Advances in Experimental Medicine and Biology

Volume 1145

More information about this series at http://www.springer.com/series/5584

Jian Li • Roger L. Nation • Keith S. Kaye
Editors

Polymyxin Antibiotics: From Laboratory Bench to Bedside

Springer

Editors
Jian Li
Biomedicine Discovery Institute
Infection & Immunity Program and
Department of Microbiology
Monash University, Clayton Campus
Melbourne, VIC, Australia

Roger L. Nation
Drug Delivery, Disposition and Dynamics
Monash Institute of Pharmaceutical
Sciences
Monash University, Parkville Campus
Melbourne, VIC, Australia

Keith S. Kaye
Division of Infectious Diseases
University of Michigan Medical School
Ann Arbor, MI, USA

ISSN 0065-2598 ISSN 2214-8019 (electronic)
Advances in Experimental Medicine and Biology
ISBN 978-3-030-16371-6 ISBN 978-3-030-16373-0 (eBook)
https://doi.org/10.1007/978-3-030-16373-0

This Springer imprint is published by the registered company Springer Nature Switzerland AG.
The registered company address is: Gewerbestrasse 11, 6330 Cham, Switzerland

Preface

Antibiotics have been a cornerstone of modern medicine and have saved millions of lives. The use of antibiotics in the clinic has made many complicated procedures and treatment modalities possible. Unfortunately, resistance to these 'magic bullets' has become widespread and now poses a serious threat to human health on a global scale. Over the last two decades, Gram-negative 'superbugs', in particular *Acinetobacter baumannii*, *Pseudomonas aeruginosa*, *Klebsiella pneumoniae* and *Escherichia coli*, have been very successful in developing resistance to most, and in some cases, all currently available antibiotics. Very often, clinicians have no newer alternatives but an 'old' class of antibiotics, namely, polymyxins, to treat deadly infections caused by these notorious pathogens. However, polymyxins (i.e. colistin and polymyxin B) were almost abandoned soon after their approval in the late 1950s and had never been rigorously evaluated through the modern drug regulatory system. Therefore, major knowledge gaps exist and have significantly limited the optimization of their clinical use. Furthermore, the polymyxins have already been off-patent for several decades and pharmaceutical companies are not interested in redeveloping both 'old' antibiotics.

The onus lies upon a number of academic research groups who have conducted an enormous number of pharmacological, chemical, microbiological and clinical studies since the late 1990s with the funding mainly from governments. Now, clinicians are in a much better position with regard to dosing polymyxins for a variety of types of patients. Excitingly, several promising candidates are being evaluated in the drug discovery pipeline. Recognizing these achievements, three international conferences were held on polymyxins (2013, Prato, Italy; 2015, San Diego, USA; and 2018, Madrid, Spain) with international opinion leaders and attendees from a large number of countries around the world. Programmes and slides from the presentations are freely available at the website of the International Society of Antimicrobial Pharmacology (https://www.isap.org/index.php/activities/special-meetings).

Polymyxins are arguably one of the most difficult classes of antibiotics to research for several reasons, including their complex amphiphilic chemical structures, stickiness to tubes and plates, complicated product composition and confusing product labelling conventions. We believe that many researchers have met significant challenges in developing a sensitive and reliable assay for the measurement of polymyxins in different biological matrices for pharmacokinetic and pharmacodynamic studies. To promote the research and facilitate their clinical use, here, we have brought together the top researchers in the field to write the first-ever book on the polymyxins. This book comprises chapters on all major topics in the polymyxin field and reviews the current progress that has been made in our understanding of the chemistry, microbiology, pharmacology, clinical use and drug discovery of polymyxins. We are hopeful that this book will provide readers with a one-stop source about polymyxin research and help clinicians to improve patient care.

The world is heading towards a potential post-antibiotic era due to the rapid increase of antibiotic resistance and the lack of commercial interest in developing new antibiotics. Therefore, every effort must be made to secure the clinical utility of this last-line defence against Gram-negative pathogens.

Finally, we would acknowledge the many contributions of authors and reviewers in the creation of this polymyxin book. This book is in memory of Professors Alan Forrest and Johan Mouton, who made significant contributions to polymyxin pharmacology. We are also very grateful to our families and colleagues for their support.

Enjoy the reading!

Monash University, Clayton Campus,
Melbourne, VIC, Australia

Jian Li

Monash University, Parkville Campus,
Melbourne, VIC, Australia

Roger L. Nation

University of Michigan Medical School,
Ann Arbor, MI, USA

Keith S. Kaye

Contents

Reviving Polymyxins: Achievements, Lessons and the Road Ahead

1

Jian Li

Abstract

Antibiotic resistance has become the most significant threat to human health across the globe. Polymyxins are often used as the only available therapeutic option against Gram-negative 'superbugs', namely *Acinetobacter baumannii*, *Pseudomonas aeruginosa* and *Klebsiella pneumoniae*. The limited pharmacological and clinical knowledge on the polymyxins in the old literature substantially limited optimizing their clinical use. The current chapter provides a general introduction to this first-ever polymyxin book which comprehensively reviews the significant progress over the last two decades in the chemistry, microbiology, pharmacology, clinical use and drug discovery of polymyxins. In particular, recent pharmacological results have led to the first scientifically-based dosing recommendations and facilitated the discovery of new-generation polymyxins. Future challenges in polymyxin research are highlighted, aiming at improving the clinical utility of this last-line defence.

Keywords

Antibiotic resistance · Drug discovery · Polymyxin · Pharmacology · Clinical use

1.1 Introduction

One of the most outstanding achievements of modern medicine was the development of antibiotics for treatment of bacterial infections that were widely fatal. Antibiotics are regarded as 'miracle drugs' and have significantly decreased mortality worldwide over the last century [1]. They have made many complicated surgical procedures and treatments possible; unfortunately, an increasing number of infections (e.g. pneumonia) are becoming more and more difficult to treat, as our current antibiotics are losing their efficacy. Over the last three decades resistance to these 'magic bullets' has presented the most significant threat to human health globally. If proactive solutions are not found to prevent the widespread antibiotic resistance, it is estimated that by 2050 approximately 10 million people per year will die of antimicrobial-resistant infections, which is more than the number of people dying from any other type of disease (Fig. 1.1) [2].

Antibiotic resistance causes increased mortality, longer hospital stays and higher medical costs. Globally, the cost of antimicrobial resistance is enormous in terms of the economy and

J. Li (✉)
Biomedicine Discovery Institute, Infection & Immunity Program and Department of Microbiology, Monash University, Clayton Campus, Melbourne, VIC, Australia
e-mail: jian.li@monash.edu

J. Li et al. (eds.), *Polymyxin Antibiotics: From Laboratory Bench to Bedside*, Advances in Experimental Medicine and Biology 1145, https://doi.org/10.1007/978-3-030-16373-0_1

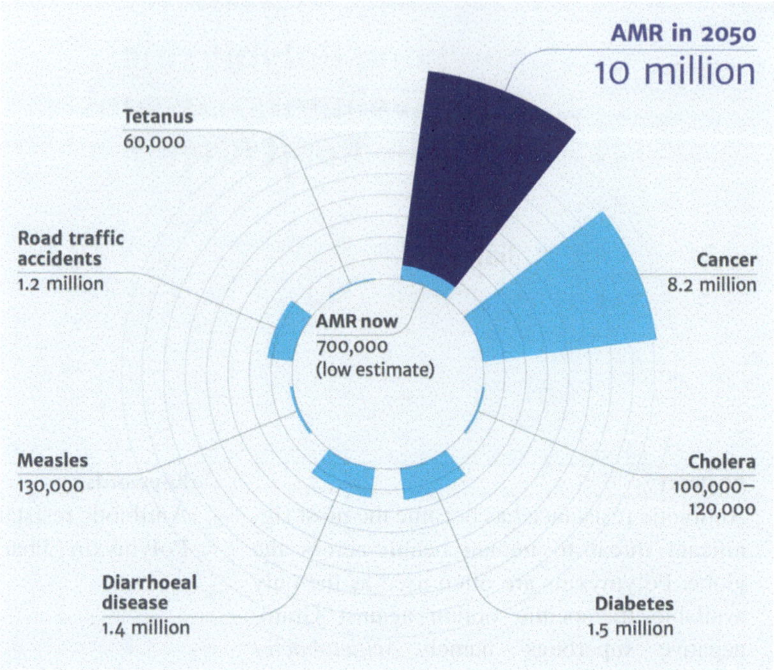

Fig. 1.1 Predicted global deaths due to antimicrobial-resistant infections every year, compared to other major diseases [2]

human health [3, 4]. It is predicted that a cumulative US$100 trillion of economic output by 2050 is at risk due to antimicrobial resistance [2]. Based upon the projections of the world economy in 2017–2050, The World Bank Group estimated that antibiotic resistance could cost the world economy $1 trillion every year by 2050 [5]. A recent study showed that the total economic cost of antibiotic resistance in *Staphylococcus aureus, Escherichia coli, Klebsiella pneumoniae, Acinetobacter baumannii* and *Pseudomonas aeruginosa* reached $2.8 billion per year in the US [6]. Furthermore, antibiotic resistance increases poverty worldwide and affects poorest countries the most [5].

Worryingly, many large pharmaceutical companies have left the antibiotic market, because the development of new antibiotics is scientifically challenging and not as profitable as for drugs used to treat chronic conditions and lifestyle issues [7–9]. A recent report reviewed the major pharmaceutical launches between 2014 and 2016 across a range of therapeutic areas [9]. It is evident that in anti-cancer drugs the risk of losing $450 million on a new molecular entity is easily offset by $8,200 million expected net present value. However, developing new antibiotics is astonishingly less attractive, as the expected net present value for the new antibiotics launched during 2014–2016 is −$100 million while with a financial risk of $500 million [9]. The World Health Organization (WHO) has urged all government sectors and society to act on antibiotic resistance. In 2017, WHO identified a list of priority pathogens which are resistant to the majority of currently available antibiotics and urgently require new therapeutic options (Fig. 1.2) [10].

As shown in the WHO Priority Pathogen List, carbapenem-resistant Gram-negative *A. baumannii, P. aeruginosa* and *K. pneumoniae* are particularly problematic, as efficacious therapeutic options are quickly diminishing against life-threatening infections caused by these pathogens [10]. All three 'superbugs' can develop resistance to almost all major classes of antibiotics via multiple mechanisms. A recent study investigated antibiotic resistance in *A. baumannii* infections with inpatients or outpatients from 54 studies (35 from the Organisation for Economic Co-operation and Development [OECD] countries with 57,188 bacterial isolates and 19 from non-OECD countries with 7,395 isolates) by searching Medline,

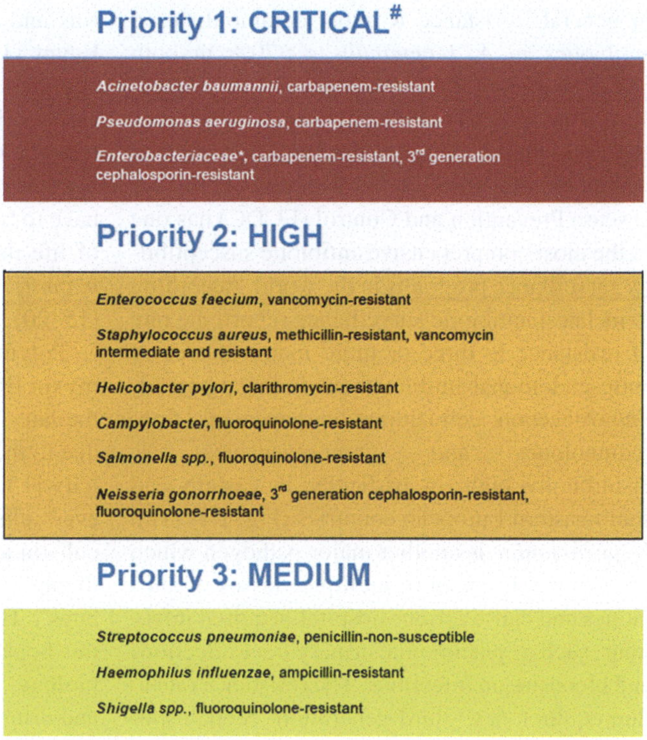

Fig. 1.2 A list of priority pathogens identified by WHO for research and development of new antibiotics [10]

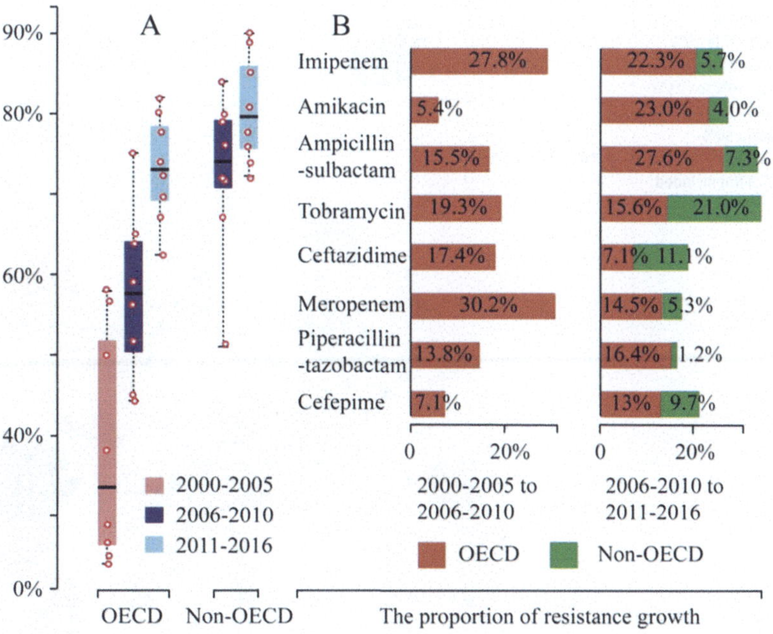

Fig. 1.3 Antibiotic resistance in *A. baumannii*. (**a**) Prevalence of multidrug-resistance to major antibiotics except colistin and tigecycline during 2000 and 2016 in the Organisation for Economic Co-operation and Development (OECD) and non-OECD countries. (**b**) Increasing antibiotic resistance in OECD and non-OECD countries between 2000 and 2016 [11]. http://creativecommons.org/licenses/by/4.0/

Embase, Web of Science, and Cochrane databases [11]. Strikingly, a high prevalence of multidrug-resistance in *A. baumannii* infections is evident in both OECD and non-OECD countries, and a faster increase was clearly shown in OECD countries over the last decade (Fig. 1.3).

In general, resistance to most commonly used antibiotics in *A. baumannii* is >70% in both OECD and non-OECD countries [11]. *P. aeruginosa* is intrinsically resistant to many antibiotics and is a major cause of healthcare-associated infections globally. The European Centre for Disease Prevention and Control (ECDC) has one of the most comprehensive antibiotic susceptibility surveillance programs in the world. According to its latest antibiotic surveillance report, the rate of resistance to three or more major classes of antipseudomonal antibiotics (including piperacillin/tazobactam, ceftazidime, carbapenems, fluoroquinolones and aminoglycosides) is disturbingly high, in particular in eastern and south-eastern European countries (Fig. 1.4) [12]. *K. pneumoniae* is another major pathogen which can become resistant to multiple classes of antibiotics and cause serious hospital-acquired infections, such as pneumonia, urinary tract infections and bloodstream infections. The resistance rate to fluoroquinolones, third-generation cephalosporins and aminoglycosides has reached >50% in Egypt [13] and a number of eastern European countries (Fig. 1.5) [12]. Sadly, few novel antibiotics will become available for these very problematic Gram-negative pathogens in the near future [2, 14, 15]. In many cases, polymyxins have to be used as the last resort for the treatment of life-threatening infections caused by *A. baumannii*, *P. aeruginosa* and *K. pneumoniae* [15–20].

Polymyxins (i.e. colistin [also known as polymyxin E] and polymyxin B) entered the clinic in the late 1950s, but their use waned in the 1970s due to the potential nephrotoxicity and neurotoxicity [15, 16, 18, 21, 22]. Since the 2000s, however, clinicians have had to increasingly use colistin and polymyxin B as one of the very few therapeutic options for Gram-negative 'superbugs'. This chapter serves as an introduction to this book and provides an overview of the microbiology, chemistry, pharmacology, clinical use, and drug discovery of polymyxins.

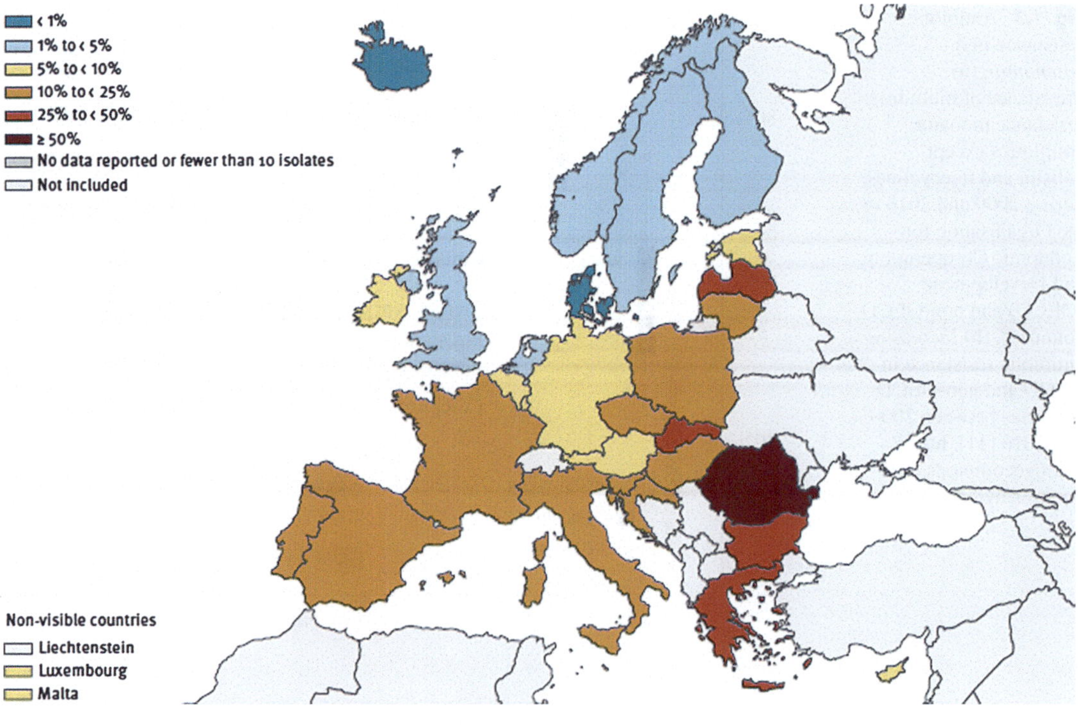

Fig. 1.4 Antibiotic resistance in *P. aeruginosa* to three or more classes among piperacillin/tazobactam, ceftazidime, carbapenem, fluoroquinolones and aminoglycosides in Europe in 2017 [12]

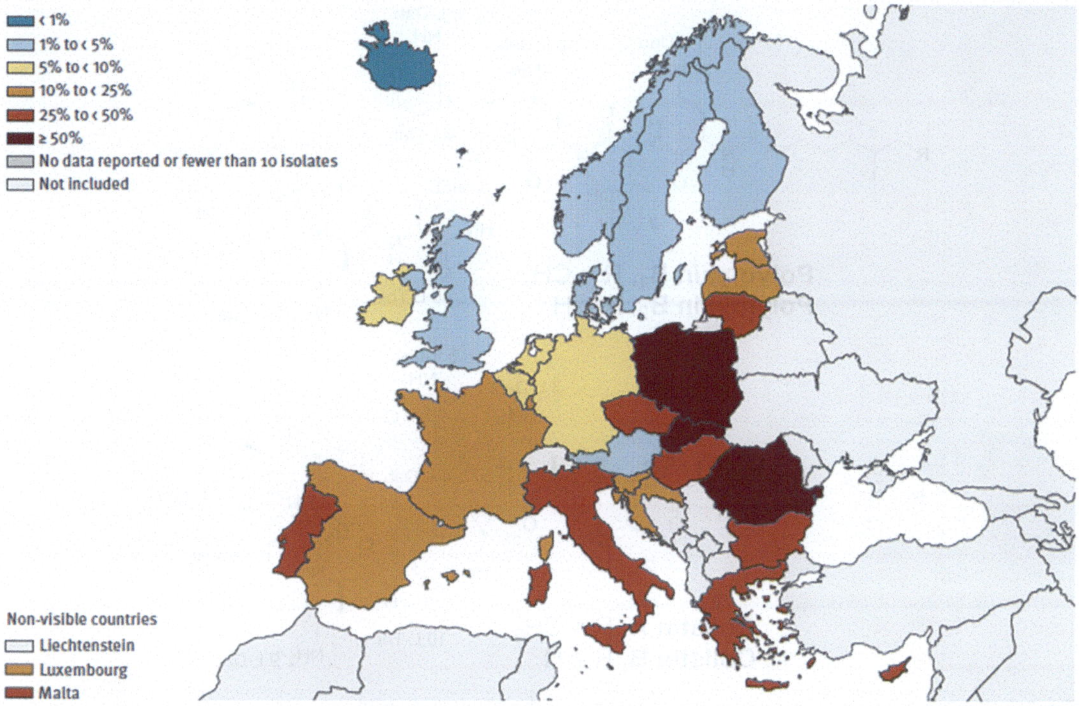

Fig. 1.5 Percentage (%) of *K. pneumoniae* isolates resistant to fluoroquinolones, third-generation cephalosporins and aminoglycosides in Europe in 2017 [12]

1.2 Polymyxins: A New 'Old' Class of Antibiotics

Colistin and polymyxin B (Fig. 1.6) were approved for clinical use in the late 1950s and were not subject to contemporary drug development evaluations and regulatory scrutiny. As polymyxins have been off patent for many years and were not widely used between the 1970s and 1990s, most pharmacological information on colistin and polymyxin B is from the studies conducted in academic research groups over the last two decades. Figure 1.7 clearly shows that polymyxins have attracted significant research and clinical interest since the early 2000s, due to increasing need to use them against multidrug-resistant Gram-negative pathogens. A number of major achievements have been made in the polymyxin field over the last two decades, including (1) a better understanding of the chemistry, structure-activity-toxicity relationships, and mechanisms of antibacterial activity, resistance and toxicity of polymyxins; (2) the first scientifically-based dosing recommendations for intravenous colistin based on the latest preclinical and clinical pharmacokinetic, pharmacodynamic and toxicodynamic information; and (3) the development of new-generation polymyxins informed by the newest chemical and pharmacological results. In this book, we invited international experts to provide comprehensive reviews on all of these major topics on polymyxins, with the aim of assembling the information needed to facilitate optimizing their clinical use and the development of novel, safer polymyxins.

This book starts with a comprehensive review on multidrug-resistance in Gram-negative 'superbugs' (Chap. 2) which highlights the urgent need to optimize the clinical use of both polymyxins and minimize any potential emergence of resistance. An in-depth introduction on the history, antibacterial spectrum and chemistry of polymyxins (Chap. 3) provides key information to understand how polymyxins kill bacterial cells (Chap. 4) and how bacteria develop resistance (Chap. 5). To optimize the dosage regimens of polymyxins, it is essential to develop sensitive and accurate analytical methods (Chap. 6), investigate the pharma-

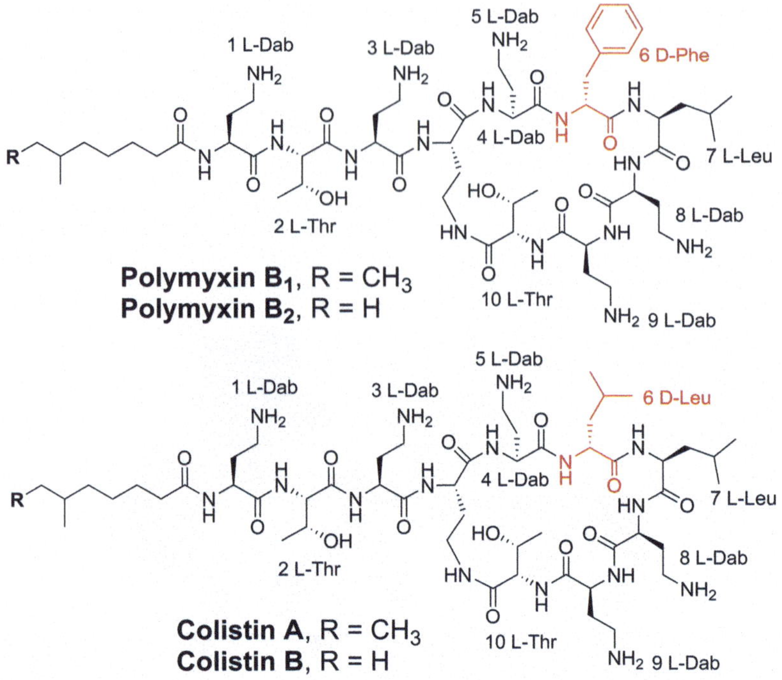

Fig. 1.6 Structures of polymyxin B and colistin. Dab, α,γ-diaminobutyric acid

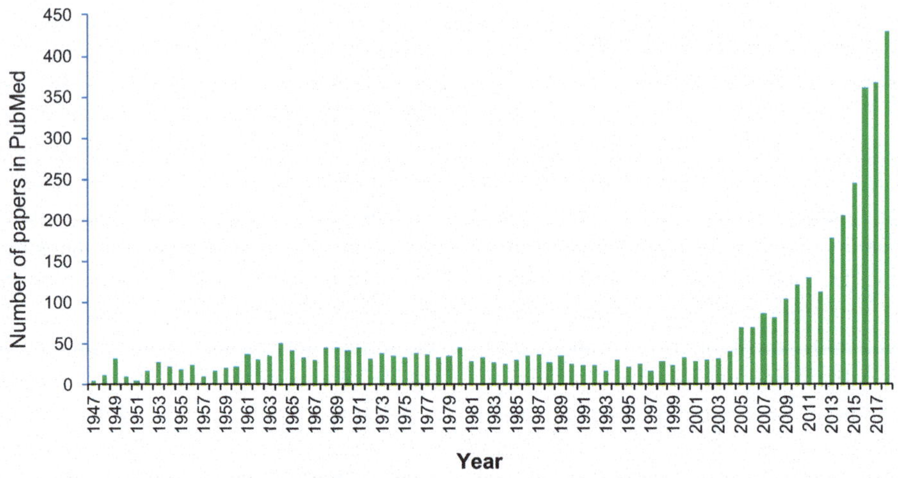

Fig. 1.7 Number of papers on polymyxins in PubMed by searching with "colistin[title] or polymyxin[title]" on 6 June 2019.

cokinetics in animals (Chap. 7), characterize the pharmacodynamics using *in vitro* and animal models (Chap. 8), and determine antibacterial susceptibility breakpoints (Chap. 9). The two different conventions used to describe the dose of colistin, the complex composition of polymyxin products, and the different pharmacopoeial standards have caused considerable confusion in different parts of the world, and together with the outdated product information can affect the ability of clinicians to optimize the use of polymyxins in patients (Chap. 10). Chapters 11, 12, 13, 14, 15, and 16 review the latest achievements in improving the use of colistin, polymyxin B and potential synergistic combinations in the clinic, which is a major focus of this polymyxin book. As polymyxins have a narrow therapeutic window and nephrotoxicity is the major dose-limiting factor [17, 22, 23], understanding their toxicities (Chap. 17) and mechanisms (Chap. 18) are crucial to ensuring their optimum and safe use in the clinic. In addition, the anti-endotoxin effect of polymyxins has been extensively evaluated for the treatment of severe sepsis and septic shock (Chap. 19). Finally, Chap. 20 reviews the latest progress in developing new-generation polymyxins with better antibacterial activity and safety profiles, which is informed by the modern polymyxin pharmacology research.

A number of major challenges and gaps in knowledge have been identified in this book. There is an imperative to systematically evaluate the clinical efficacy of intravenous colistimethate (an inactive prodrug of colistin, see Chap. 3) and polymyxin B against different types of infections (e.g. blood and urinary tract infections) [24, 25]. A large clinical PK/PD/TD study on intravenous polymyxin B is being conducted in critically-ill patients funded by the National Institutes of Health (ClinicalTrials.gov Identifier: NCT02682355). Hopefully, scientifically-based dosing recommendations will be available in the near future for intravenous polymyxin B in different types of patients. Considering the narrow therapeutic window, prospective studies with therapeutic drug monitoring and adaptive feedback control are needed for optimizing the use of both polymyxins in patients. For the treatment of MDR Gram-negative respiratory tract infections, inhalation of polymyxins is very likely a better option than intravenous administration, because of the PK/PD considerations. However, the current dosage regimens of inhaled colistin and polymyxin B are empirical and not based on PK/PD/TD information. The literature on polymyxin combination therapy *versus* monotherapy is confusing. Most clinical studies evaluating the efficacy of polymyxin combinations in the literature have overlooked the significant PK/PD issues due to the limited polymyxin exposure in the lungs after intravenous administration. PK/PD/TD principles must be considered when optimizing polymyxin combination therapy, as it is not as simple as dosing multiple antibiotics together. To achieve synergistic killing *in vivo*, all drugs should achieve optimal exposure at the infection site at the right timing; otherwise, polymyxin 'combination' therapy is essentially monotherapy. As nephrotoxicity can occur in patients receiving intravenous polymyxins, innovative approaches are warranted to increase their therapeutic indices, thereby improving the efficacy. Development of new-generation polymyxins is challenging due to the narrow chemical space and the complex relations between the chemical structure, antibacterial activity, different resistance mechanisms, toxicity and PK (e.g. plasma protein binding). Taken together, collective efforts are essential to address these challenges in the coming years.

1.3 Summary

Almost 60 years after polymyxins were approved for clinical use, clinicians are now in a much better position to determine appropriate dosage regimens for intravenous polymyxins in patients, which is the result of extensive preclinical and clinical pharmacological investigations over the last two decades. Since 2013, three international conferences have been held with the contributions of distinguished speakers worldwide, most of whom are authors in this book, the first-ever on polymyxins. It is very encouraging that substantial progress has been made across all major areas of polymyxin research, and the list of high-priority issues and challenges identified at the international polymyxin conferences becomes shorter. In this 'Bad Bugs, No Drugs' era, polymyxins will continue to play an important role in the treatment of life-threatening infections caused by Gram-negative 'superbugs'.

References

1. Armstrong GL, Conn LA, Pinner RW (1999) Trends in infectious disease mortality in the United States during the 20th century. JAMA 281:61–66
2. O'neill J (2015) Tackling a global health crisis: initial steps. Wellcome Trust, London. https://amr-review.org/sites/default/files/Report-52.15.pdf. Last accessed 12 April 2019
3. Hofer U (2019) The cost of antimicrobial resistance. Nat Rev Microbiol 17(1):3
4. Organisation for Economic Co-operation and Development (OECD) (2018) Stemming the superbug tide. http://www.oecd.org/els/health-systems/Stemming-the-Superbug-Tide-Policy-Brief-2018.pdf. Last accessed 12 April 2019
5. World Bank Group (2016) By 2050, drug-resistant infections could cause global economic damage on par with 2008 financial crisis. http://www.worldbank.org/en/news/press-release/2016/09/18/by-2050-drug-resistant-infections-could-cause-global-economic-damage-on-par-with-2008-financial-crisis. Last accessed 12 April 2019
6. Shrestha P, Cooper BS, Coast J et al (2018) Enumerating the economic cost of antimicrobial resistance per antibiotic consumed to inform the evaluation of interventions affecting their use. Antimicrob Resist Infect Control 7:98
7. Cooper MA, Shlaes D (2011) Fix the antibiotics pipeline. Nature 472:32
8. Infectious Diseases Society of America (IDSA) (2004) Bad bugs, no drugs. Infectious Diseases Society of America, Alexandria, VA. https://www.idsociety.org/globalassets/idsa/policy%2D%2Dadvocacy/current_topics_and_issues/antimicrobial_resistance/10x20/statements-manually-added/070104-as-antibiotic-discovery-stagnates-a-public-health-crisis-brews.pdf. Last accessed 12 April 2019
9. The Boston Consulting Group (BCG) for the German Federal Ministry of Health (2017) Breaking through the wall: a call for concerted action on antibiotics research and development. Follow-up report for the German GUARD initiative. https://www.bundesgesundheitsministerium.de/fileadmin/Dateien/5_Publikationen/Gesundheit/Berichte/GUARD_Follow_Up_Report_Full_Report_final.pdf. Last accessed 12 April 2019
10. World Health Organization (WHO) (2017) WHO priority pathogens list for R&D of new antibiotics. https://www.who.int/medicines/publications/WHO-PPL-Short_Summary_25Feb-ET_NM_WHO.pdf. Last accessed 12 April 2019
11. Xie R, Zhang XD, Zhao Q et al (2018) Analysis of global prevalence of antibiotic resistance in Acinetobacter baumannii infections disclosed a faster increase in OECD countries. Emerg Microbes Infect 7:31
12. European Centre for Disease Prevention and Control (ECDC) (2018) Surveillance of antimicrobial resistance in Europe – annual report of the European antimicrobial resistance surveillance network (EARS-net) 2017. ECDC, Stockholm
13. World Health Organization (WHO) (2017) Global antimicrobial resistance surveillance system (GLASS) report: early implementation 2016–2017. Geneva. ISBN: 978-92-4-151344-9. http://apps.who.int/iris/bitstream/handle/10665/259744/9789241513449-eng.pdf;jsessionid=9F6FFF4F4DA947B2346DFEAD91880B04?sequence=1. Last accessed 12 April 2019
14. Bush K, Courvalin P, Dantas G et al (2011) Tackling antibiotic resistance. Nat Rev Microbiol 9:894–896
15. Landman D, Georgescu C, Martin DA et al (2008) Polymyxins revisited. Clin Microbiol Rev 21:449–465
16. Li J, Nation RL, Turnidge JD et al (2006) Colistin: the re-emerging antibiotic for multidrug-resistant gram-negative bacterial infections. Lancet Infect Dis 6:589–601
17. Nation RL, Garonzik SM, Thamlikitkul V et al (2017) Dosing guidance for intravenous colistin in critically-ill patients. Clin Infect Dis 64:565–571
18. Poirel L, Jayol A, Nordmann P (2017) Polymyxins: antibacterial activity, susceptibility testing, and resistance mechanisms encoded by plasmids or chromosomes. Clin Microbiol Rev 30:557–596
19. Velkov T, Roberts KD, Nation RL et al (2013) Pharmacology of polymyxins: new insights into an 'old' class of antibiotics. Future Microbiol 8:711–724
20. Zavascki AP, Goldani LZ, Li J et al (2007) Polymyxin B for the treatment of multidrug-resistant pathogens: a critical review. J Antimicrob Chemother 60:1206–1215
21. Li J, Nation RL, Milne RW et al (2005) Evaluation of colistin as an agent against multi-resistant gram-negative bacteria. Int J Antimicrob Agents 25:11–25
22. Nation RL, Li J, Cars O et al (2015) Framework for optimisation of the clinical use of colistin and polymyxin B: the Prato polymyxin consensus. Lancet Infect Dis 15:225–234
23. Nation RL, Velkov T, Li J (2014) Colistin and polymyxin B: are they like peas in a pod or chalk and cheese? Clin Infect Dis 59:88–94
24. Cheah SE, Wang J, Nguyen VT et al (2015) New pharmacokinetic/pharmacodynamic studies of systemically administered colistin against Pseudomonas aeruginosa and Acinetobacter baumannii in mouse thigh and lung infection models: smaller response in lung infection. J Antimicrob Chemother 70:3291–3297
25. Landersdorfer CB, Wang J, Wirth V et al (2018) Pharmacokinetics/pharmacodynamics of systemically administered polymyxin B against Klebsiella pneumoniae in mouse thigh and lung infection models. J Antimicrob Chemother 73:462–468

Multidrug-Resistant Gram-Negative Pathogens: The Urgent Need for 'Old' Polymyxins

2

David L. Paterson and Robert A. Bonomo

Abstract

Antibiotic resistance has presented a major health challenge in the world and many isolates of Enterobacteriaceae, *Acinetobacter baumannii* and *Pseudomonas aeruginosa* become resistant to almost all current antibiotics. This chapter provides an overview on the mechanisms of antibiotic resistance in these Gram-negative pathogens and outlines the formidable problem of the genetics of bacterial resistance. Prevalent multidrug-resistance in Gram-negative bacteria underscores the need for optimizing the clinical use of the last-line polymyxins.

Keywords

Antibiotic resistance · Enterobacteriaceae · *Acinetobacter baumannii* · *Pseudomonas aeruginosa* · polymyxin

D. L. Paterson (✉)
University of Queensland Centre for Clinical Research, Brisbane, QLD, Australia
e-mail: paterson1@uq.edu.au

R. A. Bonomo
Louis Stokes Cleveland VA Medical Center, Cleveland, OH, USA

Penicillin-resistant bacteria were detected within the first decade of use of this antibiotic. More than 70 years later, antibiotic use for hospitalized patients has switched to agents such as carbapenems, quinolones, aminoglycosides and tigecycline. Unfortunately, the epidemiology of infections has changed so that bacteria resistant to some or all of these antibiotics are now commonplace in many institutions. Typically, units with compromised patients and heavy antibiotic use, such as intensive care units, hematology and transplant wards, and long-term stay units are those in which multidrug-resistant bacteria are most common. Although attention in the past focused on antibiotic-resistant Gram-positive organisms (such as methicillin-resistant *Staphylococcus aureus* or vancomycin-resistant *Enterococcus faecium*), the problem bacteria today are the Gram-negative bacteria. The Enterobacteriaceae, *Acinetobacter* spp. and *Pseudomonas aeruginosa* are common causes of healthcare-acquired infections [1] and are frequently resistant to commonly used antibiotics.

The purpose of this chapter is to outline the mechanisms of resistance in these Gram-negative pathogens, as a means of outlining the formidable problem of the genetics of bacterial resistance. It underscores the need for polymyxins, since this class of pathogens is not susceptible to beta-lactams and related resistance mechanisms so frequently seen today.

J. Li et al. (eds.), *Polymyxin Antibiotics: From Laboratory Bench to Bedside*, Advances in Experimental Medicine and Biology 1145, https://doi.org/10.1007/978-3-030-16373-0_2

2.1 Enterobacteriaceae

Carbapenem antibiotics have typically been regarded as highly stable to beta-lactamases, for example via extended-spectrum beta-lactamases (ESBLs). However, over the last decade, many of the Enterobacteriaceae have become resistant to carbapenems by way of production of carbapenemase enzymes [2].

Early reports of carbapenem-resistant Enterobacteriaceae were as a result of expression of AmpC type beta-lactamases or ESBLs plus loss of outer-membrane proteins in *K. pneumoniae* [3]. However, carbapenem resistance in the Enterobacteriaceae has emerged over the last 15–20 years due to production of beta-lactamases known as carbapenemases [4]. In the United States and some parts of Europe, the most frequently observed type of carbapenemase is the KPC type [5, 6]. KPC-producing strains are typically multidrug-resistant, being resistant to carbapenems, penicillins, cephalosporins, fluoroquinolones and aminoglycosides [6]. Therefore, polymyxins are one of the few options available for use against KPC producers. A single clone of KPC-producing *K. pneumoniae* (known as ST 258 by multilocus sequence typing (MLST)), has been found in the United States, Israel and some parts of Europe (particularly Greece and Italy) [7, 8]. This indicates that a hospital-adapted clone of KPC-producing *K. pneumoniae* was transferred from person to person as a result of breakdown in infection control measures. A new beta-lactamase inhibitor, avibactam, does have activity against the KPC beta-lactamase. However, data are limited as to its clinical effectiveness against KPC producers, and it remains to be seen as to whether it will replace polymyxins as drug of choice for KPC-producing organisms.

Although KPC has been found in China and other parts of Asia, resistance of the Enterobacteriaceae to carbapenems in Asia is more frequently due to production of metallo-beta-lactamases (MBLs) than due to KPC. A variety of MBLs have been detected. Foremost amongst these is the NDM beta-lactamase [9]. Like the KPC-type beta-lactamase, producers of NDM and other MBLs are typically resistant to carbapenems, penicillins, cephalosporins, fluoroquinolones and aminoglycosides. This explains the necessity to use polymyxins in significant infections due to NDM producers.

The NDM-type β-lactamase was first isolated in 2009 from a Swedish patient returning from India [9]. The patient was infected with NDM producing *K. pneumoniae,* resistant to multiple antibiotics including all carbapenems. The bla_{NDM} gene has now spread to all inhabited continents and is carried by multiple Gram-negative species [10]. NDM producing organisms have been strongly linked with the Indian subcontinent (India, Pakistan, Bangladesh and Nepal) [11]. China is also known to be a reservoir country for NDM producers, although surveillance data is not yet complete – at this stage it does not appear that NDM producers are as widely prevalent in China as in the Indian subcontinent. The Balkan states (for example, Serbia, Montenegro, and Bosnia-Herzegovina) may also be considered as a reservoir area for bla_{NDM} acquisition since a number of cases have been reported with no travel history to Asia [10]. In general, travel appears to be the major means by which NDM producing bacteria have spread throughout the world. Europe provided the first case in 2009 in Sweden and shortly after, many other countries began reporting travel related NDM acquisition from the Indian Subcontinent or the Balkan states. Unlike the case with KPC-producing *K. pneumoniae*, various *K. pneumoniae* sequence types (STs) have been reported to harbor bla_{NDM} [10].

A variety of other MBLs have been found to lead to carbapenem resistance in the Enterobacteriaceae. These include the VIM-type (with worldwide distribution, but particularly noteworthy in Greece), the IMP-type (with IMP-4 particularly prominent in Australia) and the SPM-type (almost exclusively found in Brazil).

Standard susceptibility testing may sometimes categorize KPC- or MBL-producing Enterobacteriaceae as susceptible to carbapenems. This issue is particularly pertinent to another group of carbapenemases found in the Enterobacteriaceae – OXA-48, and related enzymes [12]. The OXA-48 like carbapenemases are widespread in North Africa, the Middle East and India [12, 13]. There has now been significant spread to Europe [12].

The genes encoding the carbapenemases frequently reside on mobile genetic elements (such as plasmids), which are capable of transferring resistance genes from one bacterial cell to another [10]. Other resistance genes which can be co-harbored on the same genetic elements as carbapenemases include ESBLs, AmpC, quinolone resistance mechanisms, aminoglycoside modifying enzymes and 16S ribosomal RNA methylases. Chromosomally encoded mechanisms may also occur in strains with mobile genetic elements. For example, quinolone resistance in Enterobacteriaceae is usually due to chromosomally encoded alterations in target enzymes (DNA gyrase and/or topoisomerase IV) or to impaired access to the target enzymes, occurring either because of changes in porin expression or because of efflux mechanisms.

The end-result of the proliferation of this multitude of resistance mechanisms is truly multidrug-resistant Enterobacteriaceae. It is not surprising, therefore, that in settings where carbapenem-resistant Enterobacteriaceae are highly prevalent, polymyxin use becomes a necessity.

2.2 *Acinetobacter* spp.

A. baumannii and newly described species, *A. pittii* and *A. nosocomialis*, also possess a wide array of antimicrobial resistance mechanisms. Intrinsically, *Acinetobacter* species are resistant to first and second generation cephalosporins, aztreonam and ertapenem, due to a combination of poor permeability to these antibiotics and intrinsic beta-lactamase production. Antibiotics with activity against wild-type strains of *Acinetobacter* include sulbactam, meropenem, ciprofloxacin, aminoglycosides, tetracyclines, tigecycline and trimethoprim/sulfamethoxazole [14]. Acquired resistance mechanisms frequently originate from *Pseudomonas* spp., *E. coli* and other Gram-negative species, and may be localized to large resistance islands [14].

Carbapenem resistance in *Acinetobacter* spp. is a common indication for polymyxin use. As is the case with the Enterobacteriaceae, carbapenem resistance is typically mediated by production of carbapenemases.

The OXA-type beta-lactamases (especially OXA-23) are the most common mechanisms of carbapenem resistance in *Acinetobacter*. Although the KPC type carbapenemases are widely found in the Enterobacteriaceae, they are rarely found in *Acinetobacter*. However, the OXA-type carbapenemases, especially OXA-23, predominate. OXA-23 was first isolated in *Acinetobacter* spp. in the United Kingdom in 1985 [14, 15]. This carbapenemase is typically found in an internationally prevalent clone termed international clone (IC) 2. OXA-27 and OXA-49 are closely related enzymes that make up the $bla_{\text{OXA-23}}$ gene cluster in *A. baumannii* [14]. Other OXA-type genes may have carbapenemase activity including $bla_{\text{OXA-24}}$- (OXA-24, -25, -26, -40) and the $bla_{\text{OXA-58}}$-like [14] carbapenemase genes. Additionally, a chromosomally encoded gene, $bla_{\text{OXA-51}}$, is intrinsic to *A. baumannii* – its contribution to carbapenem resistance is dependent on the presence of an insertion sequence, IS*Aba1* [16]. In the absence of this insertion sequence, $bla_{\text{OXA-51}}$ does not lead to carbapenem resistance [14].

MBLs have also been well described in *Acinetobacter* spp., although they are not as frequent a cause of resistance as OXA-23 [14]. However, the carbapenem hydrolyzing activity of the MBLs is typically much more potent than that of the OXA-type carbapenemases [14]. The NDM type MBLs are particularly noteworthy since they may have originated within the genus *Acinetobacter* [10]. As noted previously, the NDM enzymes are prominent in the Indian subcontinent but have now spread widely. IMP, VIM and SIM MBLs have also been found in *Acinetobacter* spp. [17]. Additionally, *Acinetobacter* isolates may co-produce both MBL and OXA type carbapenemases [14]. Most commonly, MBLs produced by *Acinetobacter* are encoded within integrons, especially class 1 integrons. These genetic structures typically encode a wide variety of resistance genes, and contribute to the multidrug resistance typical of *Acinetobacter* [14].

Carbapenemase-producing *Acinetobacter* isolates are usually resistant to all beta-lactam antibiotics. Aminoglycoside resistance is also common, via the production of aminoglycoside modifying enzymes or 16S rRNA methylases. Ciprofloxacin resistance is typically mediated by changes in the chromosomally encoded quinolone resistance determining regions (QRDRs). Upregulated efflux pumps may also contribute to aminoglycoside, quinolone, tigecycline and beta-lactam resistance. Finally, the *sul* gene may lead to sulfamethoxazole resistance. The end-result of this multiplicity of resistance genes may be *Acinetobacter* strains resistant to all antibiotics. Hence, polymyxins play a major role in the armamentarium against the carbapenem-resistant *Acinetobacter* isolates commonly observed in clinical practice. Unfortunately, an *A. baumannii* strain has now been described which is resistant to polymyxins and all commercially available antibiotics [18].

2.3 *Pseudomonas aeruginosa*

P. aeruginosa is an organism with enhanced virulence characteristics and is the sixth most commonly isolated organism responsible for hospital-acquired infections [1]. Unlike the case with Enterobacteriaceae or *Acinetobacter* spp., the most common mechanism of carbapenem resistance in *P. aeruginosa* appears to be mutational loss of the OprD porin [19]. The primary function of OprD is importation of arginine, but it is also the major entry point for carbapenems [19]. Mutations in *oprD* can result in loss of porin function. Thus, uptake of carbapenems by *P. aeruginosa* is substantially reduced and typically confers resistance to this antibiotic class. Efflux mechanisms such as MexAB-OprM, MexXY-OprM, and the regulator MexZ may play a contributory role in carbapenem resistance, as may carbapenemase production. MBLs (such as VIM or NDM) and KPC-type beta-lactamases may be produced by *P. aeruginosa*, while OXA-type beta-lactamases are rarely seen [10, 19].

Quinolone resistance in *P. aeruginosa* is typically found to be due to mutations in the chromosomally encoded QRDRs [19]. First-step mutations occurring in the *gyrA* QRDR appear to be the most significant in causing quinolone resistance, while subsequent mutations are believed to further decrease susceptibility levels. Aminoglycoside resistance in *P. aeruginosa* is associated with efflux by the MexXY-OprM transporter, aminoglycoside modifying enzymes or 16S rRNA methyltransferases. Typically, mechanisms of quinolone and aminoglycoside resistance occur in isolates with OprD loss and/or production of carbapenemases leading to multidrug or extensively drug resistant (XDR) strains. Given that tigecycline is ineffective against *P. aeruginosa* by way of intrinsic efflux mechanisms, the polymyxins are one of the very few treatment options available for multidrug resistant strains.

2.4 Conclusions

A wide variety of issues have contributed to the problem of multidrug resistance in Gram-negative bacilli. Pharmaceutical company disinvestment in antibiotic discovery has led to fewer new options for clinical use. At the same time, proliferation of genetic elements encoding resistance has continued unabated. Exacerbators of this problem have included heavy agricultural use of antibiotics, over the counter availability of antibiotics and environmental contamination by antibiotics themselves as well as antibiotic resistant organisms in water, food and hospital environments. In particular, the recent discovery of plasmid-mediated polymyxin resistance via multiple *mcr* genes indicate that polymyxin resistance exists in food animals and patients [20–22]. The polymyxins are not perfect treatment options by any means. However, they have become the only option for many patients in this current era of resistance.

References

1. Magill SS, Edwards JR, Bamberg E et al (2014) Multistate point prevalence survey of healthcare-associated infections. N Engl J Med 370:1198–1208
2. Peleg AY, Hooper DC (2010) Hospital-acquired infections due to Gram-negative bacteria. N Engl J Med 362:1804–1813
3. Bradford PA, Urban C, Mariano N, Projan SJ, Rahal JJ, Bush K (1997) Imipenem resistance in *Klebsiella pneumoniae* is associated with the combination of ACT-1, a plasmid-mediated AmpC beta-lactamase, and the loss of an outer membrane protein. Antimicrob Agents Chemother 41:563–569
4. World Health Organization (WHO) (2017) WHO priority pathogens list for R&D of new antibiotics [Online]. Available: http://www.who.int/medicines/publications/WHO-PPL-Short_Summary_25Feb-ET_NM_WHO.pdf?ua=1. Accessed 28 Apr 2019
5. Yigit H, Queenan AM, Anderson GJ et al (2001) Novel carbapenem-hydrolyzing beta-lactamase, KPC-1, from a carbapenem-resistant strain of Klebsiella pneumoniae. Antimicrob Agents Chemother 45:1151–1161
6. Tumbarello M, Losito AR, Giamarellou H (2018) Optimizing therapy in carbapenem-resistant Enterobacteriaceae infections. Curr Opin Infect Dis 31:566–577
7. Satlin MJ, Chen L, Patel G, Gomez-Simmonds A, Weston G, Kim AC, Seo SK, Rosenthal ME, Sperber SJ, Jenkins SG, Hamula CL, Uhlemann AC, Levi MH, Fries BC, Tang YW, Juretschko S, Rojtman AD, Hong T, Mathema B, Jacobs MR, Walsh TJ, Bonomo RA, Kreiswirth BN (2017) Multicenter clinical and molecular epidemiological analysis of bacteremia due to carbapenem-resistant Enterobacteriaceae (CRE) in the CRE Epicenter of the United States. Antimicrob Agents Chemother 61: pii: e02349–16
8. Navon-Venezia S, Leavitt A, Schwaber MJ et al (2009) First report on a hyperepidemic clone of KPC-3-producing Klebsiella pneumoniae in Israel genetically related to a strain causing outbreaks in the United States. Antimicrob Agents Chemother 53:818–820
9. Yong D, Toleman MA, Giske CG et al (2009) Characterization of a new metallo-β-lactamase gene, bla_{NDM-1}, and a novel erythromycin esterase gene carried on a unique genetic structure in *Klebsiella pneumoniae* sequence type 14 from India. Antimicrob Agents Chemother 53:5046–5054
10. Wailan AM, Paterson DL (2014) The spread and acquisition of NDM-1: a multifactorial problem. Expert Rev Anti-Infect Ther 12:91–115
11. Castanheira M, Deshpande LM, Farrell SE, Shetye S, Shah N, Jones RN (2013) Update on the prevalence and genetic characterization of NDM-1-producing *Enterobacteriaceae* in Indian hospitals during 2010. Diagn Microbiol Infect Dis 75:210–213
12. Poirel L, Potron A, Nordmann P (2012) OXA-48-like carbapenemases: the phantom menace. J Antimicrob Chemother 67:1597–1606
13. Zowawi HM, Balkhy HH, Walsh TR, Paterson DL (2013) β-Lactamase production in key gram-negative pathogen isolates from the Arabian Peninsula. Clin Microbiol Rev 26:361–380
14. Peleg AY, Seifert H, Paterson DL (2008) *Acinetobacter baumannii*: emergence of a successful pathogen. Clin Microbiol Rev 21:538–582
15. Paton R, Miles RS, Hood J, Amyes SG, Miles RS, Amyes SG (1993) ARI 1: beta-lactamase-mediated imipenem resistance in *Acinetobacter baumannii*. Int J Antimicrob Agents 2:81–87
16. Turton JF, Ward ME, Woodford N, Kaufmann ME, Pike R, Livermore DM, Pitt TL (2006) The role of ISAba1 in expression of OXA carbapenemase genes in *Acinetobacter baumannii*. FEMS Microbiol Lett 258:72–77
17. Walsh TR, Toleman MA, Poirel L, Nordmann P (2005) Metallo-beta-lactamases: the quiet before the storm? Clin Microbiol Rev 18:306–325
18. Gottig S, Gruber TM, Higgins PG, Wachsmuth M, Seifert H, Kempf VA (2014) Detection of pan drug-resistant *Acinetobacter baumannii* in Germany. J Antimicrob Chemother 69:2578–2579
19. Bonomo RA, Szabo D (2006) Mechanisms of multidrug resistance in *Acinetobacter* spp. and *Pseudomonas aeruginosa*. Clin Infect Dis 43(Suppl 2):S49–S56
20. Liu YY, Wang Y, Walsh TR, Yi LX, Zhang R, Spencer J, Doi Y, Tian G, Dong B, Huang X, Yu LF, Gu D, Ren H, Chen X, Lv L, He D, Zhou H, Liang Z, Liu JH, Shen J (2016) Emergence of plasmid-mediated colistin resistance mechanism MCR-1 in animals and human beings in China: a microbiological and molecular biological study. Lancet Infect Dis 16:161–168
21. Schwarz S, Johnson AP (2016) Transferable resistance to colistin: a new but old threat. J Antimicrob Chemother 71:2066–2070
22. Wise MG, Estabrook MA, Sahm DF, Stone GG, Kazmierczak KM (2018) Prevalence of *mcr*-type genes among colistin-resistant Enterobacteriaceae collected in 2014–2016 as part of the INFORM global surveillance program. PLoS One 13:e0195281

History, Chemistry and Antibacterial Spectrum

3

Tony Velkov, Philip E. Thompson, Mohammad A. K. Azad, Kade D. Roberts, and Phillip J. Bergen

Abstract

Polymyxins are naturally occurring cyclic lipopeptides that were discovered more than 60 years ago. They have a narrow antibacterial spectrum, which is mainly against Gram-negative pathogens. The dry antibiotic pipeline, together with the increasing incidence of bacterial resistance in the clinic, has been dubbed 'the perfect storm'. This has forced a re-evaluation of 'old' antibiotics, in particular the polymyxins, which retain activity against many multidrug-resistant (MDR) Gram-negative organisms. As a consequence, polymyxin B and colistin (polymyxin E) are now used as the last therapeutic option for infections caused by 'superbugs' such as *Pseudomonas aeruginosa, Acinetobacter baumannii,* and *Klebsiella pneumoniae*. This chapter covers the history, chemistry and antibacterial spectrum of these very important last-line lipopeptide antibiotics.

Keywords

Discovery of polymyxins · Aerosporin · Chemical structure · Methanesulphonate · Antibacterial spectrum

3.1 History

3.1.1 Discovery

The polymyxins are a family of chemically distinct antibiotics produced by the widely distributed Gram-positive spore-forming soil bacterium *Paenibacillus polymyxa* (previously known as *Bacillus polymyxa*) (Table 3.1). They were first identified in the 1940s simultaneously by three different research groups working independently in the field of antibiotic discovery [1–3]. Initially, Benedict and Langlykke at the Northern Regional Research Laboratories in the United States published a paper in July of 1947 describing the antbacterial properties of crude liquid cultures of *Paenibacillus polymyxa* [1]. Later that month Stansley, Shepherd and White at the Stamford Research Laboratories of the American Cyanamid Company in the United States published a paper

T. Velkov
Department of Pharmacology and Therapeutics, University of Melbourne, Melbourne, Australia

P. E. Thompson
Medicinal Chemistry, Monash Institute of Pharmaceutical Sciences, Monash University, Melbourne, Australia

M. A. K. Azad · K. D. Roberts (✉)
Monash Biomedicine Discovery Institute, Monash University, Melbourne, Australia
e-mail: kade.roberts@monash.edu

P. J. Bergen (✉)
Centre for Medicine Use and Safety, Monash Institute of Pharmaceutical Sciences, Monash University, Melbourne, Australia
e-mail: phillip.bergen@monash.edu

J. Li et al. (eds.), *Polymyxin Antibiotics: From Laboratory Bench to Bedside*, Advances in Experimental Medicine and Biology 1145, https://doi.org/10.1007/978-3-030-16373-0_3

Table 3.1 The chemical structures of the naturally occurring polymyxins

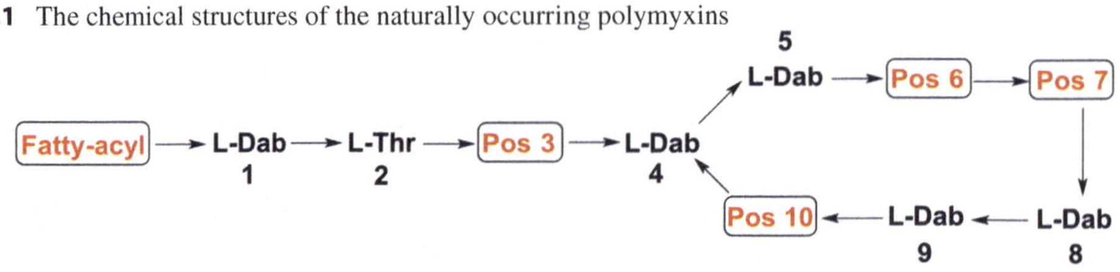

Polymyxin	Fatty-acyl group	Pos 3	Pos 6	Pos 7	Pos 10
A_1	(*S*)-6-methyloctanoyl	D-Dab	D-Leu	L-Thr	L-Thr
A_2	6-methylheptanoyl	D-Dab	D-Leu	L-Thr	L-Thr
B_1	(*S*)-6-methyloctanoyl	L-Dab	D-Phe	L-Leu	L-Thr
B_2	6-methylheptanoyl	L-Dab	D-Phe	L-Leu	L-Thr
B_3	octanoyl	L-Dab	D-Phe	L-Leu	L-Thr
B_4	heptanoyl	L-Dab	D-Phe	L-Leu	L-Thr
B_5	nonanoyl	L-Dab	D-Phe	L-Leu	L-Thr
B_6	3-hydroxy-6-methyloctanoyl[a]	L-Dab	D-Phe	L-Leu	L-Thr
B_1-Ile (Circulin A)	(*S*)-6-methyloctanoyl	L-Dab	D-Phe	L-Ile	L-Thr
B_2-Ile (Circulin A)	6-methylheptanoyl	L-Dab	D-Phe	L-Ile	L-Thr
Dab3-B_1	(*S*)-6-methyloctanoyl	D-Dab	D-Phe	L-Leu	L-Thr
Dab3-B_2	6-methylheptanoyl	D-Dab	D-Phe	L-Leu	L-Thr
$C_1^{\dagger}$	6-methyloctanoyl[b]	L/D-Dab	D-Phe	L-Thr	L-Thr
$C_2^{\dagger}$	6-methylheptanoyl	L/D-Dab	D-Phe	L-Thr	L-Thr
D_1	(*S*)-6-methyloctanoyl	D-Ser	D-Leu	L-Thr	L-Thr
D_2	6-methylheptanoyl	D-Ser	D-Leu	L-Thr	L-Thr
E_1 (Colistin A)	(*S*)-6-methyloctanoyl	L-Dab	D-Leu	L-Leu	L-Thr
E_2 (Colistin B)	6-methylheptanoyl	L-Dab	D-Leu	L-Leu	L-Thr
E_3	octanoyl	L-Dab	D-Leu	L-Leu	L-Thr
E_4	heptanoyl	L-Dab	D-Leu	L-Leu	L-Thr
E_7	7-methyloctanoyl	L-Dab	D-Leu	L-Leu	L-Thr
E_1-Ile	(*S*)-6-methyloctanoyl	L-Dab	D-Leu	L-Ile	L-Thr
E_1-Val	(*S*)-6-methyloctanoyl	L-Dab	D-Leu	L-Val	L-Thr
E_1-Nva	(*S*)-6-methyloctanoyl	L-Dab	D-Leu	L-Nva	L-Thr
E_2-Ile	6-methylheptanoyl	L-Dab	D-Leu	L-Ile	L-Thr
E_2-Val	6-methylheptanoyl	L-Dab	D-Leu	L-Val	L-Thr
E_8-Ile	7-methylnonanoyl	L-Dab	D-Leu	L-Ile	L-Thr
$F^{\dagger}$	6-methyloctanoyl[b]	L/D-Dab	D-Leu/D-Ile	L-Leu/L-Ile/L-Ser	L-Leu/L-Ile/L-Ser
$F^{\dagger}$	6-methylheptanoyl	L/D-Dab	D-Leu/D-Ile	L-Leu/L-Ile/L-Ser	L-Leu/L-Ile/L-Ser
$F^{\dagger}$	octanoyl	L/D-Dab	D-Leu/D-Ile	L-Leu/L-Ile/L-Ser	L-Leu/L-Ile/L-Ser
M_1 (Mattacin)	(*S*)-6-methyloctanoyl	D-Dab	D-Leu	L-Thr	L-Thr
M_2 (Mattacin)	6-methylheptanoyl	D-Dab	D-Leu	L-Thr	L-Thr
P_1	(*S*)-6-methyloctanoyl	D-Dab	D-Phe	L-Thr	L-Thr
P_2	6-methylheptanoyl	D-Dab	D-Phe	L-Thr	L-Thr
S_1	6-methyloctanoyl[b]	D-Ser	D-Phe	L-Thr	L-Thr
T_1	6-methyloctanoyl[b]	L-Dab	D-Phe	L-Leu	L-Leu
T_2	6-methylheptanoyl	L-Dab	D-Phe	L-Leu	L-Leu

L-Dab = L-2,4-diaminobutyric acid, D-Dab = D-2,4-diaminobutyric acid, D-Phe = D-phenylalanine, L-Leu = L-Leucine, L-Ile = L-Isoleucine, L-Val = L-Valine, L-Nva = L-Norvaline, L-Ser = L-Serine, D-Ser = D-Serine, L-Thr = L-Threonine

[a]stereochemistry at C3 and C6 not confirmed, † position of amino acid residues is speculative

[b]stereochemistry at C6 not confirmed

describing the isolation and partial purification of an antibiotic substance from *Paenibacillus polymyxa* which they designated 'Polymyxin' [2]. This organism produced on agar a wide zone of inhibition of the Gram-negative pathogen *Salmonella schottmuelleri*. The 'polymyxin' entity was unique in its remarkable specificity for Gram-negative bacteria, which distinguished it from all antibiotics previously reported. In August of 1947 Brownlee and co-workers at the Wellcome Physiological Research laboratory in England published their work on the identification of an antibiotic substance from an organism identified as *Bacillus aerosporus*, isolated from the soil of a market garden in Surry in 1946 [3]. They initially called this antibiotic 'Aerosporin' and like the antibiotic 'Polymyxin', Aerosporin had selective antimicrobial activity against Gram-negative bacteria. Brownlee and Bushby went on to further identify the chemotherapeutic and pharmacological properties of 'Aerosporin' showing that the substance they had isolated was a basic peptide [4]. Subsequently, researchers at both the Stamford and Wellcome labs determined that the three groups were working with different strains of *P. polymyxa* and that the antibiotic called 'Polymyxin' was also a basic peptide that was chemically distinct from 'Aerosporin' yet had a very similar antimicrobial spectrum and biological activity. It was concluded that the two antibiotics belonged to the same family of antibiotic compounds [5–15]. By international agreement the generic name of 'polymyxin' was adopted for all the antibiotics derived from *P. polymyxa* and a nomenclature was developed that described the chemically distinct groups of antibiotics, which comprise the polymyxin family [16, 17]. With this new nomenclature 'Aerosporin' became known as polymyxin A, while 'Polymyxin' became known as polymyxin D. Three other chemically distinct antibiotics isolated from *P. polymyxa* strains by researchers at the Wellcome labs during this period became known as polymyxin B, C and E [11]. Colistin (polymyxin E) was first described in 1950 and obtained from *Bacillus* (*Aerobacillus*) *colistinus*, a new species isolated from a soil sample in Japan [18]. Colistin was originally thought to be distinct from polymyxins, although the striking pharmacological and chemical similarities of colistin to the entire polymyxin group of antibiotics were recognized from the outset [19–21]. It was eventually determined that colistin was structurally identical to polymyxin E and that they were in fact the same compound [22–24]; colistin, however, was the name ultimately adopted in the literature. During this period the exact chemical structures of the polymyxins remained speculative [12–14, 25]. It was known that that they were peptides and possibly cyclic in nature. Individual amino acid residues had been identified and it was also established that they contained a fatty acyl group that had been identified as the *S*-6-methyloctanoyl acyl group. In 1954, Hausmann and Craig made the discovery that polymyxin B was in fact composed of two individual peptide components that differed only in the structures of the fatty-acyl groups they contained [26]. These two peptide components were labelled polymyxin B_1 and B_2 (Table 3.1). It was soon established that the presence of multiple peptide components with variations in their structures, primarily their fatty-acyl component, was a feature common to all of the polymyxin groups. In 1963, Suzuki and co-workers at the Osaka Univeristy in Japan finally determined the absolute chemical structures for polymyxin B_1, polymyxin B_2 and colistin A (polymyxin E_1) followed by colistin B (polymyxin E_2) in 1964 (Table 3.1) [27]. They went on to also confirm the structures of polymyxin D_1 and D_2 (Table 3.1) [28]. These polymyxins were all identified as being cyclic lipopeptides. Since the initial discovery of the polymyxin A, B, C, D and E groups of lipopeptides, five other groups of polymyxins containing multiple unique lipopeptide components have been identified from *P. polymyxa* strains which include the polymyxin F [29], M [30, 31], P [32, 33], S [34, 35] and T [34, 36] groups (Table 3.1). The structures and chemistry of the polymyxins are discussed in more detail in the next section of this chapter.

3.1.2 Adoption into Clinical Practice

Although the polymyxin compounds were recognized to exhibit similar antimicrobial activity, there were striking differences in their potential for eukaryotic cell toxicity [5, 6, 37–39]. For example, Brownlee et al. [37] demonstrated severe though reversible renal toxicity in rats with polymyxin A, C and D, and likewise with polymyxin A in rabbits and dogs (polymyxins C and D not tested); polymyxin B and especially colistin (polymyxin E), which were tested in all species, produced significantly less nephrotoxicity in all cases. Interestingly, in contrast to what is now known about the nephrotoxicity of both polymyxin B and colistin, the authors in that study commented that this "*lends support to the view that it [i.e. colistin] has little nephrotoxic activity*". Early reports such as this indicating substantially reduced renal toxicity from colistin and polymyxin B are likely the reason that of the five polymyxin antibiotic groups initially discovered, only these two were further developed and adopted into clinical practice. Nevertheless, while the prevailing view at the time was that colistin and polymyxin B were generally safe compounds the potential for toxicity, especially renal toxicity, was well recognized [39, 40]. Subsequently, research was undertaken to examine ways to reduce further their toxicity.

3.1.3 Sulphomethyl Derivatives

The reaction of a primary amine with an aldehyde and sodium sulphite to convert a basic substance to labile alkane sulphonic acids was introduced into drug synthesis in the early 1900s in a successful attempt to reduce the toxicity of phenetidine without loss of antipyretic activity [41]. The reaction is equally applicable to basic polypeptides such as the polymyxins (the chemistry of which is discussed later in this chapter), and the treatment of polymyxins with formaldehyde and sodium bisulphite was first reported by Stansly et al. [2]. These investigators showed that a sulphomethyl derivative of 'Polymyxin' (later shown to be polymyxin D) produced less acute toxicity than the parent antibiotic. Subsequent studies demonstrated similar results with the sulphomethylated derivatives of both colistin and polymyxin B [21, 39]. Interestingly, Stansly et al. [2] also reported substantially less painful irritation at subcutaneous or intramuscular injection sites with the sulphomethylated derivative than with the unsubstituted lipopeptide, a common problem with the polymyxins initially considered by some to be more significant than the potential renal toxicities. This is exemplified by Barnett et al. [40] who in 1964 commented that "*In the literature much value has been attached to the reduction in acute intravenous toxicity achieved by the sulphomethylation of the polymyxins, but with these antibiotics this toxicity is of no therapeutic importance because even in the unsubstituted form they have a satisfactory therapeutic index. The use of the polymyxins has, however, been much affected by the pain that develops at the site of intramuscular injection and by an undeserved reputation for nephrotoxicity. The painful reactions are undoubtedly avoided by using the sulphomethylated derivatives.*" Indeed, sulphomethylation was applied by Koyama [42] in 1957 specifically to overcome this problem with colistin. As will be discussed below colistin is still administered in the clinic intravenously as its sulphomethylated derivative.

3.1.4 Commercial Preparations

The polymyxins colistin and polymyxin B became available clinically in the late 1950s and early 1960s [43, 44]. Presently, 'colistin' is commercially available in two different forms, namely colistin sulphate [1264-72-8, CAS registry number], hereafter referred to as colistin, and its sulphomethylated derivative, sodium colistin methanesulphonate [8068-28-8] (CMS, also known as colistimethate sodium, sodium colistimethate, penta-sodium colistimethanesulphate and sulphomethyl colistin); polymyxin B is only available as polymyxin B sulphate [1405-20-5] [45]. Colistin, which is poorly absorbed from the gastrointestinal tract and through skin [21, 37, 46], has been formulated as an oral preparation (indicated for bowel decontamination) and topical preparations (indicated for bacterial skin, eye and ear infections), but is not used parenterally due to its high potential to elicit toxicity upon intravenous administration (median lethal dose (LD_{50}) = 5.46 mg/kg in mice) [21]. CMS is poorly absorbed from the adult gastrointestinal tract [47] and its sodium salt, in lyophilized form, is the form of 'colistin' that is administered parenterally, most commonly intravenously [48, 49]. However, it may also be administered intramuscularly, intrathecally, intraventricularly, and via inhalation, the latter a common route of administration for patients with cystic fibrosis. Although CMS can be administered intramuscularly at the same doses as intravenously, intramuscular administration is not commonly used in clinical practice because of variable absorption and severe pain at the injection site [50].

It is important not to use the terms colistin and CMS interchangeably, as the chemistry, antibacterial activity, toxicity and pharmacokinetics of these two entities differ substantially. Unfortunately, despite the urging of Goodwin [51] who as early as 1969 pointed out the potential confusion that may arise when the general term 'colistin' is used in reference to either colistin sulphate or CMS (as was common practice at the time; for examples, see Kunin [52], and Schwartz et al. [21]), authors to this day still occasionally report and discuss 'colistin' in generic terms which makes determination of even the preparation used (colistin sulphate or CMS) difficult. For the purposes of this and all remaining discussions, colistin sulphate will hereafter be referred to as colistin.

3.1.5 Clinical Use

In terms of their clinical use, the only difference between polymyxin B and the two commercially available forms of 'colistin' (colistin sulphate and CMS) is that polymyxin B is not indicated for oral use. Otherwise, polymyxin B sulphate can be administered via intravenous, intramuscular, inhalational, intrathecal or topical routes [45]. With the introduction of polymyxins to clinical practice, colistin was marketed as offering greater or equal antibacterial potency as compared with polymyxin B and, as the methanesulphonate (i.e. CMS), was said to lack serious toxic effect in patients [19, 21, 39, 53–57]. It was demonstrated that larger doses of CMS were required for effectiveness and thus the rate of nephrotoxicity approximated that of polymyxin B [39]; this, together with the noted reduction of pain at injection sites with the sulphomethylated derivatives, may explain why the use of CMS was adopted far more widely than polymyxin B. Interestingly, in 1961 the sodium salt of a sulphomethyl derivative of polymyxin B was administered in large doses intramuscularly and intraventricularly in five children with secondary meningitis due to *Pseudomonas pyocyanea* (now *Pseudomonas aeruginosa*) [43]. This was done in an attempt to reduce the meningeal irritant and nephrotoxic properties of polymyxin B. With all five patients cured and no toxicity observed, the authors recommended this derivative of polymyxin B for future use in the treatment of such infections. However, for reasons, which may never be known, the sulphomethylated derivative of polymyxin B was never adopted into regular clinical practice. At present there is greater worldwide use of colistin compared to polymyxin B. Notably, a survey across 56 different countries revealed formulations of polymyxins used were CMS (48.6%), colistin (sulfate) (14.1%), both (1.4%), polymyxin B (1.4%), and unknown [58]; respondents from 11 countries had no access to polymyxins. Intravenous formulations were used by 84.2% of respondents, aerosolised or nebulised colistin by 44.4%, and oral colistin for selective gut decontamination by 12.7% [58].

Despite the early belief that colistin and polymyxin B were relatively safe drugs, and the use of less toxic CMS as the parenteral form of 'colistin', clinical reports began to emerge which suggested a high incidence of nephrotoxicity and neurotoxicity following intravenous administration in a considerably large number of patients [59–67]. As a consequence, use of polymyxins declined in the 1970s with the arrival of potentially less toxic antimicrobials such as the aminoglycosides, which possessed the same or broader antibacterial spectra. However, a resurgence in their use began in the late 1980s when colistin (the most commonly used polymyxin) was reintroduced to manage infection or colonisation by *P. aeruginosa* in patients with cystic fibrosis [68]. More recently, with the emergence of multidrug-resistant (MDR) Gram-negative 'superbugs' resistant to almost all other available antibiotics [69–72], and a lack of novel antimicrobial agents in the drug development pipeline for Gram-negative infections [70–76], the place of polymyxins in therapy is presently being re-evaluated. With no new antibiotics to treat these infections to become available in the foreseeable future [71, 74], 'old' polymyxins are often the only available therapeutic options. As a consequence the use of polymyxins, especially CMS, has increased dramatically over the last decade [48, 49, 68, 77–83]. The growing importance of polymyxins as a treatment option for MDR Gram-negative infections is exemplified by the growing problem of New Delhi metallo-β-lactamase (NDM)-producing Enterobacteriaceae. Since the first identification on the Indian subcontinent in December 2009 of NDM-1-producing *Klebsiella pneumoniae* [84], NDM-producing Enterobacteriaceae (mainly *K. pneumoniae* and *E. coli*) have spread rapidly to more than 20 countries in all continents [85–87]. Many of these NDM-producing MDR isolates are only susceptible to polymyxins.

3.2 Chemistry

From a chemical perspective, the polymyxins are non-ribosomal cyclic lipopeptides and the general structure is illustrated in Table 3.1. They are decapeptides containing an intramolecular cyclic heptapeptide amide-linked loop between the amino group of the side chain of the diaminobutyric acid (Dab) residue at position 4 and the carboxyl group of the *C*-terminal threonine residue. They also have several other distinguishing structural features, which include four or five non-proteogenic Dab residues, which are charged at physiological pH. Four of these Dab residues are always found at positions 1, 5, 8 and 9 in the polymyxin scaffold and are always of the L-configuration. Position 2 of the polymyxin scaffold always contains a conserved hydrophilic L-threonine residue. Position 3 sees variation and can contain either a D or L-Dab residue or a D-serine residue. Position 6 always contains a conserved hydrophobic residue that is of the D-configuration and varies between phenyalanine, leucine. Position 7 sees the greatest variation and can either contain one of several hydrophobic residues including leucine, isoleucine, valine, norvaline or the hydrophilic residue threonine. The stereochemistry at position 7 is always of the L-configuration. Position 10 in most cases has an L-threonine residue but in at least one case contains an L-leucine residue. In regards to the *N*-terminal fatty-acyl group, six chemically distinct fatty acyl groups that vary in length from 7 to 9 carbons have been identified to date. These include (*S*)-6-methyloctanoyl, 6-methylheptanoyl, octanoyl, heptanoyl, nonanoyl and 3-hydroxy-6-methyloctanoyl. Like many other antimicrobial peptides, this mixture of lipophilic and hydrophilic groups makes them amphipathic, a chemico-physical property which is essential for their activity [88]. This also allows them to be readily water soluble (*e.g.* logP values for colistin A and colistin B are −3.15 and −3.68, respectively) [89]. The relationship between these structural features and the activity of the polymxyins is discussed in detail in Chap. 20: Discovery of Novel Polymyxin-Like Antibiotics.

Examination of the literature to date reveals that 37 unique polymyxin lipopeptides have been isolated and structurally identified from the *P. polymyxa* species [27–33, 35, 36, 90–96]. The chemical structures of these individual lipopeptides are illustrated in Table 3.1. These have been classified into 10 different groups (A, B, C, D, E, F, M, P, S and T) with each group being structurally defined and loosely classified by the presence of unique amino acid residue(s) or amino acid stereochemistry in their amino acid sequence at positions 3, 6, 7 and 10 (Table 3.1). These distinct groups of polymyxins have each been labelled with a letter. Each group can contain several individual lipopeptide components which differ from one another in the chemical structure of the fatty-acyl group they present at their *N*-terminus and in some cases the residue presented at position 7. The individual lipopeptide components of each 'polymyxin' group are labelled with a number. This nomenclature is demonstrated in Table 3.1. It is important to note here that the use of this classification system to label newly discovered polymyxins has not always been consistent as evident with the labelling of the individual components of the polymyxin E group (Table 3.1). In the case of the polymyxin C and F lipopeptides, the amino acid residue and fatty acyl composition of the lipopeptides in these two groups have been identified; however, the stereochemistry and exact positions of the amino acids are yet to be unambiguously determined. Therefore, in Table 3.1 the position of the amino acid residues for the individual lipopeptides in these two groups is speculative and based on the structural trends observed in the other polymyxin groups. To date no examples have been reported in the literature of individual polymyxin producing *P. polymyxa* strains producing 'cross mixtures' containing lipopeptides from the different polymyxin groups. Furthermore the polymyxins are always produced as mixtures of the individual lipopeptide components of that group and never as a single lipopeptide component [90, 92–95, 97]. The relative abundance of the individual components produced does vary from strain to strain and in the commercial manufacture of polymyxins from the same strain, batch-to-batch variation can be observed [92, 93, 98, 99]. Of the different 'polymyxin' groups identified to date, only the lipopeptide components of the polymyxin B and E (Colistin) groups have undergone extensive structural analysis [92–95]. This is a reflection of the fact that only 'mixtures' of individual polymyxin B lipopeptides as well as 'mixtures' of individual polymyxin E lipopeptides are used therapeutically in the clinic. The European (Ph. Eur.) and British Pharmacopeias (BP) have established limits on the minimum amount of certain components required in colistin and polymyxin B products [100, 101]. For colistin products, colistin A and B together with three minor components must constitute $\geq$ 77% of the total content; for polymyxin B products, no less than 80% of total content is to consist of polymyxin B1, B2, and two minor components. Notably, similar composition limits for colistin or polymyxin B are absent from the United States Pharmacopoeia (USP) [102]. The remaining discussion will focus only on the chemical structures of the lipopeptide components of these two groups of polymyxins.

3.2.1 Chemistry of the Polymyxin B Lipopeptides

Structurally, the lipopeptides of the polymyxin B group are generally defined by the presence of a D-phenylalanine residue at position 6, an L-leucine residue at position 7 and an L-Dab residue at position 3. To date, seven individual polymyxin B lipopeptide components have been identified (Table 3.1) [92, 95, 96]. Of these seven lipopeptides, six contain structurally different branched and non-branched *N*-terminal fatty-acyl groups varying in length from 7 to 9 carbons, which have been labelled polymyxin B_1 to B_6. The 6-methyloctanoyl fatty-acyl group of polymyxin B_1 and B_1-Ile has a stereo-centre at C6, which has been identified as being the (*S*)-configuration. Polymyxin B_6 is unique in that its fatty-acyl group contains a hydroxyl group at C3, which is not present in the fatty acyl chains of the other polymyxin B lipopeptides. This unique fatty acyl group also has two stereo-centres at C3 and C6, however the absolute stereochemistry of these two stereo-centres is yet to be reported. Interestingly, polymyxin B_1-Ile, the seventh polymyxin B lipopeptide is almost identical to polymyxin B1 except that it contains an isoleucine residue at position 7, but is still considered part of the polymyxin B group. Although isoleucine is only a structural isomer of leucine it is still a structurally distinct residue. In terms of relative abundance of individual components found in polymyxin B mixtures, polymyxin B_1 and B_2 are always the major lipopeptide components. Notably, the proportion of the different lipopeptide components in polymyxin B can vary between different brands and even between different batches from the same manufacturer [99].

In commercial preparations of polymyxin B, the lipopeptide components are always provided as their corresponding sulfate salts.

3.2.2 Chemistry of the Polymyxin E (Colistin) Lipopeptides

The polymyxin E (colistin) group of lipopeptides is generally defined by the presence of a D-leucine residue at position 6, an L-leucine residue at position 7 and an L-Dab residue at position 3 (Table 3.1). To date, 11 individual polymyxin E lipopeptide components have been identified (Table 3.1) [93, 94]. Like the polymyxin B lipopeptides the individual lipopeptide components of the polymyxin E group (polymyxin E_1, E_2, E_3, E_4, E_7 and E_8) contain structurally distinct branched and non-branched *N*-terminal fatty-acyl groups, varying in length from 7 to 9 carbons. The 6-methyloctanoyl fatty-acyl group of polymyxin E_1, E_1-Val, E_1-Ile and E_1-Nva contains a stereo-centre at C6, which has been identified as being the (*S*)-configuration. The inconsistent nature of the nomenclature used for labelling the polymyxins can be observed here with several polymyxin E lipopeptides (Polymyxin E_1-Val, E_1-Ile, E_1-Nva, E_2-Val, E_2-Ile, E_8-Ile) having structurally different amino acid residues (valine, norvaline and isoleucine) at position 7 (Table 3.1). Furthermore, no polymyxin E_5, E_6 or E_8 has been reported in the literature. In terms of relative abundance of individual components found in polymyxin E mixtures, polymyxin E_1 (colistin A) and E_2 (colistin B) are always the major lipopeptide components. Similar to commercial preparations of polymyxin B, the lipopeptide components of commercial preparations of polymyxin E are always provided as their corresponding sulfate salts.

As mentioned previously, polymyxin E (colistin) is administered intravenously as colistin methanesulphonate (CMS); CMS is an inactive prodrug of colistin [103]. CMS is chemically formed via the reaction of the amino groups of the Dab residues of polymyxin E with formaldehyde and sodium bisulphite to form sulphomethylated derivatives of each of the Dab groups (Fig. 3.1) [40, 57]. This derivatization of the amino groups of the Dab residues neutralizes the positive charge at physiological pH and imparts a negative charge through the sulphonate group, which is fully deprotonated at physiological pH. *In vivo* the sulphomethyl groups are not stable and readily undergo hydrolysis resulting in conversion back to the free amino groups to give the active form of colistin [103–114]. In the preparation of commercial CMS products, this conversion of the Dab residues to their corresponding sulphomethyled derivatives is not a complete process and as a result some of the Dab residues remain unreacted. This potentially means that even for a single polymyxin E (colistin) lipopeptide component (e.g. polymyxin E1 [colistin A]), there can be a large number of unique chemical entities in CMS, depending on the location and number (i.e. which Dab residue) of methanesulphonate groups attached. As a result commercial batches of CMS are provided as complex mixtures of fully and partially sulphomethylated derivatives [113]. Currently, no limits on the minimum or maximum amount of each potential sulphomethylated derivative within a CMS product have been established by the Ph. Eur, BP and USP [100–102].

3.2.3 Future Perspective

As we look towards the future the renewed interest in the use of polymyxins as a therapeutic option for treating MDR Gram-negative infections, alongside the constant improvement in the analytical techniques available for the identification and structural elucidation of natural products, is likely to result in the discovery of new polymyxin groups and new lipopeptide components within existing polymyxin groups. As such a more consistent use of the nomenclature for the structural classification of polymyxins is required. Therefore, the implementation of a new internationally recognised nomenclature system for structurally classifying the polymyxins is required. On a final note, an important question that still remains to be answered: *what is the physiological/biological significance of all of these individual polymyxin lipopeptides*?

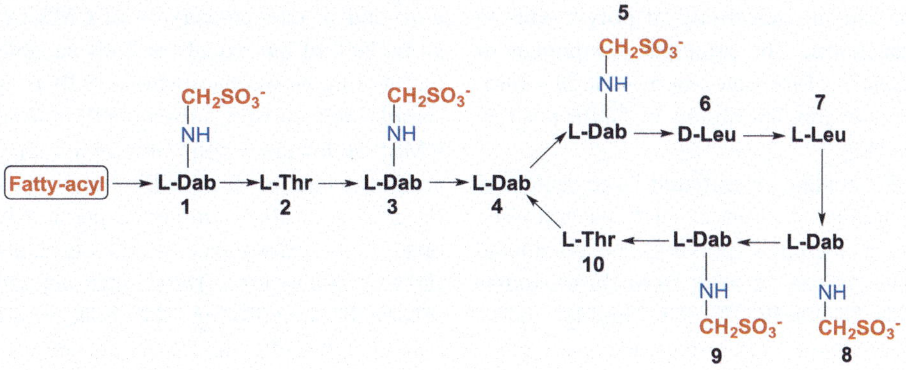

Fig. 3.1 Chemical structure of colistin methanesulphonate (CMS)

3.3 Antibacterial Spectrum

Given the structural similarities between colistin and polymyxin B as outlined above, many aspects of their antimicrobial spectrum of activity, clinical uses, toxicity and mechanism of action and resistance are shared by both [38, 45]. Both have essentially identical *in vitro* potencies (as measured by minimum inhibitory concentration [MIC]) and spectrum of activity against the commonly encountered Gram-negative organisms responsible for MDR nosocomial infections, and display a near-complete degree of cross-resistance [38, 49, 115, 116]. They exhibit a narrow antibacterial spectrum, mostly against common Gram-negative pathogens. They retain excellent bactericidal activity against most common species of Gram-negative bacilli or coccobacilli including *P. aeruginosa* [115, 117–128], *Acinetobacter* spp. [115, 117, 119, 120, 124, 125, 127, 129–131] and Enterobacteriaceae such as *Klebsiella* spp. or *E. coli* [115, 117, 119, 120, 124, 127, 132–134], the organisms against which they are most commonly used clinically. However, resistance in these and other species is increasing in some regions [119, 128, 135–147]. Interestingly, colistin-resistant isolates of several key species have been shown to be more susceptible to other antibiotics than their colistin-susceptible parent strain [128, 148, 149]. Worryingly, colistin heteroresistance (the presence of resistant subpopulations within an isolate that is susceptible based upon its MIC) has been reported in *P. aeruginosa* [150–152], *A. baumannii* [152–157], *K. pneumoniae* [144, 152, 158] and *Enterobacter cloacae* [156].

Either colistin, polymyxin B or both have also been shown to be active against *Enterobacter* spp. [117, 119, 159], *E. coli* [21, 117, 119, 124, 134, 159], *Salmonella* spp. [21, 117, 159], *Shigella* spp. [21, 117, 159], *Citrobacter* spp. [117, 159], *Haemophilus* spp. [160], *Bordetella pertussis* [40], *Legionella* spp. [161] and most *Aeromonas* species except *Ae. jandaei* (*Ae. hydrophila* has inducible resistance) [159, 162]. Polymyxins have also been reported to be potentially active against several mycobacterial species including *Mycobacterium xenopi, M. intracellulare, M. tuberculosis, M. fortuitum,* and the rapidly growing, non-pathogenic species *M. phlei and M. smegmatis* [163]. Activity against *Campylobacter* species [164, 165] and *Stenotrophomonas maltophilia* [120, 121, 166, 167] is variable, while activity against *Bartonella* species is borderline [168, 169]. Polymyxins are generally inactive against *Vibrio* spp. [159, 170], *Providentia* spp. [117, 171], *Serratia* spp. [21, 117, 124, 171, 172], *Proteus* spp. [21, 124, 171], *Morganella morganii* [173], *Helicobacter* pylori [159, 174, 175], *Neisseria* spp. (meningococci and gonococci) [21, 159, 176], *Brucella* spp. [21, 159], *Edwardsiella tarda* [177], *Burkholderia cepacia* complex [120, 178], *P. pseudomallei* [179] and *Moraxella catarrhalis* [159, 176]. Polymyxins have no significant activity against most Gram-positive bacteria, anaerobes, parasites or fungi [21, 38, 180–182]. The lack of activity against Gram-positive bacteria is likely due to the binding selectivity of polymyxins to lipopolysaccharide, the principal component of the outer leaflet of the outer membrane of Gram-negative organisms but absent in Gram-positive organisms [88].

Table 3.2 contains significant large-scale surveillance studies of antimicrobial susceptibility, which have included polymyxins conducted since 2001. As can be seen from these studies polymyxins generally remain highly active against their target Gram-negative pathogens, primarily *P. aeruginosa*, *A. baumannii* and *K. pneumoniae*. However, while the large SENTRY Antimicrobial Surveillance Program conducted between 2006 and 2009 and which contained 40,625 isolates of Gram-negative bacilli showed polymyxin-resistance generally remained stable across the collection period, a greater trend towards resistance in *Klebsiella* spp. from the Asia-Pacific and Latin American regions was noted [127]. Also noteworthy is that in the SENTRY collection, 12% of the imipenem-resistant isolates of *K. pneumoniae* were also resistant to colistin [133].

That sulphomethyl derivatives of polymyxins (including CMS) possessed substantially reduced antibacterial activity *in vitro* (as determined by MIC measurement) [21, 40, 56] and *in vivo* [21, 40] was well known from the earliest times of development. While some had speculated that the activity of sulphomethylated forms of both colistin and polymyxin B derived from unmasking of the five free amino groups present in each of the parent antibiotics, it had not been possible to determine whether any of the components had intrinsic antibacterial activity [40]. Given the sulphomethylated form of polymyxin B was never adopted into clinical practice, the uncertainty surrounding whether CMS possessed antibacterial activity in its own right persisted until recent times. Such uncertainty resulted in MIC measurements for 'colistin' having been performed using colistin [183] or CMS [117], or both [21, 136, 184]. Additionally, confusion surrounded microbiological assays used to measure 'colistin' concentrations in biological fluids. Study of the antibacterial activity of CMS, the parenteral form of colistin, had proven complicated due to the *in vitro* and *in vivo* conversion of CMS to colistin and a lack of analytical methods capable of differentiating between colistin initially present in a sample and colistin subsequently formed from CMS; on this latter point, microbiological assays are incapable of such differentiation. In 2006 Bergen et al. [103] employed previously developed high-performance liquid chromatography (HPLC) assays [185–187] which are capable of separately quantifying the concentrations of colistin and CMS (the CMS concentration determined using this approach representing the concentration of CMS (i.e. the penta-sulphomethylated species) and the numerous partially-sulphomethylated species that are intermediates in the conversion of CMS to colistin) to show that CMS may therefore be regarded as an inactive pro-drug of colistin. Additionally, this study demonstrated that the use of CMS is inappropriate for MIC measurement.

3.4 Conclusions

The polymyxins are a family of chemically distinct cyclic lipopeptide antibiotics with high specificity for Gram-negative bacteria. The chemistry of this diverse group of amphipathic compounds is complex, with each group consisting of mixtures of individual lipopeptides. Two polymyxins, polymyxin B and colistin, have been used clinically for approximately 60 years. Commercially, polymyxin B is available as polymyxin B sulphate whereas colistin is available as colistin sulphate and its sulphomethylated derivative, sodium colistin methanesulphonate (CMS); CMS is the form of 'colistin' that is administered parenterally. As polymyxins are of biological origin, the proportion of the different lipopeptide components in commercial preparations of polymyxin B or colistin vary between different brands and even between different batches from the same manufacturer. Similarly, commercial batches of CMS are provided as complex mixtures of fully and partially sulphomethylated derivatives.

Worldwide, the clinical use of colistin (predominantly as CMS) far exceeds that of poly-

Table 3.2 Summary of large-scale antimicrobial surveillance studies published from 2001–2014

Reference	Year	Polymyxin form	Species (No. of isolates)	MIC breakpoint used (mg/L)		No. of resistant isolates (%)	MIC_{50} (mg/L)	MIC_{90} (mg/L)	Range (mg/L)
				S	R				
Gales et al. [120]	2001	Colistin	*Acinetobacter* spp. (60)	≤2	≥4	2 (3.3)	≤1	2	≤1–32
			B. cepacia (12)	≤2	≥4	12 (100)	>128	>128	>128
			K. pneumoniae (9)	≤2	≥4	0 (0)	≤1	–	≤1–2
			P. aeruginosa (80)	≤2	≥4	0 (0)	≤1	2	≤1–2
			S. maltophilia (23)	≤2	≥4	17 (26.1)	≤1	8	≤1–64
			Others[a] (16)	≤2	≥4	10 (62.5)	64	>128	≤1–>128
		Polymyxin B	*Acinetobacter* spp. (60)	≤2	≥4	3 (5.0)	≤1	2	≤1–8
			B. cepacia (12)	≤2	≥4	12 (100)	>128	>128	>128
			K. pneumoniae (9)	≤2	≥4	0 (0)	≤1	–	≤1–2
			P. aeruginosa (80)	≤2	≥4	0 (0)	≤1	2	≤1–2
			S. maltophilia (23)	≤2	≥4	17 (26.1)	2	8	≤1–64
			Others[a] (16)	≤2	≥4	10 (62.5)	64	>128	≤1–>128
Fosse et al. [162]	2003	Colistin or CMS not specified	*Aeromonas* spp. (387)	≤2	≥4	83 (21.4)	ND	ND	<0.25–>256
Nicodemo et al. [167]	2004	Colistin	*S. maltophilia* (70)	≤2	≥4	17 (24.3)	2	4	0.12–32
		Polymyxin B	*S. maltophilia* (70)	≤2	≥4	16 (22.7)	2	4	0.25–16
Tan and Ng [119]	2006	Colistin	*P. aeruginosa* (33)	≤2	≥4	11 (33)	2	4	2–16
			Acinetobacter spp. (18)	≤2	≥4	0 (0)	≤1	2	<1–2
			E. coli (13)	≤2	≥4	0 (0)	≤1	≤1	<1–2
			Enterobacter spp. (8)	≤2	≥4	2 (25)	≤1	16	<1–16
			K. pneumoniae (16)	≤2	≥4	1 (6.3)	≤1	≤1	<1–4
			S. maltophilia (17)	≤2	≥4	17 (100)	128	>256	8–>256

(continued)

Table 3.2 (continued)

Reference	Year	Polymyxin form	Species (No. of isolates)	MIC breakpoint used (mg/L)		No. of resistant isolates (%)	MIC_{50} (mg/L)	MIC_{90} (mg/L)	Range (mg/L)
				S	R				
Gales et al. [159]	2006	Polymyxin B	*Acinetobacter* spp. (2621)	≤2	≥4	55 (2.1)	≤1	2	≤1–>8
			Aeromonas spp. (368)	≤2	≥4	104 (28.3)	≤1	>8	≤1–>8
			Alcaligenes spp. (121)	≤2	≥4	44 (36.4)	2	>8	≤1–>8
			B. cepacia (153)	≤2	≥4	135 (88.2)	>8	>8	0.5–8
			P. aeruginosa (8705)	≤2	≥4	113 (1.3)	≤1	2	≤1–>8
			Pseudomonas spp. (non-*aeruginosa;* 282)	≤2	≥4	33 (11.7)	≤1	4	≤1–>8
			S. maltophilia (1256)	≤2	≥4	347 (27.6)	1	8	≤0.12–>8
			Other non-enteric Gram-negative bacilli (302)	≤2	≥4	168 (55.6)	4	>4	<1–>8
			Citrobacter spp. (895)	≤2	≥4	8 (0.9)	≤1	≤1	≤1–>8
			Enterobacter spp. (4693)	≤2	≥4	784 (16.7)	≤1	>8	≤1–>8
			E. coli (18,325)	≤2	≥4	92 (0.5)	≤1	≤1	≤1–>8
			Klebsiella spp. (8188)	≤2	≥4	147 (1.8)	≤1	≤1	≤1–>8
			Indole-positive *Proteus* spp. (895)	≤2	≥4	883 (98.7)	>8	>8	≤1–>8
			P. mirabilis (1931)	≤2	≥4	1917 (99.3)	>8	>8	≤1–>8
			Salmonella spp. (2909)	≤2	≥4	698 (24.0)	≤1	4	≤1–>8
			Shigella spp. (828)	≤2	≥4	8 (1.0)	≤1	≤1	≤1–>8
			Serratia spp. (1919)	≤2	≥4	1815 (94.6)	>8	>8	0.25–>8
Yau et al. [130]	2009	Colistin	*A. baumannii* (30)	≤2	≥4	1 (3.3)	0.5	1	0.5–128

Walkty et al. [124]	2009	Colistin	*P. aeruginosa* (561)	≤2	≥4	47 (2)	2	2	0.5–>16
			E. coli (1732)	≤2	≥4	11 (0.6)	0.5	1	≤0.06–>16
			K. pneumoniae (515)	≤2	≥4	15 (2.9)	0.5	1	0.12–>16
			E. cloacae (186)	≤2	≥4	30 (16.1)	0.5	>16	0.12–>16
			P. mirabilis (119)	≤2	≥4	119 (100)	>16	>16	≥16
			S. marcescens (108)	≤2	≥4	106 (98.1)	>16	>16	1–>16
			K. oxytoca (108)	≤2	≥4	5 (4.6)	0.5	1	0.25–>16
			S. maltophilia (83)	≤2	≥4	70 (84.3)	8	>16	0.25–>16
			E. aerogenes (37)	≤2	≥4	1 (2.7)	0.5	1	0.25–4
			A. baumannii (31)	≤2	≥4	2 (6.5)	1	2	0.5–>16
Hawser et al. [132]	2010	Colistin	*K. pneumoniae* (303)	≤2	≥4	8 (2.6)	0.25	0.5	≤0.12–>4
Landman et al. [188]	2010	Polymyxin B	*K. pneumoniae* (3050)	≤2	≥4	6 (0.2)	1	1	≤0.12–>4
Lee et al. [128]	2011	Colistin	*P. aeruginosa* (215)	≤2	≥8	16 (7.4)	2	2	≤8
		Polymyxin B	*P. aeruginosa* (215)	≤2	≥8	0 (0.0)	1	2	≤2
Sader et al. [133]	2011	Colistin	*K. pneumoniae* (9774)	≤2	≥4	136 (1.4)	ND	ND	NS
		Polymyxin B	*K. pneumoniae* (9774)	≤2	≥4	156 (1.6)	ND	ND	NS
Gales et al. [115]	2011	Colistin	*Acinetobacter* spp. (4686)	≤2	≥4	42 (0.9)	≤0.5	1	NS
			E. coli (17035)	≤2	>2	34 (0.2)	≤0.5	≤0.5	NS
			Klebsiella spp. (9774)	≤2	>2	147 (1.5)	≤0.5	≤0.5	NS
			P. aeruginosa (9130)	≤2	>4	37 (0.4)	1	1	NS
		Polymyxin B	*Acinetobacter* spp. (4686)	≤2	≥4	37 (0.8)	≤0.5	≤0.5	NS
			E. coli (17035)	≤2	>2	17 (0.1)	≤0.5	≤0.5	NS
			Klebsiella spp. (9774)	≤2	>2	137 (1.4)	≤0.5	≤0.5	NS
			P. aeruginosa (9130)	≤2	>4	9 (0.1)	1	1	NS
Gales et al. [127]	2012	Colistin	*Acinetobacter* spp. (845)	≤2	≥4	10 (1.2)	≤0.5	1	≤0.5–>4
			Enterobacter spp. (451)	≤2	>2	78 (17.3)	≤0.5	>4	≤0.5–>4
			E. coli (1517)	≤2	>2	3 (0.2)	≤0.5	≤0.5	≤0.5–>4
			Klebsiella spp. (1052)	≤2	≥8	32 (3.0)	≤0.5	≤0.5	≤0.5–>4
			P. aeruginosa (1099)	≤2	≥8	1 (0.1)	1	2	≤0.5–>4

(continued)

Table 3.2 (continued)

Reference	Year	Polymyxin form	Species (No. of isolates)	MIC breakpoint used (mg/L)		No. of resistant isolates (%)	MIC_{50} (mg/L)	MIC_{90} (mg/L)	Range (mg/L)
				S	R				
Quale et al. [189]	2012	Polymyxin B	*E. coli* (3049)	≤2	≥4	6 (0.2)	1	1	≤0.12–>8
			K. pneumoniae (1155)	≤2	≥4	46 (4)	1	2	0.25–>16
			Enterobacter spp. (199)	≤2	≥4	48 (24)	1	>8	≤0.12–>16
			A. baumannii (407)	≤2	≥4	12 (3)	1	2	≤0.25–>16
			P. aeruginosa (679)	≤2	≥8	3 (0.5)	1	2	0.25–4
Queenan et al. [131]	2012	Colistin	*Acinetobacter* spp. (514)	≤2	≥4	27 (5.3)	1	2	0.12–>32
Jones et al. [126]	2013	Colistin	*P. aeruginosa* (586)	≤2	≥8	0 (0.0)	1	2	≤0.25–4
Zhanel et al. [125]	2013	Colistin	*E. coli* (5451)	–	–	–	0.25	0.5	≤0.06–>16
			ESBL *E. coli* (231)	–	–	–	0.5	1	≤0.06–4
			P. aeruginosa (2183)	≤2	≥8	24 (1.1)	2	2	≤0.06–>16
			K. pneumoniae (1659)	–	–	–	0.5	1	≤0.06–>16
			E. cloacae (637)	–	–	–	0.5	>16	≤0.06–>16
			P. mirabilis (415)	–	–	–	>16	>16	0.5–>16
			K. oxytoca (411)	–	–	–	0.5	1	0.12–>16
			S. marcescens (412)	–	–	–	>16	>16	0.5–>16
			S. maltophilia (378)	–	–	–	8	>16	0.25–>16
			E. aerogenes (163)	–	–	–	0.5	1	0.12–>16
			C. freundii (123)	–	–	–	0.5	0.5	0.12–1
			A. baumannii (104)	≤2	≥4	3 (3.2)	1	2	0.25–>16
Nakamura et al. [134]	2014	Colistin	*E. coli* (174)	≤2	>2	7 (4.0)	0.5	2	NS
			K. pneumoniae (37)	≤2	>2	5 (13.5)	0.5	2	NS

[a]*E. aerogenes* (four strains), *E. cloacae* (one strain), *M. morganii* (two strains), *P. mirabilis* (two strains), *P. rettgeri* (two strains), and *S. marcescens* (five strains)
ND not determined, *NS* not specified

myxin B. Relegated to the 'back shelf' in the 1970s due to toxicity concerns, the emergence of MDR Gram-negative 'superbugs' resistant to almost all other available antibiotics has resulted in their progressive reintroduced into clinical practice over the last two decades. Given they retain excellent bactericidal activity against most common species of Gram-negative bacilli or coccobacilli, they have become increasingly important as salvage therapy for otherwise untreatable infections caused by MDR Gram-negative organisms.

References

1. Benedict RG, Langlykke AF (1947) Antibiotic activity of Bacillus polymyxa. J Bacteriol 54(1):24
2. Stansly PG, Shepherd RG, White HJ (1947) Polymyxin: a new chemotherapeutic agent. Bull Johns Hopkins Hosp 81(1):43–54
3. Ainsworth GC, Brown AM, Brownlee G (1947) Aerosporin, an antibiotic produced by bacillus aerosporus Greer. Nature 159(4060):263
4. Brownlee G, Bushby SR (1948) Chemotherapy and pharmacology of aerosporin; a selective gram-negative antibiotic. Lancet 1(6491):127–132
5. Brownlee G, Bushby SR, Short EI (1949) Comparative biological studies of polymyxin A and polymyxin D. Ann N Y Acad Sci 51(Art. 5):891–896
6. Brownlee G, Bushby SR, Short EI (1949) The pharmacology of polymyxin A, B, and D. Ann N Y Acad Sci 51(Art. 5):952–967
7. Shepherd RG, Stansly PG et al (1948) Chemical studies on polymyxin; isolation and preliminary purification. J Am Chem Soc 70(11):3771–3774
8. Jones TS (1948) The chemical nature of aerosporin. Biochem J 42(2):xxxv
9. Brownlee G, Jones TS (1948) The polymyxins; a related series of antibiotics derived from B. polymyxa. Biochem J 43(2):xxv
10. Catch JR, Jones TS, Wilkinson S (1948) The chemistry of polymyxin A; isolation of the amino-acids D-leucine, L-threonine, L-alpha gamma-diamiobutyric acid, and an unknown fatty acid. Biochem J 43(2):xxvii
11. Jones TS (1948) The chemical basis for the classification of the polymyxins. Biochem J 43(2):xxvi
12. Bell PH, Bone JF et al (1949) Chemical studies on polymyxin; comparison with aerosporin. Ann N Y Acad Sci 51(Art. 5):897–908
13. Jones TS (1949) Chemical evidence for the multiplicity of the antibiotics produced by Bacillus polymyxa. Ann N Y Acad Sci 51(Art. 5):909–916
14. Catch JR, Jones TS, Wilkinson S (1949) The chemistry of polymyxin A. Ann N Y Acad Sci 51(Art. 5):917–923
15. White HJ, Alverson CM et al (1949) Comparative biological studies of polymyxin and aerosporin. Ann N Y Acad Sci 51(Art. 5):879–890
16. Stansly PG, Brownlee G (1949) Nomenclature of polymyxin antibiotics. Nature 163(4146):611
17. Brownlee G (1949) Antibiotics derived from bacillus polymyxa. Ann N Y Acad Sci 51(Art. 5):875–878
18. Koyama Y, Kurosasa A, Tsuchiya A, Takakuta K (1950) A new antibiotic 'colistin' produced by spore-forming soil bacteria. J Antibiot 3:457–458
19. Wright WW, Welch H (1959) Chemical, biological and clinical observations on colistin. Antibiot Annu 7:61–74
20. McCabe WR, Jackson GG, Kozij VM (1959) Clinical and laboratory observations on use of colistin in infections by gram-negative bacilli. Antibiot Annu 7:80–88
21. Schwartz BS, Warren MR, Barkley FA, Landis L (1959) Microbiological and pharmacological studies of colistin sulfate and sodium colistinmethanesulfonate. Antibiot Annu 7:41–60
22. Wilkinson S (1963) Identity of colistin and polymyxin E. Lancet 1(7287):922–923
23. Suzuki T, Hayashi K, Fujikawa K, Tsukamoto K (1965) The chemical structure of polymyxin E: the Identities of polymyxin E1 with colistin a and of polymyxin E2 with colistin B. J Biochem 57:226–227
24. Dautrevaux M, Biserte G (1961) Structure of colimycin. Bull Soc Chim Biol 43:495–504
25. Wilkinson S (1949) Crystalline derivatives of the polymyxins and the identification of the fatty acid component. Nature 164(4171):622
26. Hausmann W, Craig LC (1954) Polymyxin B1. Fractionation, molecular weight determination, amino acid and fatty acid composition. J Am Chem Soc 76(19):4892–4896
27. Hayashi K, Suzuki T (1965) Chemical structures of polymyxin series antibiotics. Bull Inst Chem Res, Kyoto Univ 43(3):259–277
28. Hayashi K, Suketa Y, Tsukamoto K, Suzuki T (1966) Chemical structure of polymyxin D1. Experientia 22(6):354–355
29. Parker WL, Rathnum ML, Dean LD, Nimeck MW, Brown WE, Meyers E (1977) Polymyxin F, a new peptide antibiotic. J Antibiot (Tokyo) 30(9):767–769
30. Wilkinson S, Lowe LA (1966) Structures of the polymyxins A and the question of identity with the polymyxins M. Nature 212(5059):311
31. Martin NI, Hu H, Moake MM, Churey JJ, Whittal R, Worobo RW, Vederas JC (2003) Isolation, structural characterization, and properties of mattacin (polymyxin M), a cyclic peptide antibiotic produced by Paenibacillus kobensis M. J Biol Chem 278(15):13124–13132. https://doi.org/10.1074/jbc.M212364200

32. Matsunaga H, Suwa K, Takeyama S, Kimura Y (1995) Polymyxin P, antibiotics from *bacillus polymyxa* t-39. Bull Mukogawa Women's Uni Nat Sci 43:37–43
33. Niu B, Vater J, Rueckert C, Blom J, Lehmann M, Ru JJ, Chen XH, Wang Q, Borriss R (2013) Polymyxin P is the active principle in suppressing phytopathogenic Erwinia spp. by the biocontrol rhizobacterium Paenibacillus polymyxa M-1. BMC Microbiol 13(1):137. https://doi.org/10.1186/1471-2180-13-137
34. Shoji J, Kato T, Hinoo H (1977) The structures of two new polymyxin group antibiotics. J Antibiot (Tokyo) 30(5):427–429
35. Shoji J, Kato T, Hinoo H (1977) The structure of polymyxin S. (Studies on antibiotics from the genus Bacillus. XXI). J Antibiot (Tokyo) 30(12):1035–1041
36. Shoji J, Kato T, Hinoo H (1977) The structure of polymyxin T1. (Studies on antibiotics from the genus Bacillus. XXII). J Antibiot (Tokyo) 30(12):1042–1048
37. Brownlee G, Bushby SR, Short EI (1952) The chemotherapy and pharmacology of the polymyxins. Br J Pharmacol Chemother 7(1):170–188
38. Storm DR, Rosenthal KS, Swanson PE (1977) Polymyxin and related peptide antibiotics. Annu Rev Biochem 46:723–763
39. Nord NM, Hoeprich PD (1964) Polymyxin B and colistin. A critical comparison. N Engl J Med 270:1030–1035
40. Barnett M, Bushby SR, Wilkinson S (1964) Sodium sulphomethyl derivatives of polymyxins. Br J Pharmacol Chemother 23:552–574
41. Lepetit R (1908-1910) Ethoxyphenylaminomethanesulphonate of sodium. Friedl 9:1969–1970
42. Koyama Y (1957) [Colimycin]. G Ital Chemioter 4(3–4):279–287
43. Clifford HE, Stewart GT (1961) Intraventricular administration of a new derivative of polymyxin B in meningitis due to Ps. pyocyanea. Lancet 2(7195):177–180
44. Ross S, Puig JR, Zaremba EA (1959) Colistin: some preliminary laboratory and clinical observations in specific gastroenteritis in infants and children. Antibiot Annu 7:89–100
45. Kwa A, Kasiakou SK, Tam VH, Falagas ME (2007) Polymyxin B: similarities to and differences from colistin (polymyxin E). Expert Rev Anti-Infect Ther 5(5):811–821. https://doi.org/10.1586/14787210.5.5.811
46. Greenwood D (2003) Miscellaneous antibacterial agents. In: Finch RG, Greenwood D, Norrby SR, Whitley RJ (eds) Antibiotics and chemotherapy: anti-infective agents and their use in therapy. Curchill Livingstone, Philadelphia, pp 409–411
47. Chamberlain G, Needham P (1976) The absorption of antibiotics from the bladder. J Urol 116(2):172–173
48. Falagas ME, Kasiakou SK (2005) Colistin: the revival of polymyxins for the management of multidrug-resistant gram-negative bacterial infections. Clin Infect Dis 40(9):1333–1341
49. Li J, Nation RL, Milne RW, Turnidge JD, Coulthard K (2005) Evaluation of colistin as an agent against multi-resistant gram-negative bacteria. Int J Antimicrob Agents 25(1):11–25
50. Bergan T, Fuglesang J (1982) Polymyxin antibiotics: chemical and pharmacokinetic properties. Antibiot Chemother 31:119–144
51. Goodwin NJ (1969) Colistin sulfate versus sodium colistimethate. Ann Intern Med 70(1):232–233
52. Kunin CM (1968) More on antimicrobials in renal failure. Ann Intern Med 69(2):397–398
53. Meleney FL, Prout GR Jr (1961) Some laboratory and clinical observations on coly-mycin (colistin) with particular reference to Pseudomonas infections. Surg Gynecol Obstet 112:211–217
54. Petersdorf RG, Hook EW (1960) The use of colistin in the treatment of infections of the urinary tract. Bull Johns Hopkins Hosp 107:133–142
55. Fekety FR Jr, Norman PS, Cluff LE (1962) The treatment of gram-negative bacillary infections with colistin. The toxicity and efficacy of large doses in forty-eight patients. Ann Intern Med 57:214–229
56. Taylor G, Allison H (1962) "Colomycin". Laboratory and clinical investigations. Br Med J 2(5298):161–163
57. Beveridge EG, Martin AJ (1967) Sodium sulphomethyl derivatives of polymyxins. Br J Pharmacol Chemother 29(2):125–135
58. Wertheim H, Van Nguyen K, Hara GL, Gelband H, Laxminarayan R, Mouton J, Cars O (2013) Global survey of polymyxin use: a call for international guidelines. J Glob Antimicrob Resistance 1(3):131–134. https://doi.org/10.1016/j.jgar.2013.03.012
59. Koch-Weser J, Sidel VW, Federman EB, Kanarek P, Finer DC, Eaton AE (1970) Adverse effects of sodium colistimethate. Manifestations and specific reaction rates during 317 courses of therapy. Ann Intern Med 72(6):857–868
60. Ryan KJ, Schainuck LI, Hickman RO, Striker GE (1969) Colistimethate toxicity. Report of a fatal case in a previously healthy child. JAMA 207(11):2099–2101
61. Brown JM, Dorman DC, Roy LP (1970) Acute renal failure due to overdosage of colistin. Med J Aust 2(20):923–924
62. Yow EM, Moyer JH (1953) Toxicity of polymyxin B. II. Human studies with particular reference to evaluation of renal function. AMA Arch Intern Med 92(2):248–257
63. Elwood CM, Lucas GD, Muehrcke RC (1966) Acute renal failure associated with sodium colistimethate treatment. Arch Intern Med 118(4):326–334
64. Duncan DA (1973) Colistin toxicity. Neuromuscular and renal manifestations. Two cases treated by hemodialysis. Minn Med 56(1):31–35

65. Price DJ, Graham DI (1970) Effects of large doses of colistin sulphomethate sodium on renal function. Br Med J 4(734):525–527
66. Lindesmith LA, Baines RD Jr, Bigelow DB, Petty TL (1968) Reversible respiratory paralysis associated with polymyxin therapy. Ann Intern Med 68(2):318–327
67. Wolinsky E, Hines JD (1962) Neurotoxic and nephrotoxic effects of colistin patients with renal disease. N Engl J Med 266:759–762
68. Nation RL, Li J (2009) Colistin in the 21st century. Curr Opin Infect Dis 22(6):535–543. https://doi.org/10.1097/QCO.0b013e328332e672
69. Jones RN (2001) Resistance patterns among nosocomial pathogens: trends over the past few years. Chest 119(2 Suppl):397S–404S
70. Talbot GH, Bradley J, Edwards JE Jr, Gilbert D, Scheld M, Bartlett JG (2006) Bad bugs need drugs: an update on the development pipeline from the antimicrobial availability task force of the infectious diseases society of America. Clin Infect Dis 42(5):657–668. (Erratum in Clin Infect Dis 2006; 2042:1065)
71. Boucher HW, Talbot GH, Benjamin DK Jr, Bradley J, Guidos RJ, Jones RN, Murray BE, Bonomo RA, Gilbert D, Infectious diseases society of A (2013) 10 x '20 progress--development of new drugs active against gram-negative bacilli: an update from the infectious diseases society of America. Clin Infect Dis 56(12):1685–1694. https://doi.org/10.1093/cid/cit152
72. Livermore DM (2004) The need for new antibiotics. Clin Microbiol Infect 10(Suppl 4):1–9. https://doi.org/10.1111/j.1465-0691.2004.1004.x
73. Spellberg B, Guidos R, Gilbert D, Bradley J, Boucher HW, Scheld WM, Bartlett JG, Edwards J Jr (2008) The epidemic of antibiotic-resistant infections: a call to action for the medical community from the infectious diseases society of America. Clin Infect Dis 46(2):155–164. https://doi.org/10.1086/524891
74. Payne DJ, Gwynn MN, Holmes DJ, Pompliano DL (2007) Drugs for bad bugs: confronting the challenges of antibacterial discovery. Nat Rev Drug Discov 6(1):29–40. https://doi.org/10.1038/nrd2201
75. Livermore DM (2003) The threat from the pink corner. Ann Med 35(4):226–234
76. Walker B, Barrett S, Polasky S, Galaz V, Folke C, Engstrom G, Ackerman F, Arrow K, Carpenter S, Chopra K, Daily G, Ehrlich P, Hughes T, Kautsky N, Levin S, Maler KG, Shogren J, Vincent J, Xepapadeas T, de Zeeuw A (2009) Environment. Looming global-scale failures and missing institutions. Science 325(5946):1345–1346. https://doi.org/10.1126/science.1175325
77. Garnacho-Montero J, Ortiz-Leyba C, Jimenez-Jimenez FJ, Barrero-Almodovar AE, Garcia-Garmendia JL, Bernabeu-Wittel IM, Gallego-Lara SL, Madrazo-Osuna J (2003) Treatment of multidrug-resistant Acinetobacter baumannii ventilator-associated pneumonia (VAP) with intravenous colistin: a comparison with imipenem-susceptible VAP. Clin Infect Dis 36(9):1111–1118. https://doi.org/10.1086/374337
78. Levin AS, Barone AA, Penco J, Santos MV, Marinho IS, Arruda EA, Manrique EI, Costa SF (1999) Intravenous colistin as therapy for nosocomial infections caused by multidrug-resistant Pseudomonas aeruginosa and Acinetobacter baumannii. Clin Infect Dis 28(5):1008–1011
79. Linden PK, Kusne S, Coley K, Fontes P, Kramer DJ, Paterson D (2003) Use of parenteral colistin for the treatment of serious infection due to antimicrobial-resistant Pseudomonas aeruginosa. Clin Infect Dis 37(11):e154–e160. https://doi.org/10.1086/379611
80. Markou N, Apostolakos H, Koumoudiou C, Athanasiou M, Koutsoukou A, Alamanos I, Gregorakos L (2003) Intravenous colistin in the treatment of sepsis from multiresistant Gram-negative bacilli in critically ill patients. Crit Care 7(5):R78–R83
81. Michalopoulos AS, Tsiodras S, Rellos K, Mentzelopoulos S, Falagas ME (2005) Colistin treatment in patients with ICU-acquired infections caused by multiresistant gram-negative bacteria: the renaissance of an old antibiotic. Clin Microbiol Infect 11(2):115–121
82. Li J, Nation RL, Turnidge JD, Milne RW, Coulthard K, Rayner CR, Paterson DL (2006) Colistin: the re-emerging antibiotic for multidrug-resistant Gram-negative bacterial infections. Lancet Infect Dis 6(9):589–601
83. Landman D, Georgescu C, Martin DA, Quale J (2008) Polymyxins revisited. Clin Microbiol Rev 21(3):449–465
84. Yong D, Toleman MA, Giske CG, Cho HS, Sundman K, Lee K, Walsh TR (2009) Characterization of a new metallo-beta-lactamase gene, bla(NDM-1), and a novel erythromycin esterase gene carried on a unique genetic structure in Klebsiella pneumoniae sequence type 14 from India. Antimicrob Agents Chemother 53(12):5046–5054. https://doi.org/10.1128/AAC.00774-09
85. Kumarasamy KK, Toleman MA, Walsh TR, Bagaria J, Butt F, Balakrishnan R, Chaudhary U, Doumith M, Giske CG, Irfan S, Krishnan P, Kumar AV, Maharjan S, Mushtaq S, Noorie T, Paterson DL, Pearson A, Perry C, Pike R, Rao B, Ray U, Sarma JB, Sharma M, Sheridan E, Thirunarayan MA, Turton J, Upadhyay S, Warner M, Welfare W, Livermore DM, Woodford N (2010) Emergence of a new antibiotic resistance mechanism in India, Pakistan, and the UK: a molecular, biological, and epidemiological study. Lancet Infect Dis 10(9):597–602. https://doi.org/10.1016/S1473-3099(10)70143-2
86. Cornaglia G, Giamarellou H, Rossolini GM (2011) Metallo-beta-lactamases: a last frontier for beta-lactams? Lancet Infect Dis 11(5):381–393. https://doi.org/10.1016/S1473-3099(11)70056-1
87. Rolain JM, Parola P, Cornaglia G (2010) New Delhi metallo-beta-lactamase (NDM-1): towards a new

pandemia? Clin Microbiol Infect 16(12):1699–1701. https://doi.org/10.1111/j.1469-0691.2010.03385.x
88. Velkov T, Thompson PE, Nation RL, Li J (2010) Structure--activity relationships of polymyxin antibiotics. J Med Chem 53(5):1898–1916. https://doi.org/10.1021/jm900999h
89. Jin L, Li J, Nation RL, Nicolazzo JA (2009) Brain penetration of colistin in mice assessed by a novel high-performance liquid chromatographic technique. Antimicrob Agents Chemother 53(10):4247–4251. https://doi.org/10.1128/AAC.00485-09
90. Kimura Y, Kitamura H, Araki T, Noguchi K, Baba M (1981) Analytical and preparative methods for polymyxin antibiotics using high-performance liquid chromatography with a porous styrene divinylbenzene paking. J Chromatogr 206:563–572
91. Terabe S, Konaka R, Inouye K (1979) Separation of some polypeptide hormones by high-performance liquid chromatography. J Chromatogr 172:163–177
92. Orwa JA, Govaerts C, Busson R, Roets E, Van Schepdael A, Hoogmartens J (2001) Isolation and structural characterization of polymyxin B components. J Chromatogr A 912(2):369–373
93. Orwa JA, Govaerts C, Busson R, Roets E, Van Schepdael A, Hoogmartens J (2001) Isolation and structural characterization of colistin components. J Antibiot (Tokyo) 54(7):595–599
94. Govaerts C, Orwa J, Van Schepdael A, Roets E, Hoogmartens J (2002) Liquid chromatography-ion trap tandem mass spectrometry for the characterization of polypeptide antibiotics of the colistin series in commercial samples. J Chromatogr A 976(1–2):65–78
95. Govaerts C, Orwa J, Van Schepdael A, Roets E, Hoogmartens J (2002) Characterization of polypeptide antibiotics of the polymyxin series by liquid chromatography electrospray ionization ion trap tandem mass spectrometry. J Pept Sci 8(2):45–55. https://doi.org/10.1002/psc.367
96. Shaheen M, Li J, Ross AC, Vederas JC, Jensen SE (2011) Paenibacillus polymyxa PKB1 produces variants of polymyxin B-type antibiotics. Chem Biol 18(12):1640–1648. https://doi.org/10.1016/j.chembiol.2011.09.017
97. Terabe S, Konaka R, Shoji J (1979) Seperation of polymyxins and octapeptins by high-performance liquid chromatography. J Chromatogr 173:313–320
98. Decolin D, Leroy P, Nicolas A, Archimbault P (1997) Hyphenated liquid chromatographic method for the determination of colistin residues in bovine tissues. J Chromatogr Sci 35(12):557–564
99. He J, Ledesma KR, Lam WY, Figueroa DA, Lim TP, Chow DS, Tam VH (2010) Variability of polymyxin B major components in commercial formulations. Int J Antimicrob Agents 35(3):308–310. https://doi.org/10.1016/j.ijantimicag.2009.11.005
100. British Pharmacopoeia Comission (2014) British pharmacopeia. Stationary Office, London
101. European Pharmacopoeia Commission (2014) European pharmacopoeia, 8th edn. Council of Europe, Strasbourg
102. United States Pharmacopeial Convention (2013) United States pharmacopeia 36: National Formulary 31. United States Pharmacopeial Convention, Inc., Rockville, MD, USA
103. Bergen PJ, Li J, Rayner CR, Nation RL (2006) Colistin methanesulfonate is an inactive prodrug of colistin against Pseudomonas aeruginosa. Antimicrob Agents Chemother 50(6):1953–1958
104. Li J, Milne RW, Nation RL, Turnidge JD, Smeaton TC, Coulthard K (2004) Pharmacokinetics of colistin methanesulphonate and colistin in rats following an intravenous dose of colistin methanesulphonate. J Antimicrob Chemother 53(5):837–840
105. Li J, Coulthard K, Milne R, Nation RL, Conway S, Peckham D, Etherington C, Turnidge J (2003) Steady-state pharmacokinetics of intravenous colistin methanesulphonate in patients with cystic fibrosis. J Antimicrob Chemother 52(6):987–992
106. Marchand S, Lamarche I, Gobin P, Couet W (2010) Dose-ranging pharmacokinetics of colistin methanesulphonate (CMS) and colistin in rats following single intravenous CMS doses. J Antimicrob Chemother 65(8):1753–1758. https://doi.org/10.1093/jac/dkq183
107. Garonzik SM, Li J, Thamlikitkul V, Paterson DL, Shoham S, Jacob J, Silveira FP, Forrest A, Nation RL (2011) Population pharmacokinetics of colistin methanesulfonate and formed colistin in critically ill patients from a multicenter study provide dosing suggestions for various categories of patients. Antimicrob Agents Chemother 55(7):3284–3294. https://doi.org/10.1128/AAC.01733-10
108. Markou N, Markantonis SL, Dimitrakis E, Panidis D, Boutzouka E, Karatzas S, Rafailidis P, Apostolakos H, Baltopoulos G (2008) Colistin serum concentrations after intravenous administration in critically ill patients with serious multidrug-resistant, gram-negative bacilli infections: a prospective, open-label, uncontrolled study. Clin Ther 30(1):143–151. https://doi.org/10.1016/j.clinthera.2008.01.015
109. Mohamed AF, Karaiskos I, Plachouras D, Karvanen M, Pontikis K, Jansson B, Papadomichelakis E, Antoniadou A, Giamarellou H, Armaganidis A, Cars O, Friberg LE (2012) Application of a loading dose of colistin methanesulfonate in critically ill patients: population pharmacokinetics, protein binding, and prediction of bacterial kill. Antimicrob Agents Chemother 56(8):4241–4249. https://doi.org/10.1128/AAC.06426-11
110. Plachouras D, Karvanen M, Friberg LE, Papadomichelakis E, Antoniadou A, Tsangaris I, Karaiskos I, Poulakou G, Kontopidou F, Armaganidis A, Cars O, Giamarellou H (2009) Population pharmacokinetic analysis of colistin methanesulphonate and colistin after intravenous administration in critically ill patients with gram-negative bacterial infec-

tions. Antimicrob Agents Chemother 53:3430–3436. https://doi.org/10.1128/AAC.01361-08
111. Lee J, Han S, Jeon S, Hong T, Song W, Woo H, Yim DS (2013) Population pharmacokinetic analysis of colistin in burn patients. Antimicrob Agents Chemother 57(5):2141–2146. https://doi.org/10.1128/AAC.00271-13
112. Imberti R, Cusato M, Villani P, Carnevale L, Iotti GA, Langer M, Regazzi M (2010) Steady-state pharmacokinetics and BAL concentration of colistin in critically Ill patients after IV colistin methanesulfonate administration. Chest 138(6):1333–1339. https://doi.org/10.1378/chest.10-0463
113. He H, Li JC, Nation RL, Jacob J, Chen G, Lee HJ, Tsuji BT, Thompson PE, Roberts K, Velkov T, Li J (2013) Pharmacokinetics of four different brands of colistimethate and formed colistin in rats. J Antimicrob Chemother 68(10):2311–2317. https://doi.org/10.1093/jac/dkt207
114. Li J, Milne RW, Nation RL, Turnidge JD, Coulthard K (2003) Stability of colistin and colistin methanesulfonate in aqueous media and plasma as determined by high-performance liquid chromatography. Antimicrob Agents Chemother 47(4):1364–1370
115. Gales AC, Jones RN, Sader HS (2011) Contemporary activity of colistin and polymyxin B against a worldwide collection of Gram-negative pathogens: results from the SENTRY Antimicrobial Surveillance Program (2006-09). J Antimicrob Chemother 66:2070–2074
116. Hirsch HA, Mc CC, Finland M (1960) Polymyxin B and colistin: activity, resistance and crossresistance in vitro. Proc Soc Exp Biol Med 103:338–342
117. Catchpole CR, Andrews JM, Brenwald N, Wise R (1997) A reassessment of the in-vitro activity of colistin sulphomethate sodium. J Antimicrob Chemother 39(2):255–260
118. Schulin T (2002) In vitro activity of the aerosolized agents colistin and tobramycin and five intravenous agents against Pseudomonas aeruginosa isolated from cystic fibrosis patients in southwestern Germany. J Antimicrob Chemother 49(2):403–406
119. Tan TY, Ng SY (2006) The in-vitro activity of colistin in gram-negative bacteria. Singap Med J 47(7):621–624
120. Gales AC, Reis AO, Jones RN (2001) Contemporary assessment of antimicrobial susceptibility testing methods for polymyxin B and colistin: review of available interpretative criteria and quality control guidelines. J Clin Microbiol 39(1):183–190. https://doi.org/10.1128/JCM.39.1.183-190.2001
121. Niks M, Hanzen J, Ohlasova D, Rovna D, Purgelova A, Szovenyiova Z, Vaculikova A (2004) Multiresistant nosocomial bacterial strains and their "in vitro" susceptibility to chloramphenicol and colistin. Klin Mikrobiol Infekc Lek 10(3):124–129
122. Bogiel T, Mikucka A, Skalski T, Gospodarek E (2010) Occurrence and susceptibility to antibiotics of carbapenem-resistant Pseudomonas aeruginosa strains between 1998 and 2009. Med Dosw Mikrobiol 62(3):221–229
123. Cernohorska L, Slavikova P (2010) Antibiotic resistance and biofilm formation in Pseudomonas aeruginosa strains isolated from patients with urinary tract infections. Epidemiol Mikrobiol Imunol 59(4):154–157
124. Walkty A, DeCorby M, Nichol K, Karlowsky JA, Hoban DJ, Zhanel GG (2009) In vitro activity of colistin (polymyxin E) against 3,480 isolates of gram-negative bacilli obtained from patients in Canadian hospitals in the CANWARD study, 2007-2008. Antimicrob Agents Chemother 53(11):4924–4926. https://doi.org/10.1128/AAC.00786-09
125. Zhanel GG, Adam HJ, Baxter MR, Fuller J, Nichol KA, Denisuik AJ, Lagace-Wiens PR, Walkty A, Karlowsky JA, Schweizer F, Hoban DJ, Canadian Antimicrobial Resistance A (2013) Antimicrobial susceptibility of 22746 pathogens from Canadian hospitals: results of the CANWARD 2007-11 study. J Antimicrob Chemother 68(Suppl 1):i7–i22. https://doi.org/10.1093/jac/dkt022
126. Jones RN, Guzman-Blanco M, Gales AC, Gallegos B, Castro AL, Martino MD, Vega S, Zurita J, Cepparulo M, Castanheira M (2013) Susceptibility rates in Latin American nations: report from a regional resistance surveillance program (2011). Braz J Infect Dis: Off Publ Braz Soc Infect Dis 17(6):672–681. https://doi.org/10.1016/j.bjid.2013.07.002
127. Gales AC, Castanheira M, Jones RN, Sader HS (2012) Antimicrobial resistance among Gram-negative bacilli isolated from Latin America: results from SENTRY antimicrobial surveillance program (Latin America, 2008–2010). Diagn Microbiol Infect Dis 73(4):354–360. https://doi.org/10.1016/j.diagmicrobio.2012.04.007
128. Lee JY, Song JH, Ko KS (2011) Identification of nonclonal Pseudomonas aeruginosa isolates with reduced colistin susceptibility in Korea. Microb Drug Resist 17(2):299–304. https://doi.org/10.1089/mdr.2010.0145
129. Kuck NA (1976) In vitro and in vivo activities of minocycline and other antibiotics against Acinetobacter (Herellea-Mima). Antimicrob Agents Chemother 9(3):493–497
130. Yau W, Owen RJ, Poudyal A, Bell JM, Turnidge JD, Yu HH, Nation RL, Li J (2009) Colistin heteroresistance in multidrug-resistant Acinetobacter baumannii clinical isolates from the Western Pacific region in the SENTRY antimicrobial surveillance programme. J Infect 58(2):138–144. https://doi.org/10.1016/j.jinf.2008.11.002
131. Queenan AM, Pillar CM, Deane J, Sahm DF, Lynch AS, Flamm RK, Peterson J, Davies TA (2012) Multidrug resistance among Acinetobacter spp. in the USA and activity profile of key agents: results from CAPITAL Surveillance 2010. Diagn Microbiol Infect Dis 73(3):267–270. https://doi.org/10.1016/j.diagmicrobio.2012.04.002

132. Hawser SP (2010) Susceptibility of Klebsiella pneumoniae clinical isolates from 2007 to 2009 to colistin and comparator antibiotics. Int J Antimicrob Agents 36(4):383–384. https://doi.org/10.1016/j.ijantimicag.2010.06.004
133. Sader HS, Farrell DJ, Jones RN (2011) Susceptibility of Klebsiella spp. to colistin and polymyxin B: results from the SENTRY antimicrobial surveillance program (2006-2009). Int J Antimicrob Agents 37(2):174–175. https://doi.org/10.1016/j.ijantimicag.2010.10.005
134. Nakamura T, Komatsu M, Yamasaki K, Fukuda S, Higuchi T, Ono T, Nishio H, Sueyoshi N, Kida K, Satoh K, Toda H, Toyokawa M, Nishi I, Sakamoto M, Akagi M, Mizutani T, Nakai I, Kofuku T, Orita T, Zikimoto T, Natsume S, Wada Y (2014) Susceptibility of various oral antibacterial agents against extended spectrum beta-lactamase producing Escherichia coli and Klebsiella pneumoniae. J Infect Chemother 20(1):48–51. https://doi.org/10.1016/j.jiac.2013.08.004
135. Johansen HK, Moskowitz SM, Ciofu O, Pressler T, Hoiby N (2008) Spread of colistin resistant non-mucoid Pseudomonas aeruginosa among chronically infected Danish cystic fibrosis patients. J Cyst Fibros 7(5):391–397. https://doi.org/10.1016/j.jcf.2008.02.003
136. Li J, Turnidge J, Milne R, Nation RL, Coulthard K (2001) In vitro pharmacodynamic properties of colistin and colistin methanesulfonate against Pseudomonas aeruginosa isolates from patients with cystic fibrosis. Antimicrob Agents Chemother 45(3):781–785
137. Falagas ME, Rafailidis PI, Matthaiou DK, Virtzili S, Nikita D, Michalopoulos A (2008) Pandrug-resistant Klebsiella pneumoniae, Pseudomonas aeruginosa and Acinetobacter baumannii infections: characteristics and outcome in a series of 28 patients. Int J Antimicrob Agents 32(5):450–454. https://doi.org/10.1016/j.ijantimicag.2008.05.016
138. Landman D, Bratu S, Alam M, Quale J (2005) Citywide emergence of Pseudomonas aeruginosa strains with reduced susceptibility to polymyxin B. J Antimicrob Chemother 55(6):954–957
139. Wang CY, Jerng JS, Chen KY, Lee LN, Yu CJ, Hsueh PR, Yang PC (2006) Pandrug-resistant Pseudomonas aeruginosa among hospitalised patients: clinical features, risk-factors and outcomes. Clin Microbiol Infect 12(1):63–68. https://doi.org/10.1111/j.1469-0691.2005.01305.x
140. Denton M, Kerr K, Mooney L, Keer V, Rajgopal A, Brownlee K, Arundel P, Conway S (2002) Transmission of colistin-resistant Pseudomonas aeruginosa between patients attending a pediatric cystic fibrosis center. Pediatr Pulmonol 34(4):257–261. https://doi.org/10.1002/ppul.10166
141. Al-Sweih NA, Al-Hubail MA, Rotimi VO (2011) Emergence of tigecycline and colistin resistance in acinetobacter species isolated from patients in Kuwait hospitals. J Chemother 23(1):13–16
142. Ko KS, Suh JY, Kwon KT, Jung SI, Park KH, Kang CI, Chung DR, Peck KR, Song JH (2007) High rates of resistance to colistin and polymyxin B in subgroups of Acinetobacter baumannii isolates from Korea. J Antimicrob Chemother 60(5):1163–1167. https://doi.org/10.1093/jac/dkm305
143. Antoniadou A, Kontopidou F, Poulakou G, Koratzanis E, Galani I, Papadomichelakis E, Kopterides P, Souli M, Armaganidis A, Giamarellou H (2007) Colistin-resistant isolates of Klebsiella pneumoniae emerging in intensive care unit patients: first report of a multiclonal cluster. J Antimicrob Chemother 59(4):786–790. https://doi.org/10.1093/jac/dkl562
144. Meletis G, Tzampaz E, Sianou E, Tzavaras I, Sofianou D (2011) Colistin heteroresistance in carbapenemase-producing Klebsiella pneumoniae. J Antimicrob Chemother 66(4):946–947. https://doi.org/10.1093/jac/dkr007
145. Mezzatesta ML, Gona F, Caio C, Petrolito V, Sciortino D, Sciacca A, Santangelo C, Stefani S (2011) Outbreak of KPC-3-producing, and colistin-resistant, Klebsiella pneumoniae infections in two Sicilian hospitals. Clin Microbiol Infect. https://doi.org/10.1111/j.1469-0691.2011.03572.x
146. Suh JY, Son JS, Chung DR, Peck KR, Ko KS, Song JH (2010) Nonclonal emergence of colistin-resistant Klebsiella pneumoniae isolates from blood samples in South Korea. Antimicrob Agents Chemother 54(1):560–562. https://doi.org/10.1128/AAC.00762-09
147. Menuet M, Bittar F, Stremler N, Dubus JC, Sarles J, Raoult D, Rolain JM (2008) First isolation of two colistin-resistant emerging pathogens, Brevundimonas diminuta and Ochrobactrum anthropi, in a woman with cystic fibrosis: a case report. J Med Case Rep 2:373. https://doi.org/10.1186/1752-1947-2-373
148. Mendes RE, Fritsche TR, Sader HS, Jones RN (2008) Increased antimicrobial susceptibility profiles among polymyxin-resistant Acinetobacter baumannii clinical isolates. Clin Infect Dis 46(8):1324–1326. https://doi.org/10.1086/533476
149. Li J, Nation RL, Owen RJ, Wong S, Spelman D, Franklin C (2007) Antibiograms of multidrug-resistant clinical Acinetobacter baumannii: promising therapeutic options for treatment of infection with colistin-resistant strains. Clin Infect Dis 45(5):594–598. https://doi.org/10.1086/520658
150. Bergen PJ, Forrest A, Bulitta JB, Tsuji BT, Sidjabat HE, Paterson DL, Li J, Nation RL (2011) Clinically relevant plasma concentrations of colistin in combination with imipenem enhance pharmacodynamic activity against multidrug-resistant Pseudomonas aeruginosa at multiple inocula. Antimicrob Agents Chemother 55(11):5134–5142. https://doi.org/10.1128/AAC.05028-11
151. Hermes DM, Pormann Pitt C, Lutz L, Teixeira AB, Ribeiro VB, Netto B, Martins AF, Zavascki AP, Barth AL (2013) Evaluation of heteroresistance to

polymyxin B among carbapenem-susceptible and -resistant Pseudomonas aeruginosa. J Med Microbiol 62. (Pt 8:1184–1189. https://doi.org/10.1099/jmm.0.059220-0
152. Vidaillac C, Benichou L, Duval RE (2012) In vitro synergy of colistin combinations against colistin-resistant acinetobacter baumannii, pseudomonas aeruginosa, and Klebsiella pneumoniae isolates. Antimicrob Agents Chemother 56(9):4856–4861. https://doi.org/10.1128/AAC.05996-11
153. Hawley JS, Murray CK, Jorgensen JH (2008) Colistin heteroresistance in acinetobacter and its association with previous colistin therapy. Antimicrob Agents Chemother 52(1):351–352. https://doi.org/10.1128/AAC.00766-07
154. Owen RJ, Li J, Nation RL, Spelman D (2007) In vitro pharmacodynamics of colistin against Acinetobacter baumannii clinical isolates. J Antimicrob Chemother 59(3):473–477. https://doi.org/10.1093/jac/dkl512
155. Tan CH, Li J, Nation RL (2007) Activity of colistin against heteroresistant Acinetobacter baumannii and emergence of resistance in an in vitro pharmacokinetic/pharmacodynamic model. Antimicrob Agents Chemother 51(9):3413–3415. https://doi.org/10.1128/AAC.01571-06
156. Lo-Ten-Foe JR, de Smet AM, Diederen BM, Kluytmans JA, van Keulen PH (2007) Comparative evaluation of the VITEK 2, disk diffusion, etest, broth microdilution, and agar dilution susceptibility testing methods for colistin in clinical isolates, including heteroresistant Enterobacter cloacae and Acinetobacter baumannii strains. Antimicrob Agents Chemother 51(10):3726–3730. https://doi.org/10.1128/AAC.01406-06
157. Li J, Rayner CR, Nation RL, Owen RJ, Spelman D, Tan KE, Liolios L (2006) Heteroresistance to colistin in multidrug-resistant Acinetobacter baumannii. Antimicrob Agents Chemother 50(9):2946–2950. https://doi.org/10.1128/AAC.00103-06
158. Poudyal A, Howden BP, Bell JM, Gao W, Owen RJ, Turnidge JD, Nation RL, Li J (2008) In vitro pharmacodynamics of colistin against multidrug-resistant Klebsiella pneumoniae. J Antimicrob Chemother 62(6):1311–1318. https://doi.org/10.1093/jac/dkn425
159. Gales AC, Jones RN, Sader HS (2006) Global assessment of the antimicrobial activity of polymyxin B against 54 731 clinical isolates of Gram-negative bacilli: report from the SENTRY antimicrobial surveillance programme (2001-2004). Clin Microbiol Infect 12(4):315–321. https://doi.org/10.1111/j.1469-0691.2005.01351.x
160. Kosakai N, Oguri T (1976) Distribution and changes of antibiotic susceptibility of genus Haemophilus (author's transl). Jpn J Antibiot 29(2):159–166
161. Thornsberry C, Baker CN, Kirven LA (1978) In vitro activity of antimicrobial agents on Legionnaires disease bacterium. Antimicrob Agents Chemother 13(1):78–80
162. Fosse T, Giraud-Morin C, Madinier I (2003) Induced colistin resistance as an identifying marker for Aeromonas phenospecies groups. Lett Appl Microbiol 36(1):25–29
163. Rastogi N, Potar MC, David HL (1986) Antimycobacterial spectrum of colistin (polymixin E). Ann Inst Pasteur Microbiol 137A(1):45–53
164. Aydin F, Atabay HI, Akan M (2001) The isolation and characterization of Campylobacter jejuni subsp. jejuni from domestic geese (Anser anser). J Appl Microbiol 90(4):637–642
165. Kiehlbauch JA, Baker CN, Wachsmuth IK (1992) In vitro susceptibilities of aerotolerant Campylobacter isolates to 22 antimicrobial agents. Antimicrob Agents Chemother 36(4):717–722
166. Felegie TP, Yu VL, Rumans LW, Yee RB (1979) Susceptibility of Pseudomonas maltophilia to antimicrobial agents, singly and in combination. Antimicrob Agents Chemother 16(6):833–837
167. Nicodemo AC, Araujo MR, Ruiz AS, Gales AC (2004) In vitro susceptibility of Stenotrophomonas maltophilia isolates: comparison of disc diffusion, Etest and agar dilution methods. J Antimicrob Chemother 53(4):604–608. https://doi.org/10.1093/jac/dkh128
168. Maurin M, Gasquet S, Ducco C, Raoult D (1995) MICs of 28 antibiotic compounds for 14 Bartonella (formerly Rochalimaea) isolates. Antimicrob Agents Chemother 39(11):2387–2391
169. Myers WF, Grossman DM, Wisseman CL Jr (1984) Antibiotic susceptibility patterns in Rochalimaea quintana, the agent of trench fever. Antimicrob Agents Chemother 25(6):690–693
170. Lesmana M, Subekti DS, Tjaniadi P, Simanjuntak CH, Punjabi NH, Campbell JR, Oyofo BA (2002) Spectrum of vibrio species associated with acute diarrhea in North Jakarta, Indonesia. Diagn Microbiol Infect Dis 43(2):91–97
171. von Graevenitz A, Nourbakhsh M (1972) Antimicrobial resistance of the genera Proteus, Providencia and Serratia with special reference to multiple resistance patterns. Med Microbiol Immunol 157(2):142–148
172. Greenfield S, Feingold DS (1970) The synergistic action of the sulfonamides and the polymyxins against Serratia marcescens. J Infect Dis 121(5):555–558
173. Bisbe J, Gatell JM, Puig J, Mallolas J, Martinez JA, Jimenez de Anta MT, Soriano E (1988) Pseudomonas aeruginosa bacteremia: univariate and multivariate analyses of factors influencing the prognosis in 133 episodes. Rev Infect Dis 10(3):629–635
174. Garcia-Rodriguez JA, Garcia-Garcia MI, Garcia-Sanchez E, Garcia-Sanchez JE, Munoz Bellido JL (1989) In vitro activity of 16 antimicrobial agents against Helicobacter (Campylobacter) pylori. Enferm Infecc Microbiol Clin 7(10):544–546
175. Glupczynski Y, Delmee M, Bruck C, Labbe M, Avesani V, Burette A (1988) Susceptibility of clinical isolates of Campylobacter pylori to 24 antimicrobial and anti-ulcer agents. Eur J Epidemiol 4(2):154–157

176. Doern GV, Morse SA (1980) Branhamella (Neisseria) catarrhalis: criteria for laboratory identification. J Clin Microbiol 11(2):193–195
177. Muyembe T, Vandepitte J, Desmyter J (1973) Natural colistin resistance in Edwardsiella tarda. Antimicrob Agents Chemother 4(5):521–524
178. Laffineur K, Janssens M, Charlier J, Avesani V, Wauters G, Delmee M (2002) Biochemical and susceptibility tests useful for identification of nonfermenting gram-negative rods. J Clin Microbiol 40(3):1085–1087
179. Dance DA, Wuthiekanun V, Naigowit P, White NJ (1989) Identification of Pseudomonas pseudomallei in clinical practice: use of simple screening tests and API 20NE. J Clin Pathol 42(6):645–648
180. Finland M, Garner C, Wilcox C, Sabath LD (1976) Susceptibility of beta-hemolytic streptococci to 65 antibacterial agents. Antimicrob Agents Chemother 9(1):11–19
181. Finland M, Garner C, Wilcox C, Sabath LD (1976) Susceptibility of pneumococci and Haemophilus influenzae to antibacterial agents. Antimicrob Agents Chemother 9(2):274–287
182. Hoeprich PD (1970) The polymyxins. Med Clin North Am 54(5):1257–1265
183. Clinical and Laboratory Standards Institute (CLSI) (2005) Performance standards for antimicrobial susceptibility testing: fifteenth informational supplement (M100-S15). CLSI, Wayne
184. Eickhoff TC, Finland M (1965) Polymyxin B and colistin: In vitro activity against Pseudomonas aeruginosa. Am J Med Sci 249:172–174
185. Li J, Milne RW, Nation RL, Turnidge JD, Coulthard K, Johnson DW (2001) A simple method for the assay of colistin in human plasma, using pre-column derivatization with 9-fluorenylmethyl chloroformate in solid-phase extraction cartridges and reversed-phase high-performance liquid chromatography. J Chromatogr B Biomed Sci Appl 761(2):167–175
186. Li J, Milne RW, Nation RL, Turnidge JD, Coulthard K, Valentine J (2002) Simple method for assaying colistin methanesulfonate in plasma and urine using high-performance liquid chromatography. Antimicrob Agents Chemother 46(10):3304–3307
187. Li J, Milne RW, Nation RL, Turnidge JD, Smeaton TC, Coulthard K (2003) Use of high-performance liquid chromatography to study the pharmacokinetics of colistin sulfate in rats following intravenous administration. Antimicrob Agents Chemother 47(5):1766–1770
188. Landman D, Urban C, Backer M, Kelly P, Shah N, Babu E, Bratu S, Quale J (2010) Susceptibility profiles, molecular epidemiology, and detection of KPC-producing Escherichia coli isolates from the New York City vicinity. J Clin Microbiol 48(12):4604–4607. https://doi.org/10.1128/JCM.01143-10
189. Quale J, Shah N, Kelly P, Babu E, Backer M, Rosas-Garcia G, Salamera J, George A, Bratu S, Landman D (2012) Activity of polymyxin B and the novel polymyxin analogue CB-182,804 against contemporary Gram-negative pathogens in New York city. Microb Drug Resist 18(2):132–136. https://doi.org/10.1089/mdr.2011.0163

Polymyxins: Mode of Action

4

Zhifeng Li and Tony Velkov

Abstract

The dry antibiotic development pipeline coupled with the emergence of multi-drug resistant Gram-negative 'superbugs' has driven the revival of the polymyxin lipopeptide antibiotics. Understanding the mode of action of antibiotics is an important precursor for optimizing their use and development. This chapter provides a concise treatise of the current knowledge-based on the primary mode of action of polymyxins as well as recent developments in understanding of bacterial cell responses and secondary modes of action.

Keywords

Polymyxin · LPS · Free radical · Type II NADH-quinone oxidoreductase · Outer membrane remodelling

Z. Li
State Key Laboratory of Microbial Technology, Shandong University, Qingdao Campus, Qingdao, Shandong, People's Republic of China

T. Velkov (✉)
Department of Pharmacology and Therapeutics, University of Melbourne, Melbourne, Australia
e-mail: Tony.Velkov@unimelb.edu.au

The outer membrane of the Gram-negative cell acts as a permeability barrier that protects the cell from various noxious substances, including numerous antimicrobials [1]. Polymyxins exert their antimicrobial action via direct interaction with the lipid A component of the lipopolysaccharide (LPS) which leads to a disruption of this critical barrier function. Accordingly, understanding the mechanism of polymyxin antibacterial activity requires a brief review of the architecture of LPS and the outer membrane. The complex asymmetrical structure of the outer membrane comprises an inner phospholipid leaflet, as well as an outer leaflet that predominantly contains LPS, proteins and phospholipids [1]. LPS is composed of three domains, a conserved inner core 2-keto-3-deoxyoctonoic acid (Kdo) bound to lipid A and a variable *O*-antigen composed of repeating units of various polysaccharides [1–4]. The structure of lipid A consists of aβ-10-6-linked D-glucosamine (GlcN) disaccharide that is phosphorylated at the 1- and 4′-positions and decorated by a variable number of saturated hydrocarbon chains, generally C10-C14 in length [2, 4]. Lipid A is intercalated within the outer leaflet, where in the saturated lipid A hydrocarbon chains are tightly packed together within the membrane through van der Waals forces, thereby acting as an anchor for the entire LPS structure [1, 5, 6]. Divalent cations (Mg^{2+} and Ca^{2+}) associate with the lipid A phosphoresters and function to bridge adjacent LPS mole-

J. Li et al. (eds.), *Polymyxin Antibiotics: From Laboratory Bench to Bedside*, Advances in Experimental Medicine and Biology 1145, https://doi.org/10.1007/978-3-030-16373-0_4

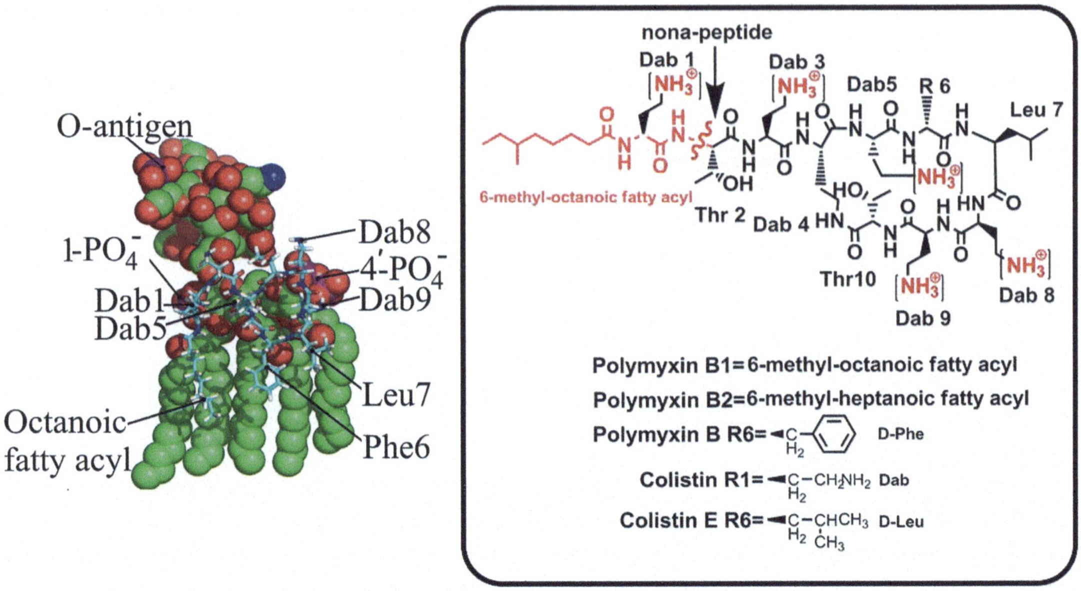

Fig. 4.1 *Left panel.* NMR-based model of polymyxin B bound to the lipid A component of *E. coli* LPS [7, 8]. *Right panel.* Chemical structures of polymyxin B, colistin and nonapeptide. Polymyxin residues: Thr: threonine; Leu: leucine; Phe: phenylalanine; Dab: α,γ-diaminobutyric acid

cules [1, 5, 6]. The barrier function of the outer membrane is further accentuated by a highly repulsive anionic charge conveyed by lipid A phosphorester moieties, as well as phosphate and carboxylate functionalities decorating the core and *O*-antigen sugars [1, 5, 6].

Amphipathicity of the polymyxins is critical for their outer membrane permeabilizing action. The conserved elements in the chemical structure of polymyxins that contribute to this amphipathicity includes the two hydrophobic domains (the *N*-terminal fatty acyl chain and the hydrophobic position 6–7 segment) separated by segments of polar (Thr) and cationic Dab side chains. The elucidation of the three-dimensional NMR solution state structure of polymyxin B in complex with LPS revealed the polymyxin B molecule is folded such that the polar and hydrophobic domains form two distinct faces, thereby conferring structural amphipathicity (Fig. 4.1) [7, 8].

4.1 Primary Mode of Action of Polymyxins

Polymyxins are believed to exert their primary antimicrobial mode of action by permeabilizing the outer membrane via a direct interaction with LPS. Polymyxins zone-into their primary cellular target, LPS through the initial electrostatic interaction of the cationic L-α,γ-diaminobutyric acid (Dab) side-chains with the phosphate groups of the lipid A component of LPS, displacing divalent cations (Ca^{2+} and Mg^{2+}) that bridge adjacent LPS molecules (Fig. 4.2, Stage 1) [1, 5, 6, 9–11]. This initial electrostatic interaction allows the *N*-terminal fatty acyl chain and hydrophobic position 6–7 motif (Polymyxin B: D-Phe^6-L-Leu^7 and colistin: D-Leu^6-L-Leu^7) of the polymyxin molecule to insert into the fatty acyl chain layer of the lipid A molecules. The insertion of the

Fig. 4.2 (continued) leads to the displacement of divalent cations that help stabilize the outer membrane structure by bridging adjacent LPS molecules. *Stage 2.* The positively charged polymyxins displace divalent cations that bridge adjacent LPS molecules. The hydrophobic insertion destabilizes the outer membrane. *Stage 3.* The polymyxin molecule penetrates into the inner membrane and inhibits the respiratory enzyme NDH-2

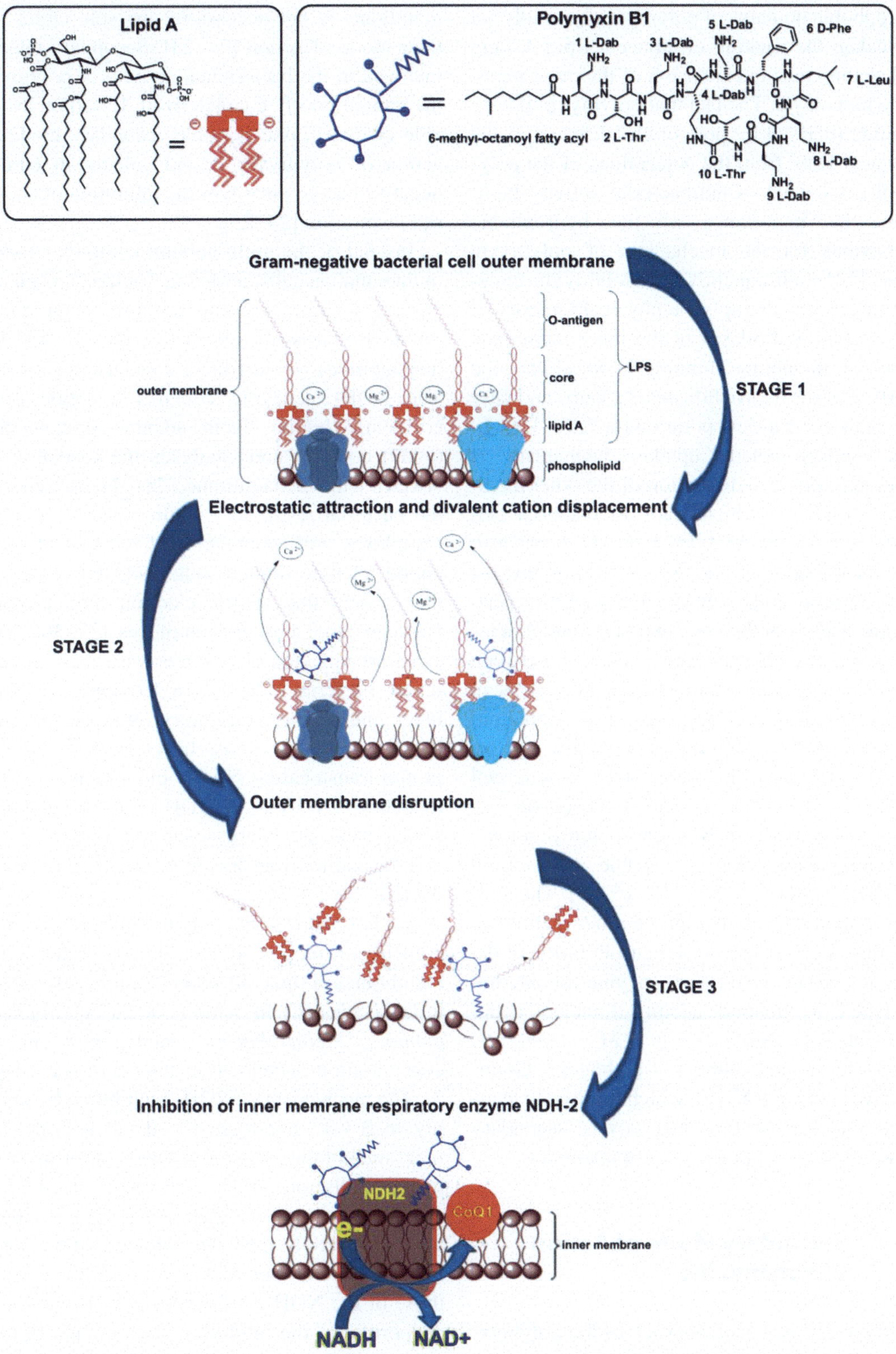

Fig. 4.2 Schematic diagram depicting the putative mode of action of polymyxins. *Stage 1.* Electrostatic attraction between the positively charges polymyxin molecule and the negatively charges bacterial outer membrane surface

hydrophobic domains of polymyxins possibly act to weaken the packing of adjacent lipid A fatty acyl chains causing expansion of the outer membrane monolayer. The fact that polymyxin B nonapeptide (derived by proteolytic cleavage of the fatty acyl-Dab[1] from the *N*-terminus of the polymyxin) is devoid of antibacterial activity highlights the importance of the hydrophobic interactions for the mechanism of polymyxin action [12]. Subsequently, the polymyxin molecule inserts and disrupts the physical integrity of the phospholipid bilayer of the inner membrane leaflet via membrane thinning by straddling the interface of the hydrophilic head groups and fatty acyl chains or transient poration [6, 7, 10, 13, 14]. This 'self-promoted' uptake mechanism is believed to produce disruption of the outer membrane structures, which leads to bacterial cell death (Fig. 4.2, Stage 2) [5, 6, 9–11]. It has also been proposed that the MOA of polymyxins involves producing contacts between the periplasmic leaflets of the inner and outer membranes that promotes phospholipid exchange between the inner and outer membrane leaflets. This in turn would result in the loss of phospholipid compositional specificity, potentially to an osmotic imbalance that contributes to lytic cell death [15, 16]. This postulate is based on evidence that polymyxin B when bound to anionic phospholipid vesicles is capable of forming vesicle-to-vesicle contacts [13, 15–17]. The ability of polymyxins to disrupt the inner membrane structure is coincident with their inhibition of the inner membrane respiratory enzyme the alternative type 2 nicotinamide adenine dinucleoti dedehydrogenase (NDH-2) in *Mycobacterium smegmatis* and in a number of pathogenic Gram-negative bacteria [18, 19], which intuitively leads us to the next section that covers secondary modes of action of polymyxin lipopeptides.

4.2 Secondary Mode of Action of Polymyxins

Although cationic peptides such as the polymyxins are traditionally thought of as outer membrane-active agents [20], the bacterial outer membrane is not necessarily the sole target for their mode of action [21–23]. Secondary targets involved in the bactericidal activity of polymyxins remain poorly characterized. Based on available evidence, one possible secondary mode of action of polymyxin B and colistin in Gram-negative bacteria involves the inhibition of bacterial respiration [24, 25].

Instead of the multi-subunit complex I found in mammalian cells, protozoa, bacteria and plants possess a single sub-unit non-proton pumping, rotenone insensitive alternative type II NADH-menaquinone oxidoreductase (NDH-2) [26–29]. The NDH-2 enzyme contains a single non-covalently bound flavin adenine dinucleotide (FAD) cofactor and catalyzes the oxidation of NADH with menaquinone [28–31]. In general, the bacterial respiratory chain consists of three complexes with quinones and reduced nicotinamide adenine dinucleotide (NADH) acting as the carriers that shuttle electrons and protons between large protein complexes [32–36]. The exact organization of enzymes varies among different bacteria [32–34]. In Complex 1, three inner membrane respiratory enzymes of the NADH oxidase family have been identified: proton-translocating NADH-quinone (Q) oxidoreductase (NDH-1), NADH-Q oxidoreductase which lacks an energy-coupling site (NDH-2) and the sodium-translocating NADH-Q oxidoreductase [32–34, 36].

We have shown that polymyxin B, B1, B2 and colistin can inhibit NDH-2 activity in the inner membranes of three different Gram-negative bacterial species (*Escherichia coli*, *Klebsiella pneumoniae*, *Acinetobacter baumannii*) in a concentration-dependent manner (Fig. 4.2, Stage 3). The mechanism of NDH-2 inhibition by polymyxin B was investigated in detail with *E. coli* inner membrane preparations and conformed to a mixed inhibition model with respect to ubiquinone-1 and a non-competitive inhibition model with respect to NADH. The structure of the polymyxins (cyclic peptides) being distinct from those of the NDH-2 substrates NADH and Q_1 is supportive of the inhibition kinetic data, in that they are unlikely to compete for the same sites on the enzyme. Our kinetic data are in line with the reported data for *Gluconobacter oxydans* which showed that the inhibition by gramicidin S and scopafungin was non-competitive with respect to NADH [37]. Scopafungin, which like polymyxin B and colistin possesses a cyclic ring and a long acyl chain in its structure, displayed a mixed inhibition mode with respect to ubiquinone, whereas gramicidin S was a competitive inhibitor [37].

The IC_{50} values for the inhibition by polymyxin B and colistin of NDH-2 activity in inner membrane of the three different Gram-negative bacterial species were in most part comparable, indicating that inter-species differences in NDH-2 do not impact the inhibitory activity of the polymyxins. Polymyxin B was a better inhibitor compared to colistin, which is in line with reported results with the Gram-positive *M. smegmatis* NDH-2 [37]. Although polymyxin B and colistin display high IC_{50} values (polymyxin B $IC_{50} = 50\ \mu M$; colistin $IC_{50} = 251\ \mu M$) for NDH-2 inhibition, under *in vivo* conditions there remains the possibility that very high local concentrations of the antibiotic can accumulate at the site of infection that fall within these IC_{50} value ranges. Coincidently, we have garnered *in vitro* evidence that suggests that polymyxins can accumulate in the inner membrane of Gram-negative bacteria [38]. Therefore, the high IC_{50} values do not dismiss the possibility that NDH-2 represents one of the secondary pathways that is targeted once the polymyxin penetrates the outer membrane. Notably also colistin inhibited NADH-quinone oxidoreductase activity in the polymyxin-susceptible strain of *K. pneumoniae* with a comparable IC_{50} to that of the polymyxin-resistant strain, suggesting polymyxin resistance in these strains is not at the level of the inner membrane respiratory enzymes. Our previous study had indicated that the resistant derivative of *K. pneumoniae* exhibited less negative charge than the wild type that lead to failure of polymyxin interaction at the outer membrane [39]. The NDH-2 activity was not inhibited by CMS, polymyxin B nonapeptide and colistin nonapeptide. The loss of inhibitory activity seen with the polymyxin nonapeptide and CMS suggests that the *N*-terminal fatty acyl chain and the positive charges of the polymyxin molecule are critical for NDH-2 inhibitory activity [40].

The fact that NDH-2 enzymes are not found in mammalian mitochondria and are mainly expressed by protozoa, bacteria and plants makes them very attractive drug targets [41]. Our data suggest that one of the secondary target sites of polymyxins is the type II NADH-quinone oxidoreductase respiratory enzyme that forms an integral part of the bacterial electron transport pathway. In view of the dry antibiotic pipe-line, together with the increasing incidence of multidrug resistant in Gram-negative bacteria, NDH-2 represents an important target that can be exploited for the development of new antibiotics against these problematic pathogens. Notably, energy metabolism and NDH-2 in particular, is emerging as an important drug target in *Mycobacterium tuberculosis* and *Plasmodium falciparum* [28, 29, 31, 42–45].

A recent preliminary biochemical study reported that rapid killing of *A. baumannii* by polymyxins is mediated by a hydroxyl radical death pathway, which although under explored, potentially represents another secondary mode of action whereby polymyxin kill Gram-negative bacterial cells [46]. Coincidently, it has been proposed that most antibiotics cause bacterial cell death via a common mechanism whereby they disrupt bacterial metabolism leading to the generation of reactive oxygen species (ROS) that eventually kills the bacterial cell [47].

4.3 Stress Responses to Polymyxin Treatment in Gram-negative Bacteria

The antibacterial activity of bactericidal antibiotic is not solely governed by its mode of action and its ability to interact with targets [47–51]. There are a number of bacterial response factors associated with exposure to sub-lethal concentrations of polymyxins which will be reviewed in this section. These factors include activation of adaptive resistance mechanisms, the stimulation of protective changes to cell physiology, and even induction of resistance mutations.

The protective stress response, also named adaptive resistance, is an auto-regulated antibiotic induced phenomenon and reversal to the sensitive phenotype in the absence of inducer [52, 53]. Extensive studies in recent years have provided significant insight into the outer membrane remodelling mechanisms responsible for adaptive resistance to polymyxins, perhaps best studied in *Salmonella* [54–57].

In response to polymyxin exposure, the outer membrane undergoes extensive remodelling of structural alterations contributing to adaptive resistance to polymyxins which is triggered by two-component systems (Fig. 4.3) [52, 54, 58]. In *Salmonella*, the response to polymyxins is mediated by the PhoPQ two-component system, where polymyxins interact with and activate the sensor PhoQ by displacing divalent cations from their metal binding sites in the sensor domain [59] and then activates the PhoP response regulator to up-regulate a variety of target genes and ultimately promote adaptation to the stress. Specific changes in OM regulated by activated PhoP include: increasing *O*-antigen chain length, acylating, inhibiting deacylating, hydroxylating lipid A, derivitizing lipid A and LPS core phosphates with cationic groups, palmitoylating OM phosphatidylglycerols, and increasing the level of OM cardiolipins [54, 56]. Therefore, upon PhoPQ activation an extensive alteration of LPS, GPLs, and proteins elaborates the OM barrier more impermeable to polymyxins, thereby promoting bacterial survival (Fig. 4.3) [54, 56].

4.4 Regulation of LPS Remodelling

Upon sensing polymyxins, PhoPQ increases transcription of *pmrD* and the *pmrCAB* operon to activate the response regulator PmrA (Fig. 4.3) [60–62]. The activated PmrA induces expression of a short membrane peptide, PmrR, which binds to and inhibits the lipid A phosphorylase LpxT, an enzyme responsible for increasing the negative charge of the outer membrane leaflet [56, 63]. The activated PmrA induces transcription of genes encoding enzymes that covalently modify lipid A and core sugar phosphates with positively charged aminoarabinose (L-Ara4N) and phosphoethanolamine (pEtN) [54]. The initial step for L-Ara4N moiety modification begins with the oxidation of UDP-glucose in the cytosol by the PhoP and PmrA-regulated UDP-glucose dehydrogenase (PagA, Ugd, or pmrE) [54, 64]. The remaining steps responsible for L-Ara4N moiety modification are encoded within an operon *pmrHFIJKLM* (also known as *arnBCADTEF*), which is activated by PhoPQ through PmrA [54, 64]. When the biosynthesis proceeds to complete UDP formylated-L-Ara4N synthesis, the formylated-L-Ara4N residue is transfer from UDP to undecaprenyl phosphate carrier lipid on the inner leaflet of the inner membrane and then deformylated to form undecaprenyl phosphate-L-Ara4N [54, 64]. Next, the undecaprenyl phosphate-L-Ara4N is flipped into the outer leaflet of the IM where the membrane protein ArnT (also known as PmrK) transfers the L-Ara4N moiety to nascent lipid A phosphates [54, 64]. Finally, *O*-antigen is loaded onto the core structure and then the assembled LPS molecules are moved from the inner membrane through the periplasm to the OM of cell surface by the lipopolysaccharide transport (Lpt) proteins complex acquired driving powers from cytoplasmic ATP hydrolysis [54, 65]. The PmrAB-controlled pEtN transferases encoded by *eptA* (also known as *lptA* or *pmrC*) and *cptA* (or *eptB*) also contribute to polymyxins resistance via their modification of lipid A and LPS core respectively, with positively charged pEtN [54, 56]. Decreased negative charge conferred by cationic groups L-Ara4N and pEtN on lipid A molecules diminishing binding sites plays a significant role in polymyxins resistance while modification of cationic groups on LPS core plays modest effect on resistance [54, 56, 66]. Though varies in details across Gram-negative species, the positively charged modification of lipid A mediated by two-component systems is the most commonly seen mechanism of polymyxins resistance [52, 56, 67, 68].

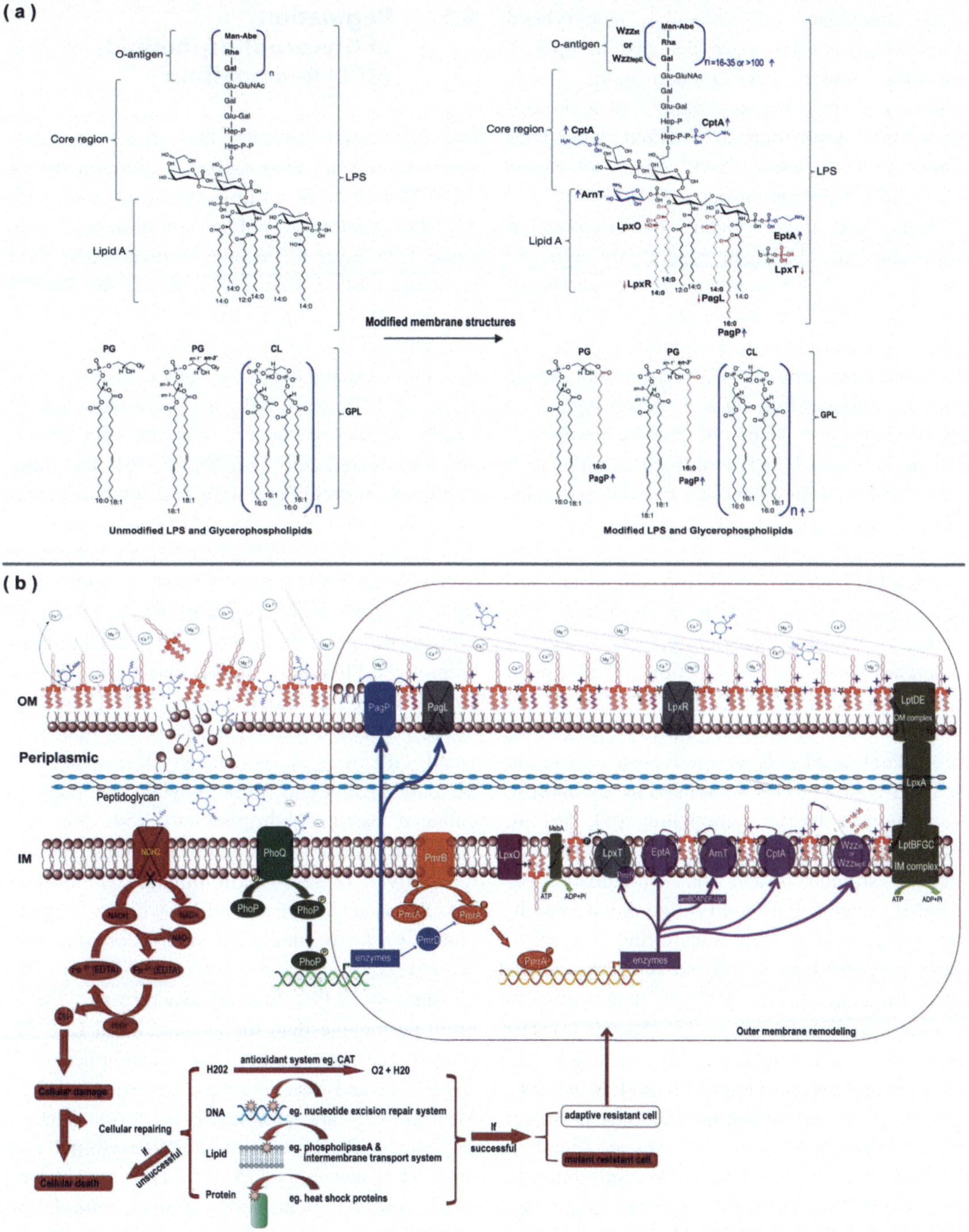

Fig. 4.3 (**a**) Glycerophospholipid and lipid A modifications that ensue with Gram-negative membrane remodelling associated with the development of polymyxin resistance. (**b**) Polymyxin-induced outer membrane remodelling, intracellular biochemical perturbations and resistance pathways. The initial outer membrane disorganisation caused by polymyxin exposure is followed by intracellular redox perturbations of NDH-2 activity. These events are accompanied by activation of repair pathways and outer membrane remodelling. The PhoP/PhoQ two component regulatory system activates the lipid A deacylase PagL and PmrD which in turn activates PmrA. PmrA activates the expression of *arnBCADTEF* which are a collective of enzymes modifying lipid A with cationic groups (e.g. 4-amino-4-deoxy-L-arabinose or phosphoethanolamine) to repel the polymyxin. PmrA also activates PmrR responsible for the repression of LpxT (phosphorylation of lipid A) and LpxR (deacylation of lipid A) genes

An abundance of uniquely hydroxylated myristoyl groups has been detected in lipid A structures under PhoPQ activation in *S. typhimurium* [69]. The modification of hydroxylation of myristoyl groups is catalyzed by the inner membrane dioxygenase, LpxO, which act as part of a coordinated stress response [54, 56, 70].

PhoPQ activation regulates the composition proportions of *O*-antigen repeats through the action of the PmrA-regulated inner membrane protein complex Wzz_{st} and Wzz_{fepE} [71–73]. Wzz_{st} selectively determines the formation of 'long' (L) *O*-chain lengths around 16–35 repeat units, while Wzz_{fepE} is responsible for the 'very long' (VL) lengths with over than one hundred subunits. Activation of pmrA increased the fraction of such L-type and VL-type *O*-antigen in LPS molecules [71–73]. Raising the proportions of L-type and VL-type *O*-antigen leads to heightened resistance to serum [71]; moreover, VL-type enhanced two-fold increase in polymyxin B resistance [73]. Thus, it is plausible that such types of *O*-antigen somehow promote barrier function and contribute to polymyxins resistance [54].

Three outer membrane enzymes PagP (or CrcA), PagL and LpxR with active-site exposed to the outer leaflet of OM are related to acylation or deacylation in barrier remodelling [54, 56]. In response to PhoPQ activation, the transcription of *pagP* is stimulated and thus upregulated the encoded proteins that modify lipid A with palmitate [74]. Palmitoyl group transferring from phospholipid (GPL) donors to lipid A occurs on the extracellular active-site of PagP and results in hepta-acylated lipid A species [54, 56, 64, 74]. The other two outer membrane enzymes, PagL and LpxR, acting as deacylation of lipid A are subjected to post-translational inhibition by L-Ara4N modified lipid A as their active sites are found on the extracellular surface and in close proximity to lipid A [56, 64]. Increased palmitoylation and inhibited deacylation enhance the hydrophobicity of the OM and prevent penetration of the amphipathic polymyxin molecules [54, 56].

4.5 Regulation of Glycerophospholipids (GPL) Remodelling

Recent research indicates that glycerophospholipids (GPL) are also regulated components of the OM barrier in Gram-negative bacteria [55, 75]. Penetration of polymyxin molecules can cause LPS layer to become displaced and shed from the outer leaflet and activated the PhoPQ system [59, 68, 75]. To maintain bilayer barrier integrity, GPL from the inner leaflet migrate into the outer leaflet of the OM to replace the breached areas of LPS with GPL as a consequence of locally weakened barrier which are only detectable in stressed cells [54, 55, 75, 76]. The outer membrane protein PagP with dual substrate specificity activated by the PhoPQ system can function as a membrane-intrinsic probe to restoration of the permeability barrier [55, 64]. In addition to lipid A, PagP transfers palmitoyl groups from GPL to the polar head group of phosphatidylglycerol (PG) that have flipped onto the surface of OM forming palmitoyl-PG [54, 55]. Therefore, once LPS layer disrupted and PhoPQ system activated, GPL may be increasingly translocated to the outer leaflet and further acylated by pagP to enhance barrier hydrophobicity [54, 55, 64]. Also, the research work detected modest yet significantly increases in cardiolipin (CL) amount of the OM on activation of PhoPQ, which was speculated to form functional micro-domains that promote OM lipid re-modelling [54, 55]. The phospholipid (PL) bilayer structure of OM is more permeable than the asymmetrical LPS-PL bilayer barrier, so the OM phospholipase A (OMPLA) and inter-membrane transport system Mla pathway are functioned to prevent surface exposure of PLs and maintain lipid asymmetry of the OM if necessary [75, 77]. Thus, coordinate regulation of LPS and GPL forms a remodelled OM barrier critical for bacterial protective responses and survival to polymyxins.

4.6 Other Responses to Polymyxin Treatment in Gram-negative Bacteria

In Salmonella, antimicrobial peptides and cations occupy the overlapping binding site of the sensor PhoQ [59]. Divalent cations are bound to the acidic surface region of PhoQ sensor under normal conditions, while the displacement of these cations by antimicrobial peptides results in a conformational change that activates PhoQ and triggers the hierarchical regulation [59]. To date, various two-component systems have been reported to associate with the adaption to sub-inhibitory concentrations of antimicrobial peptides in *Pseudomonas aeruginosa,* including the widespread PhoPQ and PmrAB systems, and the ParRS, ColRS and CprRS systems [52, 58, 78, 79]. In contrast to *Salmonella, P. aeruginosa* senses divalent cations and cationic peptides via different mechanisms. Divalent cations are detected by PhoQ and PmrB but not peptides [58, 79]. Whereas ParS and CprS can detect cationic antimicrobial peptides regardless of Mg^{2+} concentrations which are independent two-component systems that might recognize different properties of peptides or the different effects of peptides on cell at specific concentrations. As for polymyxins, both participated at all concentrations, with a greater involvement of ParRS which is likely to be the key component [78, 79]. The occurrence of at least two direct polymyxins response systems and three associated response regulatory systems in *P. aeruginosa* highlights the complexity of the adaptive resistant pathways in this organism.

Several tripartite efflux systems play considerable roles in the intrinsic and acquired resistance in *P. aeruginosa.* Each system consists of three proteins with presumed functions: a cytoplasmic membrane component of the resistance-nodulation-division family acting as a transporter, an outer membrane component forming channels, and a membrane fusion protein linking the two membranes [80, 81]. At the gene level, a constitutively expressed operon, *mexAB-oprM,* coding for an efflux system (MexAB-OprM) which contributes intrinsic resistance in *P. aeruginosa* produced at a basal level in wild-type bacteria [80]; however, it has been observed that the MexAB-OprM efflux system is overexpressed in the metabolically active subpopulations of *P. aeruginosa* biofilm, conferring an unspecific adaptive resistant phenotype to polymyxins [52, 82]. Another outer membrane efflux pump system MexXY lacking OM protein in its own operon utilises OprM of the MexAB-OprM system and forms MexXY-OprM system that has been shown to provide natural to aminoglycosides and various unrelated antibiotics [80, 81]. It has been demonstrated that polymyxins can promote expression of the *mexXY* operons besides *pmrAB* and *arnBCDTEF-ugd,* and to coordinately downregulate the *oprD* gene that promoting β-Lactams resistance through the activation of the two-component systems ParRS [78]. These researches indicate that polymyxins are able to induce multidrug adaptive resistance of cells through the activation of distinct mechanisms (efflux, porin loss, and LPS modification) in *P. aeruginosa* [78, 82].

Microarray and high-throughput RNA-seq analysis revealed a global change pattern of gene expression leading to adaptive responses. When *P. aeruginosa* cells are exposed to sub-inhibitory concentrations of colistin, approximately 0.5% of 5500 genes showed significantly changed expression levels in the colistin-treated sample. Among them, 13 were upregulated and 17 were downregulated. The upregulated genes are involved in quorum sensing (QS) and biofilm formation besides well-known LPS modification, while the downregulated genes are involved in motility (swarming and swimming motility) and osmotolerance [83]. Upon exposure to a much higher concentration of polymyxin (10 × MIC) treatment, a wider profile of global changes was identified from protective responses of *Yersinia pestis* to survive the stressful environments. A total of 291 genes were differentially expressed and 158 of them were induced. Among the 158 upregulated genes, 22 were regulatory genes including 8 two-component systems that globally or locally governing a wide set of stress-protective functions, 19 genes were involved in remodelling of cell envelope encoding membrane components or polysaccharide surface structures, 4 operons including 9 genes were essential for dissimilation of *sn*-glycerol 3-phosphate which is a direct precursor for phospholipid biosynthesis [84], 10 were heat shock proteins that play important roles in preventing aggregation of proteins and repairing misfolded or damaged proteins caused by environmental stresses, 5 were related to drug resistance (3 of tellurium resistance and 2 of multidrug transport system), 11 genes organized into four operons were components of siderophore-based iron acquisition systems which known as high-pathogenicity island shared by three pathogenic yersiniae [85]. Our recent high-throughput RNA-seq study of the transcriptomic response of *Acinetobacter baumannii* to colistin under conditions that able to kill partial cells revealed hundreds of genes differentially expressed including those of two-component systems, glycerophospholipid metabolism, lipopolysaccharide biosynthesis, biofilm synthesis, drug resistant proteins, heat shock proteins as discovered from other Gram-negative strains (data unpublished). Moreover, genes involved in nucleotide excision repair and peroxisome were also significantly induced by killing concentration of colistin. Two genes encoding the peroxisome superoxide dismutase SOD1 and catalase KatE (belonging to the antioxidant system) were upregulated after colistin treatment.

4.7 Mutations and Death

Recently, it has been demonstrated that a number of bactericidal antibiotic classes trigger the endogenous production of lethal active forms of hydroxyl radicals in bacteria through the Fenton reaction [46, 47, 51, 86] which depends on the availability of hydrogen peroxide, an iron species and reducing equivalents [86–88]. Hydrogen peroxide is generated within cells as a by-product of oxidative metabolism when molecular oxygen accidentally acquires electrons from the reduced cofactors of flavoproteins [50, 88] and bactericidal antibiotics elevated the generation of deleterious reactive species [48]. Iron is necessary for bacteria to survive and acquired from environment by biosynthetic iron chelators known as siderophores [86]. Hydrogen peroxide is capable of interacting with intracellular ferrous form iron unincorporated or associated with biological molecules including iron-sulphur-dependent dehydratases and mononuclear iron proteins, oxidizing the iron and forming hydroxyl radicals in the process [50, 86]. This is a cyclical process in vivo since intracellular reductants can reduce the oxidized iron back to ferrous form [86]. The event of iron disintegration from proteins eliminates a variety of enzyme activity and the hydroxyl radical is an extremely powerful oxidant that reacts with virtually all organic molecules including giving a wide variety of DNA damage [50]. Ultimately, the oxidative damage of hydroxyl radical to DNA, lipids, and proteins eventually reaches levels that cannot be controlled and thus contribute to cell death [47, 48].

It has been demonstrated that the rapid killing of Gram-negative species *A. baumannii*, *Escherichia coli* and *Francisella novicida* by polymyxins is in part mediated by a hydroxyl radical death pathway [46]. In addition, this mechanism of killing occurs in polymyxin-susceptible *A. baumannii* isolates including multidrug-resistant clinical isolates but this response is not induced in a polymyxin-resistant isolate [46]. The mechanism by which polymyxin treatment induces the production of hydroxyl radicals in Gram-negative bacteria is not clear yet. Polymyxin-induced hydroxyl radical death does not occur in polymyxin-resistance isolates of *A. baumannii* but in susceptible isolates [46], and polymyxin resistance in *A. baumannii* is commonly due to blocking the entries through remodelling of the outer membrane [67, 68, 89]. Therefore, it seems that the event of the hydroxyl radical death pathway induced by polymyxins happened after the initial drug-target interactions on the outer membrane. The secondary MOA of polymyxins that they can inhibit NDH-2 activity in the inner membranes of detected Gram-negative species [90] may result in accumulation of nicotinamide adenine dinucleotide (NADH) which can be utilized as a source of reducing equivalents that contribute to Fenton reaction (Fig. 4.3b, bottom left part) [88]. *In vitro* the NADH-Fe (III)-EDTA-H_2O_2 system drives an ongoing Fenton reaction and DNA ensue breaks in such system while NAD^+ is ineffective. Furthermore, the rate of DNA nicking corresponds to the rate of NADH oxidation [88].

Exposure of Gram-negative species to low concentrations of hydrogen peroxide resulted in DNA damage that causes mutagenesis and kills the cell [87, 88]. It has been proposed that antibiotic-induced ROS may provide a mechanism of acquiring beneficial mutations when stresses are small but induce lethality when stresses are large [48]. A recent study revealed that treatments with sub-lethal levels of bactericidal antibiotics resulting in up to 10 times increases in the mutation rate relative to an untreated control, a strong correlation between ROS (reactive oxygen species) formation and fold change in mutation rate, and heterogeneous increases in the minimum inhibitory concentration for a range of antibiotics irrespective of the drug target [49]. Colistin was bactericidal in a concentration-dependent manner [91]. Although colistin resistance is rare, colistin-resistant *A. baumannii, P. aeruginosa* and *K. pneumoniae* isolates have been reported worldwide. The emergence of colistin resistance or heteroresistance after colistin treatment can be easily selected *in vitro* and *in vivo* with mutations of key genes for protective responses such as the initial two-component regulatory systems [91–93]. However, a recent study of colistin mutant prevention concentrations (MPCs) for *A. baumannii*, *P. aeruginosa* and *K. pneumoniae* were shown to be very high (>64 μg/mL) [93], which recommended polymyxin combination therapy to prevent the emergency of resistant mutants and the risk of toxicity at high concentrations [93].

Accordingly, bacterial cells contain scavenging enzymes to prevent the accumulation of reactive oxygen species. Two well-known antioxidant system enzymes, catalase and SOD, can transform active radicals into oxygen and water [72, 73]. Our unpublished work of transcriptomic response of *A. baumannii* to colistin supported such protective response as 6 genes encoding peroxidases (including catalase and SOD) were significantly unregulated. Moreover, nucleotide excision repair genes for DNA damage repairing in our unpublished data, heat shock proteins [85] for repairing mis-folded or damaged proteins, iron acquisition systems [85] for recovering the function of iron disintegrated proteins were up-regulated significantly which might partially arise from the response to hydroxyl radical damage.

In summary, polymyxins are capable of inducing cell damage and death by interfering with their primary targets which trigger stress responses that induce outer membrane remodelling preventing polymyxins from entering. While downstream secondary target-polymyxin interactions might trigger redox-related physiological alterations that result in the formation of toxic reactive oxygen species as well as stimulating repairing responses which further contribute to cellular damage, mutations and death [48]. The formidable challenge of sub-lethal polymyxin treatment leading to multidrug resistance rather than completely killing the bacteria required us to expand our understanding of the polymyxin stress responses in Gram-negative bacteria on a detailed system-wide level. Such understanding helps in providing a foundation for finding key protective responses molecules. Those molecules can be utilised as therapy targets in combination with polymyxin-treatments to improve current therapeutic options and avoid resistant mutations.

4.8 Imaging Polymyxin Penetration and Localization in the Gram-negative Bacterial Outer Membrane

The unavailability of valid imaging probes with native activities is a significant barrier to examine the intra-cellular localization of polymyxins in Gram-negative bacterial cells [94]. Most reports of polymyxin probes either employed inactive nonapeptide derivatives or dansylated polymyxin B, which was derived by non-specifically reacting polymyxin B with dansyl-chloride [95–98]. We have previously highlighted the deficiencies of directly amine-coupling dansyl groups onto the Dab side chains in semi-synthetic preparations of dansyl-polymyxin B [99]. Analysis of these semi-synthetic dansyl-polymyxin B preparations revealed the existence of *mono-*, *di-*, *tri-*, and *tetra-*dansyl substituted species [99]. Furthermore, as polymyxin B is comprised of two major components (B_1 and B_2), the potential for either of these components to be substituted at any of the five Dab side chains with up to four dansyl molecules results in a highly variable mixture of dansylated derivatives [99]. Commercial preparations of 4-bora-3*a*,4*a*-diaza-*s*-indacene (BODIPY) labelled polymyxin B displayed a markedly reduced antibacterial activities compared to polymyxin B [100]. Moreover, our group previously reported a fully synthetic [dansyl-Lys][1]polymyxin B_3 probe that was devoid of antibacterial activity [99]. These findings are consistent with our understanding of polymyxin SAR wherein the dansyl modification of the Dab side chains inactivates antibacterial activity [99]. Clearly, there is very little value in using these semi-synthetic preparations as imaging probes since they lack native antibacterial activityand pharmacological properties of the parent compound, polymyxin B.

We recently reported the regio-selective modification of the polymyxin B core scaffold at the *N*-terminus with the dansyl fluorophore to generate an active probe (probe (**1**)) that mimics polymyxin B pharmacologically (Fig. 4.4, bottom left panel) [38]. The design and synthesis of a dansyl molecular probe through the regio-selective modification of the polymyxin B core structure was

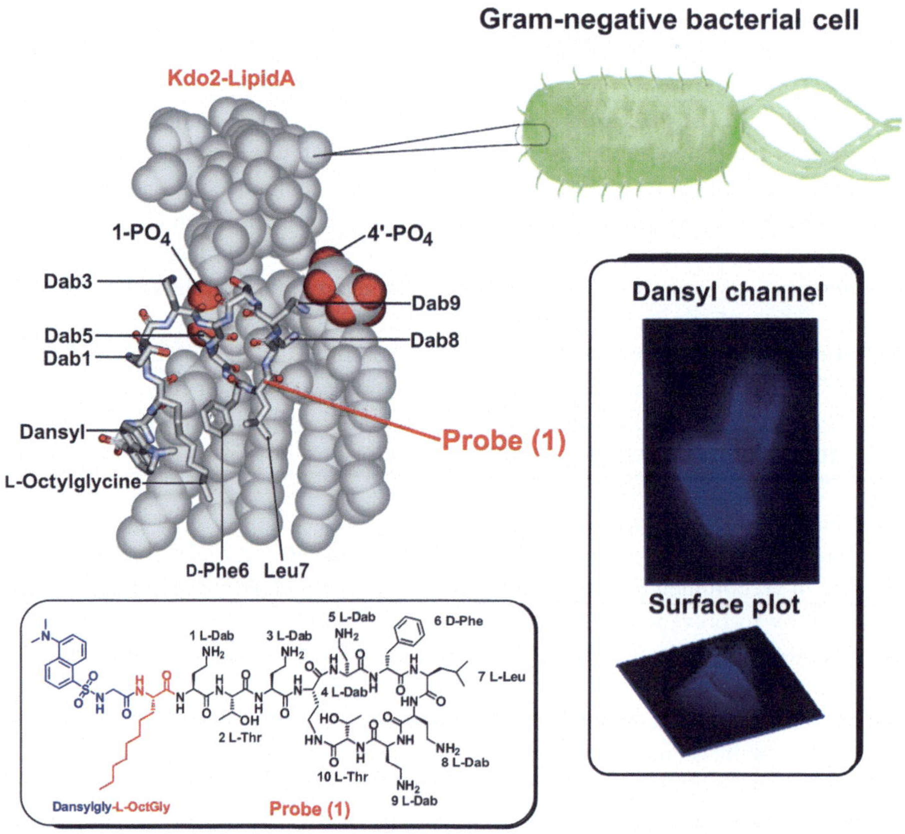

Fig. 4.4 *Bottom left panel.* Chemical structure of the probe (**1**). *Top left panel.* NMR-based model of the probe (**1**)-Kdo2 lipid A complex. *Right panel.* Laser scanning confocal microscopy image of *K. pneumoniae* ATCC 13883 cells treated with probe (**1**)

undertaken with the aforementioned lipid A interaction principles in mind; in order to mimic the polymyxin B structure as closely as possible and to maintain its native antibacterial activity. The dansyl group was utilized for the fluorescent probe as it has suitable spectral properties and its relative small size would reduce the chance of steric effects. The regio-selective incorporation of the dansyl group into the hydrophobic *N*-terminal centre of the polymyxin B core scaffold is prudent as it has a minimal impact on the native antibacterial activity of the polymyxin B scaffold. Therefore, the strategy we employed was to replace the *N*-terminal fatty acyl group of polymyxin B with the amino acid L-octylglycine, where the eight-carbon fatty acyl chain emulated the *N*-terminal fatty acyl chain of polymyxin B, whilst the N^{α}-amino group would provide a convenient point of attachment for the dansyl group (eliminating the need for additional orthogonal protection during the synthesis).

The antimicrobial activity of probe (**1**) was screened against a panel of ATCC and recent clinical isolates of polymyxin-susceptible and -resistant strains of *P. aeruginosa, A. baumannii* and *K. pneumoniae*. The probe showed antibacterial activity against polymyxin-susceptible strains (MIC 4–16 mg/L), compared to colistin (MIC 0.125–2 mg/L) and polymyxin B (MIC <0.125–2 mg/L). Probe (**1**) also displayed activity against polymyxin-resistant *P. aeruginosa* and *A. baumannii* strains (MIC 4–8 mg/L); colistin and polymyxin B (MIC >32 mg/L). It is understood that Gram-negative pathogens resist the action of polymyxins by introducing cationic modifications onto the phosphate groups on the lipid A component of LPS [101–106]. The most common mechanism involves esterification of lipid A phosphates with aminoarabinose, or ethanolamine [101–106]. The molecular tailoring serves to reduce the net negative charge of the outer membrane surface, thereby repelling the electrostatic attraction with positively charged polymyxin molecules [107]. The NMR-based molecular model of the probe (**1**)-Kdo2 Lipid A complex implies that the combination of the L-octylglycine and the dansyl substituent at the *N*-terminus provides additional hydrophobic interactive forces that compensate for the electrostatic repulsion of the aminoarabinose phosphate modifications (Fig. 4.4, top left panel). The molecular model indicates that electrostatic interactions with the 1-phosphorester group on lipid A are not hampered by the dansyl group. The model further suggests that the hydrophobic dansyl group interacts with the apolar environment formed by the fatty acyl chains of lipid A. TEM imaging of *K. pneumoniae* ATCC 13883 cells treated with probe (**1**) at 0.5 × MIC revealed the formation of numerous protrusions or blebs extending from the outer membrane of the cells that possibly represent outer membrane fragments. A similar blebbing effect was observed with Gram-negative bacterial cells treated with polymyxin B and colistin [108].

Time-lapse laser scanning confocal microscopy imaging of the penetration of probe (**1**) into *K. pneumoniae* ATCC 13883 cells revealed that the probe initially accumulates in the outer membrane and subsequently penetrates into the inner membrane and finally becomes homogenously distributed into the cytoplasm (Fig. 4.4, right panel). Intriguingly, confocal imaging and spectrophotometric lysis assay experiments with spheroplasts isolated from *K. pneumoniae* ATCC 13883 revealed that probe (**1**) also accumulated within and disrupted the inner membrane structure. Coincidently, our group has recently show that polymyxin B and colistin inhibit the NDH-2 oxidoreductase inner membrane respiratory enzyme, which also may contribute towards their bactericidal effect (*cf.* preceding discussions; Fig. 4.1) [19]. Furthermore, the imaging experiments revealed that at sub-MIC concentrations, probe (**1**) tends to accumulate on the surface of the bacterial cell and partly penetrates into the outer membrane. Whereas at <MIC concentrations, probe (**1**) accumulated on the surface of the bacterial cell and entered into the cytoplasm. Fundamentally, the localization studies from the time-lapse laser scanning confocal microscopy imaging results helps validate the mechanistic model of polymyxin action (Fig. 4.1). Based on the imaging data, it is evident that polymyxins initially accumulate in the outer membrane, fol-

lowed by a gradual penetration into the inner membrane and finally enter the cytoplasm in *K. pneumoniae*. These findings are consistent with the secondary mode of action of polymyxin B which involves an inhibitory activity against the inner membrane NDH-2 enzyme [19]. It will be most interesting to elucidate the intracellular bacterial cell target(s) for the polymyxins, which to date remains uncharacterized.

The commercial availability of this probe would greatly facilitate molecular imaging studies on both the mode of action and pharmacokinetics; and contribute towards the development of a new generation of polymyxin lipopeptides with superior activity against polymyxin-resistant Gram-negative 'superbugs'.

In summary, polymyxins are bioactive natural products with beneficial pharmacological activities and their antimicrobial properties have been investigated for decades. Albeit, their precise mode of action remains unknown and much progress has been made towards understanding their bacterial killing effect. The compendium of data has highlighted that their bacterial killing mechanism is more complex than simply the long-standing notion that they impart a membrane disorganising effect. Secondary, pathways such as the bacterial redox chain have been implicated in their killing effect. Clearly, much work needs to be done to comprehensively elucidate the precise mode-of-action of these valuable lipopeptide antibiotics.

References

1. Nikaido H (2003) Molecular basis of bacterial outer membrane permeability revisited. Microbiol Mol Biol Rev 67(4):593–656
2. Rietschel ET, Brade H (1987) Lipopolysaccharides, endotoxins and O-antigens of gram-negative bacteria: chemical structure, biologic effect and serologic properties. Infection 15(2):133–141
3. Caroff M, Karibian D (2003) Structure of bacterial lipopolysaccharides. Carbohydr Res 338(23):2431–2447
4. Rietschel ET, Brade H, Brade L, Brandenburg K, Schade U, Seydel U et al (1987) Lipid A, the endotoxic center of bacterial lipopolysaccharides: relation of chemical structure to biological activity. Prog Clin Biol Res 231:25–53
5. Hancock RE (1997) The bacterial outer membrane as a drug barrier. Trends Microbiol 5(1):37–42
6. Hancock RE (1997) Antibacterial peptides and the outer membranes of gram-negative bacilli. J Med Microbiol 46(1):1–3
7. Pristovsek P, Kidric J (2004) The search for molecular determinants of LPS inhibition by proteins and peptides. Curr Top Med Chem 4(11):1185–1201
8. Pristovsek P, Kidric J (1999) Solution structure of polymyxins B and E and effect of binding to lipopolysaccharide: an NMR and molecular modeling study. J Med Chem 42(22):4604–4613
9. Hancock RE (1997) Peptide antibiotics. Lancet 349(9049):418–422
10. Hancock RE, Lehrer R (1998) Cationic peptides: a new source of antibiotics. Trends Biotechnol 16(2):82–88
11. Velkov T, Thompson PE, Nation RL, Li J Structure--activity relationships of polymyxin antibiotics. J Med Chem 53(5):1898–1916
12. Ofek I, Cohen S, Rahmani R, Kabha K, Tamarkin D, Herzig Y et al (1994) Antibacterial synergism of polymyxin B nonapeptide and hydrophobic antibiotics in experimental gram-negative infections in mice. Antimicrob Agents Chemother 38(2):374–377
13. Clausell A, Garcia-Subirats M, Pujol M, Busquets MA, Rabanal F, Cajal Y (2007) Gram-negative outer and inner membrane models: insertion of cyclic cationic lipopeptides. J Phys Chem B 111(3):551–563
14. Powers JP, Hancock RE (2003) The relationship between peptide structure and antibacterial activity. Peptides 24(11):1681–1691
15. Cajal Y, Ghanta J, Easwaran K, Surolia A, Jain MK (1996) Specificity for the exchange of phospholipids through polymyxin B mediated intermembrane molecular contacts. Biochemistry 35(18):5684–5695
16. Cajal Y, Rogers J, Berg OG, Jain MK (1996) Intermembrane molecular contacts by polymyxin B mediate exchange of phospholipids. Biochemistry 35(1):299–308
17. Cajal Y, Berg OG, Jain MK (1995) Direct vesicle-vesicle exchange of phospholipids mediated by polymyxin B. Biochem Biophys Res Commun 210(3):746–752
18. Mogi T, Murase Y, Mori M, Shiomi K, Omura S, Paranagama MP et al (2009) Polymyxin B identified as an inhibitor of alternative NADH dehydrogenase and malate: quinone oxidoreductase from the gram-positive bacterium *Mycobacterium smegmatis*. J Biochem 146(4):491–499
19. Deris ZZ, Akter J, Sivanesan S, Roberts KD, Thompson PE, Nation RL et al (2013) A secondary mode of action of polymyxins against gram-negative bacteria involves the inhibition of NADH-quinone oxidoreductase activity. J Antibiot. 67(2):147–151
20. Hancock REW (1997) Peptide antibiotics. Lancet 349(9049):418–422
21. Hancock REW, Scott MG (2000) The role of antimicrobial peptides in animal defenses. Proc Natl Acad Sci U S A 97(16):8856–8861

22. Cho J, Kim S (2010) Non-membrane targets of antimicrobial peptides: novel therapeutic opportunities? Advances in Molecular and Cellular Microbiology No 18. CABI, Wallingford, pp 128–140
23. Otvos L (2005) Antibacterial peptides and proteins with multiple cellular targets. J Pept Sci 11(11):697–706
24. Storm DR, Rosenthal KS, Swanson PE (1977) Polymyxin and related peptide antibiotics. Annu Rev Biochem 46(1):723–763
25. Teuber M (1974) Action of polymyxin B on bacterial membranes. Arch Microbiol 100(1):131–144
26. Boshoff HI, Barry CE 3rd (2005) Tuberculosis – metabolism and respiration in the absence of growth. Nat Rev Microbiol 3(1):70–80
27. Boshoff HI, Myers TG, Copp BR, McNeil MR, Wilson MA, Barry CE 3rd (2004) The transcriptional responses of *Mycobacterium tuberculosis* to inhibitors of metabolism: novel insights into drug mechanisms of action. J Biol Chem 279(38):40174–40184
28. Weinstein EA, Yano T, Li LS, Avarbock D, Avarbock A, Helm D et al (2005) Inhibitors of type II NADH:menaquinone oxidoreductase represent a class of antitubercular drugs. Proc Natl Acad Sci U S A 102(12):4548–4553
29. Fisher N, Warman AJ, Ward SA, Biagini GA (2009) Chapter 17 Type II NADH: quinone oxidoreductases of *Plasmodium falciparum* and *Mycobacterium tuberculosis* kinetic and high-throughput assays. Methods Enzymol 456:303–320
30. Biagini GA, Viriyavejakul P, O'Neill PM, Bray PG, Ward SA (2006) Functional characterization and target validation of alternative complex I of *Plasmodium falciparum* mitochondria. Antimicrob Agents Chemother 50(5):1841–1851
31. Fisher N, Bray PG, Ward SA, Biagini GA (2007) The malaria parasite type II NADH:quinone oxidoreductase: an alternative enzyme for an alternative lifestyle. Trends Parasitol 23(7):305–310
32. Kim MS, Kim YJ (2004) Enzymatic properties of the membrane bound NADH oxidase system in the aerobic respiratory chain of *Bacillus cereus*. J Biochem Mol Biol 37(6):753–756
33. Kerscher S, Dröse S, Zickermann V, Brandt U (2008) The three families of respiratory NADH dehydrogenases. In: Schäfer G, Penefsky H (eds) Bioenergetics. Results and problems in cell differentiation, vol 45. Springer, Berlin/Heidelberg, pp 185–222
34. Yagi T, Yano T, Di Bernardo S, Matsuno-Yagi A (1998) Procaryotic complex I (NDH-1), an overview. Biochim Biophys Acta 1364(2):125–133
35. Yagi T (1991) Bacterial NADH-quinone oxidoreductases. J Bioenerg Biomembr 23(2):211–225
36. Yagi T, Matsuno-Yagi A (2003) The proton-translocating NADH-quinone oxidoreductase in the respiratory chain: the secret unlocked. Biochemistry 42(8):2266–2274
37. Mogi T, Murase Y, Mori M, Shiomi K, Ōmura S, Paranagama MP et al (2009) Polymyxin B identified as an inhibitor of alternative NADH dehydrogenase and malate: quinone oxidoreductase from the gram-positive bacterium *Mycobacterium smegmatis*. J Biochem 146(4):491–499
38. Deris ZZ, Swarbrick JD, Roberts KD, Azad MA, Akter J, Horne AS et al (2014) Probing the penetration of antimicrobial polymyxin lipopeptides into gram-negative bacteria. Bioconjug Chem 25(4):750–760
39. Velkov T, Deris ZZ, Huang JX, Azad MA, Butler M, Sivanesan S et al (2013) Surface changes and polymyxin interactions with a resistant strain of *Klebsiella pneumoniae*. Innate Immun. 20(4):350–363. https://doi.org/10.1177/1753425913493337
40. Velkov T, Thompson PE, Nation RL, Li J (2010) Structure–activity relationships of polymyxin antibiotics. J Med Chem 53(5):1898–1916
41. Melo AM, Bandeiras TM, Teixeira M (2004) New insights into type II NAD(P)H:quinone oxidoreductases. Microbiol Mol Biol Rev 68(4):603–616
42. Dong CK, Patel V, Yang JC, Dvorin JD, Duraisingh MT, Clardy J et al (2009) Type II NADH dehydrogenase of the respiratory chain of *Plasmodium falciparum* and its inhibitors. Bioorg Med Chem Lett 19(3):972–975
43. Koul A, Arnoult E, Lounis N, Guillemont J, Andries K (2011) The challenge of new drug discovery for tuberculosis. Nature 469(7331):483–490
44. Fisher N, Bray PG, Ward SA, Biagini GA (2008) Malaria-parasite mitochondrial dehydrogenases as drug targets: too early to write the obituary. Trends Parasitol 24(1):9–10
45. Biagini GA, Fisher N, Shone AE, Mubaraki MA, Srivastava A, Hill A et al (2012) Generation of quinolone antimalarials targeting the *Plasmodium falciparum* mitochondrial respiratory chain for the treatment and prophylaxis of malaria. Proc Natl Acad Sci 109:8298–8303
46. Sampson TR, Liu X, Schroeder MR, Kraft CS, Burd EM, Weiss DS (2012) Rapid killing of *Acinetobacter baumannii* by polymyxins is mediated by a hydroxyl radical death pathway. Antimicrob Agents Chemother 56(11):5642–5649
47. Kohanski MA, Dwyer DJ, Hayete B, Lawrence CA, Collins JJ (2007) A common mechanism of cellular death induced by bactericidal antibiotics. Cell 130(5):797–810
48. Dwyer DJ, Belenky PA, Yang JH, MacDonald IC, Martell JD, Takahashi N et al (2014) Antibiotics induce redox-related physiological alterations as part of their lethality. Proc Natl Acad Sci U S A 111(20):E2100–E2109
49. Kohanski MA, DePristo MA, Collins JJ (2010) Sublethal antibiotic treatment leads to multidrug resistance via radical-induced mutagenesis. Mol Cell 37(3):311–320
50. Imlay JA (2013) The molecular mechanisms and physiological consequences of oxidative stress: lessons from a model bacterium. Nat Rev Microbiol 11(7):443–454

51. Wright GD, Hung DT, Helmann JD (2013) How antibiotics kill bacteria: new models needed? Nat Med 19(5):544–545
52. Skiada A, Markogiannakis A, Plachouras D, Daikos GL (2011) Adaptive resistance to cationic compounds in *Pseudomonas aeruginosa*. Int J Antimicrob Agents 37(3):187–193
53. Poole K (2012) Stress responses as determinants of antimicrobial resistance in gram-negative bacteria. Trends Microbiol 20(5):227–234
54. Dalebroux ZD, Miller SI (2014) Salmonellae PhoPQ regulation of the outer membrane to resist innate immunity. Curr Opin Microbiol 17:106–113
55. Dalebroux ZD, Matamouros S, Whittington D, Bishop RE, Miller SI (2014) PhoPQ regulates acidic glycerophospholipid content of the *Salmonella Typhimurium* outer membrane. Proc Natl Acad Sci U S A 111(5):1963–1968
56. Needham BD, Trent MS (2013) Fortifying the barrier: the impact of lipid A remodelling on bacterial pathogenesis. Nat Rev Microbiol 11(7):467–481
57. Poole K (2012) Bacterial stress responses as determinants of antimicrobial resistance. J Antimicrob Chemother 67(9):2069–2089
58. Fernandez L, Gooderham WJ, Bains M, McPhee JB, Wiegand I, Hancock RE (2010) Adaptive resistance to the "last hope" antibiotics polymyxin B and colistin in *Pseudomonas aeruginosa* is mediated by the novel two-component regulatory system ParR-ParS. Antimicrob Agents Chemother 54(8):3372–3382
59. Bader MW, Sanowar S, Daley ME, Schneider AR, Cho U, Xu W et al (2005) Recognition of antimicrobial peptides by a bacterial sensor kinase. Cell 122(3):461–472
60. Gunn JS, Miller SI (1996) PhoP-PhoQ activates transcription of pmrAB, encoding a two-component regulatory system involved in *Salmonella typhimurium* antimicrobial peptide resistance. J Bacteriol 178(23):6857–6864
61. Luo SC, Lou YC, Rajasekaran M, Chang YW, Hsiao CD, Chen C (2013) Structural basis of a physical blockage mechanism for the interaction of response regulator PmrA with connector protein PmrD from *Klebsiella pneumoniae*. J Biol Chem 288(35):25551–25561
62. Kato A, Groisman EA (2004) Connecting two-component regulatory systems by a protein that protects a response regulator from dephosphorylation by its cognate sensor. Genes Dev 18(18):2302–2313
63. Kato A, Chen HD, Latifi T, Groisman EA (2012) Reciprocal control between a bacterium's regulatory system and the modification status of its lipopolysaccharide. Mol Cell 47(6):897–908
64. Raetz CR, Reynolds CM, Trent MS, Bishop RE (2007) Lipid A modification systems in gram-negative bacteria. Annu Rev Biochem 76:295–329
65. Okuda S, Freinkman E, Kahne D (2012) Cytoplasmic ATP hydrolysis powers transport of lipopolysaccharide across the periplasm in E. coli. Science 338(6111):1214–1217
66. Tamayo R, Choudhury B, Septer A, Merighi M, Carlson R, Gunn JS (2005) Identification of cptA, a PmrA-regulated locus required for phosphoethanolamine modification of the *Salmonella enterica serovar typhimurium* lipopolysaccharide core. J Bacteriol 187(10):3391–3399
67. Beceiro A, Llobet E, Aranda J, Bengoechea JA, Doumith M, Hornsey M et al (2011) Phosphoethanolamine modification of lipid A in colistin-resistant variants of Acinetobacter baumannii mediated by the pmrAB two-component regulatory system. Antimicrob Agents Chemother 55(7):3370–3379
68. Velkov T, Roberts KD, Nation RL, Thompson PE, Li J (2013) Pharmacology of polymyxins: new insights into an 'old' class of antibiotics. Future Microbiol 8(6):711–724
69. Guo L, Lim KB, Gunn JS, Bainbridge B, Darveau RP, Hackett M et al (1997) Regulation of lipid A modifications by *Salmonella typhimurium* virulence genes phoP-phoQ. Science 276(5310):250–253
70. Gibbons HS, Reynolds CM, Guan Z, Raetz CR (2008) An inner membrane dioxygenase that generates the 2-hydroxymyristate moiety of Salmonella lipid A. Biochemistry 47(9):2814–2825
71. Delgado MA, Mouslim C, Groisman EA (2006) The PmrA/PmrB and RcsC/YojN/RcsB systems control expression of the Salmonella O-antigen chain length determinant. Mol Microbiol 60(1):39–50
72. Larue K, Kimber MS, Ford R, Whitfield C (2009) Biochemical and structural analysis of bacterial O-antigen chain length regulator proteins reveals a conserved quaternary structure. J Biol Chem 284(11):7395–7403
73. Pescaretti Mde L, Lopez FE, Morero RD, Delgado MA (2011) The PmrA/PmrB regulatory system controls the expression of the wzzfepE gene involved in the O-antigen synthesis of Salmonella enterica serovar Typhimurium. Microbiology 157(Pt 9):2515–2521
74. Bishop RE, Gibbons HS, Guina T, Trent MS, Miller SI, Raetz CR (2000) Transfer of palmitate from phospholipids to lipid A in outer membranes of gram-negative bacteria. EMBO J 19(19):5071–5080
75. Malinverni JC, Silhavy TJ (2009) An ABC transport system that maintains lipid asymmetry in the gram-negative outer membrane. Proc Natl Acad Sci U S A 106(19):8009–8014
76. Jia W, El Zoeiby A, Petruzziello TN, Jayabalasingham B, Seyedirashti S, Bishop RE (2004) Lipid trafficking controls endotoxin acylation in outer membranes of *Escherichia coli*. J Biol Chem 279(43):44966–44975
77. Dekker N (2000) Outer-membrane phospholipase A: known structure, unknown biological function. Mol Microbiol 35(4):711–717
78. Muller C, Plesiat P, Jeannot K (2011) A two-component regulatory system interconnects

resistance to polymyxins, aminoglycosides, fluoroquinolones, and beta-lactams in *Pseudomonas aeruginosa*. Antimicrob Agents Chemother 55(3):1211–1221

79. Fernandez L, Jenssen H, Bains M, Wiegand I, Gooderham WJ, Hancock RE (2012) The two-component system CprRS senses cationic peptides and triggers adaptive resistance in *Pseudomonas aeruginosa* independently of ParRS. Antimicrob Agents Chemother 56(12):6212–6222
80. Li XZ, Nikaido H (2004) Efflux-mediated drug resistance in bacteria. Drugs 64(2):159–204
81. Masuda N, Sakagawa E, Ohya S, Gotoh N, Tsujimoto H, Nishino T (2000) Substrate specificities of MexAB-OprM, MexCD-OprJ, and MexXY-oprM efflux pumps in *Pseudomonas aeruginosa*. Antimicrob Agents Chemother 44(12):3322–3327
82. Pamp SJ, Gjermansen M, Johansen HK, Tolker-Nielsen T (2008) Tolerance to the antimicrobial peptide colistin in *Pseudomonas aeruginosa* biofilms is linked to metabolically active cells, and depends on the pmr and mexAB-oprM genes. Mol Microbiol 68(1):223–240
83. Cummins J, Reen FJ, Baysse C, Mooij MJ, O'Gara F (2009) Subinhibitory concentrations of the cationic antimicrobial peptide colistin induce the pseudomonas quinolone signal in Pseudomonas aeruginosa. Microbiology 155.(Pt 9:2826–2837
84. Yang B, Larson TJ (1996) Action at a distance for negative control of transcription of the glpD gene encoding sn-glycerol 3-phosphate dehydrogenase of *Escherichia coli* K-12. J Bacteriol 178(24):7090–7098
85. Zhou D, Han Y, Qiu J, Qin L, Guo Z, Wang X et al (2006) Genome-wide transcriptional response of Yersinia pestis to stressful conditions simulating phagolysosomal environments. Microbes Infect/Inst Pasteur 8(12–13):2669–2678
86. Yeom J, Imlay JA, Park W (2010) Iron homeostasis affects antibiotic-mediated cell death in Pseudomonas species. J Biol Chem 285(29):22689–22695
87. Imlay JA, Chin SM, Linn S (1988) Toxic DNA damage by hydrogen peroxide through the Fenton reaction in vivo and in vitro. Science 240(4852):640–642
88. Imlay JA, Linn S (1988) DNA damage and oxygen radical toxicity. Science 240(4857):1302–1309
89. Pelletier MR, Casella LG, Jones JW, Adams MD, Zurawski DV, Hazlett KR et al (2013) Unique structural modifications are present in the lipopolysaccharide from colistin-resistant strains of *Acinetobacter baumannii*. Antimicrob Agents Chemother 57(10):4831–4840
90. Deris ZZ, Akter J, Sivanesan S, Roberts KD, Thompson PE, Nation RL et al (2014) A secondary mode of action of polymyxins against gram-negative bacteria involves the inhibition of NADH-quinone oxidoreductase activity. J Antibiot 67(2):147–151
91. Li J, Rayner CR, Nation RL, Owen RJ, Tan KE, Spelman D (2006) Heteroresistance to colistin in multidrug-resistant *Acinetobacter baumannii*. Antimicrob Agents Chemother 50(9):2946–2950
92. Hawley JS, Murray CK, Jorgensen JH (2008) Colistin heteroresistance in acinetobacter and its association with previous colistin therapy. Antimicrob Agents Chemother 52(1):351–352
93. Choi MJ, Ko KS (2014) Mutant prevention concentrations of colistin for Acinetobacter baumannii, Pseudomonas aeruginosa and *Klebsiella pneumoniae* clinical isolates. J Antimicrob Chemother 69(1):275–277
94. Velkov T, Thompson PE, Nation RL, Li J (2010) Structure-activity relationships of polymyxin antibiotics. J Med Chem 53(5):1898–1916
95. Moore RA, Bates NC, Hancock RE (1986) Interaction of polycationic antibiotics with *Pseudomonas aeruginosa* lipopolysaccharide and lipid A studied by using dansyl-polymyxin. Antimicrob Agents Chemother 29(3):496–500
96. Tsubery H, Ofek I, Cohen S, Eisenstein M, Fridkin M (2002) Modulation of the hydrophobic domain of polymyxin B nonapeptide: effect on outer-membrane permeabilization and lipopolysaccharide neutralization. Mol Pharmacol 62(5):1036–1042
97. Tsubery H, Ofek I, Cohen S, Fridkin M (2000) Structure-function studies of polymyxin B nonapeptide: implications to sensitization of gram-negative bacteria. J Med Chem 43(16):3085–3092
98. Tsubery H, Ofek I, Cohen S, Fridkin M (2001) N-terminal modifications of polymyxin B nonapeptide and their effect on antibacterial activity. Peptides 22(10):1675–1681
99. Soon RL, Velkov T, Chiu F, Thompson PE, Kancharla R, Roberts K et al (2011) Design, synthesis, and evaluation of a new fluorescent probe for measuring polymyxin-lipopolysaccharide binding interactions. Anal Biochem 409(2):273–283
100. Benincasa M, Pacor S, Gennaro R, Scocchi M (2009) Rapid and reliable detection of antimicrobial peptide penetration into gram-negative bacteria based on fluorescence quenching. Antimicrob Agents Chemother 53(8):3501–3504
101. Gunn JS (2008) The Salmonella PmrAB regulon: lipopolysaccharide modifications, antimicrobial peptide resistance and more. Trends Microbiol 16(6):284–290
102. Gunn JS, Lim KB, Krueger J, Kim K, Guo L, Hackett M et al (1998) PmrA-PmrB-regulated genes necessary for 4-aminoarabinose lipid A modification and polymyxin resistance. Mol Microbiol 27(6):1171–1182
103. Gunn JS, Ryan SS, Van Velkinburgh JC, Ernst RK, Miller SI (2000) Genetic and functional analysis of a PmrA-PmrB-regulated locus necessary for lipopolysaccharide modification, antimicrobial peptide resistance, and oral virulence of *Salmonella enterica serovar typhimurium*. Infect Immun 68(11):6139–6146
104. Schurek KN, Sampaio JL, Kiffer CR, Sinto S, Mendes CM, Hancock RE (2009) Involvement of

pmrAB and phoPQ in polymyxin B adaptation and inducible resistance in non-cystic fibrosis clinical isolates of *Pseudomonas aeruginosa*. Antimicrob Agents Chemother 53(10):4345–4351
105. Shi Y, Cromie MJ, Hsu FF, Turk J, Groisman EA (2004) PhoP-regulated Salmonella resistance to the antimicrobial peptides magainin 2 and polymyxin B. Mol Microbiol 53(1):229–241
106. Trent MS, Ribeiro AA, Lin S, Cotter RJ, Raetz CR (2001) An inner membrane enzyme in Salmonella and *Escherichia coli* that transfers 4-amino-4-deoxy-L-arabinose to lipid A: induction on polymyxin-resistant mutants and role of a novel lipid-linked donor. J Biol Chem 276(46):43122–43131
107. Soon RL, Nation RL, Cockram S, Moffatt JH, Harper M, Adler B et al (2011) Different surface charge of colistin-susceptible and -resistant *Acinetobacter baumannii* cells measured with zeta potential as a function of growth phase and colistin treatment. J Antimicrob Chemother 66(1):126–133
108. Koike M, Iida K, Matsuo T (1969) Electron microscopic studies on mode of action of polymyxin. J Bacteriol 97(1):448–452

Mechanisms of Polymyxin Resistance

5

Jennifer H. Moffatt, Marina Harper, and John D. Boyce

Abstract

Polymyxin antibiotics are increasingly being used as last-line therapeutic options against a number of multidrug resistant bacteria. These antibiotics show strong bactericidal activity against a range of Gram-negative bacteria, but with the increased use of these antibiotics resistant strains are emerging at an alarming rate. Furthermore, some Gram-negative species, such as *Neisseria meningitidis*, *Proteus mirabilis* and *Burkholderia* spp., are intrinsically resistant to the action of polymyxins. Most identified polymyxin resistance mechanisms in Gram-negative bacteria involve changes to the lipopolysaccharide (LPS) structure, as polymyxins initially interact with the negatively charged lipid A component of LPS. The controlled addition of positively charged residues such as 4-amino-L-arabinose, phosphoethanolamine and/or galactosamine to LPS results in a reduced negative charge on the bacterial surface and therefore reduced interaction between the polymyxin and the LPS. Polymyxin resistant species produce LPS that intrinsically contains one or more of these additions. While the genes necessary for most of these additions are chromosomally encoded, plasmid-borne phosphoethanolamine transferases (*mcr-1* to *mcr-8*) have recently been identified and these plasmids threaten to increase the rate of dissemination of clinically relevant colistin resistance. Uniquely, *Acinetobacter baumannii* can also become highly resistant to polymyxins *via* spontaneous mutations in the lipid A biosynthesis genes *lpxA*, *lpxC* or *lpxD* such that they produce no LPS or lipid A. A range of other non-LPS-dependent polymyxin resistance mechanisms has also been identified in bacteria, but these generally result in only low levels of resistance. These include increased anionic capsular polysaccharide production in *Klebsiella pneumoniae*, expression of efflux systems such as MtrCDE in *N. meningitidis*, and altered expression of outer membrane proteins in a small number of species.

Keywords

Polymyxin · Resistance · Loss of LPS · Lipid A modification · Remodelling of outer membrane

J. H. Moffatt
Biomedicine Discovery Institute, Infection and Immunity Program and Department of Microbiology, Monash University, Clayton, Australia

M. Harper · J. D. Boyce (✉)
Biomedicine Discovery Institute, Infection and Immunity Program and Department of Microbiology, Monash University, Clayton, Australia

Australian Research Council Centre of Excellence in Structural and Functional Microbial Genomics, Monash University, Clayton, Australia
e-mail: john.boyce@monash.edu

J. Li et al. (eds.), *Polymyxin Antibiotics: From Laboratory Bench to Bedside*, Advances in Experimental Medicine and Biology 1145, https://doi.org/10.1007/978-3-030-16373-0_5

5.1 How Do Polymyxins Kill Bacteria?

The outer membrane (OM) of Gram-negative bacteria serves as a semi-permeable barrier allowing essential molecules, such as nutrients, to enter the cell while excluding toxic compounds [1]. Lipopolysaccharide (LPS) is located on the outer leaflet of the outer membrane and is the major constituent of the Gram-negative cell surface. It is composed of the hydrophobic lipid A (endotoxin), which anchors the LPS to the outer membrane, a core oligosaccharide, and in many species a repeating distal polysaccharide (*O*-antigen) [2]. The bactericidal activity of polymyxins is mediated by an initial charge-based interaction with the lipid A component of LPS. Lipid A produced by most species carries a negative charge due to the presence of free phosphate groups; the binding of positively charged, divalent cations such as Ca^{2+} and Mg^{2+} to the negatively charged phosphate groups stabilizes the LPS [3, 4]. However, polymyxins and other cationic peptides bind these negatively charged phosphate groups with higher affinity than divalent cations and as a consequence displace Ca^{2+} and Mg^{2+}, thus, destabilizing the LPS and resulting in reduced OM integrity [5]. This in turn leads to increased OM permeability, self-promoted uptake of the polymyxin into the periplasm and probable insertion of the molecule into the inner membrane. Mechanisms of killing are unknown; the insertion of polymyxins may induce mixing between the inner and outer membranes leading to overall membrane disruption [6], although there is no indication that this results in cell leakage [7, 8]. Other mechanisms which may be involved in bacterial killing by polymyxins include the formation of hydroxyl radicals [9] and/or inactivation of protein targets, such as the type II NADH-quinone oxidoreductases [10]. For a more detailed description of the mode of action of polymyxins, see Chap. 4.

5.2 Resistance Mechanisms Affecting LPS Structure

Given that the primary interaction of polymyxins with the bacterial surface is *via* the charge-based interaction with LPS, it is not surprising that the majority of resistance mechanisms involve modifications that alter LPS structure and charge (Figs. 5.1 and 5.2). These modifications include the addition of 4-amino-$_{L}$-arabinose ($_{L}$-Ara4N), phosphoethanolamine (PEtn) and/or galactosamine. These additions occur primarily to the phosphate groups of lipid A but additions can also be made to residues within the core oligosaccharide such as 3-deoxy-D-mannooctulosonic acid (KDO).

5.2.1 Addition of 4-Amino-$_{L}$-Arabinose ($_{L}$-Ara4N)

In many bacteria, including *Salmonella enterica, Escherichia coli* and *Pseudomonas aeruginosa*, the addition of the amino sugar, $_{L}$-Ara4N, to lipid A of LPS results in high level polymyxin resistance of up to 512 mg/L [11–13]. The substitution of one or more of the negatively charged phosphate groups on the lipid A with $_{L}$-Ara4N abrogates the initial charge-based interaction with the positively charged amino groups of the polymyxin.

The biosynthesis and addition of $_{L}$-Ara4N requires the co-ordinated activity of the enzymes PmrE, PmrH, PmrF, PmrI, PmrJ, PmrK, PmrL and PmrM (also known as Ugd, ArnB/PbgP, ArnC, ArnA, ArnD, ArnT, ArnE and ArnF, respectively) (Fig. 5.3) [14]. Synthesis of $_{L}$-Ara4N begins in the cytoplasm with conversion of UDP-glucose to UDP-glucuronic acid by PmrE/Ugd, followed by oxidative decarboxylation of the UDP-glucuronic acid to UDP-4-keto-pyranose by PmrI/ArnA [2]. PmrH/ArnB then converts the UDP-4-keto-pyranose to UDP-β

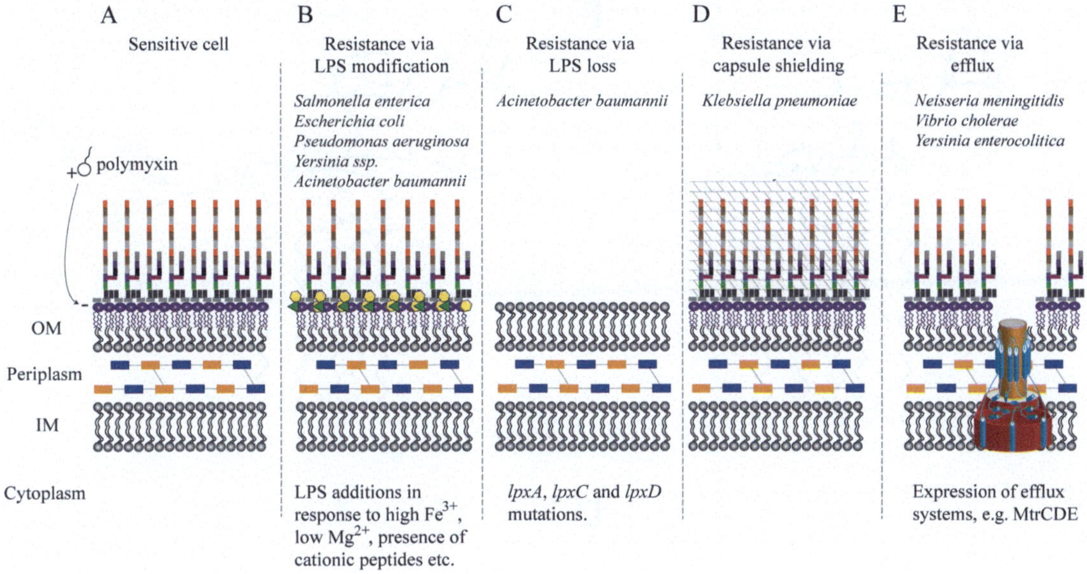

Fig. 5.1 Overview of polymyxin resistance mechanisms Schematic representations of the different polymyxin resistance mechanisms identified to date, and the species in which they have been observed. (**a**) Susceptible cell showing the inner and outer membrane and the peptidoglycan layer (yellow and blue rectangles) in the periplasm. The LPS which forms the outer leaflet of the Gram-negative cell is negatively charged and is the initial binding target of the positively charged polymyxin. (**b**) Many species, including *S. enterica*, *E. coli*, *P. aeruginosa*, *Yersinia* ssp. and *A. baumannii* can become resistant to polymyxins via modification of LPS. These changes include the addition of L-Ara4N (yellow hexagons), PEtn (green triangles) and/or galactosamine and may also include changes to the fatty acid chains. These LPS modifications are generally controlled by two component signal transduction systems such as PmrAB and PhoPQ in response to a range of conditions including, but not limited to, low Mg^{2+} high Fe^{3+} and the presence of cationic peptides (see Fig. 5.4 for more detail on regulation of gene expression). (**c**) *A. baumannii* can become resistant to polymyxins by complete loss of LPS including the lipid A anchor. Loss of LPS results from mutations within the genes *lpxA*, *lpxC* or *lpxD*. (**d**) In *K. pneumoniae* increased expression of capsule (grey hatched area) results in increased polymyxin resistance. (**e**) In *N. meningitidis* expression of the tripartite efflux system MtrCDE (orange/red/blue membrane spanning complex) results in increased polymyxin resistance. The MtrCDE structure is based on the model of Janganan et al. [95]. Polymyxin resistance has also been associated with changes in outer membrane protein expression in *Y. enterocolitica* and *V. cholera* (not shown)

-L-Ara4N, which undergoes formylation by PmrI/ArnA to generate UDP-β-L-Ara4FN [15]. The UDP-β-L-Ara4FN is then transferred to an inner membrane-located undecaprenyl phosphate carrier by the action of PmrF/ArnC, where it is then deformylated by PmrJ/ArnD and flipped across the inner membrane into the periplasm by the combined action of PmrL/ArnE and PmrM/ArnF, after which the L-Ara4N is transferred to lipid A by the glycosyltransferase PmrK/ArnT [16, 17].

In *S. enterica* and *E. coli*, L-Ara4N is preferentially added to the 4′ phosphate group (Fig. 5.2b) of lipid A by PmrK, but it can also be added to the 1 position or to both positions [18] depending on the presence/absence of the PEtn transferase PmrC/EptA (see below). The addition of L-Ara4N is highly dependent on the presence of the C14 3′-acyloxyacyl-linked myristate group on lipid A (Fig. 5.2b), which is transferred to the lipid A molecule by the myristoyl transferase LpxM. Thus, in an *lpxM* mutant, only very small amounts of L-Ara4N are added to lipid A even under inducing conditions [19].

Expression of the genes required for L-Ara4N biosynthesis and transfer to lipid A is regulated differently between species. In *S. enterica*, expression of the *pmrE* and *pmrHFIJKLM* genes is controlled both by the direct action of the two-component signal transduction system (TCSTS) PmrAB and the indirect action of the TCSTS

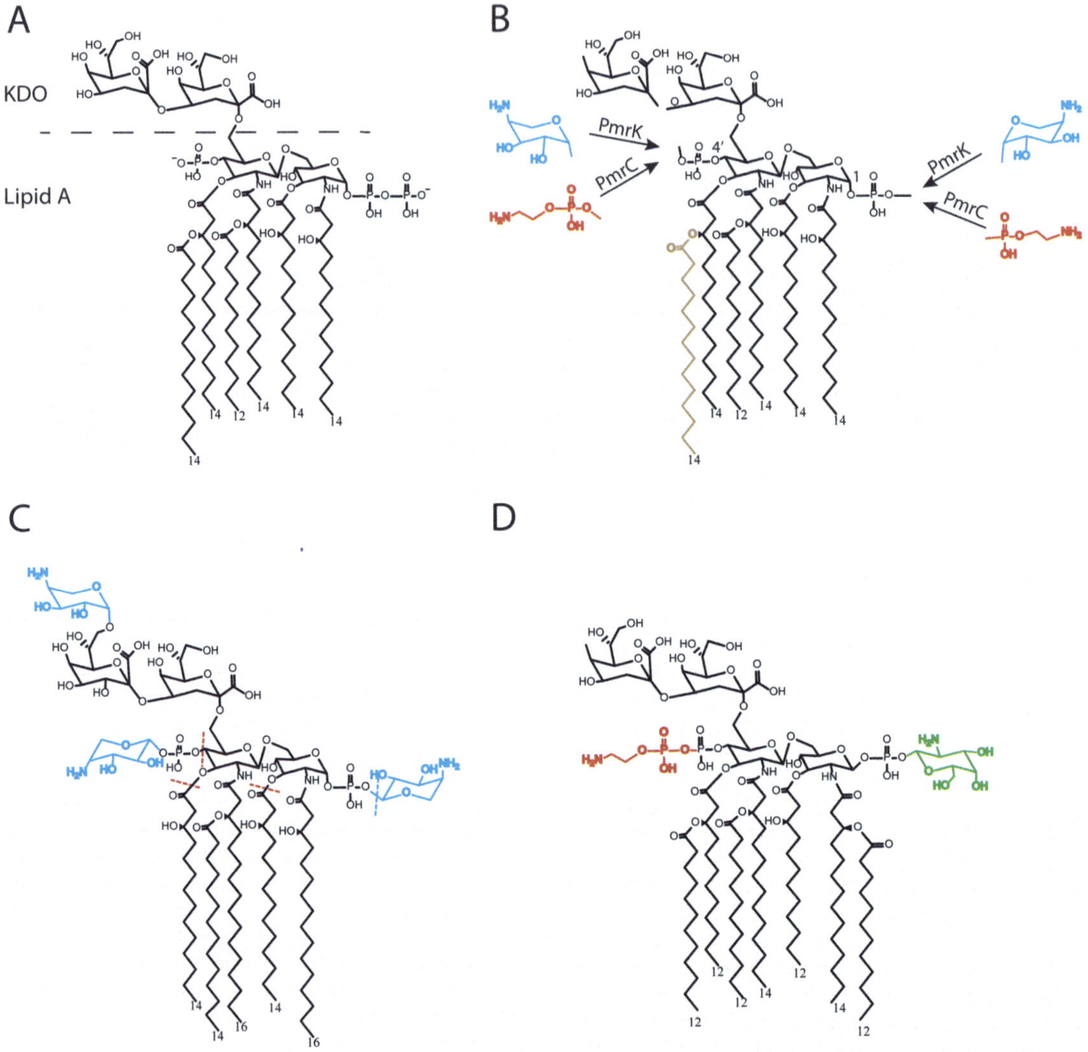

Fig. 5.2 LPS modifications that lead to polymyxin resistance
(**a**) Structure of the *E. coli* LPS isolated from strains that are susceptible to polymyxins. (**b**) Modifications of the lipid A portion of LPS that lead to polymyxin resistance [57]. The transferase required for each substitution is shown above each arrow. PEtn (shown in red) and L-Ara4N (shown in blue) are primarily added to the 1 and 4′ position of lipid A respectively although both moieties can be added to either position under certain conditions. The addition of L-Ara4N is dependent on the presence of the C14 myristate group (shown in brown), which is added during lipid A biosynthesis by the LpxM transferase. (**c**) Structure of *B. cenocepacia* LPS that gives high intrinsic polymyxin resistance to this species [54, 96]. Non-stoichiometric additions are noted by dashed red lines. (**d**) Structure of the *A. baumannii* LPS containing substitutions with both PEtn (shown in red) and galactosamine (shown in green) [75]

PhoPQ (Fig. 5.4). The membrane bound PmrB sensor kinase is activated and autophosphorylated in response to a range of stimuli including low pH, high Fe^{3+} and Al^{3+} conditions (reviewed by [20]). Activation of PmrB results in phosphate transfer to the PmrA response regulator. The phosphorylated and activated PmrA then binds to a conserved motif called the PmrA box which is located upstream of the −35 regions of a number of promoters including the *pmrHFIJKLM*, *pmrE* and *pmrCAB* promoters (Fig. 5.4) and induces increased expression of the downstream genes

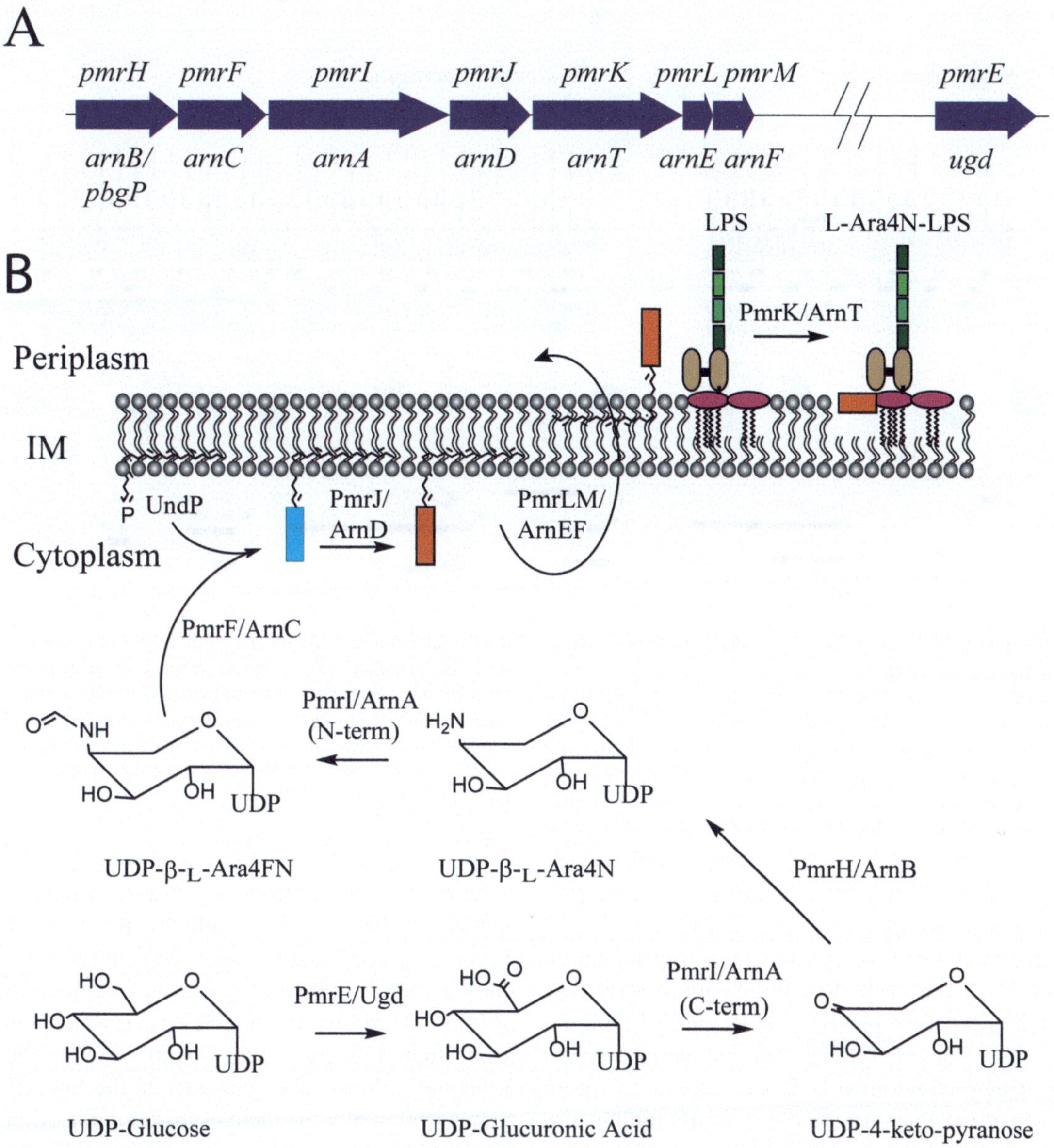

Fig. 5.3 Genetics and biochemistry of L-Ara4N addition to LPS
(**a**) Genes involved in the biosynthesis, transport and addition of L-Ara4N to LPS. (**b**) Schematic representation of the steps involved in the biosynthesis, transport and addition of L-Ara4N to the lipid A component of LPS. Enzymes required for each step are shown above or beside each arrow. UDP-glucose is first oxidised to UDP-glucuronic acid by the action of PmrE/Ugd. PmrI/ArnA then catalyzes the oxidative decarboxylation of UDP-glucuronic acid to UDP-4-keto-pyranose, PmrH/ArnB converts UDP-4-keto-pyranose to UDP-L-Ara4N and UDP-L-Ara4N undergoes formylation by PmrI/ArnA to generate UDP-L-Ara4FN. The UDP-L-Ara4FN is then transferred to the undecaprenyl phosphate carrier by the action of PmrF/ArnC. The UDP-L-Ara4FN (blue rectangle) is deformylated through the action of PmrJ/ArnD and then flipped across the inner membrane by the combined action of PmrM/ArnF and PmrL/ArnE. The L-Ara4N component of this molecule (orange rectangle) is then transferred to the lipid A of LPS by the PmrK/ArnT transferase. UndP, undecaprenyl phosphate; UDP, uridine diphosphate. Figure modified from Yan et al. [17]

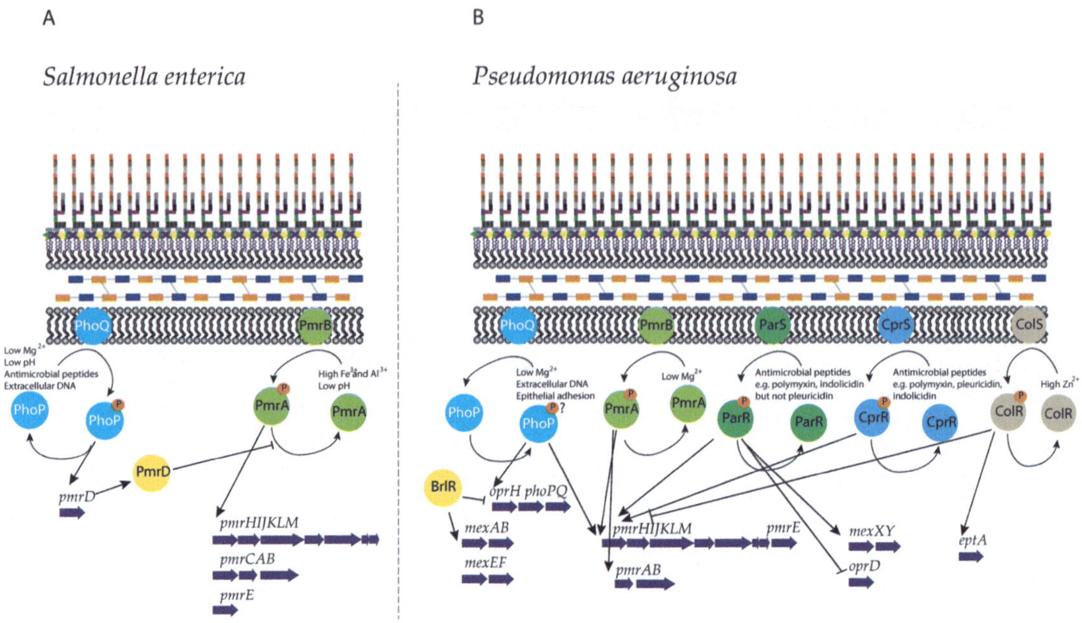

Fig. 5.4 Regulation of LPS additions that can give rise to polymyxin resistance
(**a**) The PmrAB and PhoPQ two-component systems regulate addition of L-Ara4N and PEtn to LPS in *S. enterica* in response to low Mg^{2+}, low pH, the presence of antimicrobial peptides such as polymyxins, extracellular DNA, and high levels of $Fe^{3+}$and Al^{3+}. (**b**) The PmrAB, PhoPQ, ParRS, CprRS and ColRS two component systems regulate the addition of L-Ara4N and PEtn to LPS in *P. aeruginosa* in response to a wide range of conditions. Furthermore, the ParRS system also plays a role in resistance to other antibiotics via *mexXY* and *oprD*. The question mark indicates that it is currently unclear whether the active form of PhoP is phosphorylated or unphosphorylated

[21]. The PmrA box contains a consensus sequence comprised of two YTTAAK repeats separated by 5 bp [22]. Constitutively active PmrA mutants (H81R) can give up to 3000-fold activation of the *pmrHFIJKLM* operon [21, 23].

The PhoPQ TCSTS can indirectly activate L-Ara4N addition to LPS via increased expression of the PmrD protein [24]. PmrD inhibits the dephosphorylation of the PmrA response regulator, resulting in enhanced PmrA activity and increased activation of the *pmrHFIJKLM* locus [25]. An increase in PmrD expression can occur in response to low Mg^{2+}, low pH, or in the presence of cationic peptides or extracellular DNA [26–28] (Fig. 5.4). Phosphorylated PhoP is the transcriptional activator of PmrD expression [29] and amino acid changes in PhoP (S93 N and/or Q203R) that lead to constitutive activation result in increased PmrD expression and polymyxin resistance [30] (Fig. 5.4).

In *E. coli,* the addition of L-Ara4N to lipid A appears to be controlled only by the PmrAB TCSTS. *E. coli* expresses a Mg^{2+}-responsive PhoPQ TCSTS which also activates expression of PmrD. However, the *E. coli* PmrD, which has 55% shared amino acid identity with the *Salmonella* PmrD, does not activate the PmrAB response regulator. As there is no communication between the two systems, *E. coli* is unable to modify LPS with L-Ara4N in response to low Mg^{2+} concentrations [31].

In *P. aeruginosa*, the addition of L-Ara4N to lipid A is also dependent on expression of the L-Ara4N biosynthesis and transfer genes that, unlike the situation in *Salmonella*, are organised in a single operon (*pmrHFIJKLME*). The operon is induced in response to low Mg^{2+} and in the presence of antimicrobial peptides (such as polymyxin B and LL-37 among others) or extracellular DNA [12, 32, 33]. However, the regulation of these genes appears significantly more complex in *P. aeruginosa* than in *S. enterica* (Fig. 5.4).

PmrAB is the primary TCSTS involved in controlling L-Ara4N addition to *P. aeruginosa* LPS; *pmrA* or *pmrB* mutations leading to inactivation of the PmrAB system result in strains with 2- to 16-fold increased polymyxin susceptibility [32]. Similarly, mutations that constitutively activate the sensor kinase PmrB (L243Q and A248V) result in L-Ara4N substitution of LPS and increased resistance to polymyxin [12]. In *P. aeruginosa* the PmrAB system directly responds to low Mg^{2+} but not high levels of Fe^{3+} [12]. The PhoPQ TCSTS also plays a role in polymyxin resistance, responding to low Mg^{2+}, extracellular DNA and interaction with epithelial cells [34–36]. In *P. aeruginosa* the PhoP response regulator acts directly on the *pmr* operon, unlike in *Salmonella*, where PhoP has an indirect role via up-regulation of PmrD expression [30, 37]. Unusually, the PhoQ sensor kinase appears to repress PhoP activity. Mutations that inactivate PhoQ lead to increased activity of PhoP and increased addition of L-Ara4N to LPS [35, 38]. It is unclear whether PhoQ repression of PhoP is via kinase or de-phosphorylation activity as the phosphorylation status of the active form of PhoP is not known [38]. The PhoPQ system also regulates expression of the outer membrane porin OprH, which is encoded as the first gene in the three-gene operon containing *oprH*, *phoP* and *phoQ* (Fig. 5.4) [35]. Inactivation of PhoP results in the loss of expression of *oprH, phoP* and *phoQ* and a concomitant loss in polymyxin resistance [35] supporting the hypothesis that PhoP positively regulates the L-Ara4N synthesis and transport genes.

At least three other TCSTS, designated ParRS, CprRS and ColRS, contribute to the regulation of polymyxin resistance in *P. aeruginosa* [39, 40]. In *Salmonella*, PhoQ responds to low Mg^{2+} conditions as well as the presence of antimicrobial peptides through a common binding site on PhoQ [26]. However, the *P. aeruginosa* PhoQ plays no role in the response to antimicrobial peptides. Rather, the *P. aeruginosa* ParRS and CprRS systems each activate the *pmr* operon in response to antimicrobial peptides [39, 40]. Both ParRS and CprRS respond to polymyxin as well as a range of antimicrobial peptides such as indolicidin and pleuricidin but in different ways [40]. Importantly, ParRS also controls expression of other genes involved in drug resistance, including the genes encoding the MexXY efflux system and the carbapenem-specific porin OprD [41]. Furthermore, a transposon mutagenesis screen of a *phoQ* mutant identified the TCSTS ColRS, as playing a role in polymyxin resistance [42]. ColRS has recently been shown to increase expression of the PEtn transferase PmrC ($EptA_{PA}$) but reduce expression of the L-Ara4N transferase PmrK/ArnT in response to Zn^{2+} [43]. However, this Zn^{2+}-mediated regulation of PEtn addition to LPS does not appear to affect polymyxin resistance in *P. aeruginosa*, so it is currently unclear precisely how the ColRS TCSTS directly affects polymyxin resistance. Finally, the MerR-like transcriptional regulator BrlR, which controls expression of the multidrug efflux systems MexAB-OprM and MexEF-OprN [44], also represses *phoPQ* expression and therefore the action of BrlR can increase susceptibility of *P. aeruginosa* to polymyxins [45] (Fig. 5.4).

Polymyxin-resistant *P. aeruginosa* isolates recovered from chronically infected cystic fibrosis patients have been shown to produce LPS with lipid A modifications that include L-Ara4N as well as increased addition of the C16 fatty acid palmitate [46, 47]. Furthermore, a study of polymyxin resistant *P. aeruginosa* strains that were isolated from cystic fibrosis patients who had been treated with colistin, revealed that all of the isolates contained inactivating *phoQ* mutations that resulted in increased polymyxin resistance *via* the addition of L-Ara4N to lipid A [38]. *P. aeruginosa phoQ* mutants also display novel palmitate additions to lipid A, reduced growth rate, reduced twitching motility and cytotoxicity, as well as reduced *in vivo* fitness in a rat lung infection model. However, it appears that these changes do not completely abrogate the ability of these strains to cause serious infections in cystic fibrosis patients [48].

Yersinia spp. can also express LPS with L-Ara4N substitution [49, 50]. A *Y. pestis arnT* (*pmrK*) mutant, lacking L-Ara4N substitution on the lipid A component of the LPS, was 60-fold more susceptible to polymyxin B [49]. Similarly, mutants with truncated core oligosaccharide (e.g. a *waaQ* hepIII-transferase mutant) were also highly susceptible and an LPS mutant expressing only the lipid A molecule with no L-Ara4N substitution was 250-fold more susceptible [49]. Thus, core oligosaccharide composition and/or length may play a direct role in polymyxin resistance in *Y. pestis* or may play an indirect role by inhibiting the addition of L-Ara4N to lipid A. *Y. pestis* alters its LPS composition when grown under different temperatures, resulting in changes to polymyxin resistance. When cultured at the temperature extremes of 37 °C and 6 °C, *Y. pestis* is highly susceptible to polymyxin B, but when grown at 25 °C, the addition of L-Ara4N to lipid A confers polymyxin B resistance. Modification of the LPS core with glycine, a highly uncommon LPS core component, may also play a role in polymyxin resistance in *Y. pestis* [51]. The LPS core of the *Y. pestis* strain 1146 grown at 25 °C was shown to have an increase in cationic glycine but no addition of L-Ara4N to the lipid A, suggesting that glycine alone may be responsible for the increased resistance [51].

Burkholderia cenocepacia (as well as other *Burkholderia* species) exhibits very high intrinsic polymyxin resistance, mediated by LPS that is substituted in multiple positions with L-Ara4N (Fig. 5.2c) [52]. Unusually, L-Ara4N is used by *B. cenocepacia* to modify both the lipid A and a branched D-*glycero*-D-*talo*-oct-2-ulosonic acid residue in the LPS inner core [53, 54]. Interestingly, the synthesis of L-Ara4N is essential for *B. cenocepacia* viability (Ortega at al 2007). This is because LPS is essential for *B. cenocepacia* viability and the LPS transporter, LptG, can only recognise and transport LPS molecules that are modified with L-Ara4N [52]. Other species that contain L-Ara4N as a substitution of the LPS inner core oligosaccharide, such as *Serratia*, *Proteus* and *Ralstonia* ssp., also display very high levels of polymyxin resistance [52, 55].

5.2.2 Addition of PEtn to LPS

In *S. enterica* the PmrAB TCSTS also controls the expression of genes required for PEtn addition to the lipid A component of the LPS molecule *via* the PEtn transferase PmrC (also known as EptA or PagB) [56]. PmrC is encoded by the first gene in a three-gene operon that also encodes the PmrA and PmrB TCSTS proteins (Fig. 5.4). Inactivation of only the *pmrC* gene of this operon results in mutants that lack the PEtn substitution to lipid A and are 3- to 5-fold more susceptible to killing by polymyxin B compared to the wild-type strain [56]. However, L-Ara4N substitution of the lipid A in *S. enterica* plays a greater role in polymyxin resistance; a strain unable to convert UDP-4-keto-pyranose to UDP-L-Ara4N (*pmrH*/*pbgP* mutant) was approximately 1000 times more susceptible to polymyxin B than the wild-type strain [56]. This appears to be reversed in *E. coli* as an *E. coli* L-Ara4N glycosyltransferase mutant showed an 8-fold decrease in resistance while a PEtn transferase mutant was 20-fold less resistant than the wild-type strain [57].

As noted above (Sect. 5.2.1), in *S. enterica* L-Ara4N is preferentially added to the phosphate group at the 4′ position of lipid A and PEtn is added to the 1 position [18]. However, as is the case for addition of L-Ara4N, PEtn can be added to both positions in the absence of L-Ara4N. In *S. enterica* and *E. coli*, under conditions that repress PEtn addition (e.g. high Mg^{2+}), a second phosphate group can be added to the 1 position of lipid A in place of PEtn by the LpxT transferase [57, 58]. LpxT activity is negatively regulated by the PmrAB TCSTS via induction of *pmrR*, encoding a 30 amino acid membrane peptide that directly inactivates LpxT. Thus, induction of PmrAB activates the transferases that add L-Ara4N and PEtn, and inhibits the competing LpxT transferase [57, 58].

Polymyxin resistance in *A. baumannii* can be mediated by addition of PEtn to LPS and this is dependent on the PEtn transferase PmrC as well as the TCSTS proteins PmrA and PmrB [59, 60]. As observed in *Salmonella*, activation of the

PmrAB TCSTS in *A. baumannii* results in increased transcription of the *pmrCAB* operon, with the first gene in the operon encoding the PEtn transferase PmrC. *A. baumannii* mutants with constitutively active PmrA or PmrB display increased colistin resistance of between 4- and 128-fold. Conversely, strains lacking a functional *pmrB* show 100-fold increased susceptibility to colistin [60]. Mutations in *pmrAB* associated with polymyxin resistance include point mutations leading to amino acid substitutions within the PmrA response regulator (E8D, M12I, M12K P102H) and within the PmrB sensor kinase (T13N, T13A, S14L, S17R, L87F, Y116H, M145L, M145I, A227V, R231L, T232I, P233T, P233S, T235I, A262P, R263L, R263P, R263C, G315D, N353Y, F387Y, S403F) [59–64]. Many of the amino acid substitutions within PmrB that confer polymyxin resistance fall within the predicted histidine kinase domain (amino acids 216–276); residues between 231–235 and 262–263 appear particularly important. PmrB is required for acid-induced polymyxin resistance in *A. baumannii* but not for resistance induced by high levels of Fe^{3+}, indicating that other TCSTS may also be involved in controlling polymyxin resistance in this species [61].

Polymyxin-resistant clinical isolates of *A. baumannii* have been shown to arise in patients during failed colistin treatment [62, 64–66]. Analysis of 28 *A. baumannii* isolates (14 Col^S and 14 Col^R) recovered from seven combat trauma patients before and after colistin treatment indicated that all the resistant isolates had mutations leading to amino acid changes in PmrA and/or PmrB [62]. However, the amino acid sequences of PmrA and PmrC encoded within the *pmrCAB* operon of these isolates differed sufficiently from the equivalent amino acid sequences in other *A. baumannii* strains that they were designated PmrA1 and PmrC1 respectively. Moreover, all isolates contained two additional PmrC paralogues located elsewhere on the genome, designated EptA-1 and EptA-2. An analysis of 116 *A. baumannii* genome sequences identified that 20% of the sequenced strains contained two PmrC paralogues while 4% contained three or more PmrC paralogues [62]. Importantly, transcriptional analysis by quantitative reverse transcriptase PCR indicated that in the colistin resistant, *pmrAB* TCSTS mutants, expression of each of the PEtn transferases PmrC, EptA-1 and EptA-2 was significantly increased. These data suggest that *eptA-1* and *eptA-2* are also important for polymyxin resistance and may be at least partially under the control of PmrAB [62].

Interestingly, colistin resistance due to mutations in the *A. baumannii pmrAB* genes correlates with a fitness cost [63, 64, 67, 68]. A *pmrB* mutant showed reduced *in vitro* growth and reduced competitive *in vivo* growth in a mouse systemic infection model but still caused normal levels of disease in mice [67]. Importantly, the polymyxin resistance of isolates with *pmrB* missense mutations that arose independently following colistin treatment in four patients, was rapidly lost (returned to colistin sensitivity) following termination of colistin treatment [64]. Mutations in the resistant strains all led to changes in PmrB (P233S, R263C, M145I and T13A) that likely resulted in constitutive activation [64]. One strain containing a *pmrB* mutation (leading to L271R) appeared more stable in the absence of colistin but also showed lower levels of resistance [64]. In another *pmrB* mutant (leading to P233S), isolated from a patient following cessation of colistin treatment, a reversion to polymyxin sensitivity was later observed. Genetic analyses revealed that the reversion to sensitivity was due to a secondary mutation in *pmrA* (leading to L206P) that abrogated the DNA binding ability of PmrA. Thus, in this isolate, the constitutive activation of the PmrAB TCSTS that resulted from the initial *pmrB* mutation was reversed by a second mutation in *pmrA*.

There is no structural evidence that *A. baumannii* modifies the lipid A component of the LPS with $_{L}$-Ara4N although modification with the structurally similar residue galactosamine can occur (see below). Furthermore, homologs of the PmrHFIJKLME proteins, which are essential for $_{L}$-Ara4N synthesis and attachment to LPS in *S. enterica*, *P. aeruginosa* and *E. coli* (see above), have not been identified in *A. baumannii*. In addition, there are no clear *A. baumannii* homologs of the *Salmonella* PhoPQ TCSTS proteins [60], which play important roles in controlling the addition of PEtn and $_{L}$-Ara4N to LPS in that species.

All of the above-described PEtn transferases are encoded on the bacterial genome. Thus, development of colistin resistance by these mechanisms is normally due to increased expression of the transferase genes, either as a direct response to the presence of the antibiotic or other inducers (Fig. 5.4), or following activating mutations in the controlling two component signal transduction systems (e.g. PmrAB; Fig. 5.4). However, recently some novel PEtn transferase genes, designated *mcr-1* (and highly related genes *mcr-1.2* and *mcr-1.3*) to *mcr-8*, have been identified on a number of different plasmids [69–71]. The majority of these plasmids are likely to be conjugative and some have been shown to transfer at a very high rate (10^{-1}–10^{-3} cells per recipient) [72]. This is a very worrying situation as such plasmids are likely to rapidly increase the spread of colistin resistance. Indeed, *mcr*-postive isolates have already been identified from multiple countries in Europe, Asia, Africa and the Americas [73]. At this time *mcr* plasmids have been identified mainly in members of the Enterobacteriaceae, but heterologous expression of *mcr-1* in *A. baumannii* results in PEtn addition to LPS and colistin resistance, suggesting that transfer of plasmid-mediated colistin resistance to other nosocomial pathogens is only a matter of time [74]. It is currently unknown whether expression of these plasmid-borne *mcr* genes is controlled by two-component regulatory systems in a similar way to the chromosomally-located genes.

5.2.3 Addition of Galactosamine to LPS

A. baumannii LPS can be modified by addition of galactosamine to lipid A. Galactosamine is structurally very similar to $_{L}$-Ara4N and its addition would also act to mask the negative charge on the lipid A phosphate groups (Fig. 5.2d). The *A. baumannii* colistin resistant strain, MAC204, which was selected initially by *in vitro* passage in the presence of 1 mg/L colistin, then allowed to revert to a non-resistant phenotype on normal media before final selection on 2 mg/L colistin, was shown to contain both PEtn and galactosamine modification by MALDI-TOF MS (Fig. 5.2d) [75]. The same additions of both PEtn and galactosamine have also been observed in clinical isolates recovered from patients following colistin treatment [75]. The lipid A of *Francisella tularensis* also contains galactosamine [76, 77], which is predicted to confer polymyxin resistance to this species [76].

5.2.4 Complete Loss of LPS and Lipid A

One of the most intriguing mechanisms of polymyxin resistance identified to date is the complete loss of lipopolysaccharide (LPS) from the bacterial surface. *A. baumannii* can become polymyxin resistant *via* the complete loss of LPS, including the lipid A moiety that anchors LPS to the cell surface (Fig. 5.1). This dramatic change in the cell surface results in high level resistance (>256 mg/L) to polymyxins [78]. Currently, this mechanism of polymyxin resistance has only been observed in *A. baumannii*.

Loss of LPS including the lipid A anchor in *A. baumannii* occurs following mutations in any of first three genes in the lipid A biosynthesis pathway; namely, *lpxA*, *lpxC* and *lpxD*. Analysis of 21 independent *in vitro* derived colistin resistant derivatives of the *A. baumannii* type strain ATCC 19606 showed that each contained a unique mutation in one of the first three genes in the lipid A biosynthesis pathway [78, 79]. These spontaneously occurring mutations included single base changes, large deletions, and the insertion of IS elements. Two insertion sequence elements have been identified as causing LPS loss, namely IS*Aba11* and a novel IS*4*-family element [78, 79]. Spontaneous LPS-deficient mutants may contribute to the heteroresistance phenotype observed for some strains of *A. baumannii* [78], where an apparently colistin-susceptible strain (based on MIC) harbours a small population of colistin-resistant LPS-deficient cells [80]. Heteroresistance in *A. baumannii* strains has been shown to develop into high-level colistin resistance under the selective pressure of colistin both *in vitro* [81] and *in vivo* [82].

Total loss of LPS has been observed in a small number of *A. baumannii* clinical isolates that are either colistin-resistant or heteroresistant, but recent evidence indicates that the loss of LPS results in a more significant decrease in overall bacterial fitness than does constitutive activation of PEtn addition following *pmrAB* mutations [67]. Indeed, modification of LPS by the addition of PEtn (see above) appears to be by far the more common mechanism of colistin resistance in *A. baumannii* clinical isolates [62–64].

LPS-deficient *A. baumannii* cells still elaborate an outer membrane, although, the membrane is highly permeable, allowing molecules that would typically be excluded to enter the cell [78]. Indeed, an LPS-deficient strain of *A. baumannii* displayed increased susceptibility to a variety of antibiotics, including cefepime, teicoplanin and azithromycin [78]. This increased susceptibility is likely due to the relative ease that these antibiotics can cross the compromised outer membrane. Thus, effective treatment of polymyxin resistant LPS-deficient strains can likely be accomplished by using any of a range of second antibiotics, including those to which the colistin-susceptible parent strain may have been resistant. LPS-deficient cells also show an increase in susceptibility to the human antimicrobial peptide LL-37 and this is also likely due to increased uptake across the outer membrane [83].

It is currently unclear why *A. baumannii* is able to survive without LPS while in most other species LPS appears essential for viability. Transcriptional analysis of the LPS-deficient *lpxA* mutant shows that *A. baumannii* responds to LPS loss by altering the expression of a large number of genes encoding proteins involved lipoprotein biosynthesis and transport, phospholipid transport, and production of the surface polysaccharide poly-beta-1,6-*N*-acetylglucosamine; it is likely that many of these changes are critical for its survival in the absence of LPS [84, 85].

5.3 Other Mechanisms of Polymyxin Resistance

5.3.1 Capsule Expression

The capsule of *Klebsiella pneumoniae* has been shown to contribute to polymyxin resistance (Fig. 5.1). An acapsular mutant was more susceptible to polymyxin B and displayed increased binding of polymyxin B to the bacterial surface compared to the capsulated parent [86]. Moreover, growth in the presence of polymyxin B led to an increase in the transcription of the capsule biosynthesis genes and an approximately 1.5-fold increase in the amount of capsular polysaccharide produced. An *O*-antigen deficient, LPS mutant of the same strain also showed increased susceptibility to polymyxin B. However, at low levels of polymyxin B, the acapsular strain was more susceptible than the *O*-antigen mutant [86]. In other experiments, the addition of purified capsular polysaccharide isolated from *K. pneumoniae* or *P. aeruginosa* gave increased polymyxin resistance to a susceptible un-encapsulated *K. pneumoniae* strain and purified polysaccharide from either species was found to directly bind polymyxins [87]. Taken together, these data suggest that the anionic capsular polysaccharide of *K. pneumoniae* (and perhaps other species) can bind polymyxins and physically interfere with the access of polymyxins to the outer membrane, thus abrogating their bactericidal action.

5.3.2 Outer Membrane Proteins and Efflux Systems

The *Neisseria meningitidis* Mtr efflux system (MrtCDE) is a critical efflux pump that confers a high intrinsic resistance to polymyxins (Fig. 5.1) [88]. Mutants with transposon insertions in *mtrC*, *mtrD* and *mtrE* showed 16-fold increased susceptibility to polymyxins. Mutation of *porB*, encoding an outer membrane porin, also increased polymyxin susceptibility in this species [88]. However, it is unclear whether PorB plays a role in active efflux of polymyxins alone or together with the Mtr efflux system [88]. In *K. pneumoniae*, inactivation of the KpnGH efflux system also resulted in increased susceptibility to polymyxin B [89].

In *Vibrio cholerae* strain O395, mutation of the OmpU outer membrane protein increased polymyxin B susceptibility by between 100 and 1000-fold [90]. The exact mechanism by which OmpU mediates polymyxin resistance is currently unclear; it may act directly either as a porin or part of an as yet unidentified active efflux system. However, OmpU has a known role in the regulation of the sigma factor, sigma E, and may be part of a sensor system that detects OM stress [91]. Thus, the involvement of OmpU in polymyxin resistance may be indirect *via* the appropriate activation of other resistance genes controlled by sigma E. An *ompU* mutant of *Vibrio splendidus* showed only a very low level (2-fold) of increased susceptibility to polymyxin B [92].

In *Yersinia enterocolitica*, the RosA/RosB efflux pump/potassium antiporter system has been identified as having a role in polymyxin resistance. A *Y. enterocolitica* mutant lacking functional RosA and RosB proteins was significantly more susceptible to polymyxin B; this sensitivity could be rescued by complementation with a functional copy of *rosA* [65]. Moreover, an increase in polymyxin susceptibility was observed when wild-type and *rosB* mutant cells were treated with the proton motive force uncoupler, 2-carbonyl cyanide *m*-chlorophenylhydrazone (CCCP). Thus, it is likely that RosA plays a role in direct efflux of polymyxins from the cytoplasm. It is of interest that acidification of the bacterial cytoplasm, either *via* the action of RosA/RosB or by treatment with the weak acid sodium acetate, also increased polymyxin resistance, although it is unclear whether intracellular pH plays a direct or indirect role in polymyxin resistance [65]. It should be noted that RosA/RosB are involved in the temperature- and pH-dependent regulation of *O*-antigen biosynthesis in *Y. enterocolitica* [93] and mutations affecting LPS core oligosaccharide can alter polymyxin resistance [49]. Thus, the role of the RosA/B efflux pump/potassium antiporter system in polymyxin resistance may be via its role in the regulation of LPS biosynthesis.

5.3.3 Inactivation of Polymyxins

To date, there is only a single report of a bacterial enzyme that can inactivate colistin. The Gram-positive bacterium *Paenibacillus (bacillus) polymyxa* produces both colistin and a putative serine protease, designated colistinase, that appears able to degrade colistin *via* cleavage of the DAB-DAB bond at the side chain-cyclic peptide boundary (Fig. 1.6) [94]. Why *Pa. polymyxa* has evolved an apparently secreted enzyme that inactivates colistin remains an intriguing question, as this organism is a Gram-positive bacterium so lacks the LPS that is the primary surface target of colistin. However, recent evidence suggests that polymyxins may induce the production of toxic free radicals or damage certain intracellular enzymes [9, 10]. Therefore, as the bacterium also produces colistin, it is possible that the production of colistinase is necessary for survival in the presence of polymyxin synthesized by this species. There has been no recent work to identify the gene encoding this putative colistinase or to explore its activity in more detail. It should be noted that colistin degradation has never been associated with colistin resistance in other organisms.

5.4 Conclusions

Polymyxins are currently crucial last-line treatments for infections caused by multi-drug resistant Gram-negative bacteria. However, as their clinical use has increased, so too has the isolation of resistant strains. A wide variety of resistance mechanisms has been identified, indicating that polymyxin resistance has evolved multiple times independently. The initial binding target of polymyxins is the negatively charged lipid A component of LPS on the Gram-negative bacterial surface and the most common resistance mechanisms involve modifications to the LPS that reduce the negative charge; these include the addition of L-Ara4N, PEtn and galactosamine. Other resistance mechanisms include the production of capsular polysaccharides, expression of efflux systems, and even the complete loss of LPS production. It is likely that future work will define new mechanisms of resistance and elucidate more precisely how expression of the adaptive mechanisms is regulated. As the genetics involved in many of these resistance mechanisms is now well established, it is feasible that specific molecular diagnostic approaches could be used to rapidly identify polymyxin resistance strains and their resistance mechanisms in clinical settings. Furthermore, it is hoped that a detailed understanding of these resistance mechanisms, and how certain mechanisms are favoured under particular conditions, will allow the optimization of polymyxin treatment regimens to reduce resistance development. Such optimization may prolong the useful lifespan of polymyxins as last-line treatment agents.

References

1. Nikaido H (2003) Molecular basis of bacterial outer membrane permeability revisited. Microbiol Mol Biol Rev 67(4):593–656
2. Raetz CR, Whitfield C (2002) Lipopolysaccharide endotoxins. Annu Rev Biochem 71:635–700. https://doi.org/10.1146/annurev.biochem.71.110601.135414
3. Levy SB, Leive L (1968) An equilibrium between two fractions of lipopolysaccharide in *Escherichia coli*. Proc Natl Acad Sci USA 61(4):1435–1439
4. Schnaitman CA (1971) Effect of ethylenediaminetetraacetic acid, Triton X-100, and lysozyme on the morphology and chemical composition of isolate cell walls of *Escherichia coli*. J Bacteriol 108(1):553–563
5. Hancock RE (1997) Peptide antibiotics. Lancet 349(9049):418–422. https://doi.org/10.1016/S0140-6736(97)80051-7
6. Dixon RA, Chopra I (1986) Polymyxin B and polymyxin B nonapeptide alter cytoplasmic membrane permeability in *Escherichia coli*. J Antimicrob Chemother 18(5):557–563
7. Clausell A, Garcia-Subirats M, Pujol M, Busquets MA, Rabanal F, Cajal Y (2007) Gram-negative outer and inner membrane models: insertion of cyclic cationic lipopeptides. J Phys Chem B 111(3):551–563. https://doi.org/10.1021/jp064757+
8. Wiese A, Gutsmann T, Seydel U (2003) Towards antibacterial strategies: studies on the mechanisms of interaction between antibacterial peptides and model membranes. J Endotoxin Res 9(2):67–84. https://doi.org/10.1179/096805103125001441
9. Sampson TR, Liu X, Schroeder MR, Kraft CS, Burd EM, Weiss DS (2012) Rapid killing of *Acinetobacter baumannii* by polymyxins is mediated by a hydroxyl radical death pathway. Antimicrob Agents Chemother 56(11):5642–5649. https://doi.org/10.1128/AAC.00756-12
10. Deris ZZ, Akter J, Sivanesan S, Roberts KD, Thompson PE, Nation RL, Li J, Velkov T (2014) A secondary mode of action of polymyxins against gram-negative bacteria involves the inhibition of NADH-quinone oxidoreductase activity. J Antibiot (Tokyo) 67(2):147–151. https://doi.org/10.1038/ja.2013.111
11. Moskowitz SM, Brannon MK, Dasgupta N, Pier M, Sgambati N, Miller AK, Selgrade SE, Miller SI, Denton M, Conway SP, Johansen HK, Hoiby N (2012) PmrB mutations promote polymyxin resistance of *Pseudomonas aeruginosa* isolated from colistin-treated cystic fibrosis patients. Antimicrob Agents Chemother 56(2):1019–1030. https://doi.org/10.1128/AAC.05829-11
12. Moskowitz SM, Ernst RK, Miller SI (2004) PmrAB, a two-component regulatory system of *Pseudomonas aeruginosa* that modulates resistance to cationic antimicrobial peptides and addition of aminoarabinose to lipid A. J Bacteriol 186(2):575–579
13. Nummila K, Kilpelainen I, Zahringer U, Vaara M, Helander IM (1995) Lipopolysaccharides of polymyxin B-resistant mutants of *Escherichia coli* are extensively substituted by 2-aminoethyl pyrophosphate and contain aminoarabinose in lipid A. Mol Microbiol 16(2):271–278
14. Gunn JS, Lim KB, Krueger J, Kim K, Guo L, Hackett M, Miller SI (1998) PmrA-PmrB-regulated genes necessary for 4-aminoarabinose lipid A modification and polymyxin resistance. Mol Microbiol 27(6):1171–1182
15. Breazeale SD, Ribeiro AA, McClerren AL, Raetz CR (2005) A formyltransferase required for polymyxin resistance in *Escherichia coli* and the modification of lipid A with 4-Amino-4-deoxy-L-arabinose. Identification and function oF UDP-4-deoxy-4-formamido-L-arabinose. J Biol Chem 280(14):14154–14167. https://doi.org/10.1074/jbc.M414265200
16. Trent MS, Ribeiro AA, Lin S, Cotter RJ, Raetz CR (2001) An inner membrane enzyme in *Salmonella* and *Escherichia coli* that transfers 4-amino-4-deoxy-L-arabinose to lipid A: induction on polymyxin-resistant mutants and role of a novel lipid-linked donor. J Biol Chem 276(46):43122–43131. https://doi.org/10.1074/jbc.M106961200. [pii]
17. Yan A, Guan Z, Raetz CR (2007) An undecaprenyl phosphate-aminoarabinose flippase required for poly-

myxin resistance in *Escherichia coli*. J Biol Chem 282(49):36077–36089. https://doi.org/10.1074/jbc.M706172200

18. Zhou Z, Ribeiro AA, Lin S, Cotter RJ, Miller SI, Raetz CR (2001) Lipid A modifications in polymyxin-resistant *Salmonella typhimurium*: PmrA-dependent 4-amino-4-deoxy-L-arabinose, and phosphoethanolamine incorporation. J Biol Chem 276(46):43111–43121. https://doi.org/10.1074/jbc.M106960200. [pii]
19. Tran AX, Lester ME, Stead CM, Raetz CR, Maskell DJ, McGrath SC, Cotter RJ, Trent MS (2005) Resistance to the antimicrobial peptide polymyxin requires myristoylation of *Escherichia coli* and *Salmonella typhimurium* lipid A. J Biol Chem 280(31):28186–28194. https://doi.org/10.1074/jbc.M505020200
20. Chen HD, Groisman EA (2013) The biology of the PmrA/PmrB two-component system: the major regulator of lipopolysaccharide modifications. Annu Rev Microbiol 67:83–112. https://doi.org/10.1146/annurev-micro-092412-155751
21. Gunn JS, Ryan SS, Van Velkinburgh JC, Ernst RK, Miller SI (2000) Genetic and functional analysis of a PmrA-PmrB-regulated locus necessary for lipopolysaccharide modification, antimicrobial peptide resistance, and oral virulence of *Salmonella enterica* serovar typhimurium. Infect Immun 68(11):6139–6146
22. Aguirre A, Lejona S, Vescovi EG, Soncini FC (2000) Phosphorylated PmrA interacts with the promoter region of *ugd* in *Salmonella enterica* serovar typhimurium. J Bacteriol 182(13):3874–3876
23. Roland KL, Martin LE, Esther CR, Spitznagel JK (1993) Spontaneous *pmrA* mutants of *Salmonella typhimurium* LT2 define a new two-component regulatory system with a possible role in virulence. J Bacteriol 175(13):4154–4164
24. Kox LF, Wosten MM, Groisman EA (2000) A small protein that mediates the activation of a two-component system by another two-component system. EMBO J 19(8):1861–1872. https://doi.org/10.1093/emboj/19.8.1861
25. Fu W, Yang F, Kang X, Zhang X, Li Y, Xia B, Jin C (2007) First structure of the polymyxin resistance proteins. Biochem Biophys Res Commun 361(4):1033–1037. S0006-291X(07)01637-3 [pii] https://doi.org/10.1016/j.bbrc.2007.07.144
26. Bader MW, Sanowar S, Daley ME, Schneider AR, Cho U, Xu W, Klevit RE, Le Moual H, Miller SI (2005) Recognition of antimicrobial peptides by a bacterial sensor kinase. Cell 122(3):461–472. https://doi.org/10.1016/j.cell.2005.05.030
27. Bearson BL, Wilson L, Foster JW (1998) A low pH-inducible, PhoPQ-dependent acid tolerance response protects *Salmonella typhimurium* against inorganic acid stress. J Bacteriol 180(9):2409–2417
28. Johnson L, Horsman SR, Charron-Mazenod L, Turnbull AL, Mulcahy H, Surette MG, Lewenza S (2013) Extracellular DNA-induced antimicrobial peptide resistance in *Salmonella enterica* serovar Typhimurium. BMC Microbiol 13(1):115. https://doi.org/10.1186/1471-2180-13-115
29. Shin D, Groisman EA (2005) Signal-dependent binding of the response regulators PhoP and PmrA to their target promoters *in vivo*. J Biol Chem 280(6):4089–4094. https://doi.org/10.1074/jbc.M412741200
30. Gunn JS, Ernst RK, McCoy AJ, Miller SI (2000) Constitutive mutations of the *Salmonella enterica* serovar typhimurium transcriptional virulence regulator *phoP*. Infect Immun 68(6):3758–3762
31. Winfield MD, Groisman EA (2004) Phenotypic differences between *Salmonella* and *Escherichia coli* resulting from the disparate regulation of homologous genes. Proc Natl Acad Sci USA 101(49):17162–17167. https://doi.org/10.1073/pnas.0406038101
32. McPhee JB, Lewenza S, Hancock RE (2003) Cationic antimicrobial peptides activate a two-component regulatory system, PmrA-PmrB, that regulates resistance to polymyxin B and cationic antimicrobial peptides in *Pseudomonas aeruginosa*. Mol Microbiol 50(1):205–217
33. Mulcahy H, Charron-Mazenod L, Lewenza S (2008) Extracellular DNA chelates cations and induces antibiotic resistance in *Pseudomonas aeruginosa* biofilms. PLoS Pathog 4(11):e1000213. https://doi.org/10.1371/journal.ppat.1000213
34. Gellatly SL, Needham B, Madera L, Trent MS, Hancock RE (2012) The *Pseudomonas aeruginosa* PhoP-PhoQ two-component regulatory system is induced upon interaction with epithelial cells and controls cytotoxicity and inflammation. Infect Immun 80(9):3122–3131. https://doi.org/10.1128/IAI.00382-12
35. Macfarlane EL, Kwasnicka A, Ochs MM, Hancock RE (1999) PhoP-PhoQ homologues in *Pseudomonas aeruginosa* regulate expression of the outer-membrane protein OprH and polymyxin B resistance. Mol Microbiol 34(2):305–316
36. McPhee JB, Bains M, Winsor G, Lewenza S, Kwasnicka A, Brazas MD, Brinkman FS, Hancock RE (2006) Contribution of the PhoP-PhoQ and PmrA-PmrB two-component regulatory systems to Mg2+−induced gene regulation in *Pseudomonas aeruginosa*. J Bacteriol 188(11):3995–4006. https://doi.org/10.1128/JB.00053-06
37. Gunn JS, Hohmann EL, Miller SI (1996) Transcriptional regulation of *Salmonella* virulence: a PhoQ periplasmic domain mutation results in increased net phosphotransfer to PhoP. J Bacteriol 178(21):6369–6373
38. Miller AK, Brannon MK, Stevens L, Johansen HK, Selgrade SE, Miller SI, Hoiby N, Moskowitz SM (2011) PhoQ mutations promote lipid A modification and polymyxin resistance of *Pseudomonas aeruginosa* found in colistin-treated cystic fibrosis patients. Antimicrob Agents Chemother 55(12):5761–5769. https://doi.org/10.1128/AAC.05391-11
39. Fernandez L, Gooderham WJ, Bains M, McPhee JB, Wiegand I, Hancock RE (2010) Adaptive resis-

tance to the "last hope" antibiotics polymyxin B and colistin in *Pseudomonas aeruginosa* is mediated by the novel two-component regulatory system ParR-ParS. Antimicrob Agents Chemother 54(8):3372–3382. https://doi.org/10.1128/AAC.00242-10
40. Fernandez L, Jenssen H, Bains M, Wiegand I, Gooderham WJ, Hancock RE (2012) The two-component system CprRS senses cationic peptides and triggers adaptive resistance in *Pseudomonas aeruginosa* independently of ParRS. Antimicrob Agents Chemother 56(12):6212–6222. https://doi.org/10.1128/AAC.01530-12
41. Muller C, Plesiat P, Jeannot K (2011) A two-component regulatory system interconnects resistance to polymyxins, aminoglycosides, fluoroquinolones, and beta-lactams in *Pseudomonas aeruginosa*. Antimicrob Agents Chemother 55(3):1211–1221. https://doi.org/10.1128/AAC.01252-10
42. Gutu AD, Sgambati N, Strasbourger P, Brannon MK, Jacobs MA, Haugen E, Kaul RK, Johansen HK, Hoiby N, Moskowitz SM (2013) Polymyxin resistance of *Pseudomonas aeruginosa phoQ* mutants is dependent on additional two-component regulatory systems. Antimicrob Agents Chemother 57(5):2204–2215. https://doi.org/10.1128/AAC.02353-12
43. Nowicki EM, O'Brien JP, Brodbelt JS, Trent MS (2015) Extracellular zinc induces phosphoethanolamine addition to *Pseudomonas aeruginosa* lipid A via the ColRS two-component system. Mol Microbiol 97(1):166–178. https://doi.org/10.1111/mmi.13018
44. Liao J, Schurr MJ, Sauer K (2013) The MerR-like regulator BrlR confers biofilm tolerance by activating multidrug efflux pumps in *Pseudomonas aeruginosa* biofilms. J Bacteriol 195(15):3352–3363. https://doi.org/10.1128/JB.00318-13
45. Chambers JR, Sauer K (2013) The MerR-like regulator BrlR impairs *Pseudomonas aeruginosa* biofilm tolerance to colistin by repressing PhoPQ. J Bacteriol 195(20):4678–4688. https://doi.org/10.1128/JB.00834-13
46. Ernst RK, Yi EC, Guo L, Lim KB, Burns JL, Hackett M, Miller SI (1999) Specific lipopolysaccharide found in cystic fibrosis airway *Pseudomonas aeruginosa*. Science 286(5444):1561–1565
47. Ernst RK, Moskowitz SM, Emerson JC, Kraig GM, Adams KN, Harvey MD, Ramsey B, Speert DP, Burns JL, Miller SI (2007) Unique lipid a modifications in *Pseudomonas aeruginosa* isolated from the airways of patients with cystic fibrosis. J Infect Dis 196(7):1088–1092. https://doi.org/10.1086/521367
48. Gooderham WJ, Gellatly SL, Sanschagrin F, McPhee JB, Bains M, Cosseau C, Levesque RC, Hancock RE (2009) The sensor kinase PhoQ mediates virulence in *Pseudomonas aeruginosa*. Microbiology 155(Pt 3):699–711. https://doi.org/10.1099/mic.0.024554-0
49. Dentovskaya SV, Anisimov AP, Kondakova AN, Lindner B, Bystrova OV, Svetoch TE, Shaikhutdinova RZ, Ivanov SA, Bakhteeva IV, Titareva GM, Knirel AY (2011) Functional characterization and biological significance of *Yersinia pestis* lipopolysaccharide biosynthesis genes. Biochemistry (Mosc) 76(7):808–822. https://doi.org/10.1134/S0006297911070121
50. Reines M, Llobet E, Llompart CM, Moranta D, Perez-Gutierrez C, Bengoechea JA (2012) Molecular basis of *Yersinia enterocolitica* temperature-dependent resistance to antimicrobial peptides. J Bacteriol 194(12):3173–3188. https://doi.org/10.1128/JB.00308-12
51. Anisimov AP, Dentovskaya SV, Titareva GM, Bakhteeva IV, Shaikhutdinova RZ, Balakhonov SV, Lindner B, Kocharova NA, Senchenkova SN, Holst O, Pier GB, Knirel YA (2005) Intraspecies and temperature-dependent variations in susceptibility of *Yersinia pestis* to the bactericidal action of serum and to polymyxin B. Infect Immun 73(11):7324–7331. https://doi.org/10.1128/IAI.73.11.7324-7331.2005
52. Hamad MA, Di Lorenzo F, Molinaro A, Valvano MA (2012) Aminoarabinose is essential for lipopolysaccharide export and intrinsic antimicrobial peptide resistance in *Burkholderia cenocepacia*. Mol Microbiol 85(5):962–974. https://doi.org/10.1111/j.1365-2958.2012.08154.x
53. Isshiki Y, Kawahara K, Zahringer U (1998) Isolation and characterisation of disodium (4-amino-4-deoxy-beta-L- arabinopyranosyl)-(1-->8)-(D-glycero-alpha-D-talo-oct-2-ulopyranosylona te)- (2-->4)-(methyl 3-deoxy-D-manno-oct-2-ulopyranosid)onate from the lipopolysaccharide of *Burkholderia cepacia*. Carbohydr Res 313(1):21–27
54. Silipo A, Molinaro A, Cescutti P, Bedini E, Rizzo R, Parrilli M, Lanzetta R (2005) Complete structural characterization of the lipid A fraction of a clinical strain of *B. cepacia* genomovar I lipopolysaccharide. Glycobiology 15(5):561–570. https://doi.org/10.1093/glycob/cwi029
55. Silipo A, Molinaro A (2010) The diversity of the core oligosaccharide in lipopolysaccharides. Subcell Biochem 53:69–99. https://doi.org/10.1007/978-90-481-9078-2_4
56. Lee H, Hsu FF, Turk J, Groisman EA (2004) The PmrA-regulated *pmrC* gene mediates phosphoethanolamine modification of lipid A and polymyxin resistance in *Salmonella enterica*. J Bacteriol 186(13):4124–4133. https://doi.org/10.1128/JB.186.13.4124-4133.2004
57. Herrera CM, Hankins JV, Trent MS (2010) Activation of PmrA inhibits LpxT-dependent phosphorylation of lipid A promoting resistance to antimicrobial peptides. Mol Microbiol 76(6):1444–1460. https://doi.org/10.1111/j.1365-2958.2010.07150.x
58. Kato A, Chen HD, Latifi T, Groisman EA (2012) Reciprocal control between a bacterium's regulatory system and the modification status of its lipopolysaccharide. Mol Cell 47(6):897–908. https://doi.org/10.1016/j.molcel.2012.07.017
59. Arroyo LA, Herrera CM, Fernandez L, Hankins JV, Trent MS, Hancock RE (2011) The *pmrCAB* operon mediates polymyxin resistance in *Acinetobacter baumannii* ATCC 17978 and clinical isolates through phosphoethanolamine modification of lipid

A. Antimicrob Agents Chemother 55(8):3743–3751. https://doi.org/10.1128/AAC.00256-11

60. Beceiro A, Llobet E, Aranda J, Bengoechea JA, Doumith M, Hornsey M, Dhanji H, Chart H, Bou G, Livermore DM, Woodford N (2011) Phosphoethanolamine modification of lipid A in colistin-resistant variants of *Acinetobacter baumannii* mediated by the pmrAB two-component regulatory system. Antimicrob Agents Chemother 55(7):3370–3379. https://doi.org/10.1128/AAC.00079-11
61. Adams MD, Nickel GC, Bajaksouzian S, Lavender H, Murthy AR, Jacobs MR, Bonomo RA (2009) Resistance to colistin in *Acinetobacter baumannii* associated with mutations in the PmrAB two-component system. Antimicrob Agents Chemother 53(9):3628–3634. https://doi.org/10.1128/AAC.00284-09
62. Lesho E, Yoon EJ, McGann P, Snesrud E, Kwak Y, Milillo M, Onmus-Leone F, Preston L, St Clair K, Nikolich M, Viscount H, Wortmann G, Zapor M, Grillot-Courvalin C, Courvalin P, Clifford R, Waterman PE (2013) Emergence of colistin-resistance in extremely drug-resistant *Acinetobacter baumannii* containing a novel *pmrCAB* operon during colistin therapy of wound infections. J Infect Dis 208(7):1142–1151. https://doi.org/10.1093/infdis/jit293
63. Lopez-Rojas R, McConnell MJ, Jimenez-Mejias ME, Dominguez-Herrera J, Fernandez-Cuenca F, Pachon J (2013) Colistin resistance in a clinical *Acinetobacter baumannii* strain appearing after colistin treatment: effect on virulence and bacterial fitness. Antimicrob Agents Chemother 57(9):4587–4589. https://doi.org/10.1128/AAC.00543-13
64. Snitkin ES, Zelazny AM, Gupta J, Program NCS, Palmore TN, Murray PR, Segre JA (2013) Genomic insights into the fate of colistin resistance and *Acinetobacter baumannii* during patient treatment. Genome Res 23(7):1155–1162. https://doi.org/10.1101/gr.154328.112
65. Bengoechea JA, Skurnik M (2000) Temperature-regulated efflux pump/potassium antiporter system mediates resistance to cationic antimicrobial peptides in *Yersinia*. Mol Microbiol 37(1):67–80
66. Kim Y, Bae IK, Lee H, Jeong SH, Yong D, Lee K (2014) *In vivo* emergence of colistin resistance in *Acinetobacter baumannii* clinical isolates of sequence type 357 during colistin treatment. Diagn Microbiol Infect Dis 79(3):362–366. https://doi.org/10.1016/j.diagmicrobio.2014.03.027
67. Beceiro A, Moreno A, Fernandez N, Vallejo JA, Aranda J, Adler B, Harper M, Boyce JD, Bou G (2014) Biological cost of different mechanisms of colistin resistance and their impact on virulence in *Acinetobacter baumannii*. Antimicrob Agents Chemother 58(1):518–526. https://doi.org/10.1128/AAC.01597-13
68. Hraiech S, Roch A, Lepidi H, Atieh T, Audoly G, Rolain JM, Raoult D, Brunel JM, Papazian L, Bregeon F (2013) Impaired virulence and fitness of a colistin-resistant clinical isolate of *Acinetobacter baumannii* in a rat model of pneumonia. Antimicrob Agents Chemother 57(10):5120–5121. https://doi.org/10.1128/AAC.00700-13
69. Ye H, Li Y, Li Z, Gao R, Zhang H, Wen R, Gao GF, Hu Q, Feng Y (2016) Diversified *mcr-1*-harbouring plasmid reservoirs confer resistance to colistin in human gut microbiota. mBio 7(2):e00177. https://doi.org/10.1128/mBio.00177-16
70. Xavier BB, Lammens C, Ruhal R, Kumar-Singh S, Butaye P, Goossens H, Malhotra-Kumar S (2016) Identification of a novel plasmid-mediated colistin-resistance gene, *mcr-2*, in *Escherichia coli*, Belgium, June 2016. Euro Surveill 21(27). https://doi.org/10.2807/1560-7917.ES.2016.21.27.30280
71. Wang Q, Li Z, Lin J, Wang X, Deng X, Feng Y (2016) Complex dissemination of the diversified *mcr-1*-harbouring plasmids in *Escherichia coli* of different sequence types. Oncotarget 7(50):82112–82122. https://doi.org/10.18632/oncotarget.12621
72. Liu YY, Wang Y, Walsh TR, Yi LX, Zhang R, Spencer J, Doi Y, Tian G, Dong B, Huang X, Yu LF, Gu D, Ren H, Chen X, Lv L, He D, Zhou H, Liang Z, Liu JH, Shen J (2016) Emergence of plasmid-mediated colistin resistance mechanism MCR-1 in animals and human beings in China: a microbiological and molecular biological study. Lancet Infect Dis 16(2):161–168. https://doi.org/10.1016/S1473-3099(15)00424-7
73. Jeannot K, Bolard A, Plesiat P (2017) Resistance to polymyxins in gram-negative organisms. Int J Antimicrob Agents. https://doi.org/10.1016/j.ijantimicag.2016.11.029
74. Liu YY, Chandler CE, Leung LM, McElheny CL, Mettus RT, Shanks RM, Liu JH, Goodlett DR, Ernst RK, Doi Y (2017) Structural modification of lipopolysaccharide conferred by *mcr-1* in Gram-negative ESKAPE pathogens. Antimicrob Agents Chemother. https://doi.org/10.1128/AAC.00580-17
75. Pelletier MR, Casella LG, Jones JW, Adams MD, Zurawski DV, Hazlett KR, Doi Y, Ernst RK (2013) Unique structural modifications are present in the lipopolysaccharide from colistin-resistant strains of *Acinetobacter baumannii*. Antimicrob Agents Chemother 57(10):4831–4840. https://doi.org/10.1128/AAC.00865-13
76. Wang X, Ribeiro AA, Guan Z, Raetz CR (2009) Identification of undecaprenyl phosphate-beta-D-galactosamine in *Francisella novicida* and its function in lipid A modification. Biochemistry 48(6):1162–1172. https://doi.org/10.1021/bi802211k
77. Phillips NJ, Schilling B, McLendon MK, Apicella MA, Gibson BW (2004) Novel modification of lipid A of *Francisella tularensis*. Infect Immun 72(9):5340–5348. https://doi.org/10.1128/IAI.72.9.5340-5348.2004
78. Moffatt JH, Harper M, Harrison P, Hale JD, Vinogradov E, Seemann T, Henry R, Crane B, St Michael F, Cox AD, Adler B, Nation RL, Li J, Boyce JD (2010) Colistin resistance in *Acinetobacter baumannii* is mediated by complete loss of lipopolysac-

charide production. Antimicrob Agents Chemother 54(12):4971–4977. https://doi.org/10.1128/AAC.00834-10
79. Moffatt JH, Harper M, Adler B, Nation RL, Li J, Boyce JD (2011) Insertion sequence IS*Aba11* is involved in colistin resistance and loss of lipopolysaccharide in *Acinetobacter baumannii*. Antimicrob Agents Chemother 55(6):3022–3024. https://doi.org/10.1128/AAC.01732-10
80. Li J, Rayner CR, Nation RL, Owen RJ, Spelman D, Tan KE, Liolios L (2006) Heteroresistance to colistin in multidrug-resistant *Acinetobacter baumannii*. Antimicrob Agents Chemother 50(9):2946–2950. https://doi.org/10.1128/AAC.00103-06
81. Tan CH, Li J, Nation RL (2007) Activity of colistin against heteroresistant *Acinetobacter baumannii* and emergence of resistance in an in vitro pharmacokinetic/pharmacodynamic model. Antimicrob Agents Chemother 51(9):3413–3415. https://doi.org/10.1128/AAC.01571-06
82. Rodriguez CH, Bombicino K, Granados G, Nastro M, Vay C, Famiglietti A (2009) Selection of colistin-resistant *Acinetobacter baumannii* isolates in postneurosurgical meningitis in an intensive care unit with high presence of heteroresistance to colistin. Diagn Microbiol Infect Dis 65(2):188–191. https://doi.org/10.1016/j.diagmicrobio.2009.05.019
83. Moffatt JH, Harper M, Mansell A, Crane B, Fitzsimons TC, Nation RL, Li J, Adler B, Boyce JD (2013) Lipopolysaccharide-deficient *Acinetobacter baumannii* shows altered signaling through host toll-like receptors and increased susceptibility to the host antimicrobial peptide LL-37. Infect Immun 81(3):684–689. https://doi.org/10.1128/IAI.01362-12
84. Henry R, Crane B, Powell D, Deveson Lucas D, Li Z, Aranda J, Harrison P, Nation RL, Adler B, Harper M, Boyce JD, Li J (2015) The transcriptomic response of *Acinetobacter baumannii* to colistin and doripenem alone and in combination in an *in vitro* pharmacokinetics/pharmacodynamics model. J Antimicrob Chemother. https://doi.org/10.1093/jac/dku536
85. Henry R, Vithanage N, Harrison P, Seemann T, Coutts S, Moffatt JH, Nation RL, Li J, Harper M, Adler B, Boyce JD (2012) Colistin-resistant, lipopolysaccharide-deficient *Acinetobacter baumannii* responds to lipopolysaccharide loss through increased expression of genes involved in the synthesis and transport of lipoproteins, phospholipids, and poly-beta-1,6-N-acetylglucosamine. Antimicrob Agents Chemother 56(1):59–69. https://doi.org/10.1128/AAC.05191-11
86. Campos MA, Vargas MA, Regueiro V, Llompart CM, Alberti S, Bengoechea JA (2004) Capsule polysaccharide mediates bacterial resistance to antimicrobial peptides. Infect Immun 72(12):7107–7114. https://doi.org/10.1128/IAI.72.12.7107-7114.2004
87. Llobet E, Tomas JM, Bengoechea JA (2008) Capsule polysaccharide is a bacterial decoy for antimicrobial peptides. Microbiology 154(Pt 12):3877–3886. https://doi.org/10.1099/mic.0.2008/022301-0
88. Tzeng YL, Ambrose KD, Zughaier S, Zhou X, Miller YK, Shafer WM, Stephens DS (2005) Cationic antimicrobial peptide resistance in *Neisseria meningitidis*. J Bacteriol 187(15):5387–5396. https://doi.org/10.1128/JB.187.15.5387-5396.2005
89. Srinivasan VB, Singh BB, Priyadarshi N, Chauhan NK, Rajamohan G (2014) Role of novel multidrug efflux pump involved in drug resistance in *Klebsiella pneumoniae*. PLoS One 9(5):e96288. https://doi.org/10.1371/journal.pone.0096288
90. Mathur J, Waldor MK (2004) The *Vibrio cholerae* ToxR-regulated porin OmpU confers resistance to antimicrobial peptides. Infect Immun 72(6):3577–3583. https://doi.org/10.1128/IAI.72.6.3577-3583.2004
91. Mathur J, Davis BM, Waldor MK (2007) Antimicrobial peptides activate the *Vibrio cholerae* sigmaE regulon through an OmpU-dependent signalling pathway. Mol Microbiol 63(3):848–858. https://doi.org/10.1111/j.1365-2958.2006.05544.x
92. Duperthuy M, Binesse J, Le Roux F, Romestand B, Caro A, Got P, Givaudan A, Mazel D, Bachere E, Destoumieux-Garzon D (2010) The major outer membrane protein OmpU of *Vibrio splendidus* contributes to host antimicrobial peptide resistance and is required for virulence in the oyster *Crassostrea gigas*. Environ Microbiol 12(4):951–963. https://doi.org/10.1111/j.1462-2920.2009.02138.x
93. Bengoechea JA, Zhang L, Toivanen P, Skurnik M (2002) Regulatory network of lipopolysaccharide O-antigen biosynthesis in *Yersinia enterocolitica* includes cell envelope-dependent signals. Mol Microbiol 44(4):1045–1062
94. Ito-Kagawa M, Koyama Y (1980) Selective cleavage of a peptide antibiotic, colistin by colistinase. J Antibiot (Tokyo) 33(12):1551–1555
95. Janganan TK, Bavro VN, Zhang L, Borges-Walmsley MI, Walmsley AR (2013) Tripartite efflux pumps: energy is required for dissociation, but not assembly or opening of the outer membrane channel of the pump. Mol Microbiol 88(3):590–602. https://doi.org/10.1111/mmi.12211
96. Gronow S, Noah C, Blumenthal A, Lindner B, Brade H (2003) Construction of a deep-rough mutant of *Burkholderia cepacia* ATCC 25416 and characterization of its chemical and biological properties. J Biol Chem 278(3):1647–1655. https://doi.org/10.1074/jbc.M206942200

Bioanalysis and Stability of Polymyxins

6

Robert W. Milne

Abstract

Clinical use of the polymyxin antibiotics began approximately 10 years after their discovery in the late 1940s. Their concentrations in biological fluids were measured using microbiological methods. These methods were reasonably accurate for measuring the active polymyxin base, such as polymyxin B and colistin (polymyxin E), but were used inappropriately for measuring the concentrations of "colistin" in humans or animals following the administration of colistimethate, also known as colistin methanesulphonate (CMS). The use of polymyxins for systemic infections waned in the 1970s because of their toxicity and the preference for other antibiotics, but their value for treating infections caused by several important Gram-negative pathogens becoming resistant to other antibiotics was realized in the mid-1990s. The lack of adequate pharmacokinetic and pharmacodynamic knowledge spurred the development of methods more specific for measuring polymyxin B and colistin after their administrations as sulphate salts, and of colistin and CMS after the administration of CMS sodium. These methods have been based on high-performance liquid chromatography, detection and quantification of fluorescent derivatives of the polymyxin bases, or of the bases themselves with detection and quantification by mass spectrometry.

Keywords

Colistin · Colistimethate · Polymyxin B · Quantitative assays · Biological matrices · Nonspecific binding · Stability

6.1 Introduction

The two polymyxins used in clinical practice are colistin (polymyxin E) and polymyxin B. Their historical use, chemistry and antimicrobial activity have been reviewed in Chap. 3. Briefly, the two polymyxin bases are used as a mixture primarily of colistin A and B (or polymyxin E1 and E2) or as a mixture primarily of polymyxin B1 and B2. The two polymyxin forms denoted as 1 and 2 differ only in their alkanoic acid moiety by a methylene group. Colistin and polymyxin B differ by only one amino acid in the cyclic peptide; colistin contains D-Leu while polymyxin B contains D-Phe.

Colistin for parenteral use has been administered most commonly as its methanesulphonate derivative. This is formed by derivatizing the five available amino groups of the L-diaminobutyric

R. W. Milne (✉)
School of Pharmacy and Medical Sciences, University of South Australia, Adelaide, Australia
e-mail: robert.milne@unisa.edu.au

J. Li et al. (eds.), *Polymyxin Antibiotics: From Laboratory Bench to Bedside*, Advances in Experimental Medicine and Biology 1145, https://doi.org/10.1007/978-3-030-16373-0_6

acid moieties with methanesulphonic acid. Therapeutic use commenced in the mid-1950s as the sulphate salts of the bases, and as the sodium salt of the methanesulphonate of colistin (CMS or colistin methanesulphonate; also known as colistimethate) for systemic administration in the United States in the 1960s. When administered, CMS may not be fully derivatized with methanesulphonate; it may be present in a dosage form as a mix of full and partial derivatives. Likewise, as will become apparent from the chapters covering pharmacokinetics and pharmacodynamics, there will be a complex mix of full and partial derivatives in samples of biological fluids or other aqueous media from experiments evaluating the fate or antimicrobial effectiveness of CMS *in vivo* or *in vitro*. This is because of a gradual loss of the methanesulphonate groups over time. Measuring the individual derivatives in such fluids or media has not been achieved. Doing so would be extremely complex, and of questionable value, given that colistin alone is deemed to possess antimicrobial activity [1]. Therefore, a more recent approach has been to perform two measurements on a sample [2, 3]. Firstly, the concentration of "total colistin" is measured, it being the sum of all methanesulphonate derivatives converted to colistin during processing of a sample plus pre-existing colistin in the sample; and, secondly, a measure of the concentration of colistin. When measuring the latter, one should be mindful of the instability of the methanesulphonate derivatives, ensuring appropriate storage and processing of samples under conditions which minimize conversion of any derivatives to colistin [1, 4]. Therefore, researchers should assure themselves that there is no conversion of methanesulphonate derivatives to colistin once a sample has been collected; for example, while separating plasma from a sample of blood, while stored pending analysis, after thawing, during repeated thawing and freezing, during processing and while awaiting chromatographic analysis. The difference between the two concentrations represents the concentration of all methanesulphonate derivatives (designated as CMS) in the sample. In addition, polymyxins are highly surface active, and their adsorption from aqueous solutions onto the surfaces of apparatus used during collection and processing of samples may have an impact on recovery and sensitivity. Generally, this has been minimized by including a cosolvent in stock solutions and either a cosolvent, protein (such as drug-free human plasma) or surfactant is added during the processing of samples of urine or bacterial broth [4–6].

This chapter will review the range of methods that have been used for measuring the concentrations of polymyxins in different biological fluids, and will do so in a chronological order that reflects the gradual advances in techniques that have enabled improvements in sensitivity and, more importantly, in specificity and the ease with which they are performed. It will describe methods used for pretreatment of samples, including the important issues raised above regarding stability and adsorption to surfaces, along with the methods for quantifying the concentration of polymyxin in the sample.

6.2 Microbiological Methods

The first method reported for measuring polymyxins in biological fluids appears to have been a microbiological assay for polymyxin (identified later as polymyxin D; [7]) in blood and urine based on its activity towards *Brucella bronchiseptica*. There was no apparent interference from other components of blood from the mouse, dog and human, and the lowest concentration on the calibration curve using 0.02 mL of blood was 0.25 mg/L [8, 9]. The author described measurement of the concentrations of polymyxin in pooled samples of blood after a single dose of polymyxin hydrochloride to mice (1 mg/kg, s.c.) and a dog (5 mg/kg, i.v.). Replicate analyses of a sample from the dog [8] produced coefficients of variation ranging from 7% to 12% (no concentration was provided). While the method potentially lacks specificity in the presence of other antimicrobial agents co-administered during *in vivo* studies, be they in animals or humans, it is a measure of antimicrobially-active polymyxin in a biological fluid such as blood to a level of sensitivity comparable to that achieved with HPLC of fluorescent derivatives of colistin (*vide infra*).

The limitations of nephro- and neuro-toxicity during clinical use in the early 1970s and prior led to an investigation of the binding of polymyxins to tissues following the systemic administration of polymyxin B and CMS to rabbits [10]. Concentrations were measured *via* their inhibition of *B. bronchiseptica* using the respective compounds as calibration standards. While the measurements for polymyxin B are likely to be reasonable estimates of its concentrations, those for "CMS" will be complicated by the issues of instability mentioned above and discussed subsequently in this chapter.

Later work described use of the same bacterial test organism to measure polymyxin B in bovine fluids and tissues [11, 12] following parenteral administration of polymyxin B sulphate. Two years previous, this group published values for the concentrations of polymyxin B, colistin and CMS in serum from calves administered daily doses of polymyxin B sulphate, colistin sulphate, or CMS, respectively, via the intramuscular route for three successive days [13]. The microbiological assay used by this group reported similar values of maximal serum concentration (C_{max}) and terminal half-life for polymyxin and colistin. However, greater and lesser values, respectively, were obtained for "CMS" after the administration of CMS. As will become apparent from Chap. 7 describing the pharmacokinetics of colistin in rats administered colistin sulphate and of colistin and CMS after dosing with CMS, the values for "CMS" after administration of CMS are likely to be composites derived from measuring a mix of colistin and CMS in ratios that change over time from the dose; an observation possible only when concentrations in plasma are measured more specifically by chromatographic methods (*vide infra*).

Even more recently, a microbiological method was described for measuring CMS in serum (and urine) from humans, seemingly with the intention of correlating levels of CMS in blood with toxicity; the method was advocated because it avoided more complex and instrumentally-demanding methods based on liquid chromatography [14]. This method, along with the other microbiological methods discussed above for measuring "CMS", measures antimicrobial activity that can only be attributed to colistin (assuming there are no other antimicrobially-active interferences). However, there are important issues that should be appreciated: (1) activity was relative to calibration standards prepared using CMS, which has minimal if any inherent antimicrobial activity [15]; (2) it is likely that CMS would degrade during the assay to antimicrobially-active colistin via partially sulphomethylated intermediates [15, 16], and the relative proportions of sulphomethylated colistin to colistin in the calibration standards exposed to the test organism during the assay may be quite different from those in the biological samples. Another report described the pharmacokinetics of "colistin" in a patient using concentrations measured in serum against *Acinetobacter baumannii* [17]. Further reports have appeared describing microbiological methods for measuring "colistin" following the administration of CMS to a patient [18] and in a study of the pharmacokinetics and dynamics of CMS in a pneumonia model with mice [19]; the same limitations apply to these reports. An extensive pharmacokinetic/pharmacodynamic study in mice used a microbiological method to measure "colistin" in plasma from neutropenic mice administered "colistin", but it was not clear whether the sulphate salt or CMS was administered and measured [20].

Nevertheless, microbiological assays for polymyxins are legitimate for any analysis of exposure when the polymyxin itself (as a salt of the base) rather than any prodrug (such as CMS) is being administered and the polymyxin is being used as a reference for the preparation of calibration standards and quality controls. Published examples since 2000 include assessment of the exposure to colistin after administration of colistin sulphate to piglets [21], and to polymyxin B following its intravenous administration (presumably as the sulphate) to a young male [22]. The former reported a "detection" range from 0.005 to 3 mg/L [21] and, while the lower value in this range compares favourably with limits of quantification achieved using chromatographic methods (*vide infra*), no validation was provided for a limit of quantification. Also, no details were provided by the latter report [22]. The majority of methods beyond 2000 have employed chromatographic separation and quantification of fluorescent derivatives or chromatography with quantification by mass spectrometry.

The clinical use of colistin, as CMS, began to increase in the mid-1990s in response to an increasing incidence of infections in patients with cystic fibrosis caused by bacteria resistant to the usually available antibiotics. The authors of one important study at this time concluded that intravenous colistin (as CMS) may be valuable therapy for acute respiratory exacerbations and that the risks of renal toxicity could be minimized with careful monitoring [23]. This group measured the concentrations of "colistin" in blood at steady-state by microbiological assay. The increased use of CMS in response to this and other reports of its use occurred at a time when it was recognized that previous systemic doses may have been excessive. However, the increase was at a time when there was very little information available on its pharmacokinetics and pharmacodynamics that might guide the selection of appropriate doses [24].

6.3 Chromatographic Methods

Given the limitations of microbiological methods for measuring "CMS" and "colistin" in plasma after the administration of CMS, reports began to appear of chromatographic methods being developed that were selective for colistin and CMS (the latter including partially sulphomethylated forms of colistin), with detection appropriate for the required levels of quantification of both these and polymyxin B. These methods have now been established as the most suitable for measuring the concentrations of polymyxin B and colistin in biological fluids following administration of their sulphate salts to animals or humans, or of colistin and CMS after administration of the sodium salt of CMS. The required levels of sensitivity have been achieved by detecting and quantifying fluorescent derivatives of the polymyxin base or by using mass spectrometry; fluorescent derivatives because of a lack of native ultraviolet absorbance sufficient for quantifying clinically relevant concentrations. The concentration of CMS in a sample is determined after hydrolysis of CMS to colistin and quantification of the latter as "total colistin" (from CMS and pre-existing colistin) and, after accounting for differences in molecular weight, subtracting values for the concentration of colistin measured separately from the "total". When measuring the colistin alone, one must be careful to minimize hydrolysis of CMS.

Le Brun was one of the first to report a liquid chromatographic method for measuring "colistin" in biological fluids from humans; serum, urine and sputum [25]. They adapted a method described 3 years previously for measuring residual colistin in bovine tissues [26]. The latter researchers formed a fluorescent derivative (λ_{Ex} 340 nm, λ_{Em} 440 nm) from reaction of the primary amines of colistin with *o*-phthalaldehyde. They were measuring colistin in farmed animals most likely exposed to colistin salts, using colistin sulphate and the summed responses from colistin A and B as a reference. However, Le Brun et al. used the method to measure "colistin" in patients who had inhaled CMS, seemingly using CMS as a calibration reference. This group mixed serum or sputum with methanol/trichloroacetic acid to precipitate protein, followed by reaction with *o*-phthalaldehyde and chromatographic analysis. They did not assess whether processing of samples with such an acidic mix may have facilitated partial or complete conversion of CMS to colistin, but later work (*vide infra*) indicated that complete conversion is unlikely [3]. The risk of conversion *in vitro* of methanesulphonate derivatives to colistin during processing of samples containing CMS and colistin will apply to any method [27] that purports to measure the concentrations of "colistin" in samples from humans or animals administered CMS. There is no problem when the biological samples being measured are from subjects (be they animals or humans) administered colistin sulphate or polymyxin B sulphate and calibration standards are prepared using those same substances as reference standards. For example, the formation of a derivative with *o*-phthalaldehyde was used to measure colistin in plasma (0.5 mL) and in the gastrointestinal contents (1.0 g samples extending from the duodenum to ileum) of pigs following the oral administration of colistin sulphate [28]. Samples were treated with trichloroacetic acid to precipitate protein prior to solid-phase extraction and formation of the derivative. The limit of quantification was 0.25 mg/L and 0.50 mg/kg, respectively.

The formation of other fluorescent derivatives for the chromatographic quantification of colistin in biological fluids has been published. Colistin A (polymyxin E1) was extracted from plasma of the rat and dog using a 96-well C_8 disk extraction plate prior to reacting it with dansyl chloride. The product was described as a penta-dansyl derivative (λ_{Ex} 344 nm, λ_{Em} 518 nm) that was "confirmed" with mass spectrometry [29], but the conditions for formation of the confirmed product were not identical to those for its formation during preparation of the biological samples for chromatography. The investigators intended using the method as part of the development of a single component of colistin (colistin A) as a potential therapeutic agent, and used colistin A to prepare calibration standards. The limit of quantification (in 0.2 mL plasma) was 0.05 mg/L.

In seeking to gain a better understanding of the pharmacokinetics and pharmacodynamics of colistin and CMS, separate and more specific methods were developed for measuring colistin in plasma from rats after the administration of colistin sulphate, and for measuring colistin and CMS after the administration of CMS [2, 3]. The methods relied on forming a fluorescent derivative from reaction of the amine of the L-diaminobutyric acid residues of colistin with 9-fluorenylmethyl chloroformate (FMOC). After adding trichloroacetic acid/methanol to samples of plasma, colistin in the supernatant was retained under basic conditions on a C_{18} solid-phase extraction cartridge. Extraneous substances were eluted and the reaction initiated by adding a small volume of a concentrated solution of FMOC into the cartridge. The derivatives of colistins A and B were eluted, chromatographed on a C_{18} analytical column, and detected and quantified from their fluorescence at 315 nm following excitation at 260 nm (Fig. 6.1).

Concentrations were calculated from a calibration curve of the ratio of the summed areas of the two chromatographic peaks for colistins A and B to an internal standard against the concentration of colistin sulphate. In samples spiked freshly with a "high" concentration of CMS

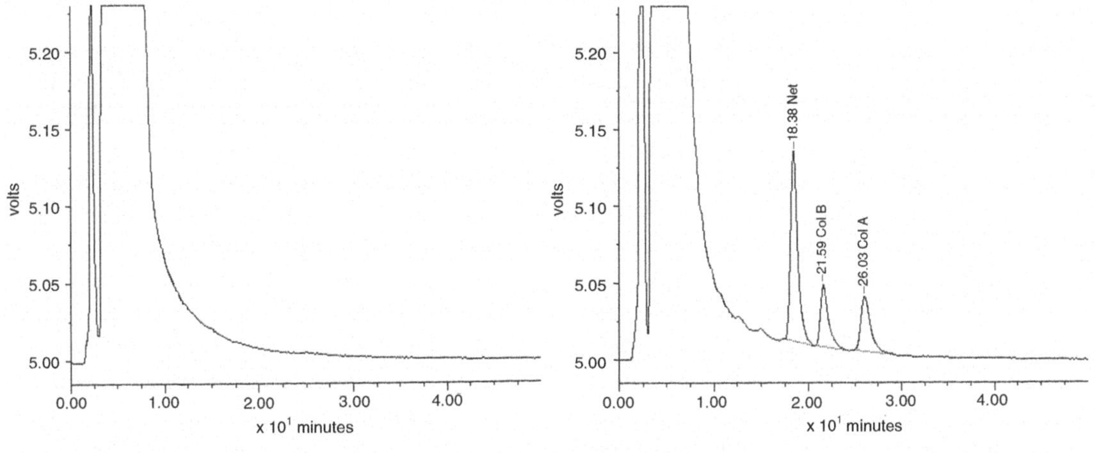

Fig. 6.1 Typical chromatograms obtained from fluorescence detection for drug-free human plasma (left) and drug-free plasma spiked with colistin sulphate (1 mg/L) (right). The fluorescent derivative of colistin A was eluted at about 26.5 min, of colistin B at about 22 min and of netilmicin, the internal standard, at about 18.5 min. [2]

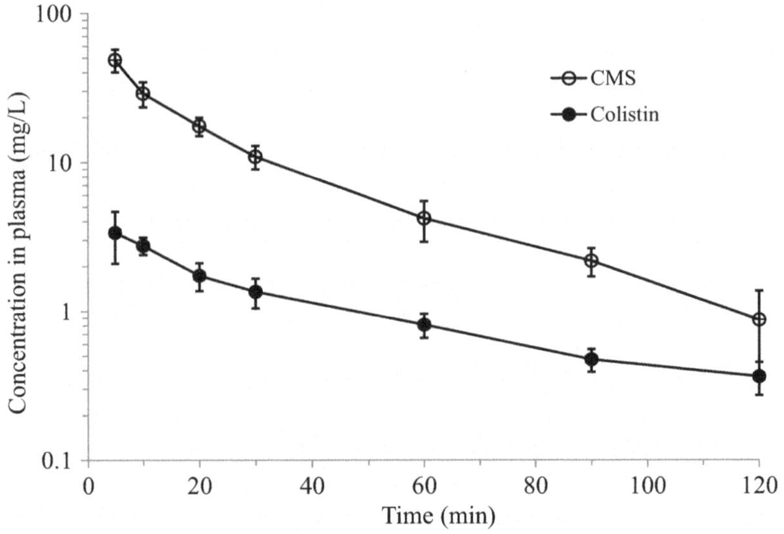

Fig. 6.2 Mean (± S.D.) concentrations of CMS and colistin in the plasma of rats following an intravenous dose of colistin methanesulphonate (Sigma, St. Louis, MO, USA; 15 mg/kg). Modified from Li et al. [31]

(10 mg/L), there was no quantifiable conversion of CMS to colistin when measured for colistin only [3]. A separate sample was treated with sulphuric acid to convert CMS and partial methanesulphonate derivatives formed *in vivo* to colistin, and then added to a cartridge for derivatizing with FMOC. The limit of quantification for colistin in plasma was 0.10 mg/L (from a 0.25 mL volume of sample) and for CMS it was 0.20 mg/L (0.15 mL) [30]. The two methods were developed for measuring both substances after administering CMS to rats (Fig. 6.2) [31] and to patients with cystic fibrosis [30]. Since then, they have been adapted/modified over the subsequent decade for measuring both substances in a more recent study of patients with cystic fibrosis [32], and after administering CMS to patients receiving continuous ambulatory peritoneal dialysis [33], and to critically-ill patients [34]; for studies with mice [35]; for measuring colistin and CMS in bronchoalveolar lavage fluid from rats after intratracheal administration of CMS [36], and for measuring both substances after administering four different brands of CMS to rats [37]. The method for colistin alone has been used also for measuring colistin in broth culture after adding colistin sulphate [38]; colistin in mouse brain after administration of colistin sulphate [39, 40]; and, for measuring colistin in plasma, urine and kidney tissue from rats administered colistin sulphate [41, 42]. Minor modifications to the volume of sample used, and the chromatographic mobile phase, and a change from trichloroacetic acid/methanol to acetonitrile for precipitating protein, have been made since initial publication of the two methods. Concentrations of CMS in plasma after administration of CMS *in vivo* (or in other media during experiments conducted, for example, with CMS *in vitro*) were calculated from the difference between "total colistin" (i.e. colistin plus methanesulphonate derivatives converted to colistin during incubation with sulphuric acid [3]), and colistin measured separately [2].

Later, the method for colistin [2] was applied to measuring the concentrations of polymyxin B in plasma from humans administered polymyxin B sulphate [43] and of a congener of polymyxin B, NAB 7061 (one of the amino acids in polymyxin B replaced with another) in plasma and urine of rats [44]. The limit of quantification was 0.125 mg/L with 0.10 mL of plasma; identical to the original method [2]. The method was applied subsequently for measuring polymyxin B in critically ill patients administered intravenous polymyxin B sulphate [45–47], some of whom were receiving continuous renal replacement or intermittent haemodialysis.

Subsequent reports from other research groups have applied the two methods [2, 3], with or without slight modifications, for measuring colistin in serum after the intravenous administration of CMS to critically ill patients [48]; colistin in serum and cerebrospinal fluid after the administration of CMS [49, 50]; colistin in plasma and bronchoalveolar lavage fluid from humans after administration of CMS [51, 52]; colistin in plasma and tissues (liver, muscle and kidney) from ducks administered colistin sulphate intramuscularly or in their feed [53]; and, colistin in plasma from pigs administered colistin sulphate [54].

A subsequent pharmacokinetic study applied the method above [2] to examine the pharmacokinetics of “colistin” in humans after inhalation of CMS [55]. These investigators claimed to have measured colistin A (polymyxin E1) but it was not clear which substance had indeed been used for preparation of the calibration standards: colistin A, colistin sulphate, or CMS. From chromatographic analysis of the calibration standards, only the peak for colistin A was used to construct a calibration curve. The sulphate salts of colistins A and B account for more than 85% of colistin sulphate and the ratios of the two can differ considerably between batches of the raw material [26] and, hence, between batches of CMS manufactured from colistin. It is important to include the peak responses for colistins A and B when constructing a calibration curve, plus the responses from the two species in biological fluids following administration of CMS. The ratios of the two components can be established by direct chromatographic analysis of the raw material, and quantified by either UV absorption [2] or mass spectrometry [4]. The validity of these two methods for assessing relative content of the components has been confirmed by their quantification in chromatographic eluate using evaporative light scattering [56].

Greater access to mass spectrometry for detection and quantification has produced a number of reports of well-described methods for measuring CMS and colistin after a dose of CMS, of colistin after colistin sulphate, and of polymyxin B after dosing with polymyxin B sulphate.

A method developed for measuring colistin in perfusate and urine collected from experiments examining the fate of colistin in the isolated perfused rat kidney also described measuring the substance in human plasma and urine [57]. Protein precipitation was achieved by mixing the samples (0.2 mL) with trichloroacetic acid/methanol followed by further clean-up using solid-phase extraction, with a portion of the eluate subjected to LC-MS/MS. Extraction was deemed necessary to maintain sufficient and consistent sensitivity. Summed intensities of the product ions from the two transitions each for colistin A and colistin B relative to an internal standard, polymyxin B1, were used to construct calibration curves. Prior to this, the proportions of colistin A and B in the reference material were established. Limits of quantification were 0.028–0.056 mg/L for colistin A and 0.016–0.032 for colistin B, depending on the biological fluid. Interestingly, this level of sensitivity was not able to be achieved when similar methods were used for preparing samples of bovine milk and tissue [58] for chromatography, despite the larger sample sizes and a more sensitive model of mass spectrometer. It is likely that the lower limits claimed by Ma et al. [57] could be extended with a more sensitive mass spectrometer. The method [57] is suitable for measuring colistin after the administration of colistin sulphate and could also be adapted for measuring polymyxin after polymyxin sulphate. However, the authors did not establish its suitability for measuring colistin in the presence of CMS.

A well-described method for measuring colistin A and B plus the concentrations of their respective methanesulphonate prodrugs in human plasma (0.10 mL) used only one step, precipitation of protein with 0.1% trifluoroacetic acid in acetonitrile, prior to chromatography connected to tandem mass spectrometry [5]. It described chilling of collected blood, separating plasma from red blood cells soon afterwards, thawing of previously frozen and stored samples of plasma in an ice bath, rapid processing of them in small batches, and storage of the supernatant at 4 °C prior to chromatography. The limits of quantification for colistin A and B were 0.019 and

0.010 mg/L, respectively.. The concentrations of CMS were calculated by difference [2, 3]. Unfortunately, the authors did not present data that validated their method for measuring CMS [5]. The method has been used by these Swedish and Greek collaborators for measuring colistin and CMS in a number of studies with intensive care patients administered CMS [59–61]. The precautions they describe in the preparation of samples, or variations of them, are not exclusive to these reports, but are necessary to minimize conversion of methanesulphonate derivatives to colistin when measuring colistin alone (see further discussion on stability below). In some instances, while the procedures described would appear to minimize conversion, it has not always been proven unequivocally [62].

Likewise, three subsequent methods based on liquid chromatography-mass spectrometry achieved comparable limits of quantification with the same (0.1 mL) [56, 63] or double the volume of plasma [64], but they also lack data validating the methods for measuring CMS in spiked samples of plasma. Data were provided with respect to the stability of colistin in plasma kept at room temperature for up to 12 h [63], but none could be identified that demonstrated lack of conversion of CMS to colistin during processing of samples. The first and third methods [56, 64] improved efficiency and accuracy by automated processing; samples of plasma thawed to 4 °C for measuring colistin were added directly into 96-well solid-phase extraction plates. With this procedure, it is quite likely that there was minimal conversion of CMS to colistin during processing of the samples, but no data in either publication [56, 64] could be identified to confirm this.

A method for measuring polymyxin B1 and B2 and colistin A and B (as well as vancomycin) in 0.5 mL of plasma from humans claimed an advantage of not requiring "a long and expensive procedure of SPE" (solid-phase extraction). However, while the authors [65] used polymyxin B sulphate as a reference for preparing calibration standards of polymyxin B1 and B2, they appear to have used colistin methanesulphonate (incorrectly) as reference standards for polymyxin E1 and E2. In contrast, another method published in the same year is comprehensive [4]. It describes limits of quantification similar to those described previously for colistin A and B [5], albeit using 2.5-times the volume of plasma, but also provides validated limits for CMS A (0.029 mg/L) and B (0.01 mg/L). The method described the processing of calibration standards and quality controls containing CMS in plasma by solid-phase extraction after conversion of the prodrug to colistin with sulphuric acid [3]. For measuring colistin alone in samples from patients administered CMS, conversion of CMS to colistin was minimized by processing previously frozen samples within 1 h of being thawed and simply diluting them with water prior to solid-phase extraction, rather than adding acetonitrile/acid to precipitate protein prior to extraction [5]. Figure 6.3 demonstrates application of the method for measuring concentrations of CMS and colistin in a subject administered CMS [4].

The method was also applied to measuring colistin and CMS in human urine; 0.2 mL of urine was mixed with half its volume of drug-free plasma "to avoid the loss of colistin by adsorption" to the 5 mL polypropylene tubes used [4]. This procedure was also found necessary by others for urine [64] and haemodiafiltrate [62], although one other group overcame the loss of polymyxin B by adding 0.5% of 3-[(3-cholamidopropyl)dimethylammonio]-1-propanesulphonate (CHAPS; a surfactant) to the sample after collection [6]. A shorter chromatographic time than described in a previous method [5] allowed the processing of larger batch sizes of samples for storage at 4 °C in an autosampler prior to analysis by liquid chromatography-mass spectrometry [4].

In 2015, a further improvement on this method was achieved with the same solid-phase extraction material but with a 96-well system and, more importantly, a chromatographic column containing an ethylene-bridged hybrid material with bound amide functional groups [66]. It is apparent from a visual comparison of chromatograms [4, 66] that the improved chromatographic efficiency provided a greater sensitivity and slightly lower limits of quantification; meanwhile

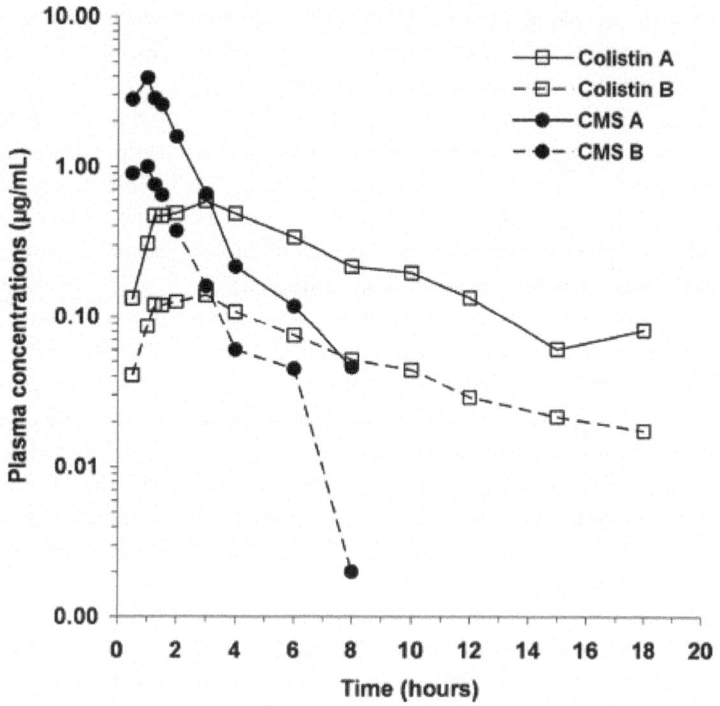

Fig. 6.3 Concentrations of colistins A and B (as the free base) and CMSs A and B (as CMS without the sodium ion) in plasma versus time from a human volunteer administered a single intravenous infusion of 80 mg CMS (Colimycin for injection, Sanofi-Aventis). From Gobin et al. [4]

using a slightly lesser volume of sample than the earlier method (plasma of 0.18 mL [66] rather than 0.25 mL [4]). However, the method also lacks data validating the measurement of CMS in plasma.

This was followed a year later by a method for measuring polymyxins B1, B2 and B1–1 (an isomer of B1) in human plasma and urine (described for measuring this group of polymyxins in the latter medium for the first time) [6]. The polymyxins were extracted in an automated manner from plasma using reversed-phase C8 HLB sorbent and from urine (to which surfactant had been added, *vide supra*) with a reversed-phase/weak cation-exchange sorbent (both Oasis®, from Waters). These authors achieved limits for the quantification of all three polymyxins in plasma (0.1 mL) and urine (0.2 mL) of 0.005 mg/L. They are superior to values reported by Thomas et al. [67] of 0.1 mg/L for polymyxins B1 and B2 using 0.25 mL of plasma although, as the authors rightly state, inspection of their chromatograms would suggest an order of magnitude lower could be achievable. Interestingly, the former authors' measurements of the three polymyxins in plasma and urine, albeit in only one subject, suggests differences in renal clearance between them [6].

Other methods based on liquid chromatography – mass spectrometry suffer from descriptions that are not clear or are incomplete. It is difficult sometimes to ascertain limits of quantification, volumes of sample used, and whether conversion of CMS to colistin in samples has been minimized and/or evaluated after collection of the samples and during their processing for quantifying colistin in studies where CMS has been administered.

For example, an appreciation of the sensitivity of the method for polymyxin B is difficult to ascertain because the volume of sample, detail of the method and quantifiable limits were not provided [68]; a limit of 0.25 mg/L can be construed from data for intra-/inter-day variations (CVs) of less than 8% for concentrations spanning the calibration range of 0.25–10.0 mg/L. Members of the same group subsequently described use of the same method for examining the pharmacokinetics of polymyxin B1, isoleucine-polymyxin B1 and summed polymyxins B2 and B3 (the two were not resolved chromatographically and the concentrations of B3 considered negligible in most samples) and their concentrations in renal tissue and urine after a single intravenous dose of polymyxin B sulphate to rats [69]. This was followed by another report of a study examining the pharmacokinetics and efficacy of polymyxin B sulphate after it had been encapsulated into a liposomal delivery system and administered intravenously to mice [70]. The concentrations of the four components of polymyxin B in 0.20 mL of serum and epithelial lining fluid were determined using UPLC-MS/MS. Trichloroacetic acid in an organic solvent was added to precipitate proteins, and the dried extract from the supernatant following centrifugation reconstituted in mobile phase (formic acid, acetonitrile and water) for chromatography. The authors had separated and purified the four components previously using preparative liquid chromatography [71], and confirmed their identity with mass spectrometry. It is assumed, therefore, that calibration curves for calculating their individual concentrations in the two fluids [72] were prepared using the purified components. The initial work from data on reproducibility for the calibration standards suggests a limit of quantification of 0.25 mg/L from an unknown volume of sample [68]; the final publication reports a limit of 0.006 mg/L for all four polymyxin B compounds (B1, isoleu-B1, B2 and B3) with 0.2 mL of serum or epithelial fluid [70].

6.4 Stability of CMS and Colistin

As noted previously, CMS is converted to colistin *in vivo* after the administration of CMS. The conversion occurs also *in vitro* in biological samples collected from studies where CMS has been administered (e.g. a pharmacokinetic study) and in studies assessing antimicrobial activity with CMS. Apparent from Fig. 6.2 are the considerably higher concentrations of CMS compared to colistin in plasma from a pharmacokinetic study in rats after an intravenous dose of CMS [31], and the higher concentrations also in humans [4, 30, 34, 59, 61], especially during the first 4 h after a dose of CMS (Fig. 6.3). Therefore, it is critical for ensuring accurate measurement of the concentrations of colistin in such studies that there is minimal conversion of CMS to colistin in the time between collection of the sample and measuring colistin.

The method for measuring colistin in plasma [2] was used for an extensive assessment of its stability when stored in a range of aqueous media (water, plasma and isotonic phosphate buffer, 0.067 mol/L, pH 7.4) and its formation from CMS stored separately in identical media plus Meuller-Hinton broth. The presence of CMS remaining in water was also examined qualitatively using strong anion exchange chromatography [16]. The levels of colistin A and B in water remained unchanged after storage at 4 °C for 60 days and at 37 °C for 120 h. When stored in the buffer (approximately 1.5 pH units higher than the solution of colistin sulphate in water) and human plasma at 37 °C, its stability was reduced markedly; more so in plasma than in the buffer. After incubation of CMS in water for 12 h at this temperature, there were clear qualitative changes in the chromatogram for CMS, suggesting partial conversion to products derivatized to a lesser degree with methanesulphonate. Between 10% and 15% of CMS in buffer and plasma had degraded to colistin within 2 h, irrespective of the source of CMS raw material. Interestingly, later work found that the stability of CMS was greater at a higher concentration in plasma (30 mg/L vs 2 mg/L); an observation made also in aqueous solutions of CMS for administration to patients [73]. It was attributed to the formation of micelles by CMS, which protected the prodrug from conversion to colistin [74].

Colistin was reported to be stable in plasma stored at −20 °C and −80 °C for up to 2 months [4]. No data was provided but, from the limits of quantification for CMS and data for the storage of plasma spiked with CMS under the same conditions and period, it can be estimated that there was no more than 1% conversion to colistin. This supports previous data [1]: CMS and colistin in plasma stored at −80 °C were stable for 4 months and 6–8 months, respectively. The "loss by adsorption", alluded to above and previously [5], was reported later [75] to be significant when dilutions of stock solutions of colistin were made using test-tubes made of soda lime glass, polystyrene and polypropylene. The least degree of loss was from low protein-binding microtubes. Although no quantitative data could be located in support, the usual procedure for minimizing adsorption is to add human plasma to those samples lacking protein prior to processing them for chromatographic analysis [4, 5]. Alternatively, it is evident from a more recent publication that the addition of a surfactant to urine after collection achieves almost 100% recovery of polymyxins B1, B1-1 and B2 [6].

These observations highlight the need for careful handling of biological samples collected from, for example, studies examining the pharmacokinetics of CMS and colistin after the administration of CMS. It is proposed that any method to be used for such studies should have conducted experiments to validate the handling of samples after collection, their storage, and their handling during processing of samples prior to forming a fluorescent derivative for chromatography or during processing prior to direct chromatographic analysis with mass spectrometric detection.

6.5 Conclusions

In summary, there have been three predominant approaches for measuring the concentrations of colistin and polymyxin B, and the prodrug of colistin (CMS), in biological fluids: microbiological assay, liquid chromatography with detection and quantification of fluorescent derivatives, and liquid chromatography with detection and quantification by mass spectrometry. The second and third approaches have facilitated rapid advances in understanding the preclinical and clinical pharmacology of polymyxins (and their relevant prodrugs) over the last 15 years. They are capable of achieving the sensitivity required to measure concentrations in samples from clinical and pharmacokinetic studies in humans, and pharmacokinetic and pharmacodynamic studies in animals, and some of the methods using these approaches have been well validated. Of the three, the most appropriate and convenient for the majority of research laboratories would be liquid chromatography in combination with triple quadrupole mass spectrometry; even a single quadrupole may be sufficient [76] and could be adapted for clinical samples. The processing of samples is generally relatively simple, but one must ensure that there is insignificant conversion of CMS to colistin when quantifying the latter in samples where CMS is present also. The only limitation is access to a mass spectrometer. The formation of fluorescent derivatives has sufficient sensitivity but does require the additional step of forming the derivative during processing of the samples. These two approaches are designed to quantify the polymyxin base. If samples are from subjects or animals administered CMS, the concentrations of the base are determined before and after forced conversion of the prodrug to the base. From these separate determinations, the concentration of prodrug in biological fluid can be calculated. Microbiological methods have, in general, suffered from insufficient validation. Potentially, such methods possess sufficient sensitivity for measuring the concentrations of polymyxin B after therapeutic doses (and of colistin after administering colistin sulphate; available in China), but they are time-consuming. Often, they are described as being used to measure "colistin" in studies where CMS is investigated without taking any account of the presence of its prodrug, the lack of antimicrobial activity of that prodrug, and its potential conversion to colistin in both the samples and calibration standards to differing degrees during incubation. Some chromatography-based methods also suffer from this shortcoming.

References

1. Dudhani RV, Nation RL, Li J (2010) Evaluating the stability of colistin and colistin methanesulphonate in human plasma under different conditions of storage. J Antimicrob Chemother 65(7):1412–1415
2. Li J, Milne RW, Nation RL, Turnidge JD, Coulthard K, Johnson DW (2001) A simple method for the assay of colistin in human plasma, using pre-column derivatization with 9-fluorenylmethyl chloroformate in solid-phase extraction cartridges and reversed-phase high-performance liquid chromatography. J Chromatogr B Biomed Sci Appl 761(2):167–175
3. Li J, Milne RW, Nation RL, Turnidge JD, Coulthard K, Valentine J (2002) Simple method for assaying colistin methanesulfonate in plasma and urine using high-performance liquid chromatography. Antimicrob Agents Chemother 46(10):3304–3307
4. Gobin P, Lemaitre F, Marchand S, Couet W, Olivier JC (2010) Assay of colistin and colistin methanesulfonate in plasma and urine by liquid chromatography-tandem mass spectrometry. Antimicrob Agents Chemother 54(5):1941–1948
5. Jansson B, Karvanen M, Cars O, Plachouras D, Friberg LE (2009) Quantitative analysis of colistin A and colistin B in plasma and culture medium using a simple precipitation step followed by LC/MS/MS. J Pharm Biomed Anal 49(3):760–767
6. Meng M, Wang L, Liu S, Jaber O, Gao L, Chevrette L, Reuschel S (2016) Simultaneous quantitation of polymyxin B1, polymyxin B2 and polymyxin B1-1 in human plasma and treated human urine using solid phase extraction and liquid chromatography-tandem mass spectrometry. J Chromatogr B 1012–1013:23–36
7. Brownlee G, Bushby SRM, Short EI (1949) The pharmacology of polymyxin A, B, and D. Ann NY Acad Sci 51(5):952–967
8. Stansly PG (1949) A simple method for the assay of polymyxin in blood and urine. Ann NY Acad Sci 51:980–981
9. Stansly PG (1948) Studies on polymyxin: an assay method for blood and urine. Proc Soc Exp Biol Med 68:301–304
10. Craig WA, Kunin CM (1973) Dynamics of binding and release of the polymyxin antibiotics by tissues. J Pharmacol Exp Ther 184(3):757–765
11. Ziv G, Schultze WD (1982) Pharmacolunetics of polymyxin B administered via the bovine mammary gland. J Vet Pharmacol Ther 5:123–129
12. Ziv G, Nouws JF, van Ginneken CA (1982) The pharmacokinetics and tissue levels of polymyxin B, colistin and gentamicin in calves. J Vet Pharmacol Ther 5(1):45–58
13. Ziv C, Wanner M, Nicolet J (1980) Clinical pharmacology of polymyxin B, colistin and colistimethate in young calves. J Vet Pharmacol Ther 3:87–94
14. Wootton M, Holt HA, MacGowan AP (2005) Development of a novel assay method for colistin sulphomethate. Clin Microbiol Infect 11:243–244
15. Bergen PJ, Li J, Rayner CR, Nation RL (2006) Colistin methanesulfonate is an inactive prodrug of colistin against Pseudomonas aeruginosa. Antimicrob Agents Chemother 50(6):1953–1958
16. Li J, Milne RW, Nation RL, Turnidge JD, Coulthard K (2003) Stability of colistin and colistin methanesulfonate in aqueous media and plasma as determined by high-performance liquid chromatography. Antimicrob Agents Chemother 47(4):1364–1370
17. Healy DP, Sombun AD, Gardner JC, Good K, Durkee PJ, Toner L, Rieman MT, Neely AN, Kagan RJ (2011) Pharmacokinetics of colistin in an adolescent boy with extensive burn injury. J Burn Care Res 32(1):e7–e11
18. Jimenez-Mejias ME, Pichardo-Guerrero C, Marquez-Rivas FJ, Martin-Lozano D, Prados T, Pachon J (2002) Cerebrospinal fluid penetration and pharmacokinetic/pharmacodynamic parameters of intravenously administered colistin in a case of multidrug-resistant acinetobacter baumannii meningitis. Eur J Clin Microbiol Infect Dis 21(3):212–214
19. Aoki N, Tateda K, Kikuchi Y, Kimura S, Miyazaki C, AIshii Y, Tanabe Y, Gejyo F, Yamaguchi K (2009) Efficacy of colistin combination therapy in a mouse model of pneumonia caused by multidrug-resistant Pseudomonas aeruginosa. J Antimicrob Chemother 63(3):534–542
20. Hengzhuang W, Wu H, Ciofu O, Song Z, Hoiby N (2012) In vivo pharmacokinetics/pharmacodynamics of colistin and imipenem in Pseudomonas aeruginosa biofilm infection. Antimicrob Agents Chemother 56(5):2683–2690
21. Lin B, Zhang C, Xiao X (2005) Toxicity, bioavailability and pharmacokinetics of a newly formulated colistin sulfate solution. J Vet Pharmacol Ther 28(4):349–354
22. Sarria JC, Angulo-Pernett F, Kimbrough RC, McVay CS, Vidal AM (2004) Use of intravenous polymyxin B during continuous venovenous hemodialysis. Eur J Clin Microbiol Infect Dis 23:340–341
23. Conway SP, Pond MN, Watson A, Etherington C, Robey HL, Goldman MH (1997) Intravenous colistin sulphomethate in acute respiratory exacerbations in adult patients with cystic fibrosis. Thorax 52:987–993
24. Hermsen ED, Sullican CJ, Rotschafer JC (2003) Polymyxins: pharmacology, pharmacokinetics, pharmacodynamics, and clinical applications. Infect Dis Clin North Am 17:545–562
25. Le Brun PP, de Graaf AI, Vinks AA (2000) High-performance liquid chromatographic method for the determination of colistin in serum. Ther Drug Monit 22(5):589–593
26. Decolin D, Leroy P, Nicolas A, Archimbault P (1997) Hyphenated liquid chromatographic method for the determination of colistin residues in bovine tissues. J Chromatogr Sci 35(12):557–564
27. Reed MD, Stern RC, O'Riordan MA, Blumer JL (2001) The pharmacokinetics of colistin in patients with cystic fibrosis. J Clin Pharmacol 41(6):645–654
28. Guyonnet J, Manco B, Baduel L, Kaltsatos V, Aliabadi MH, Lees P (2010) Determination of a dos-

age regimen of colistin by pharmacokinetic/pharmacodynamic integration and modeling for treatment of G.I.T. disease in pigs. Res Vet Sci 88(2):307–314
29. Gmur DJ, Bredl CR, Steele SJ, Cai S, VanDevanter DR, Nardella PA (2003) Determination of polymyxin E1 in rat plasma by high-performance liquid chromatography. J Chromatogr B 789(2):365–372
30. Li J, Coulthard K, Milne R, Nation RL, Conway S, Peckham D, Etherington C, Turnidge J (2003) Steady-state pharmacokinetics of intravenous colistin methanesulphonate in patients with cystic fibrosis. J Antimicrob Chemother 52(6):987–992
31. Li J, Milne RW, Nation RL, Turnidge JD, Smeaton TC, Coulthard K (2004) Pharmacokinetics of colistin methanesulphonate and colistin in rats following an intravenous dose of colistin methanesulphonate. J Antimicrob Chemother 53(5):837–840
32. Yapa SWS, Li J, Patel K, Wilson JW, Dooley MJ, George J, Clark D, Poole S, Williams E, Porter CJ, Nation RL, McIntosh MP (2014) Pulmonary and systemic pharmacokinetics of Inhaled and Intravenous colistin methanesulfonate in cystic fibrosis patients: targeting advantage of inhalational administration. Antimicrob Agents Chemother 58(8):2570–2579
33. Koomanachai P, Landersdorfer CB, Chen G, Lee HJ, Jitmuang A, Wasuwattakul S, Sritippayawan S, Li J, Nation RL, Thamlikitkula V (2014) Pharmacokinetics of colistin methanesulfonate and formed colistin in end-stage renal disease patients receiving continuous ambulatory peritoneal dialysis. Antimicrob Agents Chemother 58(1):440–446
34. Garonzik SM, Li J, Thamlikitkul V, Paterson DL, Shoham S, Jacob J, Silveira FP, Forrest A, Nation RL (2011) Population pharmacokinetics of colistin methanesulfonate and formed colistin in critically ill patients from a multicenter study provide dosing suggestions for various categories of patients. Antimicrob Agents Chemother 55(7):3284–3294
35. Dudhani RV, Turnidge J, Coulthard K, Milne R, Rayner CR, Li J, Nation RL (2010) Elucidation of the pharmacokinetic/pharmacodynamic determinant of colistin activity against Pseudomonas aeruginosa in murine thigh and lung infection models. Antimicrob Agents Chemother 54(3):1117–1124
36. Yapa SWS, Li J, Porter CJ, Nation RL, Patel K, McIntosh MP (2013) Population pharmacokinetics of colistin methanesulfonate in rats: achieving sustained lung concentrations of colistin for targeting respiratory infections. Antimicrob Agents Chemother 57(10):5087–5095
37. He H, Li JC, Nation RL, Jacob J, Chen G, Lee HJ, Tsuji BT, Thompson PE, Roberts KD, Velkov T, Li J (2013) Pharmacokinetics of four different brands of colistimethate and formed colistin in rats. J Antimicrob Chemother 68(10):2311–2317
38. Bergen PJ, Li J, Nation RL, Turnidge JD, Coulthard K, Milne RW (2008) Comparison of once-, twice- and thrice-daily dosing of colistin on antibacterial effect and emergence of resistance: studies with Pseudomonas aeruginosa in an in vitro pharmacodynamic model. J Antimicrob Chemother 61(3):636–642
39. Jin L, Li J, Nation RL, Nicolazzo JA (2009) Brain penetration of colistin in mice assessed by a novel high-performance liquid chromatographic technique. Antimicrob Agents Chemother 53(10):4247–4251
40. Jin L, Li J, Nation RL, Nicolazzo JA (2012) Effect of systemic infection induced by Pseudomonas aeruginosa on the brain uptake of colistin in mice. Antimicrob Agents Chemother 56(10):5240–5246
41. Yousef JM, Chen G, Hill PA, Nation RL, Li J (2012) Ascorbic acid protects against the nephrotoxicity and apoptosis caused by colistin and affects its pharmacokinetics. J Antimicrob Chemother 67(2):452–459
42. Yousef JM, Chen G, Hill PA, Nation RL, Li J (2011) Melatonin attenuates colistin-induced nephrotoxicity in rats. Antimicrob Agents Chemother 55(9):4044–4049
43. Cao G, Ali F, Chiu F, Alexandre P, Zavascki A, Nation R, Li J (2008) Development and validation of a reversed-phase high-performance liquid chromatography assay for polymyxin B in human plasma. J Antimicrob Chemother 62(5):1009–1014
44. Ali FE, Cao G, Poudyal A, Vaara T, Nation RL, Vaara M, Li J (2009) Pharmacokinetics of novel antimicrobial cationic peptides NAB 7061 and NAB 739 in rats following intravenous administration. J Antimicrob Chemother 64(5):1067–1070
45. Zavascki AP, Li J, Goldani LZ, Cao G, Superti SV, Lutz L, Barth AL, Ramos F, Boniatti MM, Nation RL (2008) Pharmacokinetics of intravenous polymyxin B in critically-ill patients. Clin Infect Dis 47:1298–1304
46. Sandri AM, Landersdorfer CB, Jacob J, Boniatti MM, Dalarosa MG, Falci DR, Behle TF, Bordinhão RC, Wang J, Forrest A, Nation RL, Li J, Zavascki AP (2013) Population pharmacokinetics of intravenous polymyxin B in critically ill patients: Implications for selection of dosage regimens. Clin Infect Dis 57:524–531
47. Sandri AM, Landersdorfer CB, Jacob J, Boniatti MM, Dalarosa MG, Falci DR, Behle TF, Saitovitch D, Wang J, Forrest A, Nation RL, Zavascki AP, Li J (2013) Pharmacokinetics of polymyxin B in patients on continuous venovenous haemodialysis. J Antimicrob Chemother 68(3):674–677
48. Markou N, Markantonis SI, Dimitrakis E, Panidis D, Boutzouka E, Karatzas S, Rafailidis P, Apostolakos H, Baltopoulos G (2008) Colistin serum concentrations after intravenous administration in critically ill patients with serious multidrug-resistant, gram-negative bacilli infections: A prospective, open-label, uncontrolled study. Clin Ther 30(1):143–151
49. Markantonis SL, Markou N, Fousteri M, Sakellaridis N, Karatzas S, Alamanos I, Dimopoulou E, Baltopoulos G (2009) Penetration of colistin into cerebrospinal fluid. Antimicrob Agents Chemother 53(11):4907–4910
50. Ziaka M, Markantonis SL, Fousteri M, Zygoulis P, Panidis D, Karvouniaris M, Makris D, Zakynthinos E (2013) Combined intravenous and intraventricular

administration of colistin methanesulfonate in critically ill patients with central nervous system infection. Antimicrob Agents Chemother 57(4):1938–1940
51. Imberti R, Cusato M, Villani P, Carnevale L, Iotti GA, Langer M, Regazzi M (2010) Steady-state pharmacokinetics and BAL concentration of colistin in critically Ill patients after IV colistin methanesulfonate administration. Chest 138(6):1333–1339
52. Athanassa ZE, Markantonis SL, Fousteri MZ, Myrianthefs PM, Boutzouka EG, Tsakris A, Baltopoulos GJ (2012) Pharmacokinetics of inhaled colistimethate sodium (CMS) in mechanically ventilated critically ill patients. Intensive Care Med 38(11):1779–1786
53. Zeng Z, Wu J, Yang G, Chen Z, Huang X, Ding H (2010) Study of colistin depletion in duck tissues after intramuscular and oral administration. J Vet Pharmacol Ther 33(4):408–410
54. He J, Tang S, Li L, Zhang C, Li X, Xia X, Xiao X (2011) Pharmacokinetics of a novel amoxicillin/colistin suspension after intramuscular administration in pigs. J Vet Pharmacol Ther 34(1):42–50
55. Ratjen F, Rietschel E, Kasel D, Schwiertz R, Starke K, Beier H, van Koningsbruggen S, Grasemann H (2006) Pharmacokinetics of inhaled colistin in patients with cystic fibrosis. J Antimicrob Chemother 57(2):306–311
56. Dotsikas Y, Markopoulou CK, Koundourellis JE, Loukas YL (2011) Validation of a novel LC-MS/MS method for the quantitation of colistin A and B in human plasma. J Sep Sci 34(1):37–45
57. Ma Z, Wang J, Gerber JP, Milne RW (2008) Determination of colistin in human plasma, urine and other biological samples using LC-MS/MS. J Chromatogr B Anal Technol Biomed Life Sci 862(1–2):205–212
58. Sin DW, Ho C, Wong Y, Ho S, Ip AC (2005) Analysis of major components of residual bacitracin and colistin in food samples by liquid chromatography tandem mass spectrometry. Anal Chim Acta 535(1–2):23–31. https://doi.org/10.1016/j.aca.2004.11.063
59. Plachouras D, Karvanen M, Friberg LE, Papadomichelakis E, Antoniadou A, Tsangaris I, Karaiskos I, Poulakou G, Kontopidou F, Armaganidis A, Cars O, Giamarellou H (2009) Population pharmacokinetic analysis of colistin methanesulfonate and colistin after intravenous administration in critically ill patients with infections caused by gram-negative bacteria. Antimicrob Agents Chemother 53(8):3430–3436
60. Mohamed AF, Karaiskos I, Plachouras D, Karvanen M, Pontikis K, Jansson B, Papadomichelakis E, Antoniadou A, Giamarellou H, Armaganidis A, Cars O, Friberg LE (2012) Application of a loading dose of colistin methanesulfonate in critically ill patients: population pharmacokinetics, protein binding, and prediction of bacterial kill. Antimicrob Agents Chemother 56(8):4241–4249
61. Karvanen M, Plachouras D, Friberg LE, Paramythiotou E, Papadomichelakis E, Karaiskos I, Tsangaris I, Armaganidis A, Cars O, Giamarellou H (2013) Colistin methanesulfonate and colistin pharmacokinetics in critically ill patients receiving continuous venovenous hemodiafiltration. Antimicrob Agents Chemother 57(1):668–671
62. Leporati M, Ornella Bua RO, Mariano F, Carignano P, Stella M, Biancone L, Marco Vincenti M (2014) Determination by LC–MS/MS of Colistins A and B in plasma and ultrafiltrate from critically ill patients undergoing continuous venovenous hemodiafiltration. Ther Drug Monit 36:182–191
63. Gikas E, Bazoti FN, Katsimardou M, Anagnostopoulos D, Papanikolaou K, Inglezos I, Skoutelis A, Daikos GL, Tsarbopoulos A (2013) Determination of colistin A and colistin B in human plasma by UPLC–ESI high resolution tandem MS: application to a pharmacokinetic study. J Pharm Biomed Anal 83:228–236
64. Zhao M, Wu XJ, Fan YX, Guo BN, Zhang J (2016) Development and validation of a UHPLC–MS/MS assay for colistin methanesulphonate (CMS) and colistin in human plasma and urine using weak-cation exchange solid-phase extraction. J Pharm Biomed Anal 124:303–308
65. Cheng C, Liu S, Xiao D, Hollembaek J, Yao L, Lin J, Hansel S (2010) LC–MS/MS method development and validation for the determination of polymyxins and vancomycin in rat plasma. J Chromatogr B 878(28):2831–2838
66. Mercier T, Tissot F, Gardiol C, Corti N, Wehrli S, Guidi M, Csajka C, Buclin T, Couet W, Marchetti O, Decosterd LA (2015, 1369) High-throughput hydrophilic interaction chromatography coupled to tandem mass spectrometry for the optimized quantification of the anti-Gram-negatives antibiotic colistin A/B and its pro-drug colistimethate. J Chromatogr A:52–63
67. Thomas TA, Broun EC, Abildskov KM, Kubin CJ, Horan J, Yin MT, Cremers S (2012) High performance liquid chromatography–mass spectrometry assay for polymyxin B1 and B2 in human plasma. Ther Drug Monit 34:398–405
68. Kwa AL, Lim TP, Low JG, Hou J, Kurup A, Prince RA, Tam VH (2008) Pharmacokinetics of polymyxin B1 in patients with multidrug-resistant Gram-negative bacterial infections. Diagn Microbiol Infect Dis 60(2):163–167
69. Abdelraouf K, He J, Ledesma KR, Hu M, Tam VH (2012) Pharmacokinetics and renal disposition of polymyxin B in an animal model. Antimicrob Agents Chemother 56(11):5724–5727
70. He J, Abdelraouf K, Ledesma KR, Chow DS-L, Tam VH (2013) Pharmacokinetics and efficacy of liposomal polymyxin B in a murine pneumonia model. Int J Antimicrob Agents 42(6):559–564
71. Tam VH, Cao H, Ledesma KR, Hu M (2011) In vitro potency of various polymyxin B components. Antimicrob Agents Chemother 55(9):4490–4491
72. He J, Gao S, Hu M, Chow DS-L, Tam VH (2013) A validated ultra-performance liquid chromatography–tandem mass spectrometry method for the quantification of polymyxin B in mouse serum and epithelial

lining fluid: application to pharmacokinetic studies. J Antimicrob Chemother 68:1104–1110
73. Wallace SJ, Li J, Rayner CR, Coulthard K, Nation RL (2008) Stability of colistin methanesulfonate in pharmaceutical products and solutions for administration to patients. Antimicrob Agents Chemother 52(9):3047–3051
74. Wallace SJ, Li J, Nation RL, Prankerd RJ, Velkov T, Boyd BJ (2010) Self-assembly behavior of colistin and its prodrug colistin methanesulfonate: implications for solution stability and solubilization. J Phys Chem B Condens Matter Mater Surf Interfaces Biophysical 114(14):4836–4840
75. Karvanen M (2013) Optimization of colistin dosage in the treatment of multiresistant gram-negative infections. Doctoral thesis, Uppsala Universitet, Uppsala
76. Cheah S-E, Jurgen B, Bulitta JB, Li J, Nation RL (2014) Development and validation of a liquid chromatography–mass spectrometry assay for polymyxin B in bacterial growth media. J Pharm Biomed Anal 92:177–182

Pharmacokinetics of Polymyxins in Animals

7

Sandrine Marchand, Nicolas Grégoire, and William Couet

Abstract

All of the small number of studies conducted during the second half of last century to investigate the pharmacokinetics of polymyxins in animals used microbiological methods to quantify the compounds in biological fluids. Those methods generally lacked the accuracy and precision required for such investigations and, in the case of studies involving administration of colistin methanesulfonate (CMS), ongoing conversion to colistin during microbiological incubation of collected samples artifactually elevated the measured concentration of colistin. The pharmacokinetic studies reviewed in this chapter involved use of more accurate, precise and specific methods for the measurement of the relevant compounds in biological matrices. The studies have been conducted in a number of pre-clinical animal species following administration via various routes (e.g. intravenous, intrapulmonary), and have provided important insights into not only the global pharmacokinetics as viewed from plasma but also the tissue distribution and handling by key organs particularly the kidneys.

Keywords

Colistimethate · Colistin · Polymyxin B · Animals · Global pharmacokinetics · Tissue distribution · Mechanisms involved in renal elimination

Polymyxin derivatives used in clinical practice correspond to complex mixtures of structurally related but distinct chemical entities obtained by fermentation. This raises a number of issues such as purity or differences in composition between brands and even between batches. But because of that, expressing polymyxin doses or concentrations is quite complex and may become confusing [1], in particular in the case of polymyxin E or colistin, the latter administered as a prodrug. Initial pharmacokinetic (PK) studies on colistin methansulfonate (CMS), colistin or polymyxin B in animals were conducted between 1970 and 2000 in various species (rabbits, dogs, ewes, calves) [2–8] using microbiological assays (*i.e.* bioassays) for measuring concentrations. However, as discussed in Chap. 6, these techniques are non-specific and cannot distin-

S. Marchand · N. Grégoire · W. Couet (✉)
Inserm U1070, Pôle Biologie Santé, Poitiers Cedex, France

Service de Toxicologie-Pharmacocinétique, CHU de Poitiers, Poitiers, France

Université de Poitiers, UFR Médecine-Pharmacie, Poitiers, France
e-mail: william.couet@univ-poitiers.fr

J. Li et al. (eds.), *Polymyxin Antibiotics: From Laboratory Bench to Bedside*, Advances in Experimental Medicine and Biology 1145, https://doi.org/10.1007/978-3-030-16373-0_7

guish between CMS and colistin or between colistin and co-administered antibiotics cross-reacting with the selected test strain [9]. Therefore, only PK studies in animals conducted with chromatographic assays, including high performance liquid chromatography (HPLC) coupled with fluorimetric detection [10], tandem mass spectrometry (LC-MS/MS) [11] or ultraperformance liquid chromatography-tandem mass spectrometry (UPLC-MS/MS) [12] will be reviewed. Furthermore, polymyxins are occasionally administered orally for local decontamination, but since their oral absorption is virtually negligible, this specific situation will not be covered in this review.

7.1 Pharmacokinetics of Colistin after Parenteral Administration

The first pharmacokinetic study of colistin using a chromatographic assay was published by Li et al. in 2003 [13]. Colistin sulfate was administered to rats as a single 1 mg/kg intravenous (IV) bolus dose and a specific and sensitive HPLC assay with fluorimetric detection after derivatization, previously developed and validated by the same group, was used [10]. Colistin clearance (CL) and volume of distribution at steady state (Vss) were respectively estimated at 5.2 ± 0.4 mL/min/kg and 0.50 ± 0.06 L/kg. This low Vss value indicates a limited extravascular distribution, in agreement with physico-chemical characteristics of colistin including a large molecular weight and the presence of positive charges (5 amine functions with a pKa close to 10) at physiological pH. In this study, protein binding of colistin was determined in spiked plasma by equilibrium dialysis at three concentrations (4, 8 and 12 mg/L) leading to an average unbound fraction (fu) equal to 43–45% that was independent of the concentration. Noticeably, colistin A was more extensively bound (mean fu of 36%) than colistin B (mean fu of 52%) [13]. The initially reported elimination half-life of colistin ($t_{1/2}$ = 74.6 ± 13.2 min) [13] was confirmed a few years later by Marchand et al. ($t_{1/2}$ = 75.4 ± 14.1 min) [14] after subcutaneous administration of colistin 1.5 mg/kg, although estimates of clearance and volume (CL = 8.5 ± 1.0 mL/min/kg, Vss = 0.94 ± 0.25 L/kg) in this new study [14] were somewhat higher (up to two fold for Vss) than previously reported; it should be noted that the later study involved subcutaneous administration of colistin [14]. The virtually similar $t_{1/2}$ estimate between studies suggests that colistin disposition is not rate limited by its absorption after subcutaneous administration.

Urine samples were also collected in the Li et al. study, and only 0.2% of the colistin dose was recovered unchanged in urine [13], with a corresponding renal clearance (CLr = 0.010 ± 0.008 mL/min/kg) much lower than renal clearance by glomerular filtration; estimated at 2.3 mL/min/kg under the assumption that colistin unbound fraction was equal to 0.44 [13] and glomerular filtration rate (GFR) was 5.2 mL/min/kg in rats [15]. This observation suggested an extensive tubular reabsorption of colistin, which was then confirmed by this group using an *in vitro* isolated perfused rat kidney model [16]. The extensive renal tubular reabsorption of colistin would be expected to enhance its accumulation in kidney tissue, which may have implications for its renal toxicity [16]. Active transport systems such as organic cation transporters (mainly OCTN1) and polypeptide transporter 2 (PEPT2) were proposed to be involved in the reabsorption of colistin [16, 17]. Because of this extensive tubular reabsorption, colistin elimination is mostly extra-renal. Yet mechanisms responsible for colistin elimination are mostly unknown and no degradation products have been identified. Furthermore *in vitro* degradation studies at 37 °C demonstrate that the rate of colistin disappearance is virtually similar in homogenates from various tissues such as liver, kidney, muscle or brain, but also not much different than in plasma or phosphate buffer at pH 7.4, suggesting that enzymes may not be involved in colistin elimination [18].

7.2 Pharmacokinetics of Colistin Methanesulfonate (CMS) and Colistin after Parenteral Administration of CMS

7.2.1 Pharmacokinetics of CMS and Colistin in Rats

The first pharmacokinetic study conducted after intravenous administration of CMS, the inactive prodrug of colistin used in clinical practice, in rats, was also conducted by Li et al. [19] at a dose of 15 mg/kg of CMS corresponding to ~6.3 mg/kg of colistin base activity (CBA) [1]. CMS and colistin were assayed by HPLC [20]. Noticeably as for any other chromatographic assays, CMS concentrations were not measured directly, but obtained by difference between colistin concentrations measured after and before CMS hydrolysis in plasma samples. Therefore, partially sulfomethylated CMS derivatives, that may be present within samples, cannot not be distinguished from CMS. Accordingly the authors clearly stated that estimated CMS pharmacokinetic parameters may only be considered as hybrid parameters for CMS [19]. With this limitation in mind, CMS volume of distribution (Vss) was estimated at 0.30 ± 0.06 L/kg; that is about 30% lower than that of colistin and therefore still close to the extracellular fluid volume [15]. Total CMS clearance was found equal to 11.7 ± 1.8 mL/min/kg which is about twice that of colistin [13]. Urinary recovery experiments showed that 61% ± 14% of the dose administered was recovered, mostly as unchanged CMS (2/3) and as colistin (1/3) [19]. However considering the much lower urinary recovery of colistin (0.2%) previously observed after its direct administration [13], it was concluded that post-excretion hydrolysis of CMS in urine was probably mainly responsible for this high recovery of colistin in urine after CMS administration. Consequently, a more reliable estimate of CMS renal clearance was obtained by assuming that the sum of CMS and colistin amounts recovered in urine was actually excreted as CMS. This CMS renal clearance estimate was equal to 7.2 ± 2.2 mL/min/kg, which is slightly higher than GFR in rats (5.2 mL/min/kg) [15]. Li et al. did not measure the plasma protein binding of CMS, but even if the unbound fraction was one (i.e. no binding in plasma) the relative magnitude of the renal clearance of CMS and of GFR is consistent with net tubular secretion [19]. However, it should again be remembered that CMS parameters correspond to hybrid values that are difficult to interpret. It was also possible during this study [19] to estimate that only 6.8% of the CMS dose was converted systemically into colistin in rats, and another interesting observation was that the elimination half-life of formed colistin (55.7 ± 19.3 min) was about twice that of CMS (23.6 ± 3.9 min) indicating that the disposition of formed colistin is not rate-limited by its formation.

A few years after this initial study, a dose-ranging pharmacokinetic study was conducted in rats by Marchand et al. [21], using the largest possible range of CMS doses (5–120 mg/kg of CMS base corresponding to ~2.1 to ~50 mg/kg of CBA), considering the limit of quantification of the LC-MS/MS analytical assay (0.078 µg/mL) and drug toxicity. No trend of non-linearity was observed and pharmacokinetic parameter values were consistent with those previously published by Li et al. [19], in particular when the dose administered was the same (15 mg/kg of CMS base corresponding to ~6.3 mg/kg of CBA). It was estimated that on average 10.2% of the intravenous CMS dose was converted systemically into colistin. This fm estimate is slightly higher than that previously reported (6.8%) [19] but still relatively low. Therefore, these two studies allow concluding, that at least in rats, only a small fraction of the intravenous dose of CMS is eventually converted systemically into colistin.

Interestingly also, the unusual colistin concentration versus time profiles in rats, compared with others species (discussed below), with almost instantaneous plasma peak concentrations and a slower decay of colistin compared with CMS over time, initially observed by Li et al. [19], were confirmed during the study by Marchand et al. [21] as illustrated in Fig. 7.1. A more recent pharmacokinetic study in rats by He et al. com-

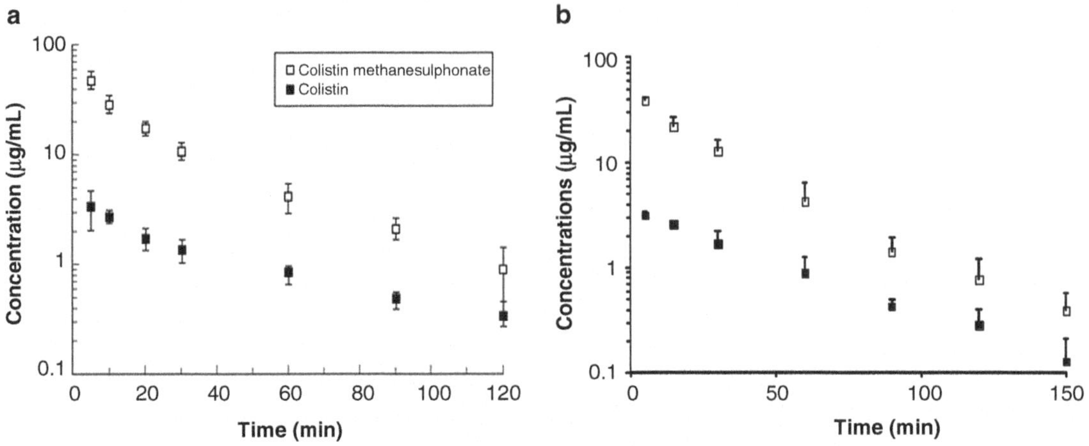

Fig. 7.1 Mean ± SD total plasma concentration versus time profiles of CMS and colistin in rats after a single 15 mg/kg IV dose of CMS (~6.3 mg/kg CBA) obtained by (**a**) Li et al., using CMS from Sigma (St Louis, USA) in 5 rats [19] or (**b**) Marchand et al. using CMS from Sanofi Aventis, (Paris, France) in 6 rats [21], by permission of Oxford University Press

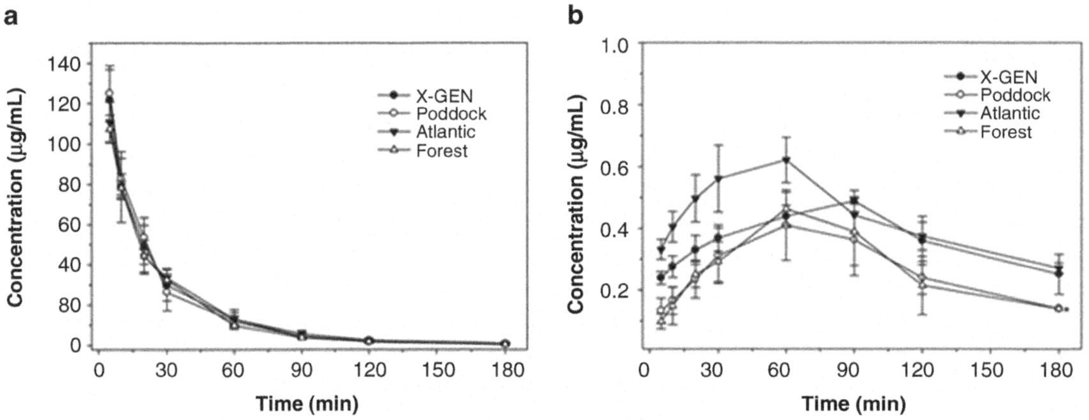

Fig. 7.2 Mean ± SD plasma concentration–time profiles of (**a**) CMS and (**b**) formed colistin in rats (n = 4) obtained by He et al., 2013, after IV administration of 28.1 mg/kg of CMS (~11.7 mg/kg CBA) from various brands, [22] by permission of Oxford University Press

paring four brands of CMS coming from various countries: Thailand (Atlantic Laboratories), United Kingdom (Forest Laboratories) and USA (X-GEN Pharmaceuticals and Paddock Laboratories) [22], may have provided an explanation for this unexpected behavior. Chemical analysis of the different brands was also performed and similar composition was observed for all brands by elementary analysis. However, chromatographic profiles of CMS showed several peaks with the Altantic CMS chromatographic profile distinct from the three other brands. This is consistent with CMS being a mixture of different fully and partially sulfomethylated derivatives that may vary between brands [19]. Following IV administration of these four brands at a dose of 28.1 mg/kg of CMS corresponding to ~11.7 mg/kg of CBA [22], CMS plasma concentrations versus time profiles (Fig. 7.2a) and CMS pharmacokinetic parameters were generally consistent with those previously described [19, 21]. Colistin elimination half-life was again longer than that of CMS, whatever the brand, confirming that colistin disposition is not limited by its formation. However, plasma concentration versus time profiles of formed colistin in the study by He et al. were different from those previously reported by Li et al. [19] and Marchand et al. [14, 21], and colistin peak concentration was considerably delayed to reach a peak after about 60 min on average (Fig. 7.2b). Similar to the results of He et al., a delay in attainment of peak plasma concentration of formed colistin following IV administration of CMS (Link Pharmaceuticals) in rats was observed by Yapa et al. [23].

Therefore the high initial concentrations of colistin after CMS injection observed by Li et al. [19] and Marchand et al. [14, 21] could be explained by the fact that in these initial studies, the administered CMS solutions may have contained a small fraction of partially sulfomethylated CMS derivatives, rapidly converted into colistin after administration. However, as discussed in the next section, this unexpected behavior was observed in rats but not in other species after IV administration of the same Sanofi-Aventis CMS brand.

Another PK study in rats was conducted to investigate the central nervous system (CNS) distribution of colistin *in vivo* and by *in situ* brain perfusion [24]. Brain-to-plasma ratios ranged between 2.1 and 3.7% and were not enhanced by co-administration of P-glycoprotein (P-gp) inhibitors (PSC833 or GF12918). Intraperitoneal injection of lipopolysaccharides to induce inflammation significantly increased brain AUC by two to threefold. In conclusion, blood brain barrier transport of colistin is negligible under healthy conditions but enhanced during systemic inflammation as might be observed in infected patients.

Bouchene et al. recently reported development of a whole-body physiologically-based pharmacokinetic (WBPBPK) model to characterize CMS and colistin distribution in various tissues of rats [25]. In order to describe the disposition of CMS and colistin, the work involved a combination of *in vitro*, *in silico* and *in vivo* data to construct the model. A key aspect of the study was the experimental determination of the tissue-to-plasma partition coefficient for CMS and colistin across 10 different tissues/organs, against which the model predictions were compared. Notably, the experimentally determined kidney-to-plasma partition coefficients for CMS (5.45) and colistin (19.7) were substantially higher than for any of the other tissues. The accumulation of CMS and colistin in kidney tissue is undoubtedly a key factor in the development of nephrotoxicity after administration of CMS. With appropriate validation studies, the WBPBPK model holds promise for inter-species extrapolations using species-specific physiological parameters [25].

7.2.2 Pharmacokinetics of CMS and Colistin in Various Animal Species

In a recent study, epithelial lining fluid (ELF) and systemic pharmacokinetics of CMS and colistin were determined in sheep after IV and pulmonary administration of both CMS (sodium) and colistin (sulfate) at respective doses of 4–8 mg/kg (corresponding to ~1.7–3.4 mg/kg CBA) and 2–3 mg/kg [26]. Concerning systemic pharmacokinetics, the maximal concentration of formed colistin was obtained at 3.13 ± 0.55 h after intravenous administration of CMS. CMS and colistin clearances were 2.29 ± 0.03 L/h (0.95 mL/min/kg) and 1.32 ± 0.23 L/h (0.55 mL/min/kg) which are at least 10 times lower than corresponding clearances in rats [13, 14, 19] and 2.5 times lower than in healthy volunteers for CMS (2.6 ± 0.3 mL/min/kg) [27]. Contrasting with previous results observed in rats and humans where colistin disposition was rate-limited by its own elimination, [19, 27] terminal half-life of formed colistin was not longer than that of CMS. The fm estimate in sheep (17.4%) is slightly higher than that previously reported in rats (6.8 and 10.2%) [19, 21] but still relatively low.

Other investigations have been conducted by our group in various species including mice (n = 36), rabbits (n = 3), baboons (n = 3) and pigs (n = 2) with the objective of developing a WBPBPK model to characterize CMS and colistin distribution in various tissues and eventually to allow between-species comparisons and extrapolations [18]. Animals received a single dose of CMS (Sanofi Aventis, Paris, France) as follows: 15 mg/kg of CMS base (~6.3 mg/kg CBA) administered subcutaneously to mice and intravenously to rabbits, on average 2.5 mg/kg of CMS base (~1.0 mg/kg CBA) infused intravenously over 10 min in baboons, and 149.5 mg of CMS base (160 mg of sodium salt) CMS per pig corresponding to ~67 mg CBA administered as a 1 h intravenous infusion. Multiple sampling of blood was conducted in all species except mice for which 4 animals were used at each of the 9 selected time points. Furthermore, colistin was also directly administered to baboons by subcutaneous 10 min infusion at doses ranging between 0.379 mg/kg and 0.485 mg/kg (of colistin sulfate), in order to estimate the fraction of the CMS dose converted into colistin in that particular species. Plasma concentrations of CMS and colistin were assayed by LC-MS/MS [11] and pharmacokinetic parameters were determined by a non-compartmental approach (WinNonLin version 6.2, Pharsight Corporation, Mountain View, California, USA). Data previously obtained in rats [14, 21] but also in healthy human volunteers [27], using the same CMS brand and analytical assay, can be used for comparison. It should be noted that different CMS batches were used, which may have an effect on concentration versus time profiles of CMS and, in particular, formed colistin. Furthermore, not the same route of administration was used for every species. Plasma concentration versus time profiles in these various species are presented in Fig. 7.3. Corresponding pharmacokinetic parameters are listed in Table 7.1 after correction of volume and clearance terms by body weight.

In humans, baboons, pigs and rabbits, CMS plasma concentrations decayed more rapidly with time than colistin (Fig. 7.3) and accordingly, CMS half-lives were shorter (Table 7.1) [18]. These profiles indicate that in these species the disposition of formed colistin is rate limited by its own elimination and not by its formation. By contrast, CMS and colistin plasma concentrations decayed in parallel with time in mice, which is typical of a formation rate limited process. However, these data must be interpreted carefully since CMS was administered subcutaneously in mice. With the exception of mice, CMS Vss varied between approximately 130 and 330 mL/kg (Table 7.1), close to the volume of extracellular fluid (ECF) [15]. Moreover, these values are in reasonable agreement with the Vss for CMS estimated in healthy volunteers (196 mL/kg) [27]. CMS CL values in rabbits, pigs and baboons were virtually identical and close to the value estimated in humans (2.6 mL/min/kg) [27] but CL estimates for CMS in rats and mice were respectively about 5 and 10 times higher than in human volunteers.

Recently, Viel et al. reported details of a WBPBPK model to characterize the disposition of CMS and colistin in pigs, especially in regard to the renal handling of these compounds [28]. A number of different pharmacokinetic studies were conducted; specifically to elucidate the plasma and kidney pharmacokinetics after single IV and IM administration of CMS as well as after repeated IM administration, and also investigations of tissue-to-plasma partition coefficients and plasma protein binding of CMS. The experimental data were subjected to WBPBPK modelling. Key findings of the experimental and modelling studies were: extensive accumulation of colistin in kidney, followed by slow elimination from that organ; very substantial contribution of tubular secretion in the renal elimination of CMS, with some conversion to colistin within tubular cells; extensive tubular reabsorption of colistin; some degradation of colistin within tubular cells; and, a plasma unbound fraction of approximately 0.4 [28].

Colistin PK-PD has been investigated *in vivo* using mice after direct subcutaneous administration of colistin sulfate to demonstrate that the AUC to MIC ratio corrected for plasma protein binding (*f*AUC/MIC) is the relevant PK-PD index to predict colistin efficacy against *Pseudomonas aeruginosa* and *Acinetobacter baumannii* [29]. Those studies were preceded by single-dose studies at 10, 20 and 40 mg/kg to elucidate the pharmacokinetics of colistin in infected neutropenic

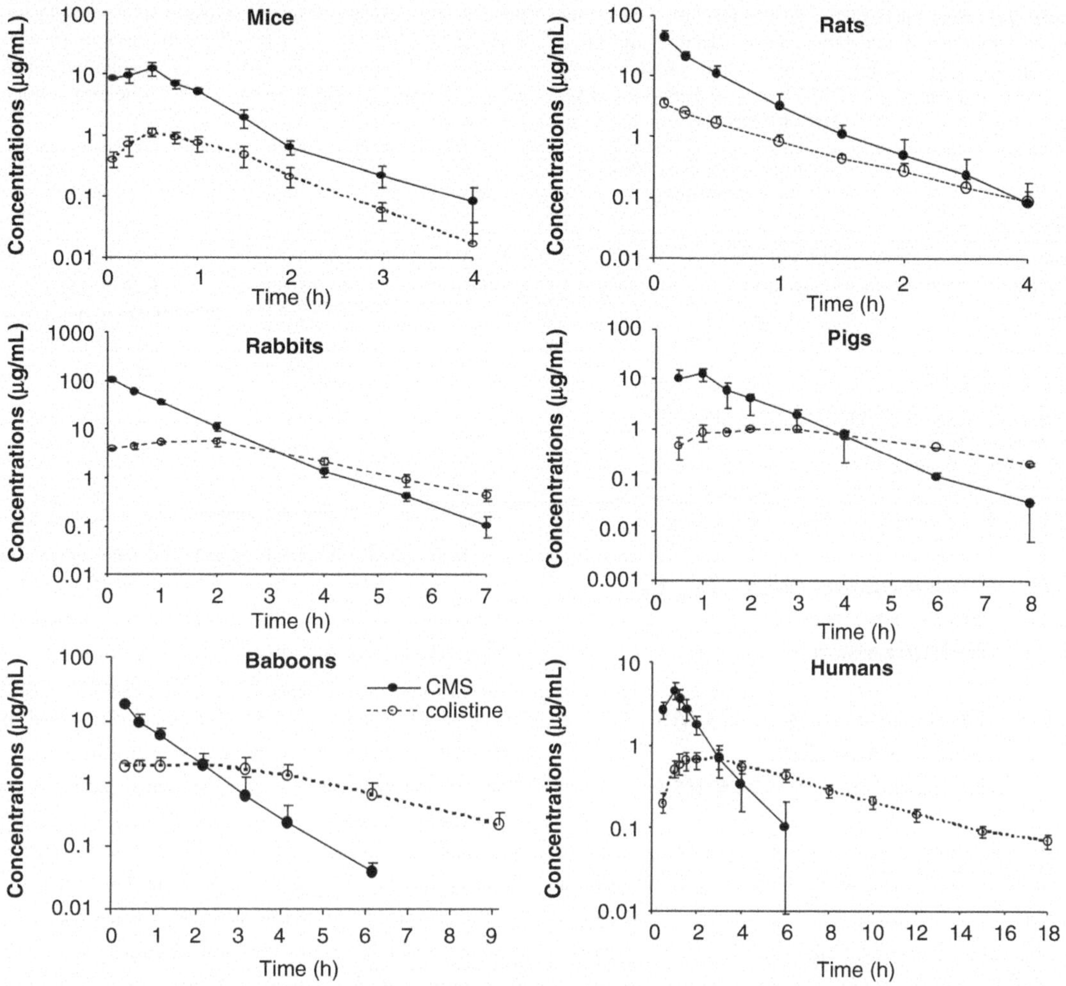

Fig. 7.3 Mean ± SD plasma concentration–time profiles of CMS (closed circles and full line) and formed colistin (open circles and dotted line) in: mice (n = 4 per time point) after subcutaneous administration of CMS base 15 mg/kg (~6.3 mg/kg CBA); rats (n = 6) after IV bolus administration of CMS base 15 mg/kg (~6.3 mg/kg CBA) [21]; rabbits (n = 3) after IV bolus administration of CMS base 15 mg/kg (~6.3 mg/kg CBA); pigs (n = 2) after 1 h infusion of 149.5 mg of CMS base (~62.3 mg CBA); baboons (n = 3) after 10 min infusion of on average 2.5 mg/kg of CMS base (~1.0 mg/kg CBA), and in healthy volunteers (n = 12) after 1 h infusion of 1MIU of CMS equivalent to 80 mg of sodium CMS (~33 mg CBA)

mice. The plasma protein binding in these animals was determined using both ultracentrifugation and rapid equilibrium dialysis, with very careful attention to minimize non-specific binding to equipment used in the measurements. Over the dose range examined, the pharmacokinetics of colistin was nonlinear; the apparent clearance decreased with increasing dose while the half-life increased. The percentage of colistin bound in plasma of infected neutropenic mice was independent of plasma concentration over a wide range (~2–50 mg/L). The average plasma unbound fraction of colistin over this range by the two methods of determination was 0.084; this was very much lower than the value of ~0.5 found for critically-ill patients and healthy humans [29]. PK differences observed between rodents and non-rodents, such as differences in plasma protein binding of colistin, should be considered before extrapolating efficacy results from rodents to humans after treatment by CMS.

Table 7.1 Pharmacokinetic parameters of CMS and colistin after IV CMS bolus administration in: mice (n = 36, 4 mice per time), rats (n = 6) and rabbits (n = 3) at a dose of 15 mg/kg of CMS base (~6.3 mg/kg CBA), after 1 h IV infusion of 2 MIU of CMS (~66 mg CBA) in pigs (n = 2), after 10 min IV infusion of 2.5 mg/kg on average of CMS (~1.0 mg/kg CBA) in baboons (n = 3) and after 1 h infusion of 1 MIU of CMS equivalent to 80 mg of sodium CMS (~33 mg CBA) in healthy human volunteers

	Pharmacokinetics parameter			
Species	CL_{CMS} (mL/min/kg)	$V_{ss,CMS}$ (mL/kg)	$t_{1/2,CMS}$ (min)	$t_{1/2,coli}$ (min)
Mice[a]	22.2	1032	33.0	33.0
Rats[b]	12.9 ± 2.5	333 ± 42	24.7 ± 3.7	32.4 ± 5.0
Rabbits	2.5 ± 0.13	133 ± 1.6	43.0 ± 3.5	80.8 ± 8.3
Pigs	2.7 ± 0.8	250 ± 88	50.9 ± 4.2	129 ± 0.4
Baboons	2.6 ± 0.8	152 ± 20	38.5 ± 7.6	130 ± 15
Healthy Volunteers[c]	2.6 ± 0.3	196 ± 26	120 ± 7.2	180 ± 34

[a]estimates based on the mean of 4 mice per time
[b]previously published (21)
[c]previously published (27)

7.3 Pharmacokinetics of CMS and Colistin after Nebulization

7.3.1 Pharmacokinetics of CMS and Colistin after Nebulization in Rodents

CMS nebulization seems very appealing for the treatment of pulmonary infections. However because CMS is an inactive prodrug, it was interesting and important to characterize its pre-systemic conversion into colistin after nebulization. More specifically, it was important to estimate how much of the CMS dose is converted to colistin pre-systemically (in the lungs) and how much of the dose is directly absorbed. This question was addressed for the first time in rats and data were published in 2010 [14]. A schema for the pharmacokinetic behavior of CMS and colistin after CMS intra-tracheal administration in rats is presented in Fig. 7.4.

In order to estimate the various pharmacokinetic parameters appearing on Fig. 7.4, CMS (Sanofi-Aventis) was administered to healthy rats at a dose equal to 15 mg/kg in base (~6.3 mg CBA), either by intra-tracheal nebulization using a microsprayer aerosolizer, model IA-1B (PennCentury INC., Pennsylvania, USA) (n = 5) or intravenous administration (n = 6) [14]. In order to complete the PK analysis, it was necessary to include another group of rats (n = 6) administered directly with colistin sulfate. The subcutaneous route was selected at a dose of 1.5 mg/kg. Blood samples were collected for concentration measurements in plasma, and broncho-alveolar lavage (BAL) was conducted in extra rats (n = 14) after intra-tracheal administration for determination of CMS and colistin in ELF, after correction by urea concentration. CMS and colistin were assayed by LC-MS/MS in plasma and BAL as well as urea in BAL. A non-compartmental pharmacokinetic analysis was conducted [14]. Using the Penn Century aerosolizer, most of the nebulized dose of CMS was eventually absorbed either directly or after pre-systemic conversion into colistin. It is important to realize that in clinical practice only a small fraction of the aerosolized dose is most likely capable of reaching the absorption/infection site, which precludes direct data extrapolation from rat to human. CMS maximum plasma concentrations were much lower after nebulization than after IV administration and the peak was delayed (Fig. 7.5b). Colistin plasma concentration versus time profiles observed after CMS intravenous administration and nebulization were distinct, with a delayed peak and sustained concentrations with time after CMS nebulization. However, plasma colistin concentrations were in the same order of magnitude independent of the route of administration (Fig. 7.5). It was estimated that after intra-tracheal administration with the

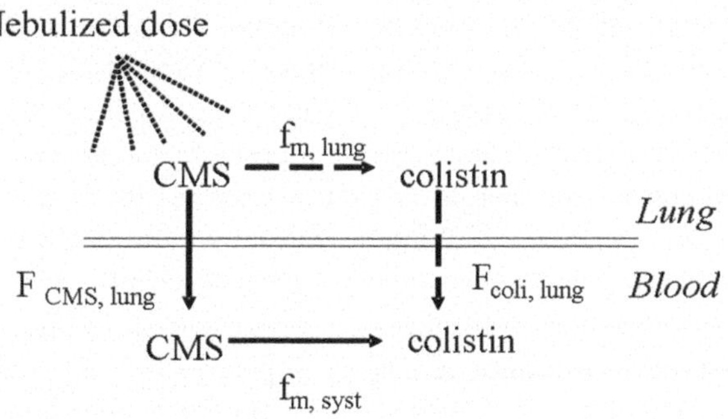

Fig. 7.4 Schema of CMS disposition after nebulization: $F_{CMS, lung}$ corresponds to the fraction of the CMS dose absorbed systematically; $f_{m,syst,}$ is the fraction of CMS converted into colistin within the systemic circulation; $f_{m,lung}$ is the fraction of the CMS dose converted into colistin pre-systematically and $F_{coli,lung}$ is the fraction of colistin which is then absorbed into the systemic circulation [14]

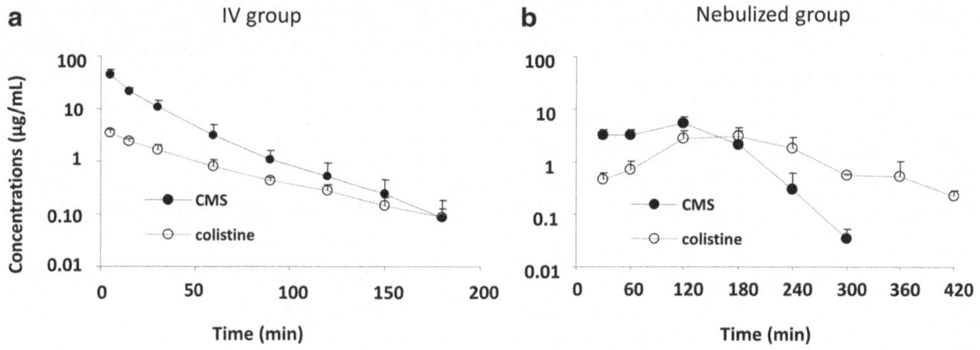

Fig. 7.5 Mean ± SD total plasma concentration-versus-time profiles of CMS and colistin after (**a**) IV administration of 15 mg/kg of CMS base (~6.3 mg/kg CBA) (n = 6), or (**b**) intra-tracheal nebulization of 15 mg/kg of CMS base (~6.3 mg/kg CBA) (n = 5). (Adapted from Marchand et al., 2010 [14])

PennCentury system, 70% of the CMS dose was directly absorbed to reach the systemic circulation and that 39% was converted pre-systematically and absorbed as colistin. These estimates reflected the experimental error and it was considered that 2/3 of the CMS was directly absorbed and 1/3 converted pre-systemically into colistin. The fraction of CMS converted into colistin after CMS intravenous administration in rats was estimated at 12.5% in the current study [14], compared with 10.2% in the previous dose-ranging study conducted by our group [21] and 6.8% in the pioneer study by Li et al. [19]. As a consequence of the relatively important CMS pre-systemic conversion in the lung, colistin systemic exposure (AUC) was about 4 times greater after CMS nebulization than after IV administration in this experimental animal model [14]. However although a significant fraction of CMS was converted into colistin pre-systemically, ELF concentrations of the active moiety were much lower (about 10 times lower at 30 and 120 min post-nebulization) than corresponding CMS concentrations, suggesting that colistin formation rate limits its absorption. In other words, colistin is rapidly absorbed after being formed within ELF, explaining that its concentrations remain always low compared with those of CMS. No modeling to better characterize these rate-limiting steps was conducted during this study.

Interestingly, these questions have been addressed in more detail by Yapa et al. [23]. In this study, CMS was provided by Link Pharmaceutical Ltd. (Auckland, New Zealand). CMS and colistin were administered to rats intra-tracheally using a 2.5 cm polyethylene tube inserted via the mouth to the tracheal carina at various doses, respectively 14 or 28 mg/kg for CMS (~5.8 and ~11.6 mg/kg CBA) and 0.41, 0.62, 0.99 and 1.49 mg/kg for colistin. They were also administered intravenously at either 14, 28 or 56 mg/kg for CMS (~5.8, ~11.6 or ~23.2 mg/kg CBA) and 0.21, 0.41 or 0.62 mg/kg for colistin. CMS and colistin concentrations were measured in plasma and BAL. Concentration versus time profiles of CMS and colistin in plasma and concentrations in ELF at predetermined sampling times after intra-tracheal administration of CMS compare favorably with those obtained by Marchand et al. at the same dose [14]. However, using the Link Pharmaceutical Ltd. CMS brand, colistin plasma concentration-time profiles after CMS IV administration, showed a delayed colistin peak as observed by He et al. [22] but not Marchand et al. [14] and Li et al. [19]. CMS and colistin concentrations in plasma and ELF were analyzed with a simultaneous population pharmacokinetic model including multi-compartments to describe a relatively complex colistin disposition within lung after intra-tracheal administration [23]. The fraction of the dose exposed to the lung after intra-tracheal administration was 40.9% for CMS and 48.5% for colistin, lower than previously described by Marchand et al. [14]. However, Yapa et al. estimated that on average 22.6% of the CMS dose was converted into colistin in ELF compared with only 3% in the systemic circulation [23], while Marchand et al. obtained respectively 39% and 10–12% for these same parameters [14, 21]. One potential explanation for the extensive CMS conversion in lungs after intra-tracheal administration, is its reduced availability for renal clearance [23]. One of the major findings in the Yapa et al. study was that intra-tracheal administration of CMS achieved much higher and sustained exposure of colistin in lungs than was possible with IV administration [23].

More recently, Gontijo et al. observed that after direct intra-tracheal administration of colistin (0.35 mg/kg), average ELF to plasma AUC ratio was equal to 1214 [30] and that colistin exposure in lung was much higher after nebulization than IV administration, confirming observations by Yapa et al. [23]. A complex absorption pattern of colistin after nebulization was again observed by Gontijo et al., but the best PK model to describe the data incorporated non-linear transfer which was further challenged by increasing the dose [30].

A recent study in neutropenic infected mice compared ELF and plasma pharmacokinetics of colistin after intra-tracheal nebulization at two doses (2.64 mg/kg and 5.28 mg/kg) versus intravenous administration (2.64 mg/kg) [31]. Whereas plasma concentrations of colistin were similar after intravenous administration and nebulization, concentrations in ELF were significantly higher than plasma concentrations after nebulization with corresponding maximum concentrations equal to 169 and 5.72 mg/L (dose of 2.64 mg/kg).

7.3.2 Pharmacokinetics of CMS and Colistin after Nebulization in Other Species

A pharmacokinetic study after CMS nebulization has been conducted in pigs [32]. In particular the aerosol delivery system used in pigs is currently used in patients and is expected to deliver much less of the dose to the pulmonary alveoli than when using the Penn Century system or direct intra-tracheal administration in rats. Furthermore, this study was conducted in piglets infected with an experimental model of pneumonia induced by *P. aeruginosa* whereas healthy rats were used previously [14, 23]. Twelve ventilated, infected piglets were included in the study, 6 received an IV infusion of CMS (3.2 mg/kg (~1.3 mg CBA) over 30 min every 8 h) and the remaining 6 received 8 mg/kg of CMS (~3.3 mg/kg CBA) over 30 min, every 12 h, by nebulization using a vibrating plate nebulizer (Aeronen Pro®, Aerogen Ltd., Galway; Ireland) [32]. Lung tissue samples were respectively obtained 1 h after the 4th intravenous infusion and the 3rd nebulization. CMS and colistin concentrations were measured in plasma and lung by HPLC [33]. After CMS IV infusions, colistin could not be detected in lung tissue whereas after aerosol delivery a median peak of colistin was estimated at 2.8 µg/g within lung tissue. The absence of colistin in lung tissue after IV administration of CMS was correlated with a lack of antimicrobial efficacy. After 24 h of treatment, 67% of pulmonary segments had bacterial counts <10^2 CFU/g following nebulization and 28% after IV administration. Therefore, these data suggest that CMS nebulization in patients could be of better value than intravenous dosing for the treatment of pulmonary infections. However measuring antibiotic concentrations in whole tissue homogenates is not recommended for reasons previously discussed [34]. ELF concentrations are probably more relevant to characterize the intrapulmonary distribution of antibiotics and predict antimicrobial efficacy against extra-cellular pathogens. Interestingly dose-normalized colistin plasma peak concentrations (Cmax) and area under concentration-time curve (AUC) were respectively 6 and 2.7 times lower after nebulization than after IV administration of CMS, suggesting that systemic side effects could be lower after CMS nebulization than IV administration [32].

Two different devices used in clinical practice, Eflow rapid® and Pari LC star® were compared by scintigraphy after nebulization of CMS at a dose of 1 MIU (corresponding approximately to 33 mg of CBA) in baboons [35]. A higher aerosol distribution into lungs was observed by imaging when nebulization was performed with a Pari LC® Star than with an Eflow Rapid® nebulizer. Accordingly, ELF concentrations simulated by PK modelling from measured plasma concentrations have confirmed a higher ELF CMS and colistin exposure after nebulization through the Pari LC® Star system than Eflow Rapid® nebulizer [35].

As partially presented above, the most recent study performed by Landersdorfer et al., has documented ELF and plasma pharmacokinetics of CMS and colistin in sheep after IV and pulmonary administration of both molecules [26]. CMS and colistin were not quantifiable in bronchoalveolar lavage fluid following intravenous CMS administration. CMS and formed colistin were not quantifiable in plasma after endotracheal nebulization of CMS at a dose of 2.6 mg/kg CBA (~6.2 mg/kg CMS sodium) whereas very high concentrations of both molecules were observed in ELF (between 1147 ± 710 mg/L and 63 ± 34 mg/L at 1 h and 24 h for CMS and between 400 ± 243 mg/L and 184 ± 190 mg/L at the same times for colistin). The therapeutic availability and the drug targeting index which characterize the targeting advantage in terms of exposure to CMS or formed colistin after nebulization were higher than 1, consequently, and in accordance with studies performed in rats [23, 26], a targeting advantage of pulmonary administration compared to intravenous administration was observed in sheep.

7.4 Pharmacokinetics of Polymyxin B in Animals

Polymyxin B and colistin are both old antibiotics used in the treatment of multidrug-resistant Gram-negative infections but not with the same clinical availability in various parts of the world. Europe and Australia have access to only colistin whereas in some other countries (*e.g.* United States, Brazil), both drugs are available. They have similar chemical structures with only one amino acid different in the ring structure [36]. However, a major difference in formulation exists since colistin is administered as an inactive prodrug (CMS), whereas polymyxin B is administered directly as its sulfate salt, which is active. Differences in terms of antibacterial concentration achievable both in plasma and in urine, in terms of toxicity but also in terms of pharmacokinetics are observable with these two molecules [36].

Due to its administration as an active form, the polymyxin B pharmacokinetics is simpler than that of colistin. The first recent modern pharmacokinetic study of polymyxin B in animals was performed in rats and relied on the measurement of four major components: polymyxin B1, isoleucine-polymyxin B1, polymyxin B2 and polymyxin B3, after IV administration of a single bolus dose equal to 4 mg/kg [37]. Since no major pharmacokinetic difference was observed between these four components, polymyxin B1, which is the most abundant component, was selected as the representative entity to describe the pharmacokinetics of polymyxin B. Polymyxin B1 pharmacokinetics was described by a one-compartment model and characterized by a clearance of 1.65 ± 0.62 mL/min and a volume of distribution of 198.1 ± 44.12 mL. Consequently the mean half-life value was estimated at 1.46 ± 0.39 h. Similar to colistin, the renal excretion of polymyxin B was negligible, with less than 1–5% of the dose recovered unchanged in urine collected up to 48 h [37, 38]. Consequently, renal insufficiency did not appear to have a significant impact on polymyxin B elimination in rats, in accordance with previous observations in patients with renal insufficiency [39, 40]. However, accumulation of polymyxin B in kidney tissue was observed, with concentrations 30 times higher in tissue than in serum in rats [41], and proximal tubular cells in the renal cortex and outer stripe of outer medulla seemed to be the main region of polymyxin accumulation [38]. Similar accumulation in renal proximal tubular cells was also observed in mice [42]. In rats, the accumulation of polymyxin B in renal tubular cells appeared to involve at least in part the transporter megalin that mediates uptake of substrates from tubular urine, and the accumulation was correlated with the onset of polymyxin B-induced nephrotoxicity [43]. Polymyxin B elimination may also occur by biliary excretion since the four major compounds of polymyxin B (B1, B2, B3 and isoleucine-polymyxin B) were detected in bile 4 hours after intravenous administration of 3 mg/kg in rats [38]. Polymyxin B concentrations were similar in muscle, heart, liver and in serum, but low concentrations were observed in brain [38]; however, whole tissue homogenate concentrations used in this study should be considered with great caution [34]. Polymyxin distribution studies in lung are limited and divergent results were found between this present study in rats which evaluated lung tissue homogenate concentrations [38] and a previous study in mice using ELF concentrations [12].

7.5 Pharmacokinetics of Novel Polymyxin-Like Compounds in Animals

Polymyxin B and colistin are used to treat serious infections caused by mutlidrug-resistant Gram-negative bacterial strains. However, both compounds are nephrotoxic which can restrict their use [44, 45]. Novel synthetic polymyxin-like antibiotics (NAB) which are potentially less toxic have been developed by Bachem AG (Budendorf, Switzerland) [46]. Among various compounds, NAB 739 and 740 present an antimicrobial activity that compares favorably with that of polymyxin B, although always slightly lower against most of the stains (*Escherichia coli* and *Klebiella pneumoniae*). Another derivative, NAB 7061, is not efficient by itself but demonstrates a strong synergism with clarithromycin and rifampicin [46, 47]. These NAB compounds present only 3 positive charges at physiological pH, compared with 5 for polymyxin B and colistin. Reducing the number of charges seems to reduce by a factor 6 to 7 the affinity of NAB compounds for isolated rat kidney brush border membrane [46] and consequently their potential renal toxicity compared with polymyxin B and colistin [48, 49]. The pharmacokinetics of NAB 739 and NAB 7061 was investigated in rats following IV bolus injection at a dose of 1 mg/kg [50]. The mean half-lives of NAB 739 and 7061 (respectively 69.0 ± 21.9 and 66.2 ± 12.3 min) were close to that of colistin (74.6 ± 13.2 min) [13] or polymyxin B (87.6 ± 23.4 min for polymyxin B1 and 79.8 ± 16.2 min for a polymyxin B2 and B3 mixture) [37]. The estimated volume of distribution of NAB 739 (222 ± 20.5 mL/kg) was slightly lower than that of NAB 7061 (339 ± 96 mL/kg), colistin (496 ± 60 mL/kg) [13] or polymyxin B (close to 800 mL/kg) [37]. Total clearance of NAB 739 (2.63 ± 0.54 mL/

min/kg) was also lower than for other polymyxins (5.22 ± 0.4 mL/min/kg for colistin and between 6.5 and 7 mL/min/kg for polymyxin B). Although higher than those of colistin or polymyxin B, urinary recoveries were still relatively low with about 20% of the dose recovered in urine for NAB 739 and 7% for NAB 7061, with corresponding renal clearances estimated at 0.53 ± 0.30 mL/min/kg and 0.28 ± 0.16 mL/min/kg. These renal clearance values are approximately 30–50 times higher than those of colistin or polymyxin B which in fact constitutes the major pharmacokinetic specificity of these NAB compounds.

7.6 Conclusions

Although polymyxin antibiotics have been commercialized more than 50 years ago, reliable PK studies based on chromatographic assays have only been conducted recently. As a consequence, polymyxin PK has been investigated almost simultaneously in animals and humans. This is a rather unusual situation since traditionally animal studies are called pre-clinical studies as they are conducted first and provide valuable information before conducting studies in humans. As a result of research conducted over the last decade or so important data concerning the effect of CMS dose, brand or route of administration on colistin PK, have been obtained in animals. Other important questions such as non-renal elimination mechanisms remain to be addressed in animals. PB-PK approaches that could be of potential value to predict the effect of infection and of major pathophysiological alterations observed in critically ill patients on polymyxin disposition will be developed and validated in humans before being extrapolated to patients. Last but not least, because of the absence of major between-species differences in polymyxin PK, animal models of infection may be used to better characterize polymyxin PK-PD administered alone or in combination, which remains a major issue to improve treatment efficacy, reduce toxicity and delay mutant selection.

References

1. Nation RL et al (2014) Consistent global approach on reporting of colistin doses to promote safe and effective use. Clin Infect Dis 58(1):139–141
2. Schwartz BS et al (1959) Microbiological and pharmacological studies of colistin sulfate and sodium colistinmethanesulfonate. Antibiot Annu 7:41–60
3. al-Khayyat AA, Aronson AL (1973) Pharmacologic and toxicologic studies with the polymyxins. II. Comparative pharmnacologic studies of the sulfate and methanesulfonate salts of polymyxin B and colistin in dogs. Chemotherapy 19(2):82–97
4. Ziv G, Sulman FG (1973) Passage of polymyxins from serum into milk in ewes. Am J Vet Res 34(3):317–322
5. Ziv G, Schultze WD (1982) Pharmacokinetics of polymyxin B administered via the bovine mammary gland. J Vet Pharmacol Ther 5(2):123–129
6. Ziv G, Nouws JF, van Ginneken CA (1982) The pharmacokinetics and tissue levels of polymyxin B, colistin and gentamicin in calves. J Vet Pharmacol Ther 5(1):45–58
7. Renard L, Sanders P, Laurentie M (1991) Pharmacokinetics of colistin sulfate administered by intravenous and intramuscular routes in the calf. Ann Rech Vet 22(4):387–394
8. Lin B, Zhang C, Xiao X (2005) Toxicity, bioavailability and pharmacokinetics of a newly formulated colistin sulfate solution. J Vet Pharmacol Ther 28(4):349–354
9. Li J et al (2005) Evaluation of colistin as an agent against multi-resistant Gram-negative bacteria. Int J Antimicrob Agents 25(1):11–25
10. Li J et al (2001) A simple method for the assay of colistin in human plasma, using pre-column derivatization with 9-fluorenylmethyl chloroformate in solid-phase extraction cartridges and reversed-phase high-performance liquid chromatography. J Chromatogr B Biomed Sci Appl 761(2):167–175
11. Gobin P et al (2010) Assay of colistin and colistin methanesulfonate in plasma and urine by liquid chromatography tandem mass spectrometry (LC-MS/MS). Antimicrob Agents Chemother 54:1941–1948
12. He J et al (2013) A validated ultra-performance liquid chromatography-tandem mass spectrometry

method for the quantification of polymyxin B in mouse serum and epithelial lining fluid: application to pharmacokinetic studies. J Antimicrob Chemother 68(5):1104–1110
13. Li J et al (2003) Use of high-performance liquid chromatography to study the pharmacokinetics of colistin sulfate in rats following intravenous administration. Antimicrob Agents Chemother 47(5):1766–1770
14. Marchand S et al (2010) Aerosol therapy with colistin methanesulfonate: a biopharmaceutical issue illustrated in rats. Antimicrob Agents Chemother 54(9):3702–3707
15. Davies B, Morris T (1993) Physiological parameters in laboratory animals and humans. Pharm Res 10(7):1093–1095
16. Ma Z et al (2009) Renal disposition of colistin in the isolated perfused rat kidney. Antimicrob Agents Chemother 53(7):2857–2864
17. Lu X et al (2016) Human oligopeptide transporter 2 (PEPT2) mediates cellular uptake of polymyxins. J Antimicrob Chemother 71(2):403–412
18. Bouchene S et al (2013) Comparison of colistin and colistimethate sodium (CMS) model-predicted whole-body distribution with measured tissue:plasma concentrations ratios in rats. In: 53rd interscience conference on antimicrobial agents and chemotherapy, Denver, CO
19. Li J et al (2004) Pharmacokinetics of colistin methanesulphonate and colistin in rats following an intravenous dose of colistin methanesulphonate. J Antimicrob Chemother 53(5):837–840
20. Li J et al (2002) Simple method for assaying colistin methanesulfonate in plasma and urine using high-performance liquid chromatography. Antimicrob Agents Chemother 46(10):3304–3307
21. Marchand S et al (2010) Dose-ranging pharmacokinetics of colistin methanesulphonate (CMS) and colistin in rats following single intravenous CMS doses. J Antimicrob Chemother 65(8):1753–1758
22. He H et al (2013) Pharmacokinetics of four different brands of colistimethate and formed colistin in rats. J Antimicrob Chemother 68(10):2311–2317
23. Yapa SWS et al (2013) Population pharmacokinetics of colistin methanesulfonate in rats: achieving sustained lung concentrations of colistin for targeting respiratory infections. Antimicrob Agents Chemother 57(10):5087–5095
24. Jin L et al (2011) Impact of p-glycoprotein inhibition and lipopolysaccharide administration on blood-brain barrier transport of colistin in mice. Antimicrob Agents Chemother 55(2):502–507
25. Bouchene S et al (2018) A whole-body physiologically based pharmacokinetic model for Colistin and Colistin Methanesulfonate in rat. Basic Clin Pharmacol Toxicol 123(4):407–422
26. Landersdorfer CB et al (2017) Substantial targeting advantage achieved by pulmonary administration of colistin methanesulfonate in a large-animal model. Antimicrob Agents Chemother 61(1):e01934-16
27. Couet W et al (2011) Pharmacokinetics of colistin and colistimethate sodium after a single 80-mg intravenous dose of CMS in young healthy volunteers. Clin Pharmacol Ther 89(6):875–879
28. Viel A et al (2018) A population WB-PBPK model of colistin and its prodrug CMS in pigs: focus on the renal distribution and excretion. Pharm Res 35(5):92
29. Cheah SE et al (2015) New pharmacokinetic/pharmacodynamic studies of systemically administered colistin against Pseudomonas aeruginosa and ACINETOBACTER baumannii in mouse thigh and lung infection models: smaller response in lung infection. J Antimicrob Chemother 70(12):3291–3297
30. Gontijo AV et al (2014) Biopharmaceutical characterization of nebulized antimicrobial agents in rats. 2. Colistin. Antimicrob Agents Chemother 58(7):3950–3956
31. Lin YW et al (2016) Pharmacokinetics/pharmacodynamics of pulmonary delivery of colistin against Pseudomonas aeruginosa in a mouse lung infection model. Antimicrob Agents Chemother 61(3):e02025-16
32. Lu Q et al (2010) Nebulized and intravenous colistin in experimental pneumonia caused by Pseudomonas aeruginosa. Intensive Care Med 36(7):1147–1155
33. Le Brun PP, de Graaf AI, Vinks AA (2000) High-performance liquid chromatographic method for the determination of colistin in serum. Ther Drug Monit 22(5):589–593
34. Mouton JW et al (2008) Tissue concentrations: do we ever learn? J Antimicrob Chemother 61(2):235–237
35. Marchand S et al (2015) Pharmacokinetics of Colistin Methansulphonate (CMS) and Colistin after CMS Nebulisation in Baboon Monkeys. Pharm Res 32(10):3403–3414
36. Nation RL, Velkov T, Li J (2014) Colistin and polymyxin B: peas in a pod, or chalk and cheese? Clin Infect Dis 59(1):88–94
37. Abdelraouf K et al (2012) Pharmacokinetics and renal disposition of polymyxin B in an animal model. Antimicrob Agents Chemother 56(11):5724–5727
38. Manchandani P et al (2016) Characterization of polymyxin B biodistribution and disposition in an animal model. Antimicrob Agents Chemother 60(2):1029–1034
39. Kwa AL et al (2011) Pharmacokinetics of polymyxin B in a patient with renal insufficiency: a case report. Clin Infect Dis 52(10):1280–1281
40. Zavascki AP et al (2008) Pharmacokinetics of intravenous polymyxin B in critically ill patients. Clin Infect Dis 47(10):1298–1304
41. Abdelraouf K et al (2012) Characterization of polymyxin B-induced nephrotoxicity: implications for dosing regimen design. Antimicrob Agents Chemother 56(9):4625–4629
42. Yun B et al (2015) Imaging the distribution of polymyxins in the kidney. J Antimicrob Chemother 70(3):827–829

43. Manchandani P et al (2017) Role of renal drug exposure in polymyxin B-induced nephrotoxicity. Antimicrob Agents Chemother 61(4):e02391-16
44. Oliveira MS et al (2009) Polymyxin B and colistimethate are comparable as to efficacy and renal toxicity. Diagn Microbiol Infect Dis 65(4):431–434
45. Akajagbor DS et al (2013) Higher incidence of acute kidney injury with intravenous colistimethate sodium compared with polymyxin B in critically ill patients at a tertiary care medical center. Clin Infect Dis 57(9):1300–1303
46. Vaara M et al (2008) Novel polymyxin derivatives carrying only three positive charges are effective antibacterial agents. Antimicrob Agents Chemother 52(9):3229–3236
47. Vaara M et al (2010) Susceptibility of carbapenemase-producing strains of Klebsiella pneumoniae and Escherichia coli to the direct antibacterial activity of NAB739 and to the synergistic activity of NAB7061 with rifampicin and clarithromycin. J Antimicrob Chemother 65(5):942–945
48. Mingeot-Leclercq MP et al (2012) Novel polymyxin derivatives are less cytotoxic than polymyxin B to renal proximal tubular cells. Peptides 35(2):248–252
49. Vaara M, Vaara T (2013) The novel polymyxin derivative NAB739 is remarkably less cytotoxic than polymyxin B and colistin to human kidney proximal tubular cells. Int J Antimicrob Agents 41(3):292–293
50. Ali FE et al (2009) Pharmacokinetics of novel antimicrobial cationic peptides NAB 7061 and NAB 739 in rats following intravenous administration. J Antimicrob Chemother 64(5):1067–1070

In vitro Pharmacodynamics and PK/PD in Animals

8

Winnie Lee, Yiying Cai, Tze-Peng Lim, Jocelyn Teo, Sonja Courtney Chua, and Andrea Lay-Hoon Kwa

Abstract

In the last decade, considerable advancements have been made to identify the pharmacokinetic/pharmacodynamic (PK/PD) index that defines the antimicrobial activity of polymyxins. Dose-fractionation studies performed in hollow-fiber models found that altering the dosing schedule had little impact on the killing or suppression of resistance emergence, alluding to AUC/MIC as the pharmacodynamic index that best describes polymyxin's activity. For *in vivo* efficacy, the PK/PD index that was the most predictive of the antibacterial effect of colistin against *P. aeruginosa* and *A. baumannii* was *f*AUC/MIC.

Keywords

Polymyxin · *In vitro* · *In vivo* · Pharmacokinetics · Pharmacodynamics

8.1 Introduction

The knowledge of pharmacokinetics (PK) and pharmacodynamics (PD) of a drug is crucial for therapeutic efficacy and minimising side effects. This is especially true with antimicrobials [1]. In a patient, the PD characteristics (i.e. antibiotic effect) have to be considered together with PK properties [2]. Many antibiotics were developed before the modern drug development process; hence, their PK/PD relationships were lacking until recently. In particular, the polymyxin class has been devoid of such information to aid clinicians in effective dosing, thus resulting in growing resistance [3–5].

Minimum inhibitory concentration (MIC) has long been the main PD endpoint for the dosing of antibacterial drugs. It was used to be just simply selecting a dose which enabled the drug plasma concentration to be higher than the MIC for as long as possible. However, in the recent decades, *in vitro* and *in vivo* studies have increased the understanding the PK and PD relationship of antibiotics [6, 7]. These studies have enabled antibiotics to be classified according to their PK/

W. Lee · Y. Cai · T.-P. Lim · J. Teo
Singapore General Hospital, Singapore

S. C. Chua
Department of Pharmacy, Faculty of Science, National University of Singapore, Singapore

A. L.-H. Kwa (✉)
Singapore General Hospital, Singapore

Department of Pharmacy, Faculty of Science, National University of Singapore, Singapore

Emerging Infectious Diseases, Duke-National University of Singapore Medical School, Singapore
e-mail: andrea.kwa@duke-nus.edu.sg

J. Li et al. (eds.), *Polymyxin Antibiotics: From Laboratory Bench to Bedside*, Advances in Experimental Medicine and Biology 1145, https://doi.org/10.1007/978-3-030-16373-0_8

PD indices which correlate to their antibacterial activity [1]. Examples of PK/PD indices are: i) fC_{max}/MIC which is the ratio of maximum unbound drug concentration to the MIC, ii) *f*AUC/MIC is the area under the unbound drug concentration curve to the MIC, and iii) $fT_{>max}$ which is the percentage of 24 h period that the unbound drug concentrations exceeds the MIC [6, 7]. The importance of PK/PD indices at the end of the day is to enable the proper identification of effective dosage regimens and to mitigate any potential risks that may occur [8].

It is only recently that research has looked into the PK/PD indices of polymyxins against various organisms, mostly using colistin. A number of recent *in vitro* and *in vivo* studies have found *f*AUC/MIC best correlates the pharmacodynamic properties of colistin [9–12]. For polymyxin B, it can be assumed that they follow similar pharmacodynamics profiles [13]. However, there has been little research done for polymyxin B, possibly because colistin is more widely used than polymyxin B.

This chapter has compiled most of the research that has been done to elucidate the PK/PD indices for colistin and polymyxin B. The results are to serve as a guide for clinicians to choose appropriate dosage regimens against specific bacterial species.

8.2 *In vitro* PD for *Pseudomonas aeruginosa*

The pharmacodynamics of the polymyxin class is most well-elucidated in *P. aeruginosa* thus far. This probably arises from the need to use more polymyxins in the treatment of multi-drug resistant Pseudomonal infections that have increased significantly in the last two decades [14, 15]. Furthermore, concerns over the drug's toxicities also prompted further studies in optimizing dosing strategies to achieve maximal antimicrobial efficacy without further increasing drug toxicity. Earlier pharmacodynamic studies on colistin were largely limited by the inability to differentiate between colistin and its inactive prodrug, colistimethate sodium (CMS). In addition, the variability in the amount of active colistin contained in the different preparations renders interpretation and application of the earlier studies difficult [15, 16].

In the last decade, considerable advancements were made to identify the PK/PD index that defines polymyxins' antimicrobial activity. One of the few *in vitro* studies performed using polymyxin B demonstrated its rapid bactericidal activity after 2 h of antibiotic exposure for standard inocula of polymyxin-susceptible *P. aeruginosa* (MIC: 0.5–1.0 mg/L) [17]. With increasing polymyxin B concentrations, the rate and extent of bacterial killing was greater. This concentration-dependent kill was also observed when tested against higher inocula (1×10^7 CFU/mL) although killing was reduced, suggesting that polymyxins have inoculum effect. Further dose-fractionation studies done with two of these strains in a hollow-fiber model found that altering the dosing schedule had little impact on the killing or suppression of resistance emergence, alluding to AUC/MIC as the pharmacodynamic index that best describes the activity of polymyxins.

In contrast to polymyxin B, more studies had been done on colistin in the past decade. Nonetheless, similar observations were made when a reference strain (PAO1; MIC 4 mg/L) and two clinical strains (MIC 1 mg/L) of *P. aeruginosa* were exposed to increasing concentrations (up to 64 times MIC) of colistin in a 24-hour time-kill experiment [18]. At standard inoculum (1×10^6 CFU/mL), up to 4 times MIC resulted in killing within 1 h and colistin concentrations more than 16 times MIC resulted in undetected CFU within 30 min. This concentration-dependent killing characteristic was replicated in several other *in vitro* PK/PD studies on colistin, using various reference and clinical strains with diverse MICs (0.5 mg/L to ≥128 mg/L) [19–23]. Not surprisingly, against colistin-non-susceptible strains, this concentration-killing profile was only noticeable when colistin was used in combination with another antibiotic [22, 23]. Most of these earlier studies employed the 24-hour one-compartment PK/PD model, with the colistin concentrations kept either constant or fluctuated to simulate its pharmacokinetic profile in humans [21].

Given the predisposition to biofilm formation in Pseudomonal infections, an *in vitro* PK/PD experiment was done on mucoid and non-mucoid biofilm-producing *P. aeruginosa* and reported that concentration-dependent activity of colistin was retained regardless. However, higher doses and longer antibiotic exposure times were needed as compared to those used for planktonic bacteria [24]. In addition, the authors also noted that mature biofilms were more resistant than young biofilms, leading to longer time to maximum kill activity (≥ 8 h versus 4 h). In a separate biofilm model study using CDC biofilm reactor, colistin monotherapy at 3.5 mg/L achieved greater and more rapid killing of biofilm-embedded bacteria compared to 1.25 mg/L. [25]

P. aeruginosa is also known to invade epithelial and phagocytic cells and hence a 24-hour *in vitro* model of THP-1 human monocytes was developed to evaluate antibiotic activity against intracellular infection [26]. From this study, it was noted that colistin exhibited concentration-dependent killing intracellularly and extracellularly, with the extent of the dose response significantly smaller against intracellular bacteria.

Quantification of the concentration-dependent activity of colistin was done using a one-compartment *in vitro* PK/PD model over 24 hours with standard inocula of ATCC reference strains (27853 and PAO1) as well as a mucoid multidrug resistant clinical strain (MIC 0.5 mg/L) [12]. Eight different dosing regimens were simulated in this time-kill model to maximally differentiate among the PK/PD indices. Early dose-dependent killing was again noted for all three strains; *f*AUC/MIC showed the best correlation with the observed killing effect ($R^2 = 0.931$) as opposed to $fT_{>MIC}$ ($R^2 = 0.785$) and fC_{max}/MIC ($R^2 = 0.868$). Additionally, *f*AUC/MIC of approximately 40, 50 and 9 achieved near maximal killing for the three strains tested.

This result corroborates with an earlier study by the same authors examining the impact of once-, twice- and thrice-daily dosage regimens of colistin, corresponding to 5 mg/kg/day, on its killing activity [27]. This dose-fractionation study found that there was no difference in bacterial killing among the different regimens when the maximum recommended daily dose was administered, strongly supporting that C_{max}/MIC is less well-correlated to killing activity than overall daily drug exposure (*f*AUC/MIC).

Similar to polymyxin B, colistin showed inoculum effect, manifested as inhibition of killing, when added to high inocula (1×10^8–1×10^9 CFU/mL) [18]. Up to 32 times MIC concentrations were required at 1×10^9 CFU/ml to achieve bactericidal activity. At high inocula, killing was also slower with nadir CFU counts at 8–12 h for PAO1. Killing rate constant was 23-fold smaller for 1×10^9 CFU/mL and sixfold small for 1×10^8 CFU/ml vs standard inoculum. The impact of inoculum load on the killing activity of colistin is well described from most *in vitro* PK/PD time-kill studies [18, 20–23]. In one study, colistin monotherapy achieved no appreciable killing within the first 6 h for colistin-susceptible isolates at a high inoculum (1×10^8 CFU/mL). The inoculum effect of colistin was postulated to be more likely due to phenotypic changes in the bacterial cells as a result of the production of hypothetical unmeasured signal molecules i.e. quorum sensing. Another hypothesis was that colistin forms mixed monolayers with phospholipids and is incorporated in micelles *in vitro*. Binding of colistin to lipopolysaccharides of killed bacteria may decrease free drug concentrations *in vitro* and thus potentiate its inoculum effect [18].

In addition, colistin was observed to possess post-antibiotic effect (PAE, between 2 and 3 h) against *P. aeruginosa*, but only at very high concentrations which are not clinically achievable after intravenous administration [19, 28].

8.3 Resistance

The phenomenon of heteroresistance (the presence of colistin-resistant subpopulations in an isolate considered susceptible by MIC breakpoint) is well studied. Heteroresistance very likely contributes to the emergence of colistin resistance, often reflected as regrowth after initial decrease in log count in *in vitro* time-kill studies [17, 18, 21–23].

When exposed to colistin monotherapy, real-time population analysis profiles (PAPs) performed at 48 h in time-kill studies demonstrated increases in colistin-resistant subpopulations at standard and high inocula. Interestingly, combining colistin with another antibiotic had minimal impact on the proportion of colistin-resistant subpopulations [22]. Another study evaluating the impact of colistin on emergency of resistance also noted that substantial increases in the proportion of colistin-resistant subpopulations at standard and high inocula and regardless of colistin dose used. However, combining colistin with doripenem resulted in significantly less colonies growing in the presence of ≥4 mg/L colistin at 96 h as compared to colistin monotherapy [23].

In an earlier study on polymyxins, standard clinical dosing of polymyxin B with standard inocula led to regrowth and emergence of resistance after 24 h in both wild-type and mutant MDR strains [17]. Interestingly, when exposed to a higher dose (8 times the clinical dose), a sustained reduction in total bacterial burden and suppression of the resistant subpopulation were achieved over 96 h for the wild-type isolate but not for the MDR isolate. Resistant isolates that emerged were found to have 8- to 16-fold rise in MIC compared to their parent strains. Subsequent serial passaging on drug-free media over 20 days led to a reversal in their susceptibility to polymyxins, which is suggestive of adaptive resistance as opposed to mutational resistance.

8.4 Rate and Extent of Bacterial Kill Against *Acinetobacter baumannii* and Enterobacteriaceae

Against *A. baumannii* colistin had been shown to exhibit rapid concentration-dependent bacterial killing in time-kill studies [29–31]. Colistin sulfate displayed rapid bactericidal killing against clinical isolates of *A. baumannii* within 20 min at concentrations of 64 times MIC; however, regrowth was observed by 3 h and continued till 24 h [29].

In an *in vitro* pharmacodynamic model, the unbound plasma concentration-time profile of colistin achieved with 3 clinically relevant intermittent regimens and a continuous infusion (5 mg/kg of body weight/day of colistin base activity) was evaluated against a clinical and ATCC *A. baumannii* isolate. Extensive killing corresponding to ≥4 $\log_{10}$ reduction in CFU/mL was observed in all regimens within 30 min with regrowth occurring at 6 h for both isolates [31].

The emergence of carbapenem resistance in Enterobacteriaceae has also resulted in the renewed interest in polymyxins for the management of such difficult-to-treat infections. Colistin exhibited rapid and extensive bacterial killing against colistin-susceptible, as well as heteroresistant clinical isolates in a time-kill experiment utilizing varying colistin concentrations ranging from 0.5 to 64X MIC. The extent of the concentration-effect could not be discerned as the killing effect was rapid even at the lowest concentrations. Regrowth was observed as early as 4 h for some isolates and occurred even at concentrations as high as 64 times MIC, a concentration much higher than those which may be achieved clinically [32]. Colistin pharmacodynamics were similar in a one-compartment *in vitro* model utilizing clinically relevant colistin regimens which mimicked the colistin pharmacokinetic profiles achieved in critically ill patients [9]. Likewise, similar observations were noted in *in vitro* PK/PD models utilizing clinically relevant polymyxin B concentrations against *Klebsiella pneumoniae,*[33] *Escherichia coli* [34] and *Enterobacter* spp. [35] There also appears to be an inoculum effect, where lower colistin concentrations exhibited no bacterial killing when a high inoculum was used [9].

8.5 Post-antimicrobial Effect

There is considerable variability in PAEs between strains [29, 36]. Significant PAEs were observed in a study on 19 clinical *A. baumannii* isolates, which demonstrated mean PAEs of 3.90 h at 1 time MIC and 4.48 h at 4 times MIC concentrations of colistin [36]. Interestingly, only a modest PAE was observed in the ATCC 19606 reference *A. baumannii* strain at ≥16X MIC, and in fact negative PAEs were observed in clinical isolates [29]. Negligible PAE (≤0.5 h) of colistin was also observed in *K. pneumoniae* clinical isolates, even with high colistin concentrations [32].

8.6 Heteroresistance

Polymyxin heteroresistance, the existence of polymyxin-resistant subpopulations (MIC ≥4 mg/L) in an otherwise susceptible population (MIC ≤2 mg/L), has been well-documented in several *in vitro* experiments.

Substantial increases in resistant subpopulations occurred by 24 h in nearly all of the colistin-susceptible *K. pneumoniae* isolates, including the ATCC reference strain after exposure to colistin. The proportion of these populations was in the order of 6×10^{-9} to 1.3×10^{-5}. Emergence of resistance was highly variable between strains, and occurred at different concentrations for different strains [32, 37]. In polymyxin B experiments, heteroresistant Enterobacteriaceae isolates showed an increase of MICs up to 32 mg/L and there was no reversal of MICs even after 20 days of passages, suggesting that these resistant phenotypes were stable [34, 35].

Heteroresistance in *A. baumannii* was demonstrated in a population analysis profile study [38]. In spite of low colistin MICs of 1 mg/L, the reference ATCC 19606 strain and a clinical strain both contained subpopulations which had the ability to grow in the presence of 10 mg/L colistin after colistin exposure. These populations had increased MICs of up to 128 mg/L. Upon further exposure to colistin-containing media (up to 200 mg/L) through passages, the proportion of the resistant subpopulations with the ability to grow in the presence of 10 mg/L colistin increased dramatically from 0.000023% to 100%.

Paradoxical effect of polymyxin B has been reported against *A. baumannii* and high drug exposure had actually amplified the resistance [39]. The following dosage regimens were simulated for polymyxin B ($t_{1/2}$ = 8 h) in the hollow-fiber infection model: non-loading dose (1.43 mg/kg of body weight every 12 h [q12h]), loading dose (2.22 mg/kg q12h for 1 dose and then 1.43 mg/kg q12h), front-loading dose (3.33 mg/kg q12h for 1 dose followed by 1.43 mg/kg q12h), burst (5.53 mg/kg for 1 dose), and supraburst (18.4 mg/kg for 1 dose). When the dose intensity of polymyxin B against two strains of *A. baumannii* was increased, a rapid initial decline in the total population was observed within the first 6 h of polymyxin exposure. The greater the polymyxin B exposure, the greater maximal killing of −1.25, −1.43, −2.84, −2.84, and −3.40 $\log_{10}$ CFU/mL within the first 6 h. However, a paradoxical effect, whereby higher polymyxin B exposures dramatically increased resistant subpopulations that grew on agar containing up to 10 mg/L of polymyxin B over 336 h, was observed. High drug exposure also proliferated polymyxin-dependent growth. The intersecting point, where the benefit of bacterial killing was equal to the cost of resistance, was an $fAUC_{0-24}$ (area under the concentration-time curve from 0 to 24 h for the unbound fraction of drug) of 38.5 mg·h/L for polymyxin B.

8.7 Compare *in vitro* PK/PD Activity of Colistin and Polymyxin B

The pharmacokinetics of the two clinically used polymyxins, polymyxin B and colistin, differ considerably, since colistin is administered as an inactive prodrug that undergoes slow conversion to colistin. However, the impact of these substantial PK differences on bacterial killing and resistance emergence is poorly understood. Cheah et al. recently assessed clinically relevant polymyxin B and colistin dosage regimens against one reference and three clinical *A. baumannii* strains in a one-compartment PK/PD model [40]. Rapid attainment of target concentrations was shown to be critical for polymyxin-induced bacterial killing. All polymyxin B regimens achieved peak concentrations of at least 1 mg/L within 1 h and caused ≥4 log10 killing at 1 h. In contrast, the slow rise of colistin concentrations to 3 mg/L over 48 h resulted in markedly reduced bacterial killing. A significant (4–6 $\log_{10}$ CFU/mL) amplification of resistant bacterial populations was common to all dosage regimens. The results also implicated adaptive polymyxin resistance as a key driver of bacterial regrowth and predicted the amplification of preexisting, highly polymyxin-resistant bacterial populations following polymyxin treatment.

8.7.1 Animal Pharmacokinetic and Pharmacodynamic Models of Polymyxins

A wide variety of animal models have been used to characterize different properties of polymyxins [10, 41–45]. Compared to *in vitro* models, animal models have the advantage of determining polymyxin efficacy at specific body sites (e.g. lung, peritoneum, brain). In addition, different host factors, such as protein binding and interaction with host immune system, can be evaluated using animal models. A variety of factors needs to be considered when evaluating *in vivo* efficacy – factors such as type of animal model, inoculum size, site of infection, drug concentrations to measure drug exposure at site of infection, presence of neutropenia and type of outcome measures need to be carefully assessed to develop meaningful conclusions.

8.7.1.1 *In Vivo* Pharmacodynamics of Polymyxins

To date, only a handful of studies have evaluated the pharmacodynamics of polymyxins using animal models. Rapid and concentration-dependent killing was exhibited against *P. aeruginosa* and *A. baumannii* in animal infection models. These studies were mainly conducted using colistin (polymyxin E). In most studies, colistin displayed variable extent of killing against susceptible clinical isolates; however, treatment with colistin usually resulted in improved survival in different animal models compared to controls [46–48]. The activity of colistin against biofilm infections was significantly poorer than against planktonic infections; in addition, while a modest PAE was observed *in vivo* against planktonic *P. aeruginosa* cells, negligible PAE was observed against *P. aeruginosa* biofilm [41]. Similar to *in vitro* studies, regrowth was observed against *A. baumannii in vivo*; this was associated with the amplification of resistant subpopulations, suggesting caution with polymyxin monotherapy against *A. baumannii* [11].

There had been almost no systematic studies that had utilised animal infection models to study the pharmacodynamics of polymyxin B, until recently [49, 50]. However, there were a few studies that had examined the therapeutic efficacy of polymyxin B in various animal infection models.

Giacometti and colleagues examined a single dose of intraperitoneal polymyxin B alone and in combination with levofloxacin in a septic shock rat infection model with the main objective to compare with novel polymyxin-like peptides against *E. coli* ATCC 25922. They found that the treatment of polymyxin B alone resulted in 86.7% survival while the addition of levofloxacin constituted 100% survival. In addition, significant decreases in plasma endotoxin and TNF-α levels were observed in the polymyxin B treatment group when compared to the untreated control group [51].

In another study of polymyxin B against IMP-type metallo-β-lactamase-producing *P. aeruginosa*, Miyajima and colleagues compared the effects of polymyxin B, colistin and other antibiotics in a murine blood infection model. They found that polymyxin B significantly increased the survival rate and decreased the bacteria inoculum in the blood in a dose-dependent manner over a range of 5–20 mg/kg. The 20 mg/kg dose achieved a survival rate of 83% and no bacteria cells were cultured at 18 h after infection [52].

In a rabbit model simulating keratitis, polymyxin B was evaluated alone and in combination against *S. aureus* ATCC 25923, *P. aeruginosa* ATCC 27853 and a clinical strain of *S. marcescens*. Polymyxin B alone did not produce any significant reduction in bacterial burden against all 3 organisms. In combination with mupirocin, a significant reduction in bacterial burden was observed when compared to each monotherapy [53].

In a novel study, Zhai and colleagues explored the anticryptococcal activity of polymyxin B in both kidney and intranasal murine infection models. In the intranasal model, polymyxin B significantly reduced fungal burden and modestly improved animal survival. However, polymyxin B modestly reduced the fungal burden of the kidney but did not improve animal survival [54].

8.7.1.2 PK/PD Index of Polymyxins

The usual methodology to determine the most optimal PK/PD target that best predicts *in vivo* efficacy is to conduct dose fractionation studies at various dosing intervals over several total doses and correlate the resulting efficacy for each regimen with the various indices. For the polymyxins, either AUC/MIC or *f*AUC/MIC was found to best correlate with optimal microbiological outcomes [10, 11, 41, 49, 50].

Polymyxin B Lin et al., recently described the PK/PD for aerosolized polymyxin B against *P. aeruginosa* in a mouse lung infection model [50]. AUC/MIC was the most predictive PK/PD index to describe the antimicrobial efficacy of aerosolized polymyxin B in treating lung infections in mice (R^2 of 0.70–0.88 for epithelial lining fluid [ELF] and 0.70–0.87 for plasma). The AUC/MIC targets associated with bacteriostasis against the three P. aeruginosa strains were 1326–1506 in ELF and 3.14–4.03 in plasma. Histopathological results showed that polymyxin B aerosol significantly reduced lung inflammation and preserved lung epithelial integrity. In addition, this study highlights the advantageous PK/PD characteristics of pulmonary delivery of polymyxin B over intravenous administration in achieving high drug exposure in ELF.

Another study assessed the PK/PD of systemically administered polymyxin B against *K. pneumoniae* in mouse thigh and lung infection models [49]. In thigh infection, antibacterial effect was well correlated with *f*AUC/MIC (R^2 = 0.89). Target values of *f*AUC/MIC for stasis and 1-$\log_{10}$ kill were 1.22–13.5 and 3.72–28.0, respectively; 2- $\log_{10}$ kill was not achieved for any strain, even at the highest tolerated dose. There was no difference ($P > 0.05$) in antibacterial activity between polymyxin B and colistin with equimolar doses. It was not possible to achieve stasis in lung infection, even at the highest dose tolerated by mice.

Colistin In the first *in vivo* study exploring the PK/PD determinant of colistin activity, Dudhani et al. demonstrated that the PK/PD index that was the most predictive of the antibacterial effect of colistin against *P. aeruginosa* was the 24 h *f*AUC/

MIC [10]. Parallel results were observed in a subsequent similar study against *A. baumannii*. To date, only one study examined the efficacy of colistin against bacterial biofilms in an animal model [11]. In the study by Wang et al., AUC/MIC was provided the best correlation with the *in vivo* efficacy of colistin. Of note, compared to planktonic *P. aeruginosa* cells, a significantly higher AUC/MIC was required to achieve stasis and bacterial killing in biofilm cells [41].

The PK/PD of pulmonary delivery of colistin in a mouse lung infection model was recently described [55]. The PK of colistin in epithelial lining fluid and plasma was determined following intratracheal delivery of a single dose of colistin solution in neutropenic lung-infected mice; the antimicrobial efficacy of intratracheal delivery of colistin against three *P. aeruginosa* strains (ATCC 27853, PAO1, and FADDI-PA022; MIC of 1 mg/L for all strains) was also examined [55]. In both ELF and plasma, *f*AUC/MIC was the PK/PD index that best described the antimicrobial effect in mouse lung infection (R^2 = 0.60–0.84 for ELF and 0.64–0.83 for plasma). The *f*AUC/MIC targets required to achieve stasis against the three strains were 684–1050 in ELF and 2.15–3.29 in plasma. The histopathological data showed that pulmonary delivery of colistin reduced infection-caused pulmonary inflammation and preserved the integrity of the lung epithelium, although colistin introduced mild pulmonary inflammation in healthy mice. This study showed pulmonary delivery of colistin provides antimicrobial effects against MDR *P. aeruginosa* lung infections superior to those of parenteral administrations. These results provide important preclinical PK/PD information for optimization of inhaled colistin therapy.

A considerable number of studies have evaluated different magnitude of AUC/MIC or *f*AUC/MIC associated with bacterial kill in different infection sites *in vivo* [10, 11, 41, 56, 57]. The results of these studies are summarised in Table 8.1. Such information provides a basis for defining the magnitude of that target required for optimal *in vivo* activity of the antibiotic, which allows us to develop clinically relevant dosing schedules for subsequent evaluation in clinical studies.

8.8 Conclusions

For the *in vitro* efficacy, dose fractionation studies performed in hollow-fiber models found that altering the dosing schedule had little impact on the killing or suppression of resistance emergence, alluding to AUC/MIC as the PK/PD index that best describes polymyxin's activity. For *in vivo* efficacy, the PK/PD index that was the most predictive of the antibacterial effect of colistin against *P. aeruginosa* and *A. baumannii* was *f*AUC/MIC. There is considerable variability in PAE between strains and bacterial species. Aerosolized polymyxins provide antimicrobial effects against MDR *P. aeruginosa* lung infections superior to those of parenteral administrations in the *in vivo* models.

Table 8.1 Summary of studies evaluating PK/PD index of colistin using animal models

Animal model	Study organism	PK/PD index	Study findings
Neutropenic mouse thigh infection model [10]	3 strains of *P. aeruginosa* – 2 reference strains and a clinical MDR strain	*f*AUC/MIC	*f*AUC/MIC required for different bacterial killing at 24 h:
			Stasis: 8.34–17.3
			1-log_{10} kill: 15.6–22.8
			2-log_{10} kill: 27.6–36.1
			3-log_{10} kill: 53.3–66.7
Neutropenic mouse lung infection model [10]	3 strains of *P. aeruginosa* – 2 reference strains and a clinical MDR strain	*f*AUC/MIC	*f*AUC/MIC required for different bacterial killing at 24 h:
			Stasis: 4.07–6.43
			1-log_{10} kill: 12.2–16.7
			2-log_{10} kill: 36.9–45.9
			3-log_{10} kill: 105–141
Neutropenic mouse thigh infection model [11]	3 strains of *A. baumannii* – A reference strain and two clinical MDR strains (including 2 colistin heteroresistant strains)	*f*AUC/MIC	*f*AUC/MIC required for different bacterial killing at 24 h:[a]
			Stasis: 1.89–7.41
			1-log_{10} kill: 6.98–13.6
			2-log_{10} kill: 17.5–43.0
Neutropenic mouse lung infection model [11]	3 strains of *A. baumannii* – A reference strain and two clinical MDR strains (including 2 colistin hetero-resistant strains)	*f*AUC/MIC	*f*AUC/MIC required for different bacterial killing at 24 h:[a]
			Stasis: 1.57–6.52
			1-log_{10} kill: 8.18–42.1
			2-log_{10} kill: 22.5–95.0[b]
Immunocompetent mouse lung infection model [57]	4 clinical strains of MDR *A. baumannii*	AUC/MIC	An AUC/MIC of 52.84 decreased the bacterial lung concentration (log_{10} CFU/g) compared to controls at 72 h (6.82 ± 3.4 versus 10.6 ± 0.27)
Immunocompetent rabbit meningitis infection model [57]	4 clinical strains of MDR *A. baumannii*	AUC/MIC	An AUC/MIC of 83.44 resulted only in a slight reduction in median CSF bacterial concentration compared to controls at 6 h (4.3 log_{10} CFU/ml versus 5.5 log_{10} CFU/mL)
Immunocompetent mouse lung infection model [56]	2 clinical NDM-1-*producing K. pneumoniae* and 1 *E. coli* NDM strains	AUC/MIC	An AUC/MIC of 158.5 resulted only in a slight reduction in bacterial lung concentration (log10 CFU/g) compared to control at 72 h for *K. pneumoniae* (8.44 ± 2.88 versus 9.60 ± 1.19) and *E. coli* (8.39 ± 3.09 versus 10.62 ± 1.35)
Neutropenic mouse lung infection model [41]	Wild type *P. aeruginosa* (PAO1) as planktonic bacteria and biofilm bacteria	AUC/MIC	Overall, higher AUC/MIC was required to achieve stasis and bacterial kill in biofilm cells at 24 h
			Stasis: AUC/$MIC_{planktonic}$ = 167, AUC/$MIC_{biofilm}$ = 500
			1-log_{10} kill: AUC/$MIC_{planktonic}$ = 297, AUC/$MIC_{biofilm}$ = 867
			2-log_{10} kill: (AUC/$MIC_{planktonic}$ = 433, AUC/$MIC_{biofilm}$ = 1033

[a]Amplification of polymyxin-resistant *A. baumannii* subpopulations was observed in all strains with all dosing regimens

[b]2-log10 kill was not achieved for one MDR strain in the lung infection model

References

1. Nielsen EI, Cars O, Friberg LE (2011) Pharmacokinetic/pharmacodynamic (PK/PD) indices of antibiotics predicted by a semimechanistic PKPD model: a step toward model-based dose optimization. Antimicrob Agents Chemother 55(10):4619–4630
2. Holford NH, Sheiner LB (1982) Kinetics of pharmacologic response. Pharmacol Ther 16(2):143–166
3. Li J, Nation RL, Turnidge JD et al (2006) Colistin: the re-emerging antibiotic for multidrug-resistant Gram-negative bacterial infections. Lancet Infect Dis 6(9):589–601
4. Loutet SA, Mussen LE, Flannagan RS, Valvano MA (2011) A two-tier model of polymyxin B resistance in Burkholderia cenocepacia. Environ Microbiol Rep 3(2):278–285
5. Bergen PJ, Li J, Rayner CR, Nation RL (2006) Colistin methanesulfonate is an inactive prodrug of colistin against Pseudomonas aeruginosa. Antimicrob Agents Chemother 50(6):1953–1958
6. Craig WA (1998) Pharmacokinetic/pharmacodynamic parameters: rationale for antibacterial dosing of mice and men. Clin Infect Dis 26(1):1–10. quiz 11-12
7. MacGowan A, Bowker K (2002) Developments in PK/PD: optimising efficacy and prevention of resistance. A critical review of PK/PD in in vitro models. Int J Antimicrob Agents 19(4):291–298
8. Ambrose PG, Bhavnani SM, Ellis-Grosse EJ, Drusano GL (2010) Pharmacokinetic-pharmacodynamic considerations in the design of hospital-acquired or ventilator-associated bacterial pneumonia studies: look before you leap! Clin Infect Dis 51(Suppl 1):S103–S110
9. Deris ZZ, Yu HH, Davis K et al (2012) The combination of colistin and doripenem is synergistic against Klebsiella pneumoniae at multiple inocula and suppresses colistin resistance in an in vitro pharmacokinetic/pharmacodynamic model. Antimicrob Agents Chemother 56(10):5103–5112
10. Dudhani RV, Turnidge JD, Nation RL, Li J (2010) fAUC/MIC is the most predictive pharmacokinetic/pharmacodynamic index of colistin against Acinetobacter baumannii in murine thigh and lung infection models. J Antimicrob Chemother 65(9):1984–1990
11. Dudhani RV, Turnidge JD, Coulthard K et al (2010) Elucidation of the pharmacokinetic/pharmacodynamic determinant of colistin activity against Pseudomonas aeruginosa in murine thigh and lung infection models. Antimicrob Agents Chemother 54(3):1117–1124
12. Bergen PJ, Bulitta JB, Forrest A, Tsuji BT, Li J, Nation RL (2010) Pharmacokinetic/pharmacodynamic investigation of colistin against Pseudomonas aeruginosa using an in vitro model. Antimicrob Agents Chemother 54(9):3783–3789
13. Hagihara M, Housman ST, Nicolau DP, Kuti JL (2014) In vitro pharmacodynamics of polymyxin B and tigecycline alone and in combination against carbapenem-resistant Acinetobacter baumannii. Antimicrob Agents Chemother 58(2):874–879
14. Falagas ME, Kasiakou SK (2005) Colistin: the revival of polymyxins for the management of multidrug-resistant gram-negative bacterial infections. Clin Infect Dis 40(9):1333–1341
15. Nation RL, Li J (2009) Colistin in the 21st century. Curr Opin Infect Dis 22(6):535–543
16. Al-Khayyat AA, Aronson AL (1973) Pharmacologic and toxicologic studies with the polymyxins. II. Comparative pharmnacologic studies of the sulfate and methanesulfonate salts of polymyxin B and colistin in dogs. Chemotherapy 19(2):82–97
17. Tam VH, Schilling AN, Vo G et al (2005) Pharmacodynamics of polymyxin B against Pseudomonas aeruginosa. Antimicrob Agents Chemother 49(9):3624–3630
18. Bulitta JB, Yang JC, Yohonn L et al (2010) Attenuation of colistin bactericidal activity by high inoculum of Pseudomonas aeruginosa characterized by a new mechanism-based population pharmacodynamic model. Antimicrob Agents Chemother 54(5):2051–2062
19. Li J, Turnidge J, Milne R, Nation RL, Coulthard K (2001) In vitro pharmacodynamic properties of colistin and colistin methanesulfonate against Pseudomonas aeruginosa isolates from patients with cystic fibrosis. Antimicrob Agents Chemother 45(3):781–785
20. Ly NS, Yang J, Bulitta JB, Tsuji BT (2012) Impact of two-component regulatory systems PhoP-PhoQ and PmrA-PmrB on colistin pharmacodynamics in Pseudomonas aeruginosa. Antimicrob Agents Chemother 56(6):3453–3456
21. Rao GG, Ly NS, Haas CE et al (2014) New dosing strategies for an old antibiotic: pharmacodynamics of front-loaded regimens of colistin at simulated pharmacokinetics in patients with kidney or liver disease. Antimicrob Agents Chemother 58(3):1381–1388
22. Bergen PJ, Forrest A, Bulitta JB et al (2011) Clinically relevant plasma concentrations of colistin in combination with imipenem enhance pharmacodynamic activity against multidrug-resistant Pseudomonas aeruginosa at multiple inocula. Antimicrob Agents Chemother 55(11):5134–5142
23. Bergen PJ, Tsuji BT, Bulitta JB et al (2011) Synergistic killing of multidrug-resistant Pseudomonas aeruginosa at multiple inocula by colistin combined with doripenem in an in vitro pharmacokinetic/pharmacodynamic model. Antimicrob Agents Chemother 55(12):5685–5695
24. Hengzhuang W, Wu H, Ciofu O, Song Z, Hoiby N (2011) Pharmacokinetics/pharmacodynamics of colistin and imipenem on mucoid and nonmucoid Pseudomonas aeruginosa biofilms. Antimicrob Agents Chemother 55(9):4469–4474
25. Lora-Tamayo J, Murillo O, Bergen PJ et al (2014) Activity of colistin combined with doripenem at

clinically relevant concentrations against multidrug-resistant Pseudomonas aeruginosa in an in vitro dynamic biofilm model. J Antimicrob Chemother 69(9):2434–2442
26. Buyck JM, Tulkens PM, Van Bambeke F (2013) Pharmacodynamic evaluation of the intracellular activity of antibiotics towards Pseudomonas aeruginosa PAO1 in a model of THP-1 human monocytes. Antimicrob Agents Chemother 57(5):2310–2318
27. Bergen PJ, Li J, Nation RL, Turnidge JD, Coulthard K, Milne RW (2008) Comparison of once-, twice- and thrice-daily dosing of colistin on antibacterial effect and emergence of resistance: studies with Pseudomonas aeruginosa in an in vitro pharmacodynamic model. J Antimicrob Chemother 61(3):636–642
28. Bozkurt-Guzel C, Gerceker AA (2012) In vitro pharmacodynamic properties of colistin methanesulfonate and amikacin against Pseudomonas aeruginosa. Indian J Med Microbiol 30(1):34–38
29. Owen RJ, Li J, Nation RL, Spelman D (2007) In vitro pharmacodynamics of colistin against Acinetobacter baumannii clinical isolates. J Antimicrob Chemother 59(3):473–477
30. Kroeger LA, Hovde LB, Mitropoulos IF, Schafer J, Rotschafer JC (2007) Colistin methanesulfonate against multidrug-resistant Acinetobacter baumannii in an in vitro pharmacodynamic model. Antimicrob Agents Chemother 51(9):3431–3433
31. Tan CH, Li J, Nation RL (2007) Activity of colistin against heteroresistant Acinetobacter baumannii and emergence of resistance in an in vitro pharmacokinetic/pharmacodynamic model. Antimicrob Agents Chemother 51(9):3413–3415
32. Poudyal A, Howden BP, Bell JM et al (2008) In vitro pharmacodynamics of colistin against multidrug-resistant Klebsiella pneumoniae. J Antimicrob Chemother 62(6):1311–1318
33. Lim TP, Cai Y, Teo J, et al (2013) In vitro pharmacodynamics of tigecycline alone and in combination with meropenem against extreme drug-resistant *Klebsiella pneumoniae* with decreased susceptibility to polymyxin B in a hollow fibre infection model. In: 53rd interscience conference on antimicrobial agents and chemotherapy, Denver, Colorado, United States
34. Teo J, Cai Y, Lim TP, et al (2014) Addition of tigecycline to polymyxin B suppresses polymyxin B-resistant subpopulations of extreme drug-resistant *Escherichia coli* in a hollow fiber infection model. In: 54th Interscience conference on antimicrobial agents and chemotherapy, Washington, DC, United States
35. Cai Y, Lim TP, Teo J, et al (2013) In vitro pharmacodynamics of polymyxin B alone and in combination with tigecycline against extreme drug-resistant *Enterobacter cloacae* with decreased susceptibility to polymyxin B in a hollow fibre infection model. In: 23rd European congress of clinical microbiology and infectious diseases, Berlin, Germany
36. Plachouras D, Giamarellos-Bourboulis EJ, Kentepozidis N, Baziaka F, Karagianni V, Giamarellou H (2007) In vitro postantibiotic effect of colistin on multidrug-resistant Acinetobacter baumannii. Diagn Microbiol Infect Dis 57(4):419–422
37. Meletis G, Tzampaz E, Sianou E, Tzavaras I, Sofianou D (2011) Colistin heteroresistance in carbapenemase-producing Klebsiella pneumoniae. J Antimicrob Chemother 66(4):946–947
38. Li J, Rayner CR, Nation RL et al (2006) Heteroresistance to colistin in multidrug-resistant Acinetobacter baumannii. Antimicrob Agents Chemother 50(9):2946–2950
39. Tsuji BT, Landersdorfer CB, Lenhard JR et al (2016) Paradoxical effect of Polymyxin B: high drug exposure amplifies resistance in Acinetobacter baumannii. Antimicrob Agents Chemother 60(7):3913–3920
40. Cheah SE, Li J, Tsuji BT, Forrest A, Bulitta JB, Nation RL (2016) Colistin and Polymyxin B dosage regimens against Acinetobacter baumannii: differences in activity and the emergence of resistance. Antimicrob Agents Chemother 60(7):3921–3933
41. Hengzhuang W, Wu H, Ciofu O, Song Z, Hoiby N (2012) In vivo pharmacokinetics/pharmacodynamics of colistin and imipenem in Pseudomonas aeruginosa biofilm infection. Antimicrob Agents Chemother 56(5):2683–2690
42. Lu Q, Girardi C, Zhang M et al (2010) Nebulized and intravenous colistin in experimental pneumonia caused by Pseudomonas aeruginosa. Intensive Care Med 36(7):1147–1155
43. Petrosillo N, Ioannidou E, Falagas ME (2008) Colistin monotherapy vs. combination therapy: evidence from microbiological, animal and clinical studies. Clin Microbiol Infect 14(9):816–827
44. Betts JW, Phee LM, Hornsey M, Woodford N, Wareham DW (2014) In vitro and in vivo activities of tigecycline-colistin combination therapies against carbapenem-resistant enterobacteriaceae. Antimicrob Agents Chemother 58(6):3541–3546
45. Hornsey M, Phee L, Longshaw C, Wareham DW (2013) In vivo efficacy of telavancin/colistin combination therapy in a Galleria mellonella model of Acinetobacter baumannii infection. Int J Antimicrob Agents 41(3):285–287
46. Mutlu Yilmaz E, Sunbul M, Aksoy A, Yilmaz H, Guney AK, Guvenc T (2012) Efficacy of tigecycline/colistin combination in a pneumonia model caused by extensively drug-resistant Acinetobacter baumannii. Int J Antimicrob Agents 40(4):332–336
47. Pantopoulou A, Giamarellos-Bourboulis EJ, Raftogannis M et al (2007) Colistin offers prolonged survival in experimental infection by multidrug-resistant Acinetobacter baumannii: the significance of co-administration of rifampicin. Int J Antimicrob Agents 29(1):51–55
48. Song JY, Cheong HJ, Lee J, Sung AK, Kim WJ (2009) Efficacy of monotherapy and combined antibiotic therapy for carbapenem-resistant Acinetobacter baumannii pneumonia in an immunosuppressed mouse model. Int J Antimicrob Agents 33(1):33–39
49. Landersdorfer CB, Wang J, Wirth V et al (2017) Pharmacokinetics/pharmacodynamics of systemi-

cally administered polymyxin B against Klebsiella pneumoniae in mouse thigh and lung infection models. J Antimicrob Chemother 73(2):462–468
50. Lin YW, Zhou Q, Onufrak NJ et al (2017) Aerosolized Polymyxin B for treatment of respiratory tract infections: determination of pharmacokinetic-pharmacodynamic indices for aerosolized Polymyxin B against Pseudomonas aeruginosa in a mouse lung infection model. Antimicrob Agents Chemother 61(8):pii: e00211-17
51. Giacometti A, Cirioni O, Ghiselli R et al (2002) Therapeutic efficacy of intraperitoneal polymyxin B and polymyxin-like peptides alone or combined with levofloxacin in rat models of septic shock. J Antimicrob Chemother 49(1):193–196
52. Miyajima Y, Hiramatsu K, Mizukami E et al (2008) In vitro and in vivo potency of polymyxin B against IMP-type metallo-beta-lactamase-producing Pseudomonas aeruginosa. Int J Antimicrob Agents 32(5):437–440
53. Moreau JM, Conerly LL, Hume EB et al (2002) Effectiveness of mupirocin and polymyxin B in experimental Staphylococcus aureus, Pseudomonas aeruginosa, and Serratia marcescens keratitis. Cornea 21(8):807–811
54. Zhai B, Lin X (2013) Evaluation of the anticryptococcal activity of the antibiotic polymyxin B in vitro and in vivo. Int J Antimicrob Agents 41(3):250–254
55. Lin YW, Zhou QT, Cheah SE et al (2017) Pharmacokinetics/pharmacodynamics of pulmonary delivery of colistin against Pseudomonas aeruginosa in a mouse lung Infection model. Antimicrob Agents Chemother 61(3):pii: e02025-16
56. Docobo-Perez F, Nordmann P, Dominguez-Herrera J et al (2012) Efficacies of colistin and tigecycline in mice with experimental pneumonia due to NDM-1-producing strains of Klebsiella pneumoniae and Escherichia coli. Int J Antimicrob Agents 39(3):251–254
57. Pachon-Ibanez ME, Docobo-Perez F, Lopez-Rojas R et al (2010) Efficacy of rifampin and its combinations with imipenem, sulbactam, and colistin in experimental models of infection caused by imipenem-resistant Acinetobacter baumannii. Antimicrob Agents Chemother 54(3):1165–1172

Polymyxin Susceptibility Testing and Breakpoint Setting

9

John Turnidge, Katherine Sei, and Johan Mouton

Abstract

Susceptibility testing of polymyxins has been subject to intensive review and revision in recent years. A joint working group was established by the Clinical and Laboratory Standards Institute and the European Committee on Antimicrobial Susceptibility Testing to establish a reference method. Issues examined included the effects of divalent cations, binding to laboratory materials, and addition of polysorbate 80. The working group recommended the use of broth microdilution without the addition of polysorbate 80 as the reference method. Published studies have shown that other testing methods, including agar dilution, disk diffusion and gradient diffusion, have unacceptably high levels of very major errors compared to the reference method, and are not recommended for routine laboratory use. Most data were for the testing of colistin; less information was available for polymyxin B. The joint working group was also asked to consider the setting of clinical breakpoints for relevant pathogens. This task involved examination of the available pharmacokinetic-pharmacodynamic, pharmacokinetic-toxicodynamic and population clinical pharmacokinetic data. All current pharmacokinetic-pharmacodynamic targets are based on MICs generated using the reference broth dilution procedure.

Keywords

Colistin · Polymyxin B · Susceptibility testing methods · Need for stringent control of conditions · Clinical breakpoints

Given that the polymyxin class has been in clinical use for more than 50 years, it would normally be assumed that susceptibility testing and the associated breakpoints would have been adequately resolved many years ago. However, because this class fell into disuse for many years, scant attention was paid to either susceptibility testing or clinical breakpoints. The resurgence in use driven by the emergence and spread of multiresistant Gram-negative bacteria has resulted in a critical re-appraisal of susceptibility testing, the standardization thereof and clinical breakpoints. Thus, the last decade has seen a substantial upgrade in the types of data and methods to be applied in setting clinical breakpoints.

J. Turnidge (✉)
Adelaide Medical School, University of Adelaide, Adelaide, SA, Australia
e-mail: jturnidge@gmail.com

K. Sei
Microscan, West Sacramento, CA, USA

J. Mouton (deceased)
Department of Medical Microbiology and Infectious Diseases, Erasmus Medical Centre, Rotterdam, The Netherlands

J. Li et al. (eds.), *Polymyxin Antibiotics: From Laboratory Bench to Bedside*, Advances in Experimental Medicine and Biology 1145, https://doi.org/10.1007/978-3-030-16373-0_9

Importantly, a clear distinction is now being made between breakpoints that distinguish the natural population without resistance mechanisms ('wild type') and those that distinguish between a high probability of cure and a low probability of cure. The former is now defined as an epidemiological cut-off value (ECOFF), is independent of any therapeutic intervention, and involves the application of *in vitro* phenotypic data only. The latter are called "breakpoints" or "clinical breakpoints" (the words are used synonymously), and are set only when there are sufficient in vitro, animal model and human pharmacokinetic-pharmacodynamic (PK-PD) and clinical outcome data [1]. Clinical breakpoints are used in the clinical laboratory to indicate whether treatment with an antibiotic is feasible or not, provided that the dosing regimen given to the patient is adequate.

A Joint Working Group was established by the European Committee on Antimicrobial Susceptibility Testing (EUCAST) and the Clinical and Laboratory Standards Institute (CLSI) to determine the most appropriate reference method for susceptibility testing and the setting of clinical breakpoints for the polymyxins.

9.1 In Vitro Susceptibility Testing

A standard reference method for susceptibility testing of antimicrobial agents is only of recent origin and was published in 2006. The accepted reference method for MIC testing of polymyxins is broth microdilution (BMD), as described in ISO 20776-1 [2], and is by and large the same as those described by CLSI [3] and EUCAST [4] with subtle differences. The vast majority of published MIC distributions have used this method, although a variety of other methods including agar dilution and Etest® have been used as well (see below for comments). Below, a number of issues are described that have a direct effect on the MIC value of polymyxins, indicating that standardization of testing is extremely important, as the conclusions that can be drawn from the test results would otherwise be invalid and highly misleading.

9.1.1 Formulations of Test Compounds

It is important to realize that for testing of colistin (polymyxin E) the colistin sulfate salt is used as the test reagent. The parenteral methanesulfonate formulation has almost no activity on its own; it is a prodrug [5]. The use of this prodrug will therefore give misleadingly high MIC values. Since slow conversion of colistin methanesulfonate to free colistin in aqueous solution does occur, MICs will be dependent on the circumstances that result in more or less conversion [6]. There are no such issues for polymyxin B, because the injectable product and the test reagent are the same, namely the sulfate salt.

9.1.2 Effects of Components of Polymyxins

Both polymyxin B and colistin contain mixtures of components. Colistin is predominantly a mixture of colistin A (E1) and colistin B (E2), which differ only in the length of the fatty acyl tail (by one carbon) [7]. Polymyxin B is also predominantly a mixture of polymyxin B1 and polymyxin B2, with two other components accounting for around 12% [8].

9.1.2.1 Colistin

Very recent work at MicroScan Microbiology systems, Beckman Coulter Inc. in California has shown that the USP standard is predominantly colistin A, while that of the Sigma-Aldrich chemical supply company, the most widely used reagent for in vitro studies, is predominantly colistin B [Sei, personal communication]. This suggests that there are substantial differences between manufacturers in the balance between the two major components of colistin and these in turn may result in different MIC values, although recent data suggest that these differences are likely to be negligible [9].

9.1.2.2 Polymyxin B

Tam et al. [10] have shown that MICs obtained against a variety of Gram-negative bacteria using Sigma-Aldrich brand polymyxin B components do not differ significantly from those observed with the USP standard. These investigators also reported only modest differences in the MICs of polymyxin B1 and B2. This was confirmed by the more recent work of Roberts et al. [9].

9.1.3 Influence of Cations

Calcium ions were shown many years ago to reduce the in vitro activity of colistin and polymyxin B against *Pseudomonas aeruginosa* [11–13]. Magnesium and some other divalent cations (strontium, barium) share this property, at least for polymyxin B [11, 13]. Added Ca^{++} and Mg^{++} also reduced the activity of polymyxin B against several other species of *Pseudomonas* and *Stenotrophomonas maltophilia* [14]. Colistin activity has also been shown to be affected when adding Ca^{++} and Mg^{++} ions to Mueller-Hinton broth. The effect may result from the interaction of these divalent cations with the outer membrane of these target species [11, 13]. It is stated without proof that this effect is not observed with *Escherichia coli*, *Klebsiella* spp., or *Proteus* spp. [12]. Using different experimental conditions, however, Chen and Feingold [13] suggested that there was an observable effect against *E. coli*. There is no published work examining the effect of other cations likely or possibly present in Mueller-Hinton broth, such as iron, zinc or manganese which have been documented to vary between manufacturers [15] and been shown to affect the MIC for some other drugs such as tigecycline [16].

The documented concentration of calcium and magnesium ions known to abolish the bactericidal effect of polymyxin B is 2 mM [13]. This translates to concentrations of 80 and 24 mg/L, respectively. ISO 16782 stipulated final cation concentrations of 20–25 mg/L of calcium and 10–12.5 mg/L of magnesium in cation-adjusted Mueller-Hinton broth used for reference and routine susceptibility testing [17]. Based on these concentrations, partial inhibition of the antibacterial activity of polymyxins against *P. aeruginosa* can be expected [13]. ISO 16782 did not stipulate final concentrations of either of these cations in Mueller-Hinton agar [17].

As the concentrations of calcium and magnesium are controlled in cation-adjusted Mueller-Hinton broth, this should not present a problem for reference or routine broth susceptibility testing. If there is some antagonism of polymyxin action, then at least it will occur consistently. However, it could potentially affect susceptibility testing using agar dilution where calcium and magnesium concentrations are not controlled and where calcium and magnesium concentrations are known to vary between brands [15]. This may also pose a problem for disk diffusion testing.

9.1.4 Binding to Plastic and Other Materials

Recent evidence has emerged that many polycationic molecules, including colistin and polymyxin B, adhere to plastics and other surfaces. Because BMD susceptibility testing is conventionally conducted in 96-well plastic microtiter trays, concern has arisen that this adherence may have a variable and deleterious effect on the accuracy of MIC measurements. Microtiter trays can be made from polystyrene, polypropylene or polycarbonate, and can have their surface charge enhanced for tissue culture work by the method discussed below. Polystyrene is by far the most commonly used for MIC and commercial panel testing.

Binding to plastic and glass has been examined to some extent by investigators at Beckman Coulter Inc. (previously Siemens Healthcare Diagnostics), manufacturers of Microscan® panels [Sei, unpublished observations]. The most influential effect on binding of colistin to typical polystyrene trays is surface charge. This was examined by measuring MICs of *E. coli* ATCC 25922 and *P. aeruginosa* ATCC 29753 in plates exposed to no external charge, and two levels of "corona" (plasma, the fourth state of matter) which ionises the plate surface (Table 9.1). Exposure to "full corona" could shift MIC values upward by at least fourfold.

In another experiment by these investigators supported by the US Centers for Disease Control and Prevention, MICs of 12 strains of Gram-negative bacteria were compared in five testing formats. Geometric mean MICs for the different formats are shown in Table 9.2.

As part of this experiment, they also undertook assays of the well/tube contents at different colistin concentrations and demonstrated that binding was concentration-dependent and saturable (Fig. 9.1).

Table 9.1 Distributions of MIC[a] replicates according to surface charge (corona strength)

Test strain	Corona strength	Number at MIC (mg/L)			
		≤0.25	0.5	1	2
E. coli ATCC 25922	None	21			
	Half	5	6	6	4
	Full		4	12	5
P. aeruginosa ATCC 29753	None	18	2		
	Half	1	11	8	
	Full			5	15

[a]Replicates were done 6 or 7 times in 3 separate brands of Mueller-Hinton broth

Table 9.2 Effect of different surfaces on MICs of 12 bacterial strains

	Microscan BMD panels	Trek BMD panels	Trek BMD Panels + Polysorbate 80	Macrobroth glass tubes	Macrobroth polypropylene tubes
Geo mean	0.11	0.94	0.22	0.40	0.42
Range	0.06–2	0.5–4	0.06–8	0.125–4	0.125–4

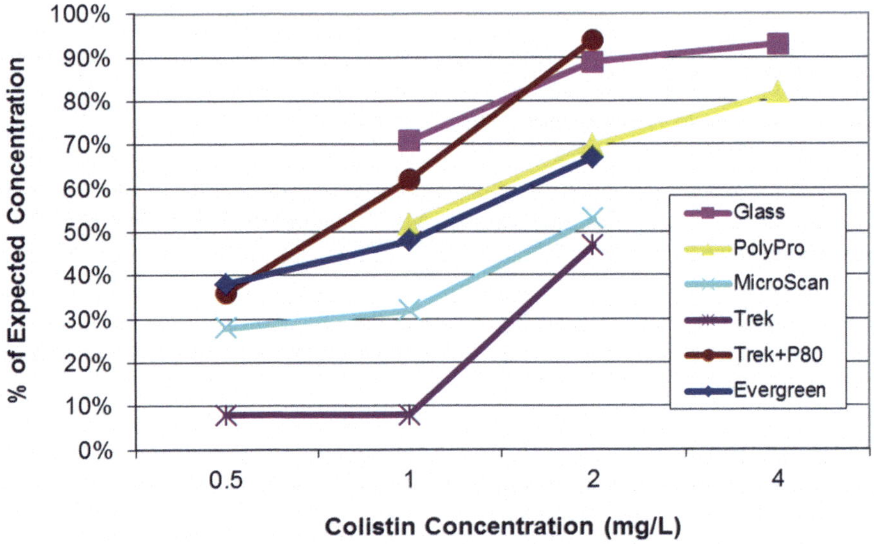

Fig. 9.1 Binding of colistin to the surface of trays and tubes made from different materials[a] and by various manufacturers[a], over a range of concentrations
[a]PolyPro = polypropylene microtiter tray; MicroScan = polystyrene in Beckman Coulter Inc. Microscan® brand tray; polystyrene in ThermoFisher's Trek brand tray; P80 = addition of polysorbate 80; Evergreen = brand of polystyrene tray

Another study from the United Kingdom examined the impact of using "tissue-culture treated" polystyrene microtiter trays (Corning brand) on colistin MICs. These trays are treated with corona discharge, resulting in a strong negative surface charge, to ensure maximum cell adhesion in cell cultures [18]. Corona treatment had a major effect on colistin binding, ranging from a 5.5-fold increase for *P. aeruginosa* to an 8.1-fold increase for Enterobacteriaceae.

Karvanen et al. examined the concentration- and time-dependent effects of binding to plastics (including polystyrene microtiter trays) and glass [19]. An exponential reduction in unbound colistin was observed, worst at the lowest concentration tested (0.125 mg/L) in glass, polypropylene and polystyrene tubes. Concentration-dependent binding was observed at 0 h and after 24 h of incubation at 37 °C; most of the loss occurred within 4–8 h (data not shown). How much this impacts on susceptibility testing is not known. Karvanen and co-workers also documented the loss of colistin during the preparation of stock solutions [19]. A drop of up to 57% was noted at the lower end of the stock solutions when prepared using the ISO-prescribed dilution method. Losses were even higher when stocks were prepared by straight serial dilution.

There is currently no information on binding to other materials such as Silastic®, rubber or other materials used in the preparation of drug stock solutions and dilutions.

9.1.5 Effect of Polysorbate 80 (P80) on 'Binding' and MICs of Polymyxins

It is known that some other antimicrobial classes have high binding to plastic and other materials, especially the lipoglycopeptides such as dalbavancin [20], and oritavancin [21]. Binding can be reduced or even eliminated by the addition of a non-ionic surfactant such as polysorbate 80 (P80, often known by one of its brand names, Tween® 80).

A number of experiments have been conducted at MicroScan Microbiology Systems, Beckman Coulter Inc. to determine the efficacy of P80 in reducing binding of colistin, the most recent of which examined this in the plates used to develop quality control ranges. P80 was included in the broth at 0.002% but not in drug dilutions. The most important features of this study were the demonstration that binding to plastic is concentration-dependent and saturable (Fig. 9.2), and influenced by brands/lots of Mueller-Hinton broth. They showed that P80 at 0.002% reduces binding but does not eliminate it.

Hindler and Humphries [22] have published a comparison of BMD MICs conducted using Evergreen brand microtiter trays with and without P80 at a concentration of 0.002%. They clearly demonstrated that the addition of P80 lowered the MICs of the 50 strains they tested, an expected effect if P80 increased free active drug in the well. They also demonstrated that the effect was dependent on the MIC in the absence of P80, with smaller differences between the MICs measured with and without P80 for less susceptible strains.

Sader et al. [23] have recently also shown the same phenomenon of reduced MICs in the presence of P80 at 0.002% for both colistin (Fig. 9.3a) and polymyxin B (Fig. 9.3b) against 124 strains of Enterobacteriaceae, 60 strains of *Acinetobacter* spp. and 63 strains of *P. aeruginosa*. Again, a concentration-dependent effect was shown.

Unpublished data have kindly been provided to the Joint EUCAST-CLSI Working Group on Polymyxins by IHMA (http://www.ihmainc.com/), the US-based company heavily involved in international surveillance programs such as the "SMART" study. They have conducted some in-house work comparing MICs generated with and without the presence of 0.002% P-80. Because of the truncation of MIC values at the lower end, most pertinent to Enterobacteriaceae, it is only possible to make general observations about their data (Table 9.3).

A recent experiment conducted at MicroScan Microbiology Systems, Beckman Coulter Inc., has thrown a cloud over all previous P80 findings, showing in fact that if anything the addition of P80 made colistin MICs higher. The only notable difference from previous experiments con-

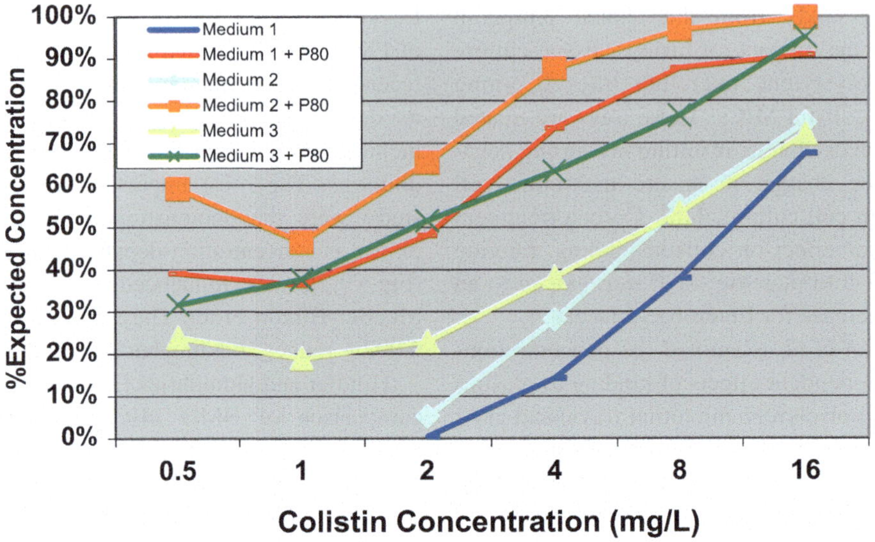

Fig. 9.2 Effect of polysorbate 80 on binding of colistin to polystyrene trays using three lots of Mueller-Hinton media

Fig. 9.3a Colistin MICs in the presence and absence of polysorbate 80 [23]

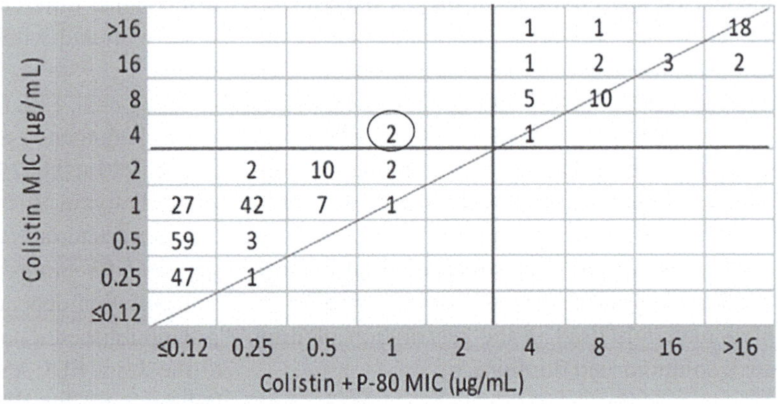

Fig. 9.3b Polymyxin B MICs in the presence and absence of polysorbate 80 [23]

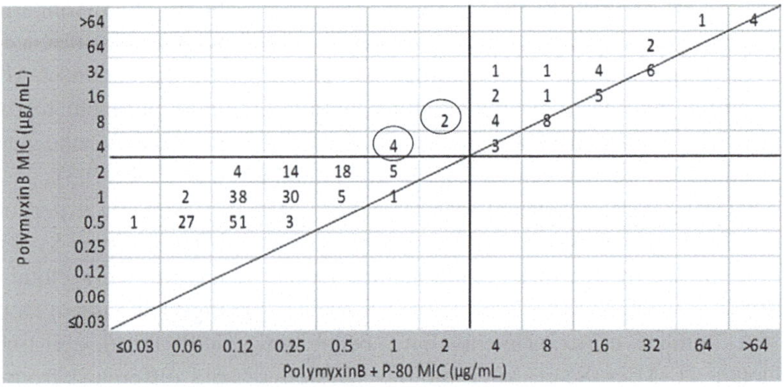

Table 9.3 Distribution of MICs with and without polysorbate 80

Species	P-80	MIC (mg/L)						
		≤0.125	0.25	0.5	1	2	4	>4
Selected Enterobacteriaceae[a]	Without	8	65	324	61	5	4	20
	With	454	2	3	4		1	18
P. aeruginosa	Without		1	5	52	98	16	
	With	54	38	39	41			
A. baumannii†	Without			44	34	8	5	4
	With	75	11	4	3		1	1

[a]*E. coli, K. pneumoniae, E. cloacae, E. aerogenes, C. freundii, C. koseri.* All other Enterobacteriaceae species had MICs >4 mg/L

Table 9.4 Effects of corona treatment and polysorbate 80 on colistin MICs determined against a range of gram-negative bacteria

Group	Geometric Mean MIC (mg/L)			
	Corona-treated without P80	Corona-treated with P80	Untreated without P80	Untreated with P80
Enterobacteriaceae (n = 29)	1.54	2.10	0.12	0.40
A. baumannii (n = 10)	2.46	3.48	0.23	0.47
P. aeruginosa (n = 11)	1.66	2.00	0.16	0.18
Other non-fermenters (n = 5)	5.28	5.28	1.74	1.74

ducted there was that the drug dilutions were dispensed into their own brand microtiter trays using stainless steel equipment and Teflon™-coated tubing, rather than Silastic® tubing. Subsequent testing suggested significant loss of colistin when run through Silastic® tubing. The effect of corona treatment was again confirmed (Table 9.4).

These findings are difficult to explain. Communications with other investigators who have generated data described above has shown that the use of Silastic® tubing for dispensing drug solutions would not explain colistin 'binding'. For instance, Hindler and Humphries [22] only added P80 to the inoculum added to the wells in the trays and not to the drug solutions.

9.1.5.1 Optimum Polysorbate 80 Concentration

There are no data on the effect of different concentrations of P80 on binding of polymyxins. However, it has been shown for dalbavancin that the optimum range of P80 concentrations, as measured by the greatest effect on lowering MICs for *S. aureus*, was 0.002–0.02% [20]. For such antimicrobials a concentration of 0.002% is the most widely used in susceptibility testing.

9.1.5.2 Micelle Formation with Polysorbate 80

P80 is known to form micelles above the critical concentration of 0.0014% [24]. Thus the typical 0.002% concentration (=20 mg/L) used in susceptibility testing is above the critical micelle concentration. How this affects drug activity is not known. Its potential importance is that micelles may sequester drug and reduce the concentration of free drug in the test system.

9.1.5.3 Quality Control Range Studies and the Influence of Polysorbate 80

At the January 2013 meeting of the CLSI Antimicrobial Susceptibility Testing Subcommittee, the results of a formal 8-laboratory

CLSI study designed to establish QC ranges for colistin and polymyxin B in BMD testing were presented [25]. The study included MIC testing in the absence and presence of P80. The important components of the study were (i) source of drug, Sigma-Aldrich; (ii) source of trays, Sarstedt brand polystyrene; (iii) source of polysorbate 80, Spectrum Chemical; (iv) addition of P80 to media, but not to drug dilutions performed before dispensing into trays; (v) dissolution and dilution of stock drugs in glass tubes, followed by filter sterilization, and dispensing into trays through Silastic® and/or rubber tubes. The inclusion of P80 resulted in ~5.5-fold reduction in MICs for *E. coli* and ~fourfold for *P. aeruginosa* (Tables 9.5 and 9.6).

The study was the first clear demonstration that Mueller-Hinton medium brand/lot could also have an impact on MICs. In 4 of 8 instances, one of the three medium lots gave significantly lower MICs than those observed with the other two medium lots (as tested by Analysis of Variance). Importantly, this was only observed with *P. aeruginosa* ATCC 27853.

Disappointingly, the addition of P80 did not reduce the overall assay variance observed in the QC study for either colistin or polymyxin B, so in this respect did not offer an advantage over performing MIC testing in the absence of P80 (Table 9.6). The QC study was performed with a single set of plastic trays of unknown surface charge.

Table 9.5 Effect of polysorbate 80 (P80) on MIC distributions from the CLSI quality control study of colistin and polymyxin B [25]

Agent	Quality control strain	Additive	MIC (mg/L)							
			0.03	0.06	0.13	0.25	0.5	1	2	4
Colistin	*E coli* ATCC 25922	P80	14	134	82	13	2			
		None			2	62	152	23	7	
	P. aeruginosa ATCC 27853	P80		12	53	155	23			
		None			1	9	93	129	11	
Polymyxin B	*E coli* ATCC 25922	P80	2	109	103	22	7			
		None				61	102	71	8	1
	P. aeruginosa ATCC 27853	P80		12	52	139	36	7		
		None				4	44	159	33	6

Table 9.6 Assay variance (SD) for QC strains from the CLSI QC study

Agent	Strain	Polysorbate 80	Geometric mean MIC (log_2)	Geometric SD (log_2)
Colistin	*E. coli* 25922	Present	−4.089	0.716
		Absent	−1.618	0.693
		Fold difference	5.54	
	P. aeruginosa 27853	Present	−2.722	0.680
		Absent	−0.924	0.660
		Fold difference	3.47	
Polymyxin B	*E. coli* 25922	Present	−3.817	0.768
		Absent	−1.381	0.837
		Fold difference	5.41	
	P. aeruginosa 27853	Present	−2.606	0.811
		Absent	−0.528	0.690
		Fold difference	4.22	

9.1.6 Adherence to Plastic or Synergy?

P80 is thought to have antibacterial properties at certain high concentrations, although concentrations of 0.05% have been shown to have no effect on the short-term viability of *P. aeruginosa* [26]. Results from the same study suggested that P80 is synergistic with polymyxin B at concentrations of P80 as low as 0.001%. However, these studies were conducted at a time when the binding to plastic and other surfaces was not appreciated, and the effect was ignored.

Recently, investigators at Microscan Microbiology Systems, Beckman Coulter Inc. undertook a novel experiment by conducting broth dilution testing in Teflon™-coated trays (mini-muffin pans for kitchen use). After confirming that there was minimal adherence of colistin to the Teflon™ surface of these trays, the investigators showed a fourfold drop in MIC in the ATCC control strains of *E. coli* and *P. aeruginosa* with the addition of 0.002% P80. This experiment was repeated and expanded in Australia using similar kitchen-use Teflon™-coated mini-muffin pans [Bell, Li and Nation, unpublished observations]. While this brand of tray did bind colistin to a small extent, about 20–30% in the presence or absence of bacteria, this amount of binding did not account for the 2- to 128-fold reduction in MICs observed with addition of 0.002% P80 to 5 of 6 strains of bacteria, the *P. aeruginosa* control strain and 2 each of clinical isolates of *K. pneumoniae* and *A. baumannii* complex.

These MIC studies confirmed that the dominant effect on MIC measurements of colistin is due to a synergistic activity of this surfactant on that antimicrobial agent. MicroScan Microbiology Systems, Beckman Coulter Inc., have shown the same effect with another non-ionic surfactant, Pluronic® P-104 [Sei, unpublished observations]. We hypothesise that the synergy of non-ionic surfactants with polymyxins is due to a direct action of the surfactants on the bacterial outer membrane.

9.1.7 Summary of Issues Relating to Broth Microdilution Testing

Colistin and polymyxin B bind to the plastics and probably other materials used in reference BMD susceptibility tests and commercial systems. The binding is dependent on surface charge, type of plastic, brand of plate, and the tubing and pipette materials used for plate preparation. The effect is concentration dependent and saturable from little binding at concentrations of 4 mg/L and higher, and up to 90% binding after 24 h at the lowest test concentrations (0.03–0.06 mg/L).

Of these factors, the most important by far appears to be surface charge, particularly the plastic in microtiter trays. The addition of P80 surfactant appears to reduce the binding of colistin and polymyxin B in most instances, but does not completely eliminate it, nor does it appear to offer a great advantage or disadvantage from the assay point of view, as it does not reduce the assay variance.

More important is the evidence that P80 synergises with colistin and polymyxin B. The effect is more potent than that of reducing binding, and called into question the value of adding P80 to MIC test systems. As a consequence, both EUCAST and CLSI have agreed that reference MIC testing of polymyxins should not include the addition of P80 or other non-ionic surfactants.

9.1.8 Disc Diffusion Testing

Soon after the introduction of polymyxins into clinical practice disc diffusion susceptibility testing was introduced because of the great popularity with the method at the time. Although still used in many laboratories, the major change over time has been that MIC microdilution testing has become the standard of susceptibility testing. The immediate consequence is that any susceptibility testing method should be referenced to the standard, including automated methods, gradient tests and disk diffusion. With respect to disk susceptibility testing of polymyxins, results have been disappointing, probably because the size and charge of the molecules results in poor diffusion through agar. Very soon after introduction of the BMD MIC method, it was shown by Matsen et al. that there was very poor correlation between disk zone size and BMD MIC [27].

In more recent years, in an extensive study involving 200 bloodstream isolates, Gales and colleagues [28] documented clearly that there is a serious problem with very major errors associated with disk diffusion testing of both colistin and polymyxin B. They concluded that clinical laboratories should exclusively use MIC methods to assist the therapeutic application of colistin or polymyxin B. These observations were later again confirmed in comparative studies by Tan and Ng [29], Lo-Ten-Foe et al. [30], Moskowitz et al. [31] and Maalej et al. [32] who all observed significant rates of false susceptibility with disc diffusion. In addition, in a recent study with 10, 25 or 50 μg colistin disks, resistant and susceptible isolates could not be reliably separated [Kahlmeter and Matuschek, personal communication]. The current view is that disk susceptibility testing is unreliable and should not be used for susceptibility testing in the clinical laboratory.

9.1.9 Gradient Diffusion Methods

Two commercial strips for gradient diffusion testing are available, those of bioMérieux (Etest®) and Liofilchem (MTS®). The Etest in particular has been compared to other testing methods by several authors and has shown conflicting results [30, 32–34], but when compared to reference BMD MIC testing, always showed significant proportions of very major errors [30, 33]. Both brands were recently compared to the standard method and both showed significant major errors in a recent comparison by the EUCAST Development Laboratory [35]. The EUCAST has placed a warning on their website in 2016 against using the gradient methods (http://www.eucast.org/ast_of_bacteria/warnings/) until such time as the manufacturers are able to address the problems (at the time of writing, one brand was still commercially available).

9.1.10 MIC Distributions and ECOFFs

Current methods to determine the susceptibility of micro-organisms are not very reproducible when compared with other clinical tests. This is due to both the inherent biological variation of micro-organisms and assay variation. MICs are normally determined using a twofold dilution series of the antimicrobial agent and the MIC distribution of the wild-type strains tested is log-normally distributed. Moreover, repeated measurement of the same strain will provide MICs that show at least a 50–100% coefficient of variation. MICs of strain collections therefore always show a log-normal distribution in the wild type, and the variation within that distribution is due to both intra- and inter-laboratory variation. MIC distributions are specific to each combination of species and antimicrobial agent. MIC distributions for a very broad range of species/antimicrobial combinations can be found at the website of EUCAST (http://mic.eucast.org/Eucast2/).

Methods have been sought to describe MIC distributions statistically, and in particular to determine whether strains are wild type or non-wildtype. This has led to the introduction of the concept of epidemiological cutoff values (ECOFFs) which are MIC values that mark the high end of the wild-type distribution.

9.1.10.1 Colistin MIC Distributions

BMD MIC distribution data and ECOFFs for colistin are on the EUCAST website: http://mic.eucast.org/Eucast2/, and were updated (February, 2016) using more stringent rules of data acceptance that are under development by EUCAST.

9.1.10.2 Polymyxin B MIC Distributions

There are few published data on MIC distributions of polymyxin B, and none currently listed in the EUCAST website. Only three publications currently provide on-scale MIC distribution data for polymyxin B by species [33, 36, 37]. The data from two of these studies, as well as data obtained from the SENTRY surveillance program [Sader, personal communication, 2015], where BMD was used, are shown in Table 9.7. Sader et al. [38] have shown that polymyxin BMD MICs tend to be higher for polymyxin B when compared directly with colistin for the wild-types of three species, *E. coli*, *K. pneumoniae* and *P. aeruginosa*.

9.2 Breakpoint Setting

Since 2000, the methods for selecting interpretive criteria (breakpoints) for susceptibility tests have undergone profound change. Prior to that time, much weight was applied to MIC distribution data, although pharmacokinetic and some pharmacodynamic data were taken into account in some European committees but not elsewhere [39]. Clinical data were used where available, although clinical trial design improved considerably after 2000. Since that time, the science of antimicrobial pharmacokinetic-pharmacodynamics (PK-PD) has come to provide a suite of tools to integrate susceptibility data with pharmacokinetics, based on knowledge of PK-PD indices (fT>MIC, $fAUC_{24}$/MIC and fC_{max}/MIC) and their respective target values associated with efficacy. PK-PD is now an integral part of the breakpoint setting standards applied by EUCAST and CLSI committees.

The process of setting clinical breakpoints involves several steps and procedures, both preclinical and clinical, as described Mouton, et al. [39]. Ideally, each step is known and taken into account when setting the clinical breakpoint, and for new drugs this information is generally available, or becomes available during the development of the drug. However, for polymyxins a substantial amount of this information is not available. At the time of registration of the polymyxins, the PK-PD of antimicrobial agents as a science did not exist and there was no reference method for susceptibility testing.

Polymyxins, like other 'old antibiotics', therefore needed redevelopment using modern standards in order to determine breakpoints. Although much information has become available in recent years, there are still many gaps that need to be filled. Below, we discuss the most important issues: the pharmacodynamic target, pharmacokinetics in patients and the modelling to determine the probability of target attainment (PTA). These processes ultimately lead to the setting of clinical breakpoints.

9.2.1 The Pharmacodynamic Target of Polymyxins

The pharmacodynamic target (PT) of an antimicrobial involves two types of studies. In the first, time-kill experiments are conducted to determine whether the drug shows primarily time-dependent

Table 9.7 Distributions of polymyxin B (sulfate) MIC

Ref	**Species**	**0.03**	**0.06**	**0.125**	**0.25**	**0.5**	**1**	**2**	**4**	**8**	**≥ 16**
a	*P. aeruginosa*				20	26	26	4		1	
Ref	**Species**		**≤ 0.5**	**1**	**2**	**4**	**8**	**16**	**32**	**64**	**>64**
b	*E. coli*		29	14	8	2	6	2			
	K. pneumoniae		39	9	2	1	0	4	6		1
	P. aeruginosa		2	29	18	4	1	2	2	1	4
Ref	**Species**			**0.125**	**0.25**	**0.5**	**1**	**2**	**4**	**8**	**≥ 16**
c	*E. coli*			4	466	4463	1340	15	20	5	
	Klebsiella spp.			1	43	2368	1561	35	30	40	100
	P. aeruginosa			3	12	95	1850	1854	4	0	3

a. van der Heijden et al. [33]
b. Vaara et al. [37]
c. Sader, personal communication, 2015

or concentration-dependent killing. Maximum kill at relatively low concentrations proceeding over time is usually associated with time-dependent effects, and efficacy thus primarily correlated with the time the concentration of the drug remains above the MIC, usually expressed as %*f*T>MIC, where "*f*" refers to the unbound fraction of drug. In contrast, increased killing as a result of increasing concentrations is usually associated with area under the time-concentration curve (AUC), most often taken over 24 h, divided by the MIC of the target organism ($f\mathrm{AUC}_{24}$/MIC). Killing curves for polymyxins show concentration-dependent killing [40], which generally predicts that bacterial killing in vivo is associated with AUC.

As suggested above, protein binding of an antimicrobial agent must be accounted for in determination of PK-PD indices. The initial experiments with protein binding indicated that it might be concentration-dependent [41]. This subsequently proved to be an artefact of the assay systems used, due to the adherence of colistin and polymyxin B to laboratory plastics and surfaces. When this process was controlled for, it was shown that protein binding was not concentration-dependent, and values were found for the percent binding in human volunteers and in infected patients, both approximately 50% [42].

Studies to determine the PK-PD indices that predict killing have been undertaken so far for colistin against *P. aeruginosa* and *Acinetobacter baumannii* in murine thigh and lung infection models [42] and against *Klebsiella pneumoniae* in an in vitro PK-PD model [43]. There is a report of a PK-PD study with polymyxin B against *P. aeruginosa* in an in vitro model [44] and a report of a study in murine thigh and lung infection models [45]. All of the above-mentioned studies indicated that the *f*AUC/MIC ratio is the PK-PD index that is most predictive of efficacy. For polymyxin B, notwithstanding the qualitative similarity in the nature of the PK-PD relationship between the two studies, there was a substantial quantitative difference. For example, over the same range of *f*AUC/MIC (exposure) values, up to six-$\log_{10}$ bacterial killing was achievable in the in vitro model [43] but less than two-$\log_{10}$ bacterial killing was possible in the murine thigh infection model and no killing was observed in the lung infection model [45]. In addition, in the latter study, there was a relatively wide range in the *f*AUC/MIC target values for stasis and one-$\log_{10}$ kill in the thigh model. Clearly, more PK-PD data are required for *K. pneumoniae* and for polymyxin B.

The most recent of the colistin studies [42], conducted in the neutropenic mouse thigh and lung models of *P. aeruginosa* and *A. baumannii* infection, established target values for $f\mathrm{AUC}_{24}$/MIC for stasis and one- and two-$\log_{10}$ killing at the site of infection. For thigh infection, a mean target $f\mathrm{AUC}_{24}$/MIC ratio of ~9 for one-$\log_{10}$ kill and ~12 for two-$\log_{10}$ kill was observed. For lung infection, target $f\mathrm{AUC}_{24}$/MIC ratios were much higher, and were considered to be unachievable in clinical practice based upon the finding that there is a substantially increased risk of nephrotoxicity in critically ill patients when the average steady-state plasma colistin concentration exceeds ~2.5 mg/L [46, 47].

9.2.2 Human Pharmacokinetics

The next step in breakpoint setting, after PK-PD targets have been established, is to choose appropriate human estimates of the pharmacokinetic parameter of interest, in this case, $f\mathrm{AUC}_{24}$. Most commonly, this is done using population PK studies, either from human volunteer studies, or preferably PK studies in patients with infections. A number of such studies have now been published [48–52]. An important feature of colistin is that the parenteral preparation is the methanesulfonate, an inactive prodrug which is cleared by the kidney and slowly broken down in plasma and tissue to the active colistin molecule. As a consequence, colistin exposure is very strongly influenced by the degree of renal function [52, 49, 51].

The most useful data for assisting in breakpoint setting that are available is from a large international multicenter trial of colistin treatment in infected intensive care patients [53, 52]. Due to the nature of the types of patients treated with colistin, the study found patients with an extremely broad range of creatinine clearances, which meant that estimates of target attainment had to be calculated across groups comprising different degrees of renal function. In order to do this, both the FDA [54] and the EMA [55] dosing recommendations for patients with different levels of renal function must be considered because they differ somewhat in their recommendations, but reflect the dosing regimens most widely used internationally. The analyses have assumed that there is 50% protein binding in humans, and that the appropriate target value for $fAUC_{24}/MIC$ is 12, based on the mouse thigh infection model study for at least two $\log_{10}$ of killing [42]. Note that the parameter chosen by Nation et al. [53, 52], namely average steady-state plasma colistin concentration of total drug, relates to $fAUC_{24}$ as follows. At 50% protein binding a $fAUC_{24}$ of 12 is equivalent to a total drug AUC_{24} of 24, which in turn is equal to an average steady-state plasma colistin concentration of total drug of 1 mg/L (i.e. 24 mg*h/L divided by 24 h). In essence, this means that the target average steady-state plasma colistin concentration is equal to the MIC of the infecting organism.

It is clear from these analyses that at a colistin MIC of 0.5 mg/L adequate target attainment (>90%) is likely to be achieved with both the FDA and the EMA dosing recommendations [53]. At a colistin MIC of 1 mg/L the attainment percentages are more variable: using FDA dosing recommendations there is low target attainment (<30%) at the lowest and highest levels of renal function, while with the EMA recommendations, low target attainment is seen only with patients having creatinine clearance >80 mL/min (<40% target attainment). For a colistin MIC of 2 mg/L, only the EMA dosing recommendations were able to achieve satisfactory target attainment for the three lowest renal function categories (i.e. patients with creatinine clearance <80 mL/min). As 2 mg/L is the epidemiological cut-off value for *A. baumannii*, the Joint CLSI-EUCAST Working Group on Polymyxins recommended a colistin breakpoint for this species of 2 mg/L, accompanied by the recommendation of using maximum recommended dose. It also led EUCAST to lower the *P. aeruginosa* breakpoint from the previous value of ≤4 mg/L, even though a small proportion of the wild-type population has an MIC of 4 mg/L. In addition, because of the less than optimum attainment for some degrees of renal function, the Working Group also recommended that there be no "Intermediate" category for the interpretive susceptibility test criteria.

In making its breakpoint recommendations to CLSI and EUCAST, the Joint Working Group was aware that these recommendations were based on murine thigh and lung infection models only, and that validation of the PK-PD targets will be required from prospective clinical studies. Furthermore, the data from the murine models would suggest that target attainment is suboptimal in pulmonary infections caused by these two species, and thus the recommended breakpoints may not apply in this setting.

9.2.3 Future Goals

There is further work to be done on breakpoint setting. For example:

- Data on clinical response rates by colistin MIC for *P. aeruginosa* and *A. baumannii* are lacking.
- More information is awaited on the PK-PD target $fAUC/MIC$ ratios for Enterobacteriaceae. In the meantime, both EUCAST and CLSI will work with an epidemiological cut-off value of 2 mg/L for this group of micro-organisms.
- There are insufficient data in all the areas for polymyxin B: MIC distributions, PK-PD target data, and human PK data.

These are all eagerly awaited.

References

1. Turnidge J, Paterson DL (2007) Setting and revising antibacterial susceptibility breakpoints. Clin Microbiol Rev 20(3):391–408. https://doi.org/10.1128/cmr.00047-06
2. ISO 20776-1 (2006) Clinical laboratory testing and in vitro diagnostic test systems – Susceptibility testing of infectious agents and evaulation of peformance of antimicrobial susceptibility test devices. Part 1: Reference method for testing the in vitro activity of antimicrobial agents against rapidly growing aerobic bacteria involved in infectious diseases. Internation Organization for Standardization, Geneva
3. Clinical and Laboratory Standards Institute (CLSI) (2015) Methods for dilution antimicrobial susceptibility tests for bacteria that grow aerobically; Approved Standard, 10th edn. Clinical and Laboratory Standards Institute, Wayne
4. EUCAST DISCUSSION DOCUMENT E. Dis 5.1 Determination of minimum inhibitory concentrations (MICs) of antibacterial agents by broth dilution (2003). Clin Microbiol Infect 9:1–7
5. Bergen PJ, Li J, Rayner CR, Nation RL (2006) Colistin methanesulfonate is an inactive prodrug of colistin against Pseudomonas aeruginosa. Antimicrob Agents Chemother 50(6):1953–1958. https://doi.org/10.1128/AAC.00035-06
6. Li J, Milne RW, Nation RL, Turnidge JD, Coulthard K (2003) Stability of colistin and colistin methanesulfonate in aqueous media and plasma as determined by high-performance liquid chromatography. Antimicrob Agents Chemother 47(4):1364–1370
7. Orwa JA, Govaerts C, Busson R, Roets E, Van Schepdael A, Hoogmartens J (2001) Isolation and structural characterization of colistin components. J Antibiot (Tokyo) 54(7):595–599
8. He J, Ledesma KR, Lam WY, Figueroa DA, Lim TP, Chow DS, Tam VH (2010) Variability of polymyxin B major components in commercial formulations. Int J Antimicrob Agents 35(3):308–310. https://doi.org/10.1016/j.ijantimicag.2009.11.005
9. Roberts KD, Azad MA, Wang J, Horne AS, Thompson PE, Nation RL, Velkov T, Li J (2015) Antimicrobial activity and toxicity of the major lipopeptide components of polymyxin B and colistin: last-line antibiotics against multidrug-resistant gram-negative bacteria. ACS Infect Dis 1(11):568–575. https://doi.org/10.1021/acsinfecdis.5b00085
10. Tam VH, Cao H, Ledesma KR, Hu M (2011) In vitro potency of various polymyxin B components. Antimicrob Agents Chemother 55(9):4490–4491. https://doi.org/10.1128/AAC.00119-11
11. Newton BA (1954) Site of action of polymyxin on Pseudomonas aeruginosa: antagonism by cations. J Gen Microbiol 10(3):491–499
12. Davis SD, Iannetta A, Wedgwood RJ (1971) Activity of colistin against Pseudomonas aeruginosa: inhibition by calcium. J Infect Dis 124(6):610–612
13. Chen CC, Feingold DS (1972) Locus of divalent cation inhibition of the bactericidal action of polymyxin B. Antimicrob Agents Chemother 2(5):331–335
14. D'Amato RF, Thornsberry C, Baker CN, Kirven LA (1975) Effect of calcium and magnesium ions on the susceptibility of Pseudomonas species to tetracycline, gentamicin polymyxin B, and carbenicillin. Antimicrob Agents Chemother 7(5):596–600
15. Girardello R, Bispo PJ, Yamanaka TM, Gales AC (2012) Cation concentration variability of four distinct Mueller-Hinton agar brands influences polymyxin B susceptibility results. J Clin Microbiol 50(7):2414–2418. https://doi.org/10.1128/JCM.06686-11
16. Veenemans J, Mouton JW, Kluytmans JA, Donnely R, Verhulst C, van Keulen PH (2012) Effect of manganese in test media on in vitro susceptibility of Enterobacteriaceae and Acinetobacter baumannii to tigecycline. J Clin Microbiol 50(9):3077–3079. https://doi.org/10.1128/jcm.01485-12
17. ISO/TS 16782 (2016) Clinical laboratory testing -- Criteria for acceptable lots of dehydrated Mueller-Hinton agar and broth for antimicrobial susceptibility testing. International Organization for Standardization, Geneva
18. Albur M, Noel A, Bowker K, Macgowan A (2014) Colistin susceptibility testing: time for a review. J Antimicrob Chemother 69(5):1432–1434. https://doi.org/10.1093/jac/dkt503
19. Karvanen M, Malmberg C, Lagerback P, Friberg LE, Cars O (2017) Colistin is extensively lost during standard in vitro experimental conditions. Antimicrob Agents Chemother 61(11). https://doi.org/10.1128/aac.00857-17
20. Rennie RP, Koeth L, Jones RN, Fritsche TR, Knapp CC, Killian SB, Goldstein BP (2007) Factors influencing broth microdilution antimicrobial susceptibility test results for dalbavancin, a new glycopeptide agent. J Clin Microbiol 45(10):3151–3154. https://doi.org/10.1128/jcm.02411-06
21. Arhin FF, Sarmiento I, Belley A, McKay GA, Draghi DC, Grover P, Sahm DF, Parr TR Jr, Moeck G (2008) Effect of polysorbate 80 on oritavancin binding to plastic surfaces: implications for susceptibility testing. Antimicrob Agents Chemother 52(5):1597–1603. https://doi.org/10.1128/aac.01513-07
22. Hindler JA, Humphries RM (2013) Colistin MIC variability by method for contemporary clinical isolates of multidrug-resistant Gram-negative bacilli. J Clin Microbiol 51(6):1678–1684. https://doi.org/10.1128/JCM.03385-12
23. Sader HS, Rhomberg PR, Flamm RK, Jones RN (2012) Use of a surfactant (polysorbate 80) to improve MIC susceptibility testing results for polymyxin B and colistin. Diagn Microbiol Infect Dis 74(4):412–414. https://doi.org/10.1016/j.diagmicrobio.2012.08.025
24. Wan LS, Lee PF (1974) CMC of polysorbates. J Pharm Sci 63(1):136–137
25. Brown S, Traczewski M (2013) Quality control limits for microdilution susceptibility tests of colistin against Escherichia coli ATCC 25922 and Pseudomonas

aeruginosa ATCC 278563. Clinical and Laboratory Standards Institute
26. Brown MR, Winsley BE (1971) Synergism between polymyxin and polysorbate 80 against Pseudomonas aeruginosa. J Gen Microbiol 68(3):367–373
27. Matsen JM, Koepcke MJ, Quie PG (1969) Evaluation of the Bauer-Kirby-Sherris-Turck single-disc diffusion method of antibiotic susceptibility testing. Antimicrob Agents Chemother (Bethesda) 9:445–453
28. Gales AC, Reis AO, Jones RN (2001) Contemporary assessment of antimicrobial susceptibility testing methods for polymyxin B and colistin: review of available interpretative criteria and quality control guidelines. J Clin Microbiol 39(1):183–190. https://doi.org/10.1128/JCM.39.1.183-190.2001
29. Tan TY, Ng LS (2006) Comparison of three standardized disc susceptibility testing methods for colistin. J Antimicrob Chemother 58(4):864–867. https://doi.org/10.1093/jac/dkl330
30. Lo-Ten-Foe JR, de Smet AM, Diederen BM, Kluytmans JA, van Keulen PH (2007) Comparative evaluation of the VITEK 2, disk diffusion, etest, broth microdilution, and agar dilution susceptibility testing methods for colistin in clinical isolates, including heteroresistant Enterobacter cloacae and Acinetobacter baumannii strains. Antimicrob Agents Chemother 51(10):3726–3730. https://doi.org/10.1128/AAC.01406-06
31. Moskowitz SM, Garber E, Chen Y, Clock SA, Tabibi S, Miller AK, Doctor M, Saiman L (2010) Colistin susceptibility testing: evaluation of reliability for cystic fibrosis isolates of Pseudomonas aeruginosa and Stenotrophomonas maltophilia. J Antimicrob Chemother 65(7):1416–1423. https://doi.org/10.1093/jac/dkq131
32. Maalej SM, Meziou MR, Rhimi FM, Hammami A (2011) Comparison of disc diffusion, Etest and agar dilution for susceptibility testing of colistin against Enterobacteriaceae. Lett Appl Microbiol 53(5):546–551. https://doi.org/10.1111/j.1472-765X.2011.03145.x
33. van der Heijden IM, Levin AS, De Pedri EH, Fung L, Rossi F, Duboc G, Barone AA, Costa SF (2007) Comparison of disc diffusion, Etest and broth microdilution for testing susceptibility of carbapenem-resistant P. Aeruginosa to polymyxins. Ann Clin Microbiol Antimicrob 6:8. https://doi.org/10.1186/1476-0711-6-8
34. Tan TY, Ng SY (2007) Comparison of Etest, Vitek and agar dilution for susceptibility testing of colistin. Clin Microbiol Infect 13(5):541–544. https://doi.org/10.1111/j.1469-0691.2007.01708.x
35. Matuschek E, Ahman J, Webster C, Kahlmeter G (2018) Antimicrobial susceptibility testing of colistin – evaluation of seven commercial MIC products against standard broth microdilution for Escherichia coli, Klebsiella pneumoniae, Pseudomonas aeruginosa, and Acinetobacter spp. Clin Microbiol Infect 24(8):865–870. https://doi.org/10.1016/j.cmi.2017.11.020
36. Hogardt M, Schmoldt S, Gotzfried M, Adler K, Heesemann J (2004) Pitfalls of polymyxin antimicrobial susceptibility testing of Pseudomonas aeruginosa isolated from cystic fibrosis patients. J Antimicrob Chemother 54(6):1057–1061. https://doi.org/10.1093/jac/dkh470
37. Vaara M, Sader HS, Rhomberg PR, Jones RN, Vaara T (2013) Antimicrobial activity of the novel polymyxin derivative NAB739 tested against Gram-negative pathogens. J Antimicrob Chemother 68(3):636–639. https://doi.org/10.1093/jac/dks438
38. Sader HS, Rhomberg PR, Farrell DJ, Jones RN (2015) Differences in potency and categorical agreement between colistin and polymyxin B when testing 15,377 clinical strains collected worldwide. Diagn Microbiol Infect Dis 83(4):379–381. https://doi.org/10.1016/j.diagmicrobio.2015.08.013
39. Mouton J, Ambrose P, Kahlmeter G, Wikler M, Craig W (2007) Applying pharmacodynamics for susceptibility breakpoint selection and susceptibility testing. In: Nightingale CAP, Drusano GL, Murakawa T (eds) Antimicrobial pharmacodynamics in theory and clinical practice, 2nd edn. Informa Healthcare/CRC Press, New York, pp 21–44
40. Li J, Turnidge J, Milne R, Nation RL, Coulthard K (2001) In vitro pharmacodynamic properties of colistin and colistin methanesulfonate against Pseudomonas aeruginosa isolates from patients with cystic fibrosis. Antimicrob Agents Chemother 45(3):781–785. https://doi.org/10.1128/AAC.45.3.781-785.2001
41. Dudhani RV, Turnidge JD, Coulthard K, Milne RW, Rayner CR, Li J, Nation RL (2010) Elucidation of the pharmacokinetic/pharmacodynamic determinant of colistin activity against Pseudomonas aeruginosa in murine thigh and lung infection models. Antimicrob Agents Chemother 54(3):1117–1124. https://doi.org/10.1128/AAC.01114-09
42. Cheah SE, Wang J, Nguyen VT, Turnidge JD, Li J, Nation RL (2015) New pharmacokinetic/pharmacodynamic studies of systemically administered colistin against Pseudomonas aeruginosa and Acinetobacter baumannii in mouse thigh and lung infection models: smaller response in lung infection. J Antimicrob Chemother 70(12):3291–3297. https://doi.org/10.1093/jac/dkv267
43. Tsala M, Vourli S, Georgiou PC, Pournaras S, Tsakris A, Daikos GL, Mouton JW, Meletiadis J (2018) Exploring colistin pharmacodynamics against Klebsiella pneumoniae: a need to revise current susceptibility breakpoints. J Antimicrob Chemother 73(4):953–961. https://doi.org/10.1093/jac/dkx522
44. Tam VH, Schilling AN, Vo G, Kabbara S, Kwa AL, Wiederhold NP, Lewis RE (2005) Pharmacodynamics of polymyxin B against Pseudomonas aeruginosa. Antimicrob Agents Chemother 49(9):3624–3630. https://doi.org/10.1128/AAC.49.9.3624-3630.2005
45. Landersdorfer CB, Wang J, Wirth V, Chen K, Kaye KS, Tsuji BT, Li J, Nation RL (2018) Pharmacokinetics/pharmacodynamics of systemically administered polymyxin B against Klebsiella pneumoniae in

mouse thigh and lung infection models. J Antimicrob Chemother 73(2):462–468. https://doi.org/10.1093/jac/dkx409
46. Sorli L, Luque S, Grau S, Berenguer N, Segura C, Montero MM, Alvarez-Lerma F, Knobel H, Benito N, Horcajada JP (2013) Trough colistin plasma level is an independent risk factor for nephrotoxicity: a prospective observational cohort study. BMC Infect Dis 13:380. https://doi.org/10.1186/1471-2334-13-380
47. Forrest A, Garonzik SM, Thamlikitkul V, Giamarellos-Bourboulis EJ, Paterson DL, Li J, Silveira FP, Nation RL (2017) Pharmacokinetic/Toxicodynamic analysis of colistin-associated acute kidney injury in critically ill patients. Antimicrob Agents Chemother 61(11). https://doi.org/10.1128/aac.01367-17
48. Plachouras D, Karvanen M, Friberg LE, Papadomichelakis E, Antoniadou A, Tsangaris I, Karaiskos I, Poulakou G, Kontopidou F, Armaganidis A, Cars O, Giamarellou H (2009) Population pharmacokinetic analysis of colistin methanesulfonate and colistin after intravenous administration in critically ill patients with infections caused by gram-negative bacteria. Antimicrob Agents Chemother 53(8):3430–3436. https://doi.org/10.1128/AAC.01361-08
49. Garonzik SM, Li J, Thamlikitkul V, Paterson DL, Shoham S, Jacob J, Silveira FP, Forrest A, Nation RL (2011) Population pharmacokinetics of colistin methanesulfonate and formed colistin in critically ill patients from a multicenter study provide dosing suggestions for various categories of patients. Antimicrob Agents Chemother 55(7):3284–3294. https://doi.org/10.1128/AAC.01733-10
50. Karaiskos I, Friberg LE, Pontikis K, Ioannidis K, Tsagkari V, Galani L, Kostakou E, Baziaka F, Paskalis C, Koutsoukou A, Giamarellou H (2015) Colistin population pharmacokinetics after application of a loading dose of 9 MU colistin methanesulfonate in critically Ill patients. Antimicrob Agents Chemother 59(12):7240–7248. https://doi.org/10.1128/aac.00554-15
51. Jacobs M, Gregoire N, Megarbane B, Gobin P, Balayn D, Marchand S, Mimoz O, Couet W (2016) Population pharmacokinetics of colistin methanesulphonate (CMS) and colistin in critically ill patients with acute renal failure requiring intermittent haemodialysis. Antimicrob Agents Chemother 60(3):1788–1793. https://doi.org/10.1128/aac.01868-15
52. Nation RL, Garonzik SM, Thamlikitkul V, Giamarellos-Bourboulis EJ, Forrest A, Paterson DL, Li J, Silveira FP (2017) Dosing guidance for intravenous colistin in critically ill patients. Clin Infect Dis 64(5):565–571. https://doi.org/10.1093/cid/ciw839
53. Nation RL, Garonzik SM, Li J, Thamlikitkul V, Giamarellos-Bourboulis EJ, Paterson DL, Turnidge JD, Forrest A, Silveira FP (2016) Updated US and European dose recommendations for intravenous colistin: how do they perform? Clin Infect Dis 62(5):552–558. https://doi.org/10.1093/cid/civ964
54. Food-and-Drug-Administration (2013) Approved drug products. Label and approval history for Coly-Mycin M, NDA 050108. http://www.accessdata.fda.gov/drugsatfda_docs/label/2013/050108s030lbl.pdf. Last accessed 16 June 2019
55. European-Medicines-Agency (2014) Assessment report on polymyxin-based products. Referral under Article 31 of Directive 2001/83/EC. http://www.ema.europa.eu/docs/en_GB/document_library/Referrals_document/Polymyxin_31/WC500179664.pdf. Last accessed 16 June 2019

Labelling Conventions and Product Package Insert of Parenteral Polymyxins: Factors Causing Potential Medication Errors and Impeding Optimal Clinical Use

10

Jian Li, Kingsley Coulthard, and Roger L. Nation

Abstract

Two different labelling conventions for the contents of colistin methanesulfonate (i.e. colistin base activity [CBA] and international unit [IU]) are used in different parts of the world, and have caused prescribing errors and patient safety issues. This chapter discusses the key issues on the conversion between CBA and IU, and highlights that in pharmacokinetic analyses only the absolute mass of the chemical colistin methanesulfonate should be employed, but not the CBA or IU values. The scientific evidence is unknown for the limits specified for the pharmacopeial standards of the major components of colistin methanesulfonate and polymyxin B. The package information of parenteral colistin methanesulfonate in Europe has now been significantly improved by incorporating the latest pharmacokinetic/pharmacodynamic data. However, the current package information of almost all different brands of parenteral polymyxin B products is substantially out of date without solid pharmacological data. Updating the package information of different products of both polymyxins requires the coordination between major regulatory authorities and will significantly facilitate the optimisation of their use in patients.

Keywords

Polymyxin B · Colistin · Labelling convention · Product information · Pharmacopeial standard

J. Li (✉)
Biomedicine Discovery Institute, Infection & Immunity Program and Department of Microbiology, Monash University, Clayton Campus, Melbourne, VIC, Australia
e-mail: jian.li@monash.edu

K. Coulthard
School Pharmacy and Medical Sciences, University of South Australia, Adelaide, SA, Australia

R. L. Nation
Drug Delivery, Disposition and Dynamics, Monash Institute of Pharmaceutical Sciences, Monash University, Parkville Campus, Melbourne, VIC, Australia

Polymyxins are one of the very few 'old' antibiotic classes that have been 'redeveloped' by academics and clinicians since the early 2000s using contemporary drug development procedures (Chap. 1). Colistin (i.e. polymyxin E) and polymyxin B are the only two polymyxins available for clinical use. Unlike polymyxin B which is used as its sulfate salt, colistin is available as an inactive prodrug colistin methanesulfonate (sodium salt; also known as colistimethate [1]) for intravenous, intrathecal and inhalation administration in patients in

J. Li et al. (eds.), *Polymyxin Antibiotics: From Laboratory Bench to Bedside*, Advances in Experimental Medicine and Biology 1145, https://doi.org/10.1007/978-3-030-16373-0_10

most countries. Unfortunately, two different and confusing conventions have been used around the world to label the contents of parenteral products of colistin methanesulfonate: *international unit* (IU, mainly in Europe) and *colistin base activity* (CBA, largely in America and Asia) [2–4]. These terms may cause prescribing errors and significantly compromise patient safety ([2–4], ISMP [5]). There is also a product of colistin sulfate available for intravenous administration in China [6]. This chapter discusses three polymyxin drugs: (1) colistin methanesulfonate, (2) polymyxin B and (3) colistin; and reviews the different labelling conventions of colistin methanesulfonate and polymyxin B products and the approaches to promote their safe use in patients.

10.1 Labelling Conventions of Colistin Methanesulfonate and Polymyxin B Products

10.1.1 Colistin Methanesulfonate Products

Products of colistin methanesulfonate are an extremely complex mixture of various methanesulfonated derivatives of colistin A, colistin B and many other components [7, 8]. It is unknown in the literature how two different antibacterial activity units were originally employed to label the vial content of colistin methanesulfonate products in different regions of the world. The labelling convention of IU per injection vial is employed mainly in Europe, the United Kingdom, and India. In North and South America, Singapore, Malaysia, New Zealand and Australia, the convention of milligrams of CBA is used to label parenteral products of colistin methanesulfonate. Unfortunately, in Australia the number of IU is also used for labelling the dose and vial contents of an inhalation product of colistin methanesulfonate named Tadim® (https://www.phebra.com/product/colistimethate-sodium/). This is very confusing and problematic for clinicians in Australia, given that milligrams of CBA is also employed in the Colistin Link Parenteral® product for intravenous use [9].

Milligrams of CBA sounds like a mass unit; however, it is very important to note that both CBA and IU are based upon the antibacterial activity measured by microbiological assays *in vitro* [3]. Such microbiological assays require overnight incubation of agar plates at 35–37 °C with bacteria treated by antibiotics, before an inhibition zone is formed. As colistin methanesulfonate is not stable under such conditions, ongoing conversion to colistin occurs in the agar plate during the overnight incubation. Hence, the units of CBA and IU obtained are only apparent values for the antibacterial activity. These values do not represent the absolute amount of the chemical colistin methanesulfonate in injection vials, nor do they indicate the amount of colistin that will be formed in an individual patient as this will be influenced by the renal function of the patient (see Chap. 15). Therefore, these values cannot be used for any pharmacokinetic/pharmacodynamic (PK/PD) purposes. Accordingly, neither CBA nor IU values should be used in pharmacokinetic analyses. For the conversion between the two very different conventions, the absolute mass of colistin methanesulfonate plays an important role, i.e. ~80 mg of the chemical colistin methanesulfonate = one million IU (MIU) = ~33.3 mg of CBA. A given number of milligrams of CBA is equivalent to approximately 2.4 times (i.e. 80÷33.3) that number of milligrams of the chemical colistin methanesulfonate. Even though expressing the dose in milligrams of the chemical colistin methanesulfonate reflects the amount of the vial content, this may further complicate the already existing two different labelling conventions (i.e. CBA and IU) used since the late 1950s. Table 10.1 shows a conversion between CBA and MIU for colistin methanesulfonate. A publicly available iPhone and iPad app was recently developed for dose calculations of intravenous colistin methanesulfonate in patients and provides the conversion between CBA and MIU (https://itunes.apple.com/us/app/colistindose/id1336806844?mt=8).

Clinical use of colistin methanesulfonate around the world is substantially confounded by the expression of doses using different conventions [3, 4]. Importantly, the different conven-

Table 10.1 Look-up table of the conversion between CBA and MIU, based on one million IU = 33.3 mg CBA

CBA mg	Million international unit (MIU)
100	3.00
110	3.30
120	3.60
130	3.90
140	4.20
150	4.50
160	4.80
170	5.10
180	5.40
190	5.70
200	6.00
210	6.30
220	6.60
230	6.90
240	7.20
250	7.50
260	7.80
270	8.10
280	8.40
290	8.70
300	9.00

Please note the factor of 33.3 may slightly vary from different brands and batches of colistin methanesulfonate products

tions have a significant impact on the ability of clinicians to optimally use this last-line antibiotic with maximal antibacterial effect. The confusing labelling can also compromise patient safety as polymyxins have a narrow therapeutic window [10, 11]. International harmonization was called for labelling colistin methanesulfonate products when the problematic labelling conventions were first identified in 2006 [2, 3]. At *the First International Conference on Polymyxins* (Prato, Italy, 2–4 May 2013), international leaders in research and clinical use of polymyxins and the attendees from 27 countries reached a consensus on reporting of colistin methanesulfonate doses [12]. Recommendations were made for publishing of papers on clinical use of colistin methanesulfonate in international journals [4]. The Prato Consensus recommends that for a clinical paper, an equivalence should be provided in the methods section for the two major conventions of labelling colistin methanesulfonate, such as that 1 million IU is equivalent to ~33.3 mg CBA [3, 4, 12]. When reporting methods used in clinical pharmacokinetic studies, it is essential to provide an equivalence to the absolute mass of the chemical colistin methanesulfonate (e.g. 1 million IU or 33.3 mg CBA is equivalent to ~80 mg colistin methanesulfonate); the CBA or IU values must not be used in pharmacokinetic analyses. In the sections of Introduction, Results and Discussion of any clinical paper, expressing doses in terms of milligrams of colistin methanesulfonate should be avoided, considering any potential dosing errors. Instead, colistin methanesulfonate doses should be expressed with the convention used in the region of the world where the study was conducted (i.e. number of IU or milligram of CBA) [4]. For example, in clinical studies from Europe, only IU should be used except in the methods section. These recommendations aim to minimize potential confusions between milligram of CBA and milligram of the chemical colistin methanesulfonate (i.e. 33.3 mg CBA is equivalent to ~80 mg colistin methanesulfonate).

10.1.2 Parenteral Products of Polymyxin B Sulfate

Unlike colistin methanesulfonate, products of polymyxin B from different regions have only one labelling convention, i.e. units; each milligram of pure polymyxin B base is equivalent to 10,000 units of polymyxin B. For polymyxin B sulfate, 1 mg of pure polymyxin B sulfate corresponds to approximately 8,300 IU [13], which is also consistent with the different molecular weights of polymyxin B base and polymyxin B sulfate. If we assume that the molar ratio of the two major polymyxin B components B1 (molecular weight = 1203.5) to B2 (molecular weight = 1189.5) is 1, the average molecular weight of polymyxin B is approximately 1196.5. Therefore, 1 mg pure polymyxin B sulfate (molecular weight = 1441.7) should contain 0.8299 mg polymyxin B base, i.e. 8,299 IU. As stated in the package insert, polymyxin B sulfate products should have a potency of ≥6,000 units per mg calculated on the anhydrous basis [14–18].

In pharmacokinetic analyses, correct dose values of polymyxin B base in mg should be employed.

10.1.3 Parenteral Products of Colistin Sulfate

To the best of our knowledge, there is a parenteral product of colistin sulfate available in China [6]. There is very limited pharmacological information in the literature on intravenous colistin sulfate in patients. Therefore, parenteral colistin sulfate will not be discussed in Sects. 10.2 and 10.3 below. Nevertheless, this parenteral product of colistin sulfate is labeled as 0.5 million IU per vial (1 mg of pure colistin base = 17,000 IU of colistin), and the recommended dosage regimen in renally healthy patients is 1–1.5 million IU per day (i.e. 58.8–88.2 mg colistin base) in 2–3 doses. For a 60-kg patient, the highest dose of intravenous colistin appears about half of that recommended for polymyxin B (Sect. 10.3).

In summary, to optimize the use of polymyxins in critically-ill patients, it is essential for clinicians to understand the different activity units and terms employed by different colistin methanesulfonate products in different regions of the world. Fortunately, the labelling conventions for the parenteral products of polymyxin B sulfate and colistin sulfate are less confusing and relatively easy to understand.

10.2 Pharmacopeial Standards of Colistin Methanesulfonate and Polymyxin B

Both polymyxins contain multiple components, colistin A (i.e. polymyxin E1) and B (i.e. polymyxin E2) for colistin, and polymyxin B1 and B2 for polymyxin B (Chap. 2). As colistin and polymyxin B are prepared by fermentation, manufacture is difficult to control compared to synthetic processes, leading to variable and less predictable product composition and impurity profiles [19]. Different brands of colistin methanesulfonate and polymyxin B may have different molar ratios of their respective major components [20]. Quality control limits for the major components of colistin and polymyxin B are missing in the US Pharmacopeia [21]. Limits for the major components of polymyxin B are provided in the European and British Pharmacopeia (i.e. sum of polymyxin B1, B2, B3 and B1-I $\geq$ 80.0%) [22, 23]; while only the latter specified the limits for the major components of colistin methanesulfonate (i.e. the colistin starting material must have colistin A 50–75% and colistin B 5–20%) [22, 24]. Furthermore, it also specified that the sum of all major methanesulfonated derivatives of colistin methanesulfonate A and B should be $\geq$77.0% [22].

The scientific evidence is unknown for the limits specified by the British Pharmacopeia (2018) and European Pharmacopoeia (2017) for colistin A, colistin B, colistin methanesulfonate A and colistin methanesulfonate B. Colistin methanesulfonate is a very complex mixture of multiple methanesulfonate intermediates (Fig. 10.1) [8]. On top of the aforementioned complexity due to multiple colistin components, the sulfomethylation of colistin also substantially increases the complexity of the product [7, 8, 22, 24–26]. Methanesulfonate groups can be coupled onto any of the five Dab residues of each colistin component [7, 22, 24, 26]; therefore, the extremely complex resulting mixture makes challenging, indeed virtually impossible, the labelling of the injection vial contents based on the chemical composition. As colistin methanesulfonate derivatives are usually not stable in aqueous solutions, it is difficult to accurately measure the proportion of each methanesulfonated derivative. It has been demonstrated that different brands of colistin methanesulfonate may have different compositions [25]. That study using high-performance liquid chromatography (HPLC) showed that four different brands of parenteral colistin methanesulfonate from Europe, Asia and North America had different HPLC fingerprints [25]. Elemental analysis results demonstrated similar elemental compositions of the four brands of colistin methanesulfonate products from three continents (Table 10.2). Furthermore, in a rat pharmacokinetic model intravenous administration of different brands of colistin

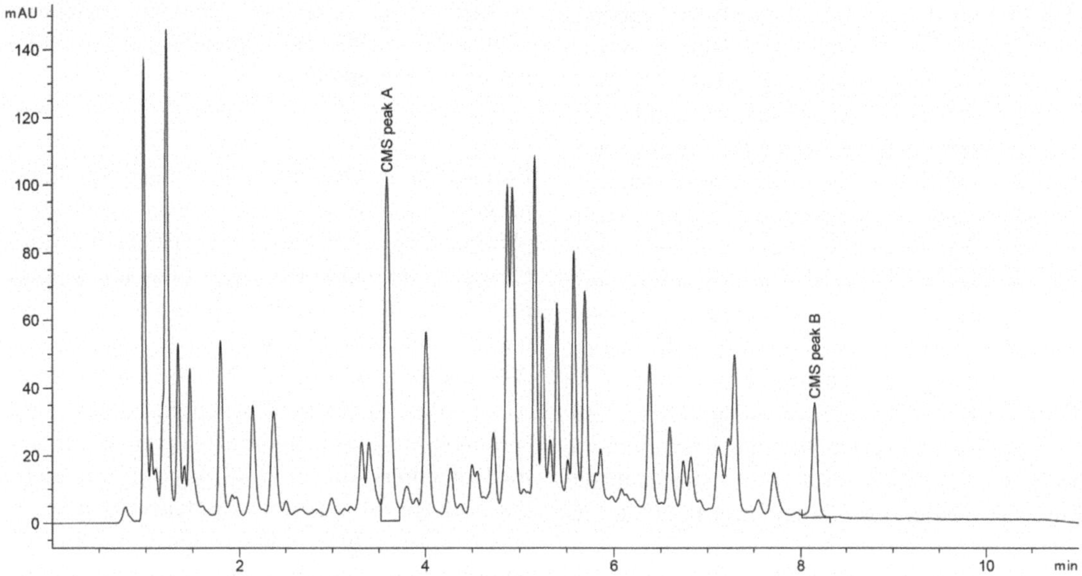

Fig. 10.1 HPLC chromatogram of CMS solution prepared from Promixin (Zambon, Bresso, Italy; a freeze-dried pharmaceutical preparation containing no excipients) [8]. Permission obtained from Elsevier

Table 10.2 CMS (sodium) contents per vial (n = 3) of four different brands and elemental analysis (%) [25]

	X-GEN (USA)	Paddock (USA)	Atlantic (Thailand)	Forest (UK)	Theoretical value
Vial label	150 mg CBA			2 million IU	–
Weight (mg/vial)	366.8 ± 0.80	340.6 ± 0.08	380.0 ± 5.97	159.3 ± 1.75	–
Carbon	34.56%	34.91%	34.41%	34.67%	39.60%
Hydrogen	5.87%	5.86%	5.80%	5.95%	6.07%
Nitrogen	10.95%	11.42%	11.22%	11.22%	12.85%
Oxygen	34.26%	36.83%	34.46%	34.91%	25.69%
Sulfur	10.35%	9.76%	8.71%	8.80%	9.19%

methanesulfonate products led to different exposures to the formed colistin, the antibacterial entity [25]. Therefore, caution is required in the interpretation of pharmacokinetic and pharmacodynamic studies of colistin methanesulfonate conducted in different parts of the world.

Recent pharmacological studies demonstrated that different components of colistin and polymyxin B may have different pharmacokinetic, activity and toxicity profiles [27, 28]. Therefore, specifying the molar ratio of major colistin and polymyxin B components in parenteral products is desirable. Unfortunately, limits of the multiple components of polymyxin products in different pharmacopeias are inconsistent. As both colistin and polymyxin B have a low therapeutic index [10–12], careful re-evaluation of the current pharmacopoeia standards is urgently required.

10.3 Product Package Inserts of Colistin Methanesulfonate and Polymyxin B

To promote optimal and safe use of both polymyxins in the clinic, it is essential that clinicians are provided with the latest pharmacological information. Unfortunately, for almost all different brands of parenteral polymyxin B sulfate, the current package information is substantially out of date with the pharmacological data obtained more than five decades ago. The European Medicines Agency rapidly responded to the Prato Consensus after the First International Conference on Polymyxins in 2013 and the package information of parenteral colistin methanesulfonate in Europe has now been substantially improved by incorporating the latest pharmacokinetic/pharmacodynamic data [29]. The product information of many different brands of colistin methanesulfonate in the other parts of the world still contains very confusing or even wrong information, in particular on their pharmacokinetics [3, 12, 30]. This section will review the major errors in the package information sheets of colistin methanesulfonate (sodium) and polymyxin B sulfate. As pharmacokinetic data for colistin (sulfate) in patients are not available in the literature, parenteral products of colistin sulfate will not be discussed below.

Colistin Methanesulfonate Except those in Europe, the package information of most colistin methanesulfonate products is based on studies conducted more than half a century ago and does not provide accurate pharmacological information to clinicians [3, 12, 30]. For example, the serum drug concentration - time profile in the package information for products available in the USA and some other regions was obtained using microbiological assays [31–33]; therefore, concentrations of CBA (labelled in terms of mg/L) in serum are misleading and not correct due to the ongoing conversion of colistin methanesulfonate to the antibacterial colistin during the microbiological assay (Sect. 10.1.1). In the recently updated package information of European products, the first scientifically-based dosing recommendations (Chaps. 12 and 15) [12] were included [29]. For example, in the package information by Teva UK Limited the dosing regimens for patients on different renal replacement therapy have been provided based on the largest population pharmacokinetic study in critically-ill patients to date [34–36]. In North America, it appears that the package information of colistin methanesulfonate products has not been substantially updated as of June 2018. Clearly, global coordination is essential among regulatory authorities in different continents, in particular between EMA and FDA.

Polymyxin B Over the last decade significant progress has also been made in understanding the pharmacology of polymyxin B [10, 37–44] (also Chaps. 12, 14 and 15). It appears that intravenous polymyxin B is increasingly used more frequently in North America, South America and Asia due to its better PK/PD characteristics compared with colistin methanesulfonate, except for certain indications including treatment of urinary tract infections [10, 12, 37, 44, 45]. Unfortunately, the package information of polymyxin B products is not helpful and has seriously limited the optimization of this important last-line antibiotic in the clinical setting.

For example, no plasma concentration - time profile is provided in product information for polymyxin B sulfate products. The Clinical Pharmacology section is superficial and lacks basic pharmacokinetic information. Furthermore, the susceptibility breakpoints require updating and the *in vitro* susceptibility measurement using disk diffusion assay has been demonstrated inappropriate for determining polymyxin MICs [46]. For intravenous administration, the recommended dosage regimen in the package information is 15,000–25,000 units/kg/day in patients with normal kidney function and the total daily dose should not exceed 25,000 units/kg/day [15, 18]. Although the package information recommends that doses should be reduced for patients with kidney impairment [14, 16, 17], from the PK/PD point of view this is not supported by recent population pharmacokinetic studies [37–41, 45] (also Chaps. 14 and 15). When acute kidney injury occurs, reducing the dose in renally impaired patients may be reasonable; nevertheless, more clinical pharmacological studies are warranted on the PK/PD and toxicodynamics of polymyxin B. For patients on renal replacement therapy, information is lacking on the optimal dosage regimens in the package insert. Based on the data from a very small number of patients thus far, it has been suggested that the recommended polymyxin B doses should not be reduced for patients on continuous venovenous haemodialysis [42, 43]. A recent animal PK/PD study revealed that, similar to colistin against *Pseudomonas aeruginosa* and *Acinetobacter baumannii* [47], *f*AUC/MIC is the most predictive PK/PD parameter for parenteral polymyxin B against *Klebsiella pneumoniae* using mouse thigh and lung infection models [48]. No significant difference was observed in the antibacterial activity between polymyxin B and colistin with equimolar doses in infected mice [48]. In summary, significant gaps in the preclinical and clinical pharmacology of polymyxin B should be addressed, and scientifically-based dosing guidelines for polymyxin B are urgently needed for different types of patients (e.g. on renal replacement therapy).

In conclusion, confusing package information of colistin methanesulfonate and polymyxin B products may cause potential medication errors and impede optimal clinical use. Updating the package information of different products is required globally based on the recent pharmacological studies; and coordination between major regulatory authorities (e.g. EMA and FDA) is crucial. Before new antibiotics become available for Gram-negative 'superbugs', we must optimize the clinical use of the last-line polymyxins, thereby maximizing antibacterial efficacy while minimizing emergence of resistance and nephrotoxicity.

References

1. Bergen PJ, Li J, Rayner CR, Nation RL (2006) Colistin methanesulfonate is an inactive prodrug of colistin against *Pseudomonas aeruginosa*. Antimicrob Agents Chemother 50:1953–1958
2. Li J, Nation RL, Turnidge JD (2006) Defining the dosage units for colistin methanesulfonate: urgent need for international harmonization. Antimicrob Agents Chemother 50: 4231. author reply 4231–4232
3. Li J, Nation RL, Turnidge JD, Milne RW, Coulthard K, Rayner CR, Paterson DL (2006) Colistin: the re-emerging antibiotic for multidrug-resistant gram-negative bacterial infections. Lancet Infect Dis 6:589–601
4. Nation RL, Li J, Cars O, Couet W, Dudley MN, Kaye KS, Mouton JW, Paterson DL, Tam VH, Theuretzbacher U, Tsuji BT, Turnidge JD (2014) Consistent global approach on reporting of colistin doses to promote safe and effective use. Clin Infect Dis 58:139–141
5. Institute for Safe Medication Practices (2011) Warning! Dosing confusion with colistimethate for injection. [Online]. Horsham, PA Available: https://www.nccmerp.org/sites/default/files/NANAlertJune2011.pdf. Last accessed 11 April 2019
6. Chen Y, Zheng Y, Hguan H, Huang J (2013) Injectable preparation and clinical applications of polymyxin E. Shanghai Med Pharm J 34:16–19
7. Li J, Milne RW, Nation RL, Turnidge JD, Coulthard K (2003) Stability of colistin and colistin methanesulfonate in aqueous media and plasma studied by high-performance liquid chromatography. Antimicrob Agents Chemother 47:1364–1370
8. Metcalf AP, Hardaker LEA, Hatley RHM (2017) A simple method for assaying colistimethate sodium in pharmaceutical aerosol samples using high performance liquid chromatography. J Pharm Biomed Anal 142:15–18
9. Link Medical Products (2014) Coly-Mycin M parenteral [package insert]. Link Medical Products, Warriewood
10. Bergen PJ, Landersdorfer CB, Zhang J, Zhao M, Lee HJ, Nation RL, Li J (2012) Pharmacokinetics and pharmacodynamics of 'old' polymyxins: what is new? Diagn Microbiol Infect Dis 74:213–223
11. Falagas ME, Kasiakou SK (2006) Toxicity of polymyxins: a systematic review of the evidence from old and recent studies. Crit Care 10:R27
12. Nation RL, Li J, Cars O, Couet W, Dudley MN, Kaye KS, Mouton JW, Paterson DL, Tam VH, Theuretzbacher U, Tsuji BT, Turnidge JD (2015) Framework for optimisation of the clinical use of colistin and polymyxin B: the Prato polymyxin consensus. Lancet Infect Dis 15:225–234
13. Lightbown JW, Thomas AH, Grab B, Outschoorn AS (1973) The second international standard for polymyxin B. Bull World Health Organ 48:85–90
14. APP (2008) Polymyxin B for injection, USP [package insert]. 2008. APP Pharmaceuticals, LLC, Schaumburg
15. Athenex Pharmaceutical Division (2016) Polymyxin B injection, powder, lyophilized, for solution [package insert]. Schaumburg
16. Bedford Laboratories (2012) Polymyxin B for injection [package insert]. Bedford laboratories™, Bedford
17. X-GEN (2006) Polymyxin B for injection USP [package insert]. X-GEN Pharmaceuticals, Northport
18. Xellia Pharmaceuticals APS (2016) Polymyxin B sulfate injection [package Insert]. 16 February 2016. Xellia Pharmaceuticals APS, Denmark
19. Brink AJ, Richards GA, Colombo G, Bortolotti F, Colombo P, Jehl F (2014) Multicomponent antibiotic substances produced by fermentation: implications

for regulatory authorities, critically ill patients and generics. Int J Antimicrob Agents 43:1–6
20. Fujihara M, Nishiyama S, Hasegawa S (1994) Effects of agars on determination of potency of polymyxin B sulfate by the agar plate diffusion method. Antimicrob Agents Chemother 38:2665–2667
21. United States Pharmacopeia (2017) Monograph for polymyxin B sulfate. In: United States Pharmacopoeia 40 2017. The United States Pharmacopeial Convention: 5750–5751, Rockville
22. British Pharmacopoeia (2018) Monographs for colistimethate sodium. In: British pharmacopoeia 2018. Stationery Office, London. I-675 to I-678
23. European Pharmacopoeia (2017) Monograph for polymyxin B sulfate. In: European Pharmacopoeia 9.0 2017. Strasbourg: Directorate for the Quality of Medicines and HealthCare of the Council of Europe: 3366–3367
24. European Pharmacopoeia (2017) Monograph for colistimethate sodium. In: European pharmacopoeia 9.0 2017. Strasbourg: Directorate for the Quality of Medicines and HealthCare of the Council of Europe: 2165–2166
25. He H, Li JC, Nation RL, Jacob J, Chen G, Lee HJ, Tsuji BT, Thompson PE, Roberts K, Velkov T, Li J (2013) Pharmacokinetics of four different brands of colistimethate and formed colistin in rats. J Antimicrob Chemother 68:2311–2317
26. Kenyon R (2015) Structural evidence for a new understanding of the nature of CMS. In: The Second International Conference on Polymyxins. 22–24 September 2015, San Diego, USA
27. Roberts KD, Azad MA, Wang J, Horne AS, Thompson PE, Nation RL, Velkov T, Li J (2015) Antimicrobial activity and toxicity of the major lipopeptide components of polymyxin B and colistin: last-line antibiotics against multidrug-resistant gram-negative bacteria. ACS Infect Dis 1:568–575
28. Sivanesan S, Roberts K, Wang J, Chea SE, Thompson PE, Li J, Nation RL, Velkov T (2017) Pharmacokinetics of the individual major components of Polymyxin B and Colistin in rats. J Nat Prod 80:225–229
29. European Medicines Agency (2014) Polymyxin-containing medicines. https://www.ema.europa.eu/en/medicines/human/referrals/polymyxin-containing-medicines. Last accessed 11 April 2019
30. Theuretzbacher U (2014) Product information for parenteral colistin varies substantially across Europe. J Antimicrob Chemother 69:1987–1992
31. JHP Pharmaceuticals (2011) Coly-Mycin M parenteral [package insert]. 2013. Rochester
32. Monarch Pharmaceuticals (2006) Coly-Mycin M parenteral [package insert]. Monarch Pharmaceuticals, Bristol
33. Xellia Pharmaceuticals APS (2016) Colistimethate injection package insert]. 12 July 2016. Xellia Pharmaceuticals APS, Denmark
34. Teva UK Limited (2016) Colomycin injection [package insert]. Teva UK Limited, Essex
35. Garonzik SM, Li J, Thamlikitkul V, Paterson DL, Shoham S, Jacob J, Silveira FP, Forrest A, Nation RL (2011) Population pharmacokinetics of colistin methanesulfonate and formed colistin in critically ill patients from a multicenter study provide dosing suggestions for various categories of patients. Antimicrob Agents Chemother 55:3284–3294
36. Nation RL, Garonzik SM, Thamlikitkul V, Giamarellos-Bourboulis EJ, Forrest A, Paterson DL, Li J, Silveira FP (2017) Dosing guidance for intravenous colistin in critically-ill patients. Clin Infect Dis 64:565–571
37. Sandri AM, Landersdorfer CB, Jacob J, Boniatti MM, Dalarosa MG, Falci DR, Behle TF, Bordinhao RC, Wang J, Forrest A, Nation RL, Li J, Zavascki AP (2013) Population pharmacokinetics of intravenous polymyxin B in critically ill patients: implications for selection of dosage regimens. Clin Infect Dis 57:524–531
38. Zavascki AP, Goldani LZ, Cao GY, Superti SV, Lutz L, Barth AL, Ramos F, Boniatti MM, Nation RL, Li J (2008) Pharmacokinetics of intravenous polymyxin B in critically-ill patients. Clin Infect Dis 47:1298–1304
39. Miglis C, Rhodes NJ, Avedissian S, Kubin CJ, Yin MT, Nelson B, Pai MP, Scheetz MH (2018) Population pharmacokinetics of polymyxin B in acutely ill adult patients. Antimicrob Agents Chemother 62:pii: e01475–17
40. Kubin CJ, Nelson BC, Miglis C, Scheetz MH, Rhodes N, Avedissian S, Cremers S, Yin MT (2018) Population pharmacokinetics of intravenous polymyxin B from clinical samples. Antimicrob Agents Chemother 62:pii: e01493-17
41. Manchandani P, Thamlikitkul V, Dubrovskaya Y, Babic JT, Lye DC, Lee LS, Tam VH (2018) Population pharmacokinetics of polymyxin B. Clin Pharmacol Ther 104:534–538
42. Sandri AM, Landersdorfer CB, Jacob J, Boniatti MM, Dalarosa MG, Falci DR, Behle TF, Saitovitch D, Wang J, Forrest A, Nation RL, Zavascki AP, Li J (2013) Pharmacokinetics of polymyxin B in patients on continuous venovenous haemodialysis. J Antimicrob Chemother 68:674–677
43. Baird JS (2014) Polymyxin B and haemofiltration in an adolescent with leukaemia. J Antimicrob Chemother 69:1434
44. Velkov T, Roberts KD, Nation RL, Thompson PE, Li J (2013) Pharmacology of polymyxins: new insights into an 'old' class of antibiotics. Future Microbiol 8:711–724
45. Nation RL, Velkov T, Li J (2014) Colistin and polymyxin B: are they like peas in a pod or chalk and cheese? Clin Infect Dis 59:88–94

46. Vasoo S (2017) Susceptibility testing for the polymyxins: two steps back, three steps forward? J Clin Microbiol 55:2573–2582
47. Cheah SE, Wang J, Nguyen VT, Turnidge JD, Li J, Nation RL (2015) New pharmacokinetic/pharmacodynamic studies of systemically administered colistin against *Pseudomonas aeruginosa* and *Acinetobacter baumannii* in mouse thigh and lung infection models: smaller response in lung infection. J Antimicrob Chemother 70:3291–3297
48. Landersdorfer CB, Wang J, Wirth V, Chen K, Kaye KS, Tsuji BT, Li J, Nation RL (2018) Pharmacokinetics/pharmacodynamics of systemically administered polymyxin B against *Klebsiella pneumoniae* in mouse thigh and lung infection models. J Antimicrob Chemother 73:462–468

Meta-analysis of Polymyxin Use in Patients

11

Mical Paul, Oren Zusman, and Leonard Leibovici

Abstract

In this chapter, we systematically reviewed studies that assessed polymyxin's effectiveness and summarized results through meta-analysis. The outcomes addressed were all-cause mortality, assuming that for patients with severe multidrug-resistant infections survival is the most important outcome, and resistance development, important for future patients. Most clinical data on polymyxins in the literature are from retrospective, observational studies at high risk of bias. The majority of clinical studies were unpowered to examine mortality controlling for other risk factors. The studies had no control of dosage regimens and treatment modifications. We identified several areas of missing data, in particular randomized controlled trials (RCTs) examining treatment options for carbapenem-resistant Gram-negative bacteria, different dosage regimens, polymyxins versus alternative antibiotics (e.g. aminoglycosides, tigecycline), and monotherapy versus specific combination therapies. Ideally, mortality and development of resistance should be examined in RCTs, with further longitudinal studies required for the latter.

Keywords

Meta-analysis · Polymyxin · Randomized controlled trial · Resistance · Combination therapy

M. Paul (✉)
Division of Infectious Diseases, Rambam Health Care Center and The Ruth and Bruce Rappaport Faculty of Medicine, Technion – Israel Institute of Technology, Haifa, Israel
e-mail: paulm@technion.ac.il

O. Zusman · L. Leibovici
Medicine E, Rabin Medical Center, Beilinson Hospital, Petah Tikva and Sackler Faculty of Medicine, Tel-Aviv University, Ramat Aviv, Tel Aviv, Israel

11.1 Why Focus on Meta-analysis

Meta-analysis is a statistical technique of combining results from different studies. In itself the term conveys little information on the methodology of a study, as the selection criteria for the studies combined are crucial to the meta-analysis results. Systematic reviews define precisely the question addressed and the studies to be included in a meta-analysis and then attempt to include each and every study that has been performed. The advantage over a narrative review is that the information contained within the summary result is transparent and highly specific. This is also the limitation of the meta-analysis result; it addresses precisely the question addressed (patient population, intervention, comparison and outcome).

Meta-analysis provides a single point estimate summarizing all known studies that is much

J. Li et al. (eds.), *Polymyxin Antibiotics: From Laboratory Bench to Bedside*, Advances in Experimental Medicine and Biology 1145, https://doi.org/10.1007/978-3-030-16373-0_11

simpler to deal with than the many separate results of the original individual studies. However, many times the pooled estimate has poor credibility because of heterogeneity in the patient populations, interventions and outcomes assessed despite the attempt to ask specific questions. For example, addressing the question of the survival benefit of colistin-meropenem combination therapy vs. colistin monotherapy among patients with bloodstream infections is seemingly highly specific. However, the studies might evaluate mortality at different time points (in-hospital, 14-day, 28-day), colistin and meropenem might be given in different doses and schedules, patients might be infected by different Gram-negative bacteria with different MICs for meropenem. Readers of meta-analyses are advised to critically examine whether the pooled effect estimate is useful. Frequently meta-analyses will examine clinical and statistical heterogeneity and might be able to point to the factors underlying differences in results.

In this chapter, we will address systematically several questions previously reviewed in the book and try to summarize results through meta-analysis.

11.2 "Effectiveness"

The only study design appropriate to examine the effectiveness of a drug is a well-powered and well-conducted randomized controlled trial (RCT), since only RCTs can achieve unbiased comparisons. There are no RCTs comparing colistin vs. another antibiotic for the treatment of severe infections. Historically, colistin has been considered as poorly effective and has been replaced by beta-lactams once broad-spectrum beta-lactams covering Gram-negative bacteria became available. Currently several studies and authors claim that colistin is "effective". Its use has certainly increased in recent years and it is a primary mode of treatment for carbapenem-resistant bacteria. The question of effectiveness is important as it should determine our inclination to use colistin empirically, before we know whether the patient is infected with carbapenem-resistant bacteria. It should also determine the selection of the antibiotic to be used against carbapenem-resistant bacteria if the isolates are susceptible in-vitro to antibiotics other than colistin (e.g. an aminoglycoside, fosfomycin, tigecycline). Contained within the question of the effectiveness of colistin is also the question of optimal dosing.

Given the lack of RCTs, we compared contemporary observational studies that assessed the effectiveness of colistin (update of a previous review [1]). The inclusion criteria were studies comparing a systemic polymyxin against a drug regimen not including a polymyxin in a comparative clinical trial, cohort (prospective or retrospective) or case-control design and reporting on mortality. We did not restrict inclusion by type of infection or bacteria.

Three studies permitted a comparison between patients given colistin vs. patients receiving inappropriate antibiotic treatment (empirical treatment) [2–4]. Mortality was higher with inappropriate antibiotics, with heterogeneity between the studies (Fig. 11.1). Adjusted analyses were not available.

Thirteen studies compared polymyxins to another antibiotic [5–17]. All studies examined patients with severe healthcare-associated infections (most commonly pneumonia and bacteremia) caused by highly-resistant bacteria. *Acinetobacter baumannii* and *Pseudomonas aeruginosa* were the common bacteria and *Klebsiella pneumoniae* was more rarely assessed. Polymyxins (colistin in all but two studies) were given to patients with carbapenemase-producing or phenotypically carbapenem-resistant Gram-negative bacteria (CRGNB). Colistin was used as monotherapy in a single study [14] and in the other studies polymyxins were most commonly given in combination with other antibiotics. The comparator arm included patients with multidrug-resistant (MDR) bacteria susceptible to the non-polymyxin comparator drug and that were treated with beta-lactams (most commonly carbapenems), tobramycin (one study [8]) or tigecycline (one study [11]). Individual study results and the pooled summary for all-cause mortality are presented in Fig. 11.2. The pooled unadjusted result

Study or Subgroup	Colistin Events	Colistin Total	Comparator Events	Comparator Total	Weight	Odds Ratio M-H, Fixed, 95% CI
Qureshi 2012	5	14	4	6	15.6%	0.28 [0.04, 2.09]
Lim 2011	11	31	15	39	37.3%	0.88 [0.33, 2.34]
Koomanachai 2007	36	78	12	15	47.1%	0.21 [0.06, 0.82]
Total (95% CI)		**123**		**60**	**100.0%**	**0.47 [0.23, 0.96]**
Total events	52		31			
Heterogeneity: $Chi^2 = 3.15$, df= 2 (P = 0.21); $I^2 = 37\%$						
Test for overall effect: Z = 2.07 (P = 0.04)						

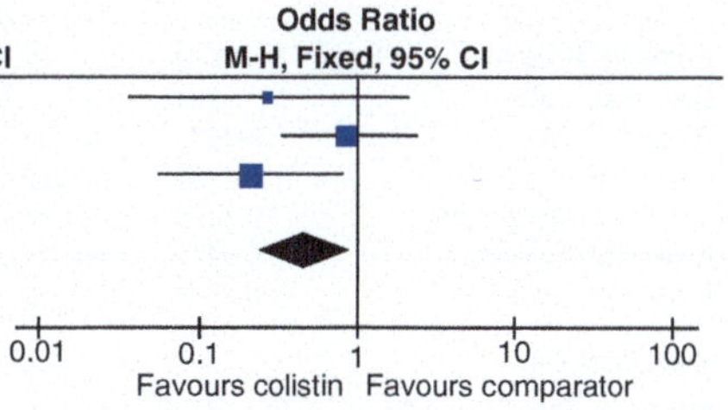

Fig. 11.1 Mortality for patients treated with colistin vs. inappropriate antibiotics

Study or Subgroup	Polymyxin Events	Polymyxin Total	Comparator Events	Comparator Total	Weight	Odds Ratio M-H, Fixed, 95% CI
1.2.1 Pseudorandomized						
Betrosian 2008	5	15	3	13	1.9%	1.67 [0.31, 8.93]
Subtotal (95% CI)		**15**		**13**	**1.9%**	**1.67 [0.31, 8.93]**
Total events	5		3			
Heterogeneity: Not applicable						
Test for overall effect: Z = 0.60 (P = 0.55)						
1.2.2 Matched retrospective						
Durakovic 2011	3	26	3	26	2.3%	1.00 [0.18, 5.48]
Kallel 2007	21	60	15	60	8.5%	1.62 [0.73, 3.56]
Subtotal (95% CI)		**86**		**86**	**10.8%**	**1.48 [0.73, 3.03]**
Total events	24		18			
Heterogeneity: $Chi^2 = 0.25$, df= 1 (P = 0.62); $I^2 = 0\%$						
Test for overall effect: Z = 1.08 (P = 0.28)						
1.2.3 Non-matched prospective						
Rigatto 2013	24	45	6	22	3.3%	3.05 [1.01, 9.21]
Garnacho-Montero 2003	13	21	9	14	3.6%	0.90 [0.22, 3.68]
Hachem 2007	19	31	30	64	6.6%	1.79 [0.75, 4.30]
Reina 2005	16	55	34	130	12.4%	1.16 [0.57, 2.34]
Paul 2010	78	200	85	295	36.3%	1.58 [1.08, 2.31]
Subtotal (95% CI)		**352**		**525**	**62.1%**	**1.56 [1.16, 2.08]**
Total events	150		164			
Heterogeneity: $Chi^2 = 0.25$, df= 1 (P = 0.62); $I^2 = 0\%$						
Test for overall effect: Z = 2.97 (P = 0.003)						
1.2.4 Non-matched retrospective						
Ku 2012	26	45	0	16	0.3%	44.85 [2.53, 793.82]
Gounden 2009	16	32	9	32	3.9%	2.56 [0.91, 7.20]
Kvitko 2011 (polyB)	30	45	25	88	4.9%	5.04 [2.32, 10.93]
Rios 2007	16	31	13	30	5.5%	1.39 [0.51, 3.83]
Oliveira 2008 (polyB)	63	82	54	85	10.7%	1.90 [0.97, 3.75]
Subtotal (95% CI)		**235**		**251**	**25.2%**	**2.96 [1.99, 4.39]**
Total events	151		101			
Heterogeneity: $Chi^2 = 9.10$, df= 4 (P = 0.06); $I^2 = 56\%$						
Test for overall effect: Z = 5.36 (P < 0.00001)						
Total (95% CI)		**688**		**875**	**100.0%**	**1.90 [1.53, 2.37]**
Total events	330		286			
Heterogeneity: $Chi^2 = 16.80$, df= 12 (P = 0.16); $I^2 = 29\%$						
Test for overall effect: Z = 5.76 (P < 0.00001)						
Test for subgroup differences: $Chi^2 = 7.07$, df = 3 (P = 0.07), $I^2 = 57.6\%$						

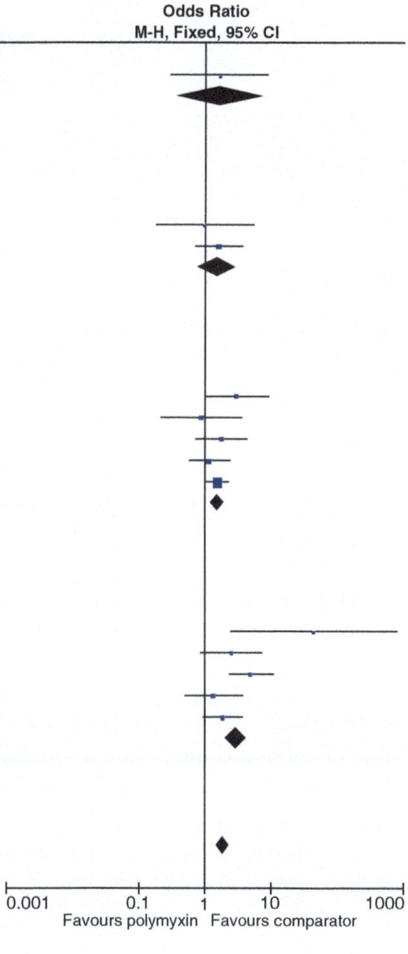

Fig. 11.2 All-cause mortality for polymyxin vs. comparator antibiotics, unadjusted results

showed nearly twice the mortality odds with polymyxins compared to comparator drugs. The study design affected results: the meta-analysis forest plot is subcategorized by study design, from the least risk of bias (top) to the highest (bottom) and odds ratios increase from top to bottom. However, a meta-analysis of adjusted odds ratios (ORs) or odds ratios from studies using matching shows also significantly higher mortality with polymyxins with no statistical heterogeneity (adjusted OR 1.79, 95% confidence intervals (CI) 1.35–2.36, Fig. 11.3). Assessment

Study or Subgroup	log[Odds Ratio]	SE	Weight	Odds Ratio IV, Fixed, 95% CI
Betrosian 2008	0.512824	0.857296	2.8%	1.67 [0.31, 8.96]
Rigatto 2013	1.360977	0.51843	7.6%	3.90 [1.41, 10.77]
Kallel 2007	0.4796	0.4027	12.6%	1.62 [0.73, 3.56]
Oliveira 2008 (polyB)	0.7275	0.3561	16.1%	2.07 [1.03, 4.16]
Kvitko 2011(polyB)	0.6471	0.3017	22.4%	1.91 [1.06, 3.45]
Paul 2010	0.3646	0.23	38.6%	1.44 [0.92, 2.26]
Total (95% CI)			**100.0%**	**1.79 [1.35, 2.36]**

Heterogeneity: $Chi^2 = 3.44$, df= 5 (P = 0.63); $I^2 = 0\%$
Test for overall effect: Z = 4.06 (P < 0.0001)

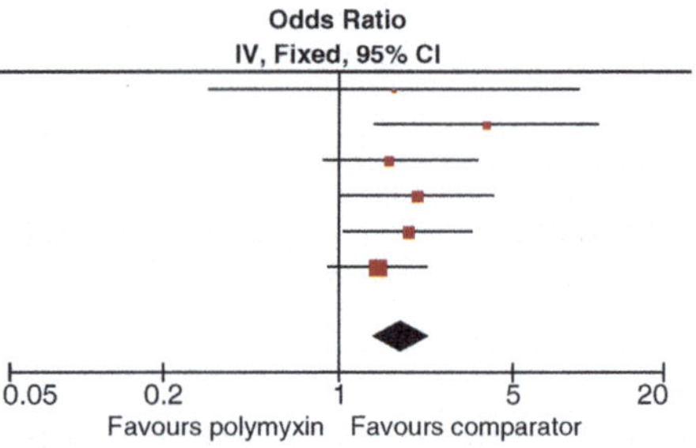

Fig. 11.3 All-cause mortality for polymyxin vs. comparator antibiotics, adjusted results

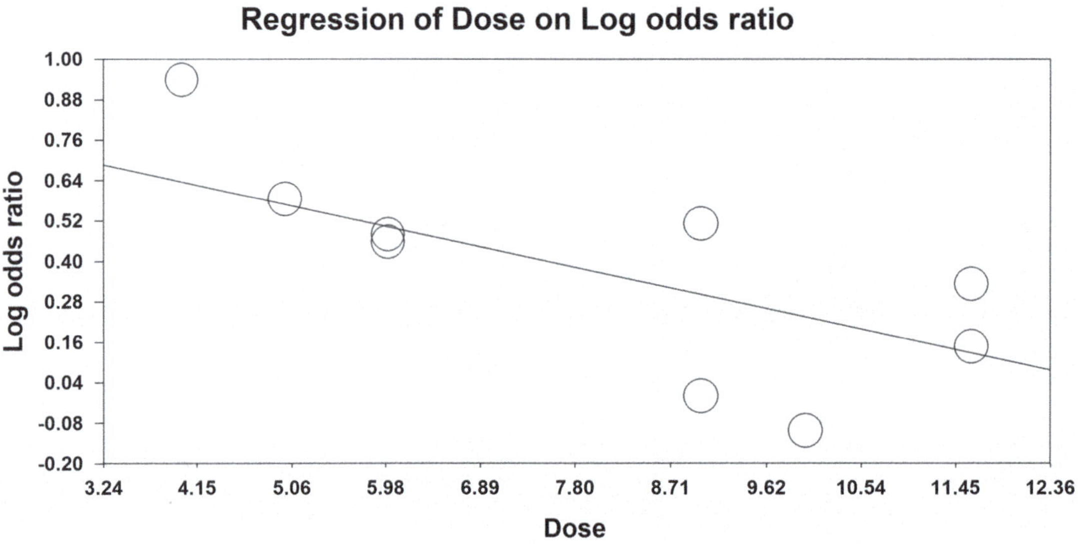

Fig. 11.4 Meta-regression of unadjusted ORs for mortality with mean colistin dose in study Colistin dose given in million international units (MIU). P for slope = 0.21

of the effect of colistin dose on results was possible in univariate analysis only including 9 studies that reported the mean colistin dose used. The meta-regression is shown in Fig. 11.4; although, not statistically significant, a trend is shown of increasing ORs (greater advantage to comparator arm) with lower colistin dosing (presented in million IUs).

Thus, the compilation of existing studies shows that polymyxins may be more effective than no antibiotics and less effective than beta-lactams. The comparison to antibiotics potentially active against CRGNB is limited to single studies. This is based on observational studies with major limitations, of which the main is that different patients are compared. Those treated with colistin have infections caused by CRGNB while those treated with comparator antibiotics usually had carbapenem-susceptible bacteria. Therefore, these studies do not assess the effectiveness of colistin (hence "effectiveness"), but its association with mortality with many limitations. Polymyxins were administered in combination, thus results are relevant to colistin combination therapy. Colistin was given in some of the studies at a lower dose than currently recommended [18, 19] and without a loading dose and lower dosing might have been associated with a larger advantage to comparator drugs. Few retrospective studies compared colistin to polymyxin B [20–23]; the cohorts were too different to allow reasonable comparisons between groups

Study or Subgroup	Polymyxin Events	Polymyxin Total	Comparator Events	Comparator Total	Weight	Odds Ratio M-H, Fixed, 95% CI
1.5.1 Colistin						
Rios 2007	0	31	0	20		Not estimable
Reina 2005	0	55	0	130		Not estimable
Kallel 2007	0	60	0	60		Not estimable
Durakovic 2011	3	26	0	26	0.8%	7.89 [0.39, 160.91]
Betrosian 2008	5	15	2	13	2.7%	2.75 [0.43, 17.49]
Lim 2011	10	20	10	35	7.0%	2.50 [0.80, 7.84]
Garnacho-Montero 2003	5	21	6	14	10.6%	0.42 [0.10, 1.79]
Hachem 2007	7	31	14	64	13.6%	1.04 [0.37, 2.92]
Paul 2010	26	168	17	244	22.6%	2.44 [1.28, 4.67]
Subtotal (95% CI)		**427**		**606**	**57.3%**	**1.84 [1.19, 2.84]**
Total events	56		49			
Heterogeneity: $Chi^2 = 7.25$, df = 5 (P = 0.20); $I^2 = 31\%$						
Test for overall effect: Z = 2.74 (P = 0.006)						
1.5.2 Polymyxin B						
Kvitko 2011 (polyB)	5	45	6	88	6.9%	1.71 [0.49, 5.94]
Rigatto 2013	9	45	4	22	8.3%	1.13 [0.30, 4.16]
Oliveira 2008 (polyB)	18	69	21	81	27.5%	1.01 [0.49, 2.10]
Subtotal (95% CI)		**159**		**191**	**42.7%**	**1.14 [0.65, 2.02]**
Total events	32		31			
Heterogeneity: $Chi^2 = 0.51$, df = 2 (P = 0.77); $I^2 = 0\%$						
Test for overall effect: Z = 0.47 (P = 0.64)						
Total (95% CI)		**586**		**797**	**100.0%**	**1.54 [1.09, 2.18]**
Total events	88		80			
Heterogeneity: $Chi^2 = 9.33$, df = 8 (P = 0.31); $I^2 = 14\%$						
Test for overall effect: Z = 2.47 (P = 0.01)						
Test for subgroup differences: $Chi^2 = 1.68$, df = 1 (P = 0.19); $I^2 = 40.6\%$						

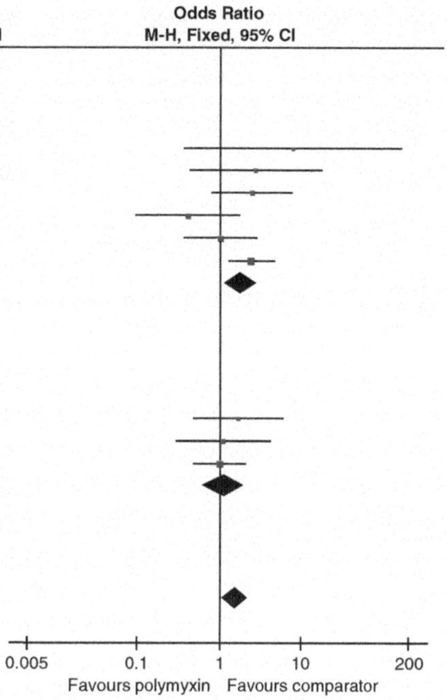

Fig. 11.5 Nephrotoxicity with polymyxins vs. comparator antibiotics, unadjusted results

for mortality and adjusted analyses for mortality were not conducted.

11.3 Nephrotoxicity

The same studies allowed the assessment of nephrotoxicity rates with polymyxins vs. non-polymyxins [3, 5–7, 9, 10, 12–17]. Nephrotoxicity was most commonly defined as at least a 1.5–2 fold increase in serum creatinine from baseline (RIFLE "risk" and above [24]). Ten studies examining colistin were identified, showing higher nephrotoxicity rates with colistin vs. comparator antibiotics, unadjusted OR 1.75 (95% CI 1.16–2.64, Fig. 11.5). Two studies examining polymyxin B did not show a significant difference vs. comparators (Fig. 11.5). None compared a polymyxin to an aminoglycoside.

Recent studies claim higher nephrotoxicity rates with colistin compared to polymyxin B [20–23]. In these studies, selection of patients depended on the type of polymyxin available (comparison between time periods or hospitals). All studies were retrospective and nephrotoxicity was similarly defined as RIFLE "risk" and above [24]. We pooled adjusted odds ratios or odds ratio reported from matched patient cohorts (non-significant univariate results taken from one study). Overall, the nephrotoxicity rate was observed to be about two-fold higher with colistin compared with polymyxin B, adjusted OR 2.12 (95% CI 1.46–3.07, Fig. 11.6).

11.4 Combination Therapy

Currently much debate surrounds the issue of polymyxin combination therapy. Empirical combination therapy is reasonable given that polymyxins are less effective than other antibiotics but more effective than no antibiotics, as shown above. The issue of debate regards combination therapy for CRGNB after receipt of the final pathogen identification and susceptibility results. Some would consider the question also pertinent for carbapenemase-producing Gram-negative bacteria that are phenotypically susceptible to carbapenems. The answer probably depends on the precise MIC of the isolate and perhaps on the type of bacterium.

Study or Subgroup	log[]	SE	Weight	IV, Fixed, 95% CI
Oliveira 2009	−0.05129	0.55333	11.6%	0.95 [0.32, 2.81]
Phe 2014	1.532557	0.514509	13.5%	4.63 [1.69, 12.69]
Tuon 2014	0.553885	0.383693	24.2%	1.74 [0.82, 3.69]
Akajagbor 2013	0.81978	0.265343	50.7%	2.27 [1.35, 3.82]
Total (95% CI)			**100.0%**	**2.12 [1.46, 3.07]**

Heterogeneity: $Chi^2 = 4.74$, df = 3 (P = 0.19); $I^2 = 37\%$
Test for overall effect: Z = 3.97 (P < 0.0001)

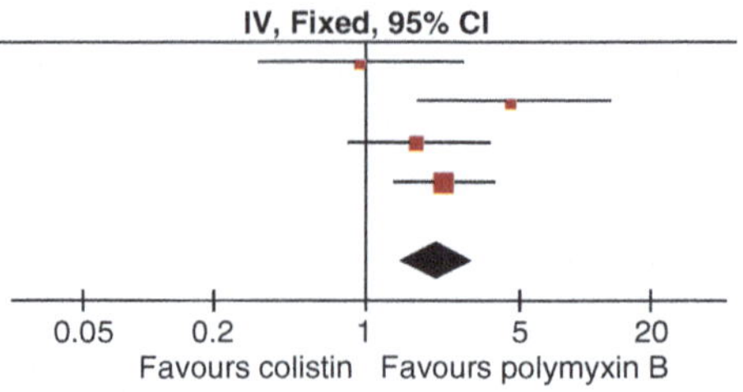

Fig. 11.6 Nephrotoxicity with colistin vs. polymyxin B, adjusted analysis

The rationale for combination therapy is based on synergy, enhanced bactericidality and prevention of polymyxin-resistance development. In a systematic review and meta-analysis we analysed in-vitro interactions between polymyxins and carabapenems for different Gram-negative bacteria [25]. Synergy rates for different carbapenems and different bacteria ranged between 24% (meropenem for *P. aeruginosa*) to 88% (doripenem for *A. baumannii*). Among all carbapenem-polymyxin combinations, synergy rates were highest for *A. baumannii*. Among all bacteria, doripenem achieved highest synergy rates with polymyxins. Antagonism rates were low; the highest value, 24%, was observed for imipenem-polymyxin against *K. pneumoniae.* Bactericidal activity of the combination was greater than that of the polymyxins in most assays, increasing from 10–26% with the polymyxin to 49–74% in different isolates. Resistance developed rapidly with polymyxins alone, whereas the combination therapy generally suppressed and delayed resistance development.

While the in-vitro data appear promising, clinical results might be very different from in-vitro interactions. We compiled all clinical studies comparing colistin administered as monotherapy vs. combination therapy including colistin for the treatment of CRGNB or carbapenemase-producing Gram-negative bacteria [26]. We included RCTs and observational studies. When the same patients were included in more than one publication, we included the publication describing the largest number of patients. The outcome assessed was all-cause mortality. Results are summarized in Fig. 11.7.

Two RCTs compared colistin alone vs. colistin-rifampin for infections caused by *A. baumannii* [27, 28], showing no survival advantage to the combination arm. In both an advantage to colistin-rifampin was shown for secondary outcomes; clinical or microbiological cure. One RCT compared colistin alone vs. colistin-meropenem combination therapy, both administered with optimized high dosing [29]. All other studies were observational (all but two retrospective) ranging from very small case series to cohort studies, the largest analysing 250 patients. Nine studies permitted the comparison between colistin alone vs. colistin-carbapenem combination therapy [4, 29–36]. No advantage was observed to combination therapy OR 0.97, 95% CI 0.69–1.35, unadjusted except for the results of the single RCT). Similarly, the comparisons between colistin monotherapy vs. colistin combined with tigecycline, sulbactam and aminoglycoside showed no significant difference between regimens [4, 11, 30, 32, 34, 37, 38]. Four studies presented a comparison between colistin monotherapy vs. "any" combination therapy, that is difficult to translate to clinical practice. Combinations frequently included three-drug regimens. In this set of studies the combination therapy was significantly associated with higher mortality (unadjusted OR 2.09, 95% CI 1.33–3.28). The risk of bias in these studies was very high, as previously discussed [26]. The main reason underlying heterogeneity in the observational studies was carbapenem MICs, with lower MICs associated with an advantage to the combination therapy.

Thus, these meta-analyses show that despite favorable in-vitro interactions for specific antibiotic combinations, clinical studies do not demonstrate an advantage to combination therapy. The only combinations that have been tested in RCTs are those of colistin-rifampin and colistin-

Study or Subgroup	Colistin mono Events	Colistin mono Total	Combi Events	Combi Total	Weight	Odds Ratio M-H, Fixed, 95% CI
1.1.1 Colistin rifampin (RCTs)						
Aydemir 2013 Ab	16	22	13	21	12.3%	1.64 [0.45, 5.94]
Durante-Mangoni 2013 Ab	45	105	45	104	87.7%	0.98 [0.57, 1.70]
Subtotal (95% CI)		**127**		**125**	**100.0%**	**1.06 [0.64, 1.76]**
Total events	61		58			
Heterogeneity: Chi² = 0.52, df = 1 (P = 0.47); I² = 0%						
Test for overall effect: Z = 0.24 (P = 0.81)						
1.1.2 Colistin carbapenem						
Tuon 2013 Kp	1	1	2	3	0.6%	1.80 [0.04, 79.42]
Qureshi 2012 Kp	4	7	1	5	0.7%	5.33 [0.38, 75.78]
Bergamasco 2011 Kp	1	3	2	3	1.9%	0.25 [0.01, 7.45]
Navarro 2013 Kp	0	1	3	4	2.1%	0.14 [0.00, 5.95]
Souli 2008 Ab	1	3	4	7	2.3%	0.38 [0.02, 6.35]
Daikos 2014 Kp	12	22	3	7	2.9%	1.60 [0.29, 8.90]
Falagas 2006 Ab Pa	0	14	21	57	12.1%	0.06 [0.00, 1.03]
Batirel 2014 Ab	16	36	30	102	12.3%	1.92 [0.88, 4.20]
Paul 2018 Ab Kp Pa (RCT)	64	198	70	208	65.2%	0.94 [0.62, 1.42]
Subtotal (95% CI)		**285**		**396**	**100.0%**	**0.97 [0.69, 1.35]**
Total events	99		136			
Heterogeneity: Chi² = 10.70, df = 8 (P = 0.22); I² = 25%						
Test for overall effect: Z = 0.19 (P = 0.85)						
1.1.3 Colistin tigecycline						
Bergamasco 2011 Kp	1	3	0	3	1.9%	4.20 [0.12, 151.97]
Qureshi 2012 Kp	4	7	0	1	2.1%	3.86 [0.12, 126.73]
Navarro 2013 Kp	0	1	7	8	12.3%	0.07 [0.00, 2.56]
Daikos 2014 Kp	12	22	5	21	14.0%	3.84 [1.04, 14.21]
Kontopidou 2013 Kp	6	26	4	9	27.5%	0.38 [0.08, 1.86]
Ku 2012 Ab	26	71	7	19	42.2%	0.99 [0.35, 2.83]
Subtotal (95% CI)		**130**		**61**	**100.0%**	**1.23 [0.64, 2.35]**
Total events	49		23			
Heterogeneity: Chi² = 8.50, df = 5 (P = 0.13); I² = 41 %						
Test for overall effect: Z= 0.62 (P = 0.54)						
1.1.4 Colistin sulbactam						
Batirel 2014 Ab	16	36	22	69	35.6%	1.71 [0.75, 3.92]
Kalin 2013 Ab	27	52	27	37	64.4%	0.40 [0.16, 0.99]
Subtotal (95% CI)		**88**		**106**	**100.0%**	**0.87 [0.48, 1.57]**
Total events	43		49			
Heterogeneity: Chi² = 5.37, df = 1 (P = 0.02); I²= 81 %						
Test for overall effect: Z = 0.48 (P= 0.63)						
1.1.5 Colistin aminoglycoside						
Navarro 2013 Kp	0	1	0	2		Not estimable
Kontopidou 2013 Kp	6	26	2	17	42.0%	2.25 [0.40, 12.75]
Daikos 2014 Kp	12	22	5	17	58.0%	2.88 [0.75, 10.99]
Subtotal (95% CI)		**49**		**36**	**100.0%**	**2.62 [0.90, 7.57]**
Total events	18		7			
Heterogeneity: Chi² = 0.05, df = 1 (P = 0.83); I² = 0%						
Test for overall effect: Z = 1.77 (P = 0.08)						
1.1.6 Mixed comparators						
Simsek 2012 Ab	10	20	10	31	15.6%	2.10 [0.66, 6.67]
Daikos 2014 Kp	12	22	28	103	17.8%	3.21 [1.25, 8.27]
Tumbarello 2012 Ab	11	22	27	79	23.4%	1.93 [0.74, 5.01]
Batirel 2014 Ab	16	36	68	214	43.2%	1.72 [0.84, 3.52]
Subtotal (95% CI)		**100**		**427**	**100.0%**	**2.09 [1.33, 3.28]**
Total events	49		133			
Heterogeneity: Chi² = 1.11, df = 3 (P=0.77); I²=0%						
Test for overall effect: Z = 3.22 (P = 0.001)						

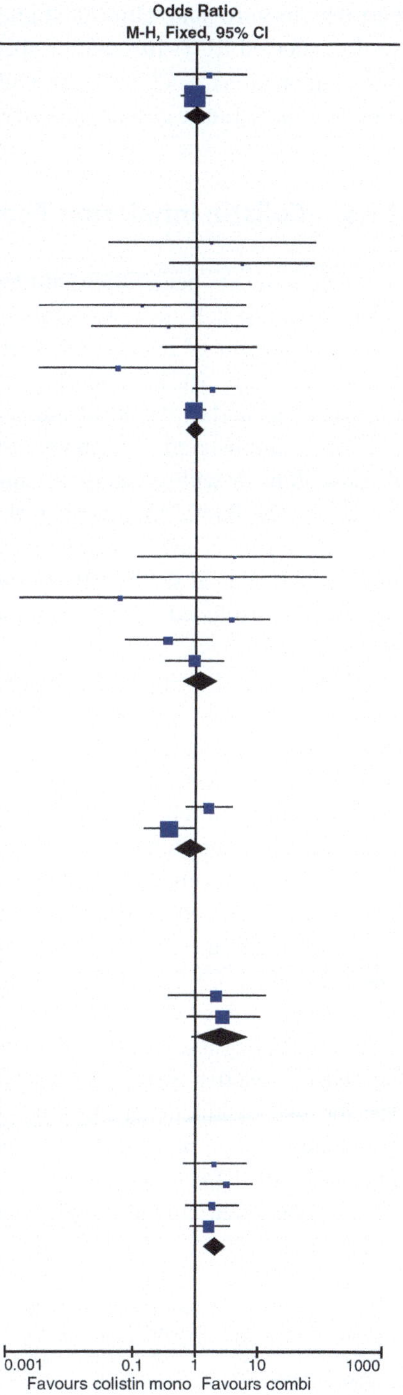

Fig. 11.7 All-cause mortality for colistin monotherapy vs. colistin combination therapy, unadjusted results

meropenem, and the results of the RCTs do not justify the use of this combination. Critical assessment of the observational studies shows very serious risk of bias and no significant survival advantage to specific polymyxin combinations. Lacking support for combination therapy for CRGNB, we believe that this practice should not be adopted as the routine. The discrepancy between in-vitro and clinical studies calls for well-conducted RCTs to examine specific antibiotic combinations. Such trials are under way and will determine future clinical practice.

11.5 Colistin Inhalation Therapy

Since polymyxins penetration into lung tissue is poor, nebulized colistin is sometimes being used for the treatment of respiratory tract infections. We searched for RCTs, cohort (prospective or retrospective) and case control studies comparing colistin administered as inhalation/nebulized therapy alone or with systemic treatment vs. systemic only antibiotic treatment in the treatment of ventilator-associated pneumonia or nosocomial pneumonia caused by MDR Gram-negative bacteria. We excluded studies examining patients with cystic fibrosis.

Three studies compared colistin inhalation alone vs. systemic antibiotic treatment for the treatment of pneumonia caused by *A. baumannii* or *P. aeruginosa* (one in neonates) [39–41]. None used matching nor reported on adjusted mortality rates. All-cause mortality was significantly lower among patients receiving colistin inhalation therapy alone compared to those treated with systemic treatment, usually polymyxins (unadjusted OR 0.37, 95% CI 0.17–0.82), with significant heterogeneity in results (Fig. 11.8).

Seven studies assessed the use of colistin inhalation as adjunctive therapy to systemic antibiotics for the treatment of *A. baumannii* (most commonly), *P. aeruginosa* or *K. pneumoniae*. One was a RCT [42], two used matching criteria for patients given colistin inhalations and those treated with systemic antibiotics alone [43, 44] and the remaining were unmatched and did not report an adjusted analysis for mortality [40, 45, 47]. The RCT showed no difference in mortality between study arms, while the observational studies showed a trend in favor of the adjunctive colistin inhalations, with heterogeneity in results (overall pooled OR 0.76, 95% CI 0.54–1.05, Fig. 11.9). A main concern with colistin inhalations is the induction of polymyxin-resistant bacteria, but the studies did not report on comparative resistance development rates. As expected, these studies show higher rates of eradication of the MDR bacteria from the respiratory tract with colistin inhalations.

These studies are suggestive of a possible benefit for colistin inhalation therapy, but these cannot form a basis for treatment recommendations. Selection bias is likely present in the analysis assessing colistin inhalations alone and this and other sources of bias affect the analysis of adjunctive colistin inhalations. The only RCT showed no advantage regarding survival for adjunctive colistin inhalations. Given the positive results of the observational studies, further RCTs are warranted and further observational studies should assess the long-term effects of colistin inhalations on the emergence of resistance.

11.6 Summary

Meta-analysis is an elegant tool to summarize outcome data gained from RCTs. Much of the data on polymyxins to date is based on observational studies at high risk of bias. The studies were unpowered to examine mortality, adjusting for all known

Study or Subgroup	Inhalation Events	Inhalation Total	IV Events	IV Total	Weight	Odds Ratio M-H, Fixed, 95% CI
Lu 2012	7	43	28	122	54.6%	0.65 [0.26, 1.63]
Naesens 2011	0	6	5	5	24.6%	0.01 [0.00, 0.41]
Nakwan 2011	1	8	5	7	20.8%	0.06 [0.00, 0.82]
Total (95% CI)		**57**		**134**	**100.0%**	**0.37 [0.17, 0.82]**
Total events	8		38			

Heterogeneity: $Chi^2 = 7.01$, df = 2 (P = 0.03); $I^2 = 71\%$
Test for overall effect: Z = 2.45 (P = 0.01)

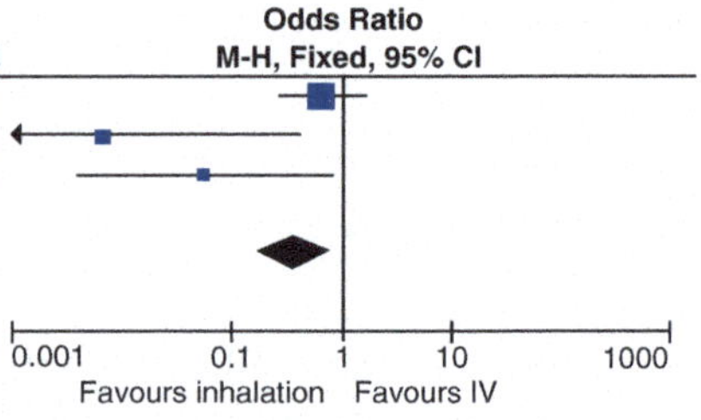

Fig. 11.8 All-cause mortality for colistin inhalations alone vs. systemic antibiotics in the treatment of pneumonia

Study or Subgroup	Inhalation+IV Events	Inhalation+IV Total	IV Events	IV Total	Weight	Odds Ratio M-H, Fixed, 95% CI
3.2.1 Randomized						
Rattanaumpawan 2010	22	51	20	49	10.7%	1.10 [0.50, 2.44]
Subtotal (95% CI)		**51**		**49**	**10.7%**	**1.10 [0.50, 2.44]**
Total events	22		20			
Heterogeneity: Not applicable						
Test for overall effect: Z = 0.24 (P = 0.81)						
3.2.2 Matched						
Kofteridis 2011	10	43	18	43	12.7%	0.42 [0.17, 1.07]
Tumbarello 2013	45	104	48	104	25.1%	0.89 [0.51, 1.54]
Subtotal (95% CI)		**147**		**147**	**37.8%**	**0.73 [0.46, 1.17]**
Total events	55		66			
Heterogeneity: $Chi^2 = 1.85$, df = 1 (P = 0.17); $I^2 = 46\%$						
Test for overall effect: Z = 1.31 (P = 0.19)						
3.2.3 Non-matched						
Bogovic 2013	6	8	17	23	2.0%	1.06 [0.17, 6.74]
Perez 2011	2	15	5	18	3.6%	0.40 [0.07, 2.45]
Kalin 2012	16	29	7	15	3.8%	1.41 [0.40, 4.91]
Naesens 2011	3	9	5	5	4.1%	0.05 [0.00, 1.17]
Amin 2013	8	28	5	12	4.6%	0.56 [0.14, 2.29]
Livermore 2010	1	8	9	15	5.0%	0.10 [0.01, 0.98]
Korbila 2010	31	78	19	43	13.6%	0.83 [0.39, 1.77]
Doshi 2013	16	44	27	51	14.7%	0.51 [0.22, 1.16]
Subtotal (95% CI)		**219**		**182**	**51.5%**	**0.60 [0.40, 0.92]**
Total events	83		94			
Heterogeneity: $Chi^2 = 8.01$, df = 7 (P = 0.33); $I^2 = 13\%$						
Test for overall effect: Z = 2.37 (P = 0.02)						
Total (95% CI)		**417**		**378**	**100.0%**	**0.70 [0.53, 0.94]**
Total events	160		180			
Heterogeneity: $Chi^2 = 11.25$, df = 10 (P = 0.34); $I^2 = 11\%$						
Test for overall effect: Z = 2.37 (P = 0.02)						
Test for subgroup differences: $Chi^2 = 1.77$, df = 2 (P = 0.41), $I^2 = 0\%$						

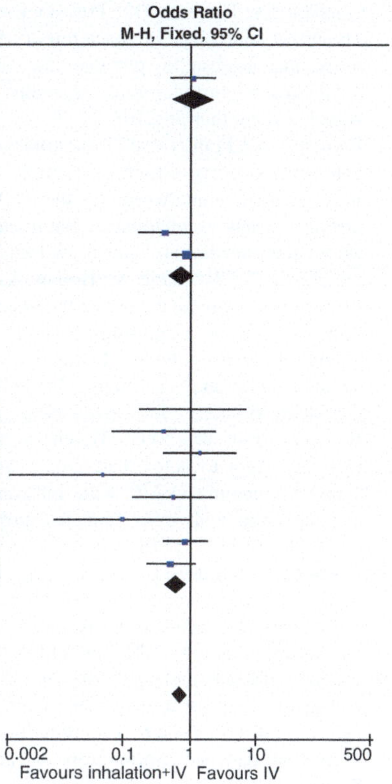

Fig. 11.9 All-cause mortality for colistin inhalations combined with systemic antibiotics vs. systemic antibiotics in the treatment of pneumonia

risk factors for mortality. Most studies were retrospective and had no control of treatment regimens and their modification during treatment. Meta-analyses of these studies suffer from the same sources of bias and only some of the biases can be accounted for by careful analysis of the methods.

We presented here only data on mortality. The original studies examined further outcomes including clinical cure and microbiological cure. We believe that for patients with severe infections caused by MDR Gram-negative bacteria survival is ultimately the only outcome that matters to the individual patient, while resistance development is relevant epidemiologically.

Systematically reviewing the evidence highlights areas of missing data. We are mostly missing RCTs examining treatment options for CRGNBs: the two polymyxins, different doses of the polymyxins, polymyxins vs. alternative antibiotics covering CRGNBs (e.g. aminoglycosides, tigecycline) and polymyxin monotherapy vs. specific combination therapies. These RCTs should examine mortality and resistance development, although the latter should also be examined in longitudinal studies befitting the timeframe of resistance development.

References

1. Yahav D, Farbman L, Leibovici L et al (2012) Colistin: new lessons on an old antibiotic. Clin Microbiol Infect 18:18–29
2. Koomanachai P, Tiengrim S, Kiratisin P et al (2007) Efficacy and safety of colistin (colistimethate sodium) for therapy of infections caused by multidrug-resistant Pseudomonas aeruginosa and Acinetobacter baumannii in Siriraj Hospital, Bangkok, Thailand. Int J Infect Dis 11:402–406
3. Lim SK, Lee SO, Choi SH et al (2011) The outcomes of using colistin for treating multidrug resistant Acinetobacter species bloodstream infections. J Korean Med Sci 26:325–331

4. Qureshi ZA, Paterson DL, Potoski BA et al (2012) Treatment outcome of bacteremia due to KPC-producing Klebsiella pneumoniae: superiority of combination antimicrobial regimens. Antimicrob Agents Chemother 56:2108–2113
5. Betrosian AP, Frantzeskaki F, Xanthaki A et al (2008) Efficacy and safety of high-dose ampicillin/sulbactam vs. colistin as monotherapy for the treatment of multidrug resistant Acinetobacter baumannii ventilator-associated pneumonia. J Infect 56:432–436
6. Durakovic N, Radojcic V, Boban A et al (2011) Efficacy and safety of colistin in the treatment of infections caused by multidrug-resistant Pseudomonas aeruginosa in patients with hematologic malignancy: a matched pair analysis. Intern Med 50:1009–1013
7. Garnacho-Montero J, Ortiz-Leyba C, Jimenez-Jimenez FJ et al (2003) Treatment of multidrug-resistant Acinetobacter baumannii ventilator-associated pneumonia (VAP) with intravenous colistin: a comparison with imipenem-susceptible VAP. Clin Infect Dis 36:1111–1118
8. Gounden R, Bamford C, van Zyl-Smit R et al (2009) Safety and effectiveness of colistin compared with tobramycin for multi-drug resistant Acinetobacter baumannii infections. BMC Infect Dis 9:26
9. Hachem RY, Chemaly RF, Ahmar CA et al (2007) Colistin is effective in treatment of infections caused by multidrug-resistant Pseudomonas aeruginosa in cancer patients. Antimicrob Agents Chemother 51:1905–1911
10. Kallel H, Hergafi L, Bahloul M et al (2007) Safety and efficacy of colistin compared with imipenem in the treatment of ventilator-associated pneumonia: a matched case-control study. Intensive Care Med 33:1162–1167
11. Ku K, Pogue JM, Moshos J et al (2012) Retrospective evaluation of colistin versus tigecycline for the treatment of Acinetobacter baumannii and/or carbapenem-resistant Enterobacteriaceae infections. Am J Infect Control 40:983–987
12. Kvitko CH, Rigatto MH, Moro AL et al (2011) Polymyxin B versus other antimicrobials for the treatment of pseudomonas aeruginosa bacteraemia. J Antimicrob Chemother 66:175–179
13. Oliveira MS, Prado GV, Costa SF et al (2008) Ampicillin/sulbactam compared with polymyxins for the treatment of infections caused by carbapenem-resistant Acinetobacter spp. J Antimicrob Chemother 61:1369–1375
14. Paul M, Bishara J, Levcovich A et al (2010) Effectiveness and safety of colistin: prospective comparative cohort study. J Antimicrob Chemother 65:1019–1027
15. Reina R, Estenssoro E, Saenz G et al (2005) Safety and efficacy of colistin in Acinetobacter and Pseudomonas infections: a prospective cohort study. Intensive Care Med 31:1058–1065
16. Rigatto MH, Ribeiro VB, Konzen D et al (2013) Comparison of polymyxin B with other antimicrobials in the treatment of ventilator-associated pneumonia and tracheobronchitis caused by Pseudomonas aeruginosa or Acinetobacter baumannii. Infection 41:321–328
17. Rios FG, Luna CM, Maskin B et al (2007) Ventilator-associated pneumonia due to colistin susceptible-only microorganisms. Eur Respir J 30:307–313
18. Couet W, Gregoire N, Marchand S et al (2012) Colistin pharmacokinetics: the fog is lifting. Clin Microbiol Infect 18:30–39
19. Plachouras D, Karvanen M, Friberg LE et al (2009) Population pharmacokinetic analysis of colistin methanesulfonate and colistin after intravenous administration in critically ill patients with infections caused by gram-negative bacteria. Antimicrob Agents Chemother 53:3430–3436
20. Akajagbor DS, Wilson SL, Shere-Wolfe KD et al (2013) Higher incidence of acute kidney injury with intravenous colistimethate sodium compared with polymyxin B in critically ill patients at a tertiary care medical center. Clin Infect Dis 57:1300–1303
21. Oliveira MS, Prado GV, Costa SF et al (2009) Polymyxin B and colistimethate are comparable as to efficacy and renal toxicity. Diagn Microbiol Infect Dis 65:431–434
22. Phe K, Lee Y, McDaneld PM et al (2014) In vitro assessment and multicenter cohort study of comparative nephrotoxicity rates associated with colistimethate versus polymyxin B therapy. Antimicrob Agents Chemother 58:2740–2746
23. Tuon FF, Rigatto MH, Lopes CK et al (2014) Risk factors for acute kidney injury in patients treated with polymyxin B or colistin methanesulfonate sodium. Int J Antimicrob Agents 43:349–352
24. Bellomo R, Ronco C, Kellum JA et al (2004) Acute renal failure – definition, outcome measures, animal models, fluid therapy and information technology needs: the second international consensus conference of the Acute Dialysis Quality Initiative (ADQI) Group. Crit Care 8:R204–R212
25. Zusman O, Avni T, Leibovici L et al (2013) Systematic review and meta-analysis of in vitro synergy of polymyxins and carbapenems. Antimicrob Agents Chemother 57:5104–5111
26. Paul M, Carmeli Y, Durante-Mangoni E et al (2014) Combination therapy for carbapenem-resistant gram-negative bacteria. J Antimicrob Chemother
27. Aydemir H, Akduman D, Piskin N et al (2013) Colistin vs. the combination of colistin and rifampicin for the treatment of carbapenem-resistant Acinetobacter baumannii ventilator-associated pneumonia. Epidemiol Infect 141:1214–1222
28. Durante-Mangoni E, Signoriello G, Andini R et al (2013) Colistin and rifampicin compared with colistin alone for the treatment of serious infections due to extensively drug-resistant Acinetobacter baumannii: a multicenter, randomized clinical trial. Clin Infect Dis 57:349–358
29. Paul M, Daikos GL, Durante-Mangoni E et al (2018) Colistin alone versus colistin plus meropenem for treatment of severe infections caused by

carbapenem-resistant Gram-negative bacteria: an open-label, randomised controlled trial., Lancet Infect Dis 18(4):391–400. https://doi.org/10.1016/S1473-3099(18)30099-9
30. Batirel A, Balkan II, Karabay O et al (2014) Comparison of colistin-carbapenem, colistin-sulbactam, and colistin plus other antibacterial agents for the treatment of extremely drug-resistant Acinetobacter baumannii bloodstream infections. Eur J Clin Microbiol Infect Dis 33(8):1311–1322. https://doi.org/10.1007/s10096-014-2070-6
31. Bergamasco MD, Barroso Barbosa M, de Oliveira Garcia D et al (2012) Infection with Klebsiella pneumoniae carbapenemase (KPC)-producing K. pneumoniae in solid organ transplantation. Transpl Infect Dis 14:198–205
32. Daikos GL, Tsaousi S, Tzouvelekis LS et al (2014) Carbapenemase-producing Klebsiella pneumoniae bloodstream infections: lowering mortality by antibiotic combination schemes and the role of Carbapenems. Antimicrob Agents Chemother 58:2322–2328
33. Falagas ME, Rafailidis PI, Kasiakou SK et al (2006) Effectiveness and nephrotoxicity of colistin monotherapy vs. colistin-meropenem combination therapy for multidrug-resistant gram-negative bacterial infections. Clin Microbiol Infect 12:1227–1230
34. Navarro-San Francisco C, Mora-Rillo M, Romero-Gomez MP et al (2013) Bacteraemia due to OXA-48-carbapenemase-producing Enterobacteriaceae: a major clinical challenge. Clin Microbiol Infect 19:E72–E79
35. Souli M, Kontopidou FV, Papadomichelakis E et al (2008) Clinical experience of serious infections caused by Enterobacteriaceae producing VIM-1 metallo-beta-lactamase in a Greek University Hospital. Clin Infect Dis 46:847–854
36. Tuon FF, Rocha JL, Arend LN et al (2013) Treatment and outcome of nine cases of KPC-producing Klebsiella pneumoniae meningitis. J Infect 67:161–164
37. Kalin G, Alp E, Akin A et al (2013) Comparison of colistin and colistin/sulbactam for the treatment of multidrug resistant Acinetobacter baumannii ventilator-associated pneumonia. Infect 42:37–42
38. Kontopidou F, Giamarellou H, Katerelos P et al (2014) Infections caused by carbapenem-resistant Klebsiella pneumoniae among patients in intensive care units in Greece: a multi-centre study on clinical outcome and therapeutic options. Clin Microbiol Infect 20:O117–O123
39. Lu Q, Luo R, Bodin L et al (2012) Efficacy of high-dose nebulized colistin in ventilator-associated pneumonia caused by multidrug-resistant Pseudomonas aeruginosa and Acinetobacter baumannii. Anesthesiology 117:1335–1347
40. Naesens R, Vlieghe E, Verbrugghe W et al (2011) A retrospective observational study on the efficacy of colistin by inhalation as compared to parenteral administration for the treatment of nosocomial pneumonia associated with multidrug-resistant Pseudomonas aeruginosa. BMC Infect Dis 11:317
41. Nakwan N, Wannaro J, Thongmak T et al (2011) Safety in treatment of ventilator-associated pneumonia due to extensive drug-resistant Acinetobacter baumannii with aerosolized colistin in neonates: a preliminary report. Pediatr Pulmonol 46:60–66
42. Rattanaumpawan P, Lorsutthitham J, Ungprasert P et al (2010) Randomized controlled trial of nebulized colistimethate sodium as adjunctive therapy of ventilator-associated pneumonia caused by gram-negative bacteria. J Antimicrob Chemother 65:2645–2649
43. Kofteridis DP, Alexopoulou C, Valachis A et al (2010) Aerosolized plus intravenous colistin versus intravenous colistin alone for the treatment of ventilator-associated pneumonia: a matched case-control study. Clin Infect Dis 51:1238–1244
44. Tumbarello M, De Pascale G, Trecarichi EM et al (2013) Effect of aerosolized colistin as adjunctive treatment on the outcomes of microbiologically documented ventilator-associated pneumonia caused by colistin-only susceptible gram-negative bacteria. Chest 144:1768–1775
45. Kalin G, Alp E, Coskun R et al (2012) Use of high-dose IV and aerosolized colistin for the treatment of multidrug-resistant Acinetobacter baumannii ventilator-associated pneumonia: do we really need this treatment? J Infect Chemother 18:872–877
46. Korbila IP, Michalopoulos A, Rafailidis PI et al (2010) Inhaled colistin as adjunctive therapy to intravenous colistin for the treatment of microbiologically documented ventilator-associated pneumonia: a comparative cohort study. Clin Microbiol Infect 16:1230–1236
47. Livermore DM, Hill RL, Thomson H et al (2010) Antimicrobial treatment and clinical outcome for infections with carbapenem- and multiply-resistant Acinetobacter baumannii around London. Int J Antimicrob Agents 35:19–24

Use of Colistin in Critically Ill Patients

12

Dror Marchaim, Donald Kaye, and Keith S. Kaye

Abstract

Due to lack of better therapeutic options, colistin use for extensively drug-resistant Gram-negative organisms was revived in the past two decades, including in patients in intensive-care units (ICU). There are multiple knowledge gaps pertaining to the clinical use and utility of colistin in critically-ill patients, but due to lack of options, it is used in these high risk patients. In this chapter, we critically review the various topics pertaining to colistin use in critically-ill patients, while highlighting the (lack of) controlled evidence supporting common current practices pertaining to colistin use by clinicians.

Keywords

ICU · Polymyxin · Critical care · Nosocomial infections · Hospital acquired infections

D. Marchaim (✉)
Unit of Infection Control, Shamir (Assaf Harofeh) Medical Center, Zerifin, Israel

Sackler School of Medicine, Tel-Aviv University, Tel-Aviv, Israel
e-mail: drormarchaim@gmail.com

D. Kaye
Drexel University College of Medicine, Philadelphia, PA, USA

K. S. Kaye
Division of Infectious Diseases, University of Michigan Medical School, Ann Arbor, MI, USA

12.1 Introduction

The polymyxins were discovered in 1947. They were widely used from 1962 until anti-*Pseudomonas* aminoglycosides became available after the middle to late 1960s. The parenterally administered polymyxins fell into disuse by the 1980s because of their nephrotoxicity.

The emergence and worldwide spread of multi-drug resistant and extensively-drug resistant gram-negative bacilli (GNB), has led to the revival of the use of polymyxins in general, and colistin in particular [1–3]. This chapter will focus on colistin (polymyxin E).

For severe bloodstream infections (BSI) caused by extensively-drug resistant (XDR) GNB, the polymyxins were frequently, up until recently, the only active therapeutic agents available other than tigecycline and certain aminoglycosides (e.g. amikacin, tobramycin) [4, 5]. However, there are important limitations to these alternatives. Tigecycline has a large volume of distribution with low serum levels, which limits its efficacy in treating BSI [6, 7]. In addition, several meta-analyses have demonstrated inferiority of tigecycline versus various comparators for certain clinical syndromes [8, 9], and there is an FDA warning of increased mortality pertaining to its use for severely septic patients [10]. Many clinicians perceive aminoglycosides as sub-optimal therapy for serious GNB infections when administered as the only active agent, particularly for

J. Li et al. (eds.), *Polymyxin Antibiotics: From Laboratory Bench to Bedside*, Advances in Experimental Medicine and Biology 1145, https://doi.org/10.1007/978-3-030-16373-0_12

invasive infections or for BSI [5, 11, 12]. Therefore, the use of the polymyxins and especially colistin has increased exponentially in parallel with the epidemic spread of XDR GNB infections. Even the recently marketed new agents, ceftolozane/tazobactam and ceftazidime/avibactam, have not contributed significantly to our available armamentarium against some XDR GNB, particularly versus *Acinetobacter baumannii* and some XDR metallo-beta-lactamase producing Enterobacteriaceae [13]. The increase in use of colistin began in the late 1990s, despite a lack of controlled clinical efficacy data for multidrug-resistant (MDR) and/or XDR-GNB infections [3, 14]. Moreover, little safety, pharmacokinetic (PK) and pharmacodynamic (PD) data were available for patients in general, and for acutely ill septic patients with XDR-GNB infections in particular [15]. Subsequently, in recent years, such data have become available.

This chapter will review the available clinical data pertaining to the use of colistin in non-cystic fibrosis, critically ill patients in the modern era. The major focus will be on parenteral administration, but topical use and use for selective decontamination of mucosal surfaces will also be discussed.

While many publications contain data pertaining to the efficacy of colistin in critically ill patients, the strengths in terms of analytic methodology vary considerably. For this reason, only the more important studies are cited and only a few will be reviewed in detail as illustrated in the following sections. In interpreting the data being reviewed, the reader should recognize the importance of certain factors (Table 12.1). These include: (1) the dose of colistin injected; (2) the patient age group being studied (adult versus pediatric patients); (3) the infectious clinical syndromes being studied (e.g., pneumonia, bloodstream infections, mixed infections, etc.); (4) the specific XDR-GNB pathogens being investigated (e.g., *Pseudomonas aeruginosa*, *A. baumannii*, carbapenem-resistant Enterobacteriaceae [CRE]); (5) renal function of the patients; (6) the concurrent administration of additional systemic therapeutics (either additional active drugs, or drugs administered as adjuncts to colistin, based on supposedly established synergistic properties); (7) the concurrent use of inhaled colistin with IV colistin for pulmonary infections; and (8) the intraventricular or intrathecal administration of colistin concurrent with intravenous colistin for central nervous system (CNS) infections. It is beyond the scope of this chapter to review the efficacy of colistin in each study based on each one of these categories, but we do categorize the available data based on some of these factors whenever sufficient data were available to do so. Emergence of resistance to colistin will be reviewed and the use of colistin as a selective digestive tract decontaminant (SDD) and/or as a selective oropharyngeal decontaminant (SOD) agent in critically ill patients will be covered.

Table 12.1 Factors affecting interpretation of studies of colistin efficacy

No.	Affecting parameter
1	Dose of colistin
2	Patient age group (adult versus pediatric patients)
3	Infectious clinical syndromes being studied (e.g., pneumonia, bloodstream infections, mixed infections, etc.)
4	Specific XDR-GNB pathogens being investigated (e.g., *Pseudomonas aeruginosa*, *Acinetobacter baumannii*, carbapenem-resistant Enterobacteriaceae [CRE])
5	Renal function of the patients
6	Concurrent administration of additional systemic therapeutics (either additional active drugs, or drugs administered as adjuncts to colistin, based on supposedly established synergistic properties)
7	Concurrent use of inhaled colistin with intravenous colistin for pulmonary infections
8	Intraventricular or intrathecal administration of colistin concurrent with intravenous colistin for central nervous system (CNS) infections

12.2 Dose

12.2.1 Lower and Higher Dose Colstin

The dosing of colistin has been confusing over the years. Publications have reported dosing schema based either on units of colistin (primarily in Europe) and milligrams of colistin base activity (primarily in the United States [US]). The conversion between these two metrics is as follows: one million units (MU) colistimethate sodium (CMS) ≈ 80 mg CMS ≈ 33 mg of colistin base activity (CBA). These dosing conversions are reviewed in detail in Chap. 10 of this book. The package inserts for colistin in both Europe and the US contain outdated information. Currently, the recommended dose in the package insert of IV colistin being distributed in the US is generally much higher than the recommended dose (per package insert) of the colistin being distributed in Europe. The recommended US dose for a patient with creatinine clearance greater than about 80 mL/min is 5 mg/kg/day of colistin base activity (CBA), which in an average 70 kg patient, is 350 mg CBA or ≈ 10.5 MU per day (which for the purposes of this chapter are considered "higher dose" as defined below). The corresponding recommended daily dose in the package insert of the product distributed in most European countries varies from 3 to 9 MU (i.e. 100 mg to 300 mg CBA) [16].

Although in many studies colistin had been dosed at levels lower than what is recommended in the US package insert [17], recent publications have virtually universally recommended the higher doses for colistin [11, 18, 19]. Due to the historical discrepancy in dosing, this chapter describes separately selected clinical experiences using "lower dose" and "higher dose" colistin. In both the US and Europe, dose reduction is suggested for patients with impaired kidney function and therefore the interpretation of results of "lower" and "higher" dose studies needs to be undertaken with caution. In addition, loading doses were used in some studies and not in others further confusing the distinction between "lower" and "higher" dose studies. The use of a loading dose is more routine in recent years, although controlled efficacy data are conflicting [20, 21]. For the purposes of this chapter, in differentiating between higher versus lower doses, the authors have used their judgment in choosing categories as sufficient information was often not available to assign precise cutoffs. For example, if a loading dose was given it was factored in to the high versus low dose categorization decision, potentially moving a study from a low to higher dose categorization.

12.2.2 Efficacy of 'Lower Dose' Colistin

With lower dose colistin, no loading dose was used, and therefore the achievement of steady state drug levels was likely delayed, in some cases by days, which might have impacted patients' outcomes and the measured efficacy of colistin [16, 22–24]. The issues pertaining to colistin dosing are discussed in detail in Chap. 15 of this book, but it is worth mentioning, that the "optimal dosing" of IV colistin is still debatable [15, 16, 25].

Historically, in Europe, low maximum daily doses of 200 mg CBA (i.e. 6 MU) and even 100 mg CBA (i.e. 3 MU) were used without a loading dose [26]. These dosing characteristics might have adversely impacted therapy and clinical outcomes. An example of a report of lower dose therapy early in the modern era of colistin use was an observational study executed at a single intensive care unit (ICU) in Athens, Greece, published in 2003 [27]. In this case-series analysis, the efficacy of intravenous (IV) colistin was evaluated as salvage therapy for critically ill patients with sepsis caused by XDR-GNB resistant to all other antibiotics available at the time. Twenty-eight relatively young (mean age 44.3 years) and severely ill patients (mean Acute Physiology and Chronic Health Evaluation II [APACHE II] score of 20.6), had received IV colistin at a dose of 3 MU (i.e. 100 mg CBA) every 8 h, adjusted for creatinine clearance. This dose was actually higher than the package insert recommendation available at that time in most European countries [16]. No loading dose was used, and compared to the US (based on the package insert), these doses would be considered as relatively 'low' doses. Sixteen of the patients had ventilator-associated pneumonia (VAP), three had a catheter-related BSI, one had post-traumatic meningitis, one had sinusitis, and one had a urinary-tract infection (UTI). For four patients, the infectious clinical syndrome was not determined. Infecting pathogens were *P. aeruginosa* (n = 20) and *A. baumannii* (n = 6). Four patients died within 48 h and were excluded from the efficacy analysis. Clinical response was observed for 73% of the treatments and survival at 30 days was 57.7%. Nephrotoxicity was observed in only 14.3% of patients and in only one case did the deterioration in renal function result in serious clinical consequences. The low numbers and the lack of a control group in this study limited the ability to extract meaningful, generalizable conclusions. However, these results were reassuring in terms of clinical outcomes and toxicity, as no other agents were available to treat these XDR-GNB infections in critically-ill patients [27].

A study from a different center in Athens was published in 2005, summarizing the data from patients enrolled from July 2001 to December 2003 [28]. The study included 43 critically ill patients with XDR-GNB infections given the same doses of colistin as in the previously described study. Various infectious syndromes were included (consisting mainly of VAP and BSI), all caused by *P. aeruginosa* and/or *A. baumannii* XDR strains. Good clinical response (cure or improvement) was again noted among the majority of patients (74.4%), and deterioration of renal function was again noted to be relatively minor, i.e. "only" among 18.6% of patients. The all-cause in-hospital mortality amounted to 27.9% among these septic individuals [28].

The same group later published a summary report pertaining to 258 patients who were enrolled at this center in Greece, over a 7-year period [23]. IV colistin was administered only for microbiologically documented XDR-GNBs: i.e. 170 (66%) *A. baumannii*, 68 (26.4%) *P. aeruginosa*, 18 (7%) *K. pneumoniae*, 1 (0.4%) *Stenotrophomonas maltophilia* and 1 (0.4%) *Enterobacter cloacae*. There were 155 cases of pneumonia (60%), 33 cases of bacteremia (13%), 22 abdominal infections (9%), 16 central venous catheter-related infections (6%) and 32 infections of other sites (12%). Cure of infection occurred in 79.1% of patients, nephrotoxicity in 10% and hospital survival in 65.1%. The average daily dose of colistin was relatively 'low': i.e. 6.1 ± 2.4 MU (i.e. 201 ± 79 mg CBA). Interestingly, in multivariable analysis, one of the independent predictors for survival was the colistin average daily dose (adjusted odds ratio for increased dose =1.22, CI-95% 1.05–1.42). The authors concluded that based on this large retrospective cohort analysis, colistin was a valuable antibiotic with acceptable nephrotoxicity and considerable effectiveness that depends on the daily dosage: i.e., higher doses are more effective than lower doses (though both 'high' and 'low' doses were relatively 'low' compared to doses used in the US) [23]. These reports resulted in a sense of security among clinicians in XDR-GNB endemic regions, who were also administering colistin (in similar dose ranges) to critically ill patients as salvage therapy.

In 2006, an additional retrospective case-series investigation was reported from Tunisia, from a single 22-bed ICU, where colistin was used as a salvage regimen for XDR *A. baumannii* or *P. aeruginosa* infections, mainly among young individuals (mean age of 48 years) [29]. The report summarized their experience with 78 patients enrolled between July 2003 and October 2004. IV colistin was administered in mean daily doses of 5.5 ± 1.1 MU/day, i.e., 182 ± 36 mg CBA (range 2–9 MU/day, 66–300 mg CBA). Most of the patients had pulmonary infection (78.2%), with the remaining having primary BSI (11.5%), UTI (7.7%) or meningitis (2.6%). No sub-analyses for specific infectious syndromes were performed. A favorable clinical response was noted in the vast majority of patients (76.9%), and nephrotoxicity was again documented among relatively few individuals (9%).

In a retrospective case-series analysis from Ankara, Turkey, 24 patients from a single center were enrolled, and the efficacy and toxicity of 'low dose' colistin was evaluated [30]. The investigators performed an interesting analysis, comparing two 'low dose' regimens of IV colistin: i.e., 3 MU per day (100 mg CBA/day) versus 6 MU per day (200 mg CBA/day). Clinical response rates were 69.2% and 72.7%, respectively ($p = 0.65$), microbiological response rates were similar ($p = 0.62$), and nephrotoxicity was revealed only in 1 of 13 patients (7.7%) for the 3 MU group and 2 of 11 patients (18.2%) in the 6 MU group ($p = 0.57$). The authors concluded that IV colistin, even in such 'low' doses, was relatively effective and non-toxic. In addition, and in contrast to other reports [23], there were no major differences in clinical outcomes between the two 'low' dose regimens [30]. A limitation of this study was that the renal function of patients receiving the 'low' dose regimens was not reported.

All these case series analyses, among many others, conducted on cohorts enrolled from the late 1990s to about 2006 were essential early reports pertaining to the use of relatively 'low doses' of colistin to treat MDR and XDR-GNB in the modern era. Colistin was mainly used as a last-resort agent in the treatment of relatively young and critically ill patients, when no other therapeutic was available. There were additional reports from the late 1990s and early 2000s that compared the efficacy of 'lower' doses of colistin versus beta-lactam agents, mainly carbapenems, for beta-lactam susceptible infections [31, 32]. However, most clinicians and researchers believe today that colistin should be used only for XDR-GNB infections, which in practice typically signifies carbapenem-resistant isolates [1, 33]. Therefore, these comparative efficacy analyses will not be further reviewed in detail. The encouraging early reports from the aforementioned "lower dose" case-series analyses soon were tempered by additional comparative studies evaluating the efficacy of colistin in the treatment of infections due to XDR-GNBs [11, 26, 34, 35].

A major event between 2006 and 2010 was recognition of the worldwide pandemic of carbapenem-resistant Enterobacteriaceae (CRE), which has substantially increased the burden of XDR-GNB infections all over the world, and the need for drugs with activity against these organisms such as colistin [36–38]. During these years tigecycline was also marketed as a theoretical alternative for the treatment of some XDR-GNB infections [39].

In 2010, two prospective controlled trials from the United Kingdom (UK) [34] and Israel [26] were published from centers still using relatively 'low doses' of colistin, without a loading dose. A clinical review of 166 consecutive patients infected or colonized with XDR *A. baumannii* was reported from 18 hospitals around London, UK [34]. Most subjects (62%) were critically ill (admitted to an ICU or high dependency unit) and 49% were carriers only (i.e., not infected). Interestingly, the survival rates among infected (68%) and colonized (67%) patients were practically the same, indicating little attributable mortality either to the pathogen or benefit to the treatment being administered. This finding of comparable outcomes between patients with *A. baumannii* respiratory infection and patients with lower respiratory colonization, was recently reported from an additional trial from Spain [40]. Of note, in the UK study, favorable outcomes (improved / resolved / cured infection) were achieved only among 50% of patients given colistin, versus 68% of patients given other agents ($P > 0.05$) [34].

The Israeli study was an observational prospective cohort investigation, conducted at a single, large tertiary centre [26]. Inclusion criteria included patients with pneumonia, UTI, surgical site infection, meningitis or bacteremia, all caused by GNB, and treated with colistin (3–6 MU or 100–200 mg CBA per day for patients with normal renal function [without a loading dose]) versus other agents. Patients were enrolled from May 2006 to July 2009. The primary outcome was 30-day mortality. Two hundred patients treated with colistin and 295 patients treated with other agents were included. Treatment with colistin was associated with older age, admission from healthcare facilities (versus home), mechanical ventilation, and lower rate of early appropriate antibiotic treatment. The 30-day mortality was 39% (78/200) for colistin versus 28.8% (85/295) for comparators (OR = 1.58, CI-95% = 1.08–2.31). Among the bacteremic patients (n = 220), the adjusted OR reflecting the inferiority of colistin increased to 1.99 (1.06–3.77). Nephrotoxicity at the end of treatment was more frequent with colistin, and treatment with colistin was associated with an increased incidence of infections due to *Proteus* species (*Proteus* species are inherently resistant to polymyxins). The authors concluded that in contrast to previous reports, where the same range of relatively 'low doses' were used, the efficacy and toxicity of colistin when studied in a prospective controlled fashion, particularly when other options (e.g., beta-lactam agents) were available, were associated with poorer survival rates and increased toxicity [26]. The Israeli study also noted a finding that was later repeated by others [11], that there were significant delays in initiation of appropriate therapy that had *in-vitro* activity against the target pathogens, and that colistin was often initiated several hours to days after ineffective empiric therapy [26]. Therefore, regardless of whether or not colistin had *in-vitro* activity against the causative pathogen, it was frequently prescribed late in the course of the acute infectious illness.

After these two prospective controlled trials were published, *in-vitro* and subsequently *in-vivo* PK/PD analyses were published, suggesting that, in most of the studies discussed so far in this chapter, colistin might have been administered in dosages that were too low [15, 41]. Reviews and editorials that summarized the data pertaining specifically to the usage of lower doses of colistin, vary considerably: i.e., from reports which claim that lower doses of colistin are relatively effective and safe [24], to reports claiming that colistin is inferior to comparators and is highly toxic [35]. These differing opinions reflect the analyses of efficacy of low dose colistin among critically ill patients in Europe over different time periods; higher doses have always been used in the US. The earlier period consisted of non-controlled case-series trials, and the later period consisted of controlled trials with a comparator arm. A recent study from Spain, in which colistin was used in lower doses (3–9 MU or 100–300 mg CBA per day, according to the discretion of the attending clinicians, without a loading dose) to treat XDR *P. aeruginosa* in 91 patients, clinical cure was observed in 72 (79%) patients. Interestingly, according to this trial, the mean colistin plasma levels at steady-state were not independently correlated with patients' clinical outcomes [42].

During the later time period, data emerged suggesting that several agents might possess synergistic activity with colistin versus certain XDR-GNB, and that colistin could be used in combination with other agents, in order to increase its efficacy and decrease the threat for emergence of resistance to colistin [43]. Thus, during the past several years, clinical research pertaining to colistin has shifted towards higher dose therapy and combination therapy. These issues are reviewed in part in a later section of this chapter and in detail in other chapters of this book.

12.2.3 Efficacy of 'Higher Dose' Colistin

The trend from use of 'lower dose' colistin in Europe to 'higher dose' use occurred recently; higher doses have always been used in the US, since its 'revival' [18, 19, 44]. Re-analyzing the pharmacologic properties of colistin in the modern era was crucial, since prior dosing was based on microbiological and PK data that were decades old (most studies were from the 1960s) [41]. In addition, dosing for critically ill patients based on PK data from non-critically ill patients (which was what was historically available with regards to colistin from the 1960s) in many instances is not appropriate [41].

In a study from Bari, Italy, published in 2012, 28 infectious episodes due to *A. baumannii* (46.4%), *K. pneumoniae* (46.4%), and *P. aeruginosa* (7.2%) XDR isolates were retrospectively analyzed [22]. Patients were given a loading dose of 9 MU colistin (300 mg CBA) and a 4.5 MU (150 mg CBA) twice-daily maintenance dose. In the presence of moderate to severe renal function, doses of 4.5 million units (150 mg CBA) were given at extended intervals, the duration based on renal function. The main types of infection were BSI (64.3%) and VAP (35.7%). Clinical cure was observed in 23 cases (82.1%), and only five patients (17.8%) developed acute kidney injury (all episodes subsided within 10 days after cessation of treatment). The authors concluded that based on this case-series analysis, a 9 MU (300 mg CBA) loading dose with 4.5 MU (150 mg CBA) twice daily and an extended-interval regimen of 4.5 MU (150 mg CBA) for renal insufficiency had good efficacy, without significant renal toxicity [22].

In a study from Turkey, the safety and efficacy of high-dose IV colistin for the treatment of VAP caused by *A. baumannii*, was retrospectively investigated at a single university hospital [45]. A total of 45 patients were enrolled: 15 patients received high-dose colistin (2.5 mg/kg CBA, i.e. 0.076 MU/kg, every 6 h), 20 patients received 'normal' doses (2.5 mg/kg CBA, 0.076 MU/kg, every 12 h), and 10 patients received lower than 'normal' doses, determined according to creatinine clearance. Of the 45 patients, 29 patients had concurrently received aerosolized colistin. Therefore, only 16 patients received parenteral colistin alone. The authors did not report a correlation between the treatment efficacies of IV colistin administered alone, versus the efficacy among patients who received in addition aerosolized colistin. According to the analysis of this entire Turkish cohort, there were no significant correlations between the dose of parenteral colistin and clinical cure, microbiological eradication rate, or mortality. Moreover, the higher doses of parenteral colistin and the addition of aerosolized colistin separately and independently increased the nephrotoxicity risk. The authors concluded that higher doses of colistin had no advantage over lower doses in treating *A. baumannii* VAP [45]. The study contained multiple potential biases and methodological flaws, but the results reported were interesting and noteworthy.

In a study from Detroit Medical Center, Michigan, USA, the efficacy of colistin (administered in high doses according to the US package insert instructions) was analyzed, compared to the efficacy of tigecycline [11]. Loading doses (of 5 mg / kg CBA, i.e. 0.152 MU/kg) were nonuniformly used throughout the study period. Adult patients with infections caused by *A. baumannii* (n = 82), CRE (n = 12), or both (n = 12) in 2009, who received ≥2 doses of colistin or tigecycline, were retrospectively studied. Seventy-one patients received colistin, 16 received tigecycline, and 19 received both colistin and tigecycline. Seven isolates were nonsusceptible to colistin and 79 to tigecycline. Patients receiving colistin alone or in combination were more likely to die during their hospitalization than patients receiving only tigecycline (P = .002). However, patients receiving colistin had higher indices of acute severity of illness and had notable delays in initiation of appropriate antimicrobial therapy ($P < 0.001$). No definite conclusions pertaining to the efficacy of (high dose) colistin vs. tigecycline could be drawn from this analysis, since the drugs were prescribed to populations with different baseline characteristics [11].

The same Israeli group that conducted the prospective single-center trial pertaining to low doses of colistin (2006–2009, 385 patients) [26], changed the practice in their hospital and conducted an additional prospective trial (2012–2015, 144 patients), comparing the outcomes between low dose and high dose regimens. The study revealed no association between high colistin dosing and all-cause mortality. However, high dosing was associated with increased nephrotoxicity [46].

12.3 Emergence of Colistin Resistance

There is a growing concern that the continually evolving pandemic of XDR-GNB infections leading to extensive use of colistin will result in the emergence and spread of colistin-resistant GNB. Non-susceptibility to colistin among Enterobacteriaceae and *A. baumannii* is defined as MIC >2 mg/L per US Clinical and Laboratory Standards Institute (CLSI) and European Committee on Antimicrobial Susceptibility Testing (EUCAST) criteria [47, 48]. Pertaining to *Pseudomonas* species, the CLSI breakpoints are similar [48], but the EUCAST criteria defines non-susceptibility to *Pseudomonas* as MIC >4 mg/L [47]. Since few alternative therapeutics for severe XDR-GNB invasive infections exist, colistin resistance could herald the spread of pan-resistant isolates [1, 49]. Initial reports of colistin resistance among clinical isolates emerged in Greece, where extensive use for treatment of XDR-GNB infections has been ongoing for several years [3, 50]. Notable, in Greece, colistin was used initially in relatively 'low' doses (as compared to current established practices) and without a loading dose [51, 52]. These factors might have contributed to or facilitated the emergence of colistin resistance among GNB [53].

A retrospective observational case-series analysis was published in 2007, pertaining to colistin-resistant *Klebsiella pneumoniae* strains isolated in a single Greek ICU during 2004–05 [51]. Overall, 18 strains isolated from 13 patients were reported, representing colonizing and/or infecting isolates. The patients' mean age was 70 years, and the mean APACHE II score upon admission to the unit was 22. All patients had a long hospitalization (median 69 days) and significant prior exposure to colistin (median 27 days). Colistin-resistant isolates were implicated as pathogens in two bacteremias, a ventilator-associated pneumonia and two soft tissue infections. Repetitive extragenic palindromic PCR (rep-PCR) identified six distinct clones, and evidence of horizontal transmission was also documented. The authors concluded (although this was not a case-control investigation) that selective pressure due to extensive colistin use might have led to the emergence of colistin resistance among *K. pneumoniae* isolates [51].

A different center from Athens, Greece, published a retrospective case-control matched analysis, pertaining to risk factors and outcomes of colistin-resistant XDR-GNB (*K. pneumoniae*, *A. baumannii*, and *P. aeruginosa*) infections, isolated from patients enrolled from 1/2006, until 4/2007 [54]. Case patients (n = 41) were those with a colistin-resistant XDR-GNB strain, and controls were selected from a pool of patients who had isolates susceptible to colistin. Controls were matched (1:1) to cases for species of GNB and for site of isolation. Various risk factors were significantly associated with the isolation of colistin-resistant isolates in bivariate analysis. However, in the multivariable model, recent prior use of colistin was identified as the only independent predictor (OR = 7.78, p = .002) [54].

A matched 1:3 case-control study was conducted in Piraeus, Greece, from 4/2008 to 6/2009 [55]. The study investigated factors predicting colistin-resistant bla_{KPC}-producing *K. pneumoniae* strain acquisition, compared to acquisition of colistin-susceptible *K. pneumoniae* strains, and the impact of colistin resistance on patient outcomes. Case patients (n = 13) were more often admitted from other institutions (p = 0.02) and had longer prior therapy with beta-lactam/beta-lactamase inhibitors (p = 0.002) compared to controls (n = 39). Colistin exposure was not a significant predictor for acquisition of colistin-resistant bla_{KPC}-producing strains. Nonetheless, two case patients carried a colistin-susceptible strain before yielding a colistin-resistant strain, and both were treated with colistin during the interval between cultures. In multivariable analysis, no parameter was significantly and independently associated with acquisition of a colistin-resistant strain. Resistance to colistin was also associated with increased mortality in bivariate analysis (p = 0.05). All 52 study isolates (13 cases and 39 matched controls) were clonally related (determined per pulsed-field gel electrophoresis [PFGE]), suggesting horizontal dissemination. The authors speculated that colistin-resistance had emerged from colistin-susceptible bla_{KPC}-producing *K. pneumoniae* isolates due to selective colistin pressure and then disseminated horizontally to other patients [55].

A cluster of five colistin-resistant bla_{KPC}-producing *K. pneumoniae* strains involving 3 healthcare facilities was reported from Detroit, Michigan, USA in 2009 [49]. An index case of colistin-resistant, bla_{KPC}-producing *K. pneumoniae*, was followed 20 days later by four additional cases occurring during a 6-day interval. All of the patients, at some point, had stayed in the same hospital and each patient had at least one opportunity for transmission with one of the other patients. Compared to 60 bla_{KPC}-producing colistin-susceptible strains of *K. pneumoniae* controls, isolated in the previous year, case patients were significantly older (p = 0.05), and the bla_{KPC}-producing *K. pneumoniae* organisms that were isolated from cases had higher MICs to imipenem than controls ($P < 0.001$). Prior colistin exposure was not a significant predictor for colistin resistance, although the index case did receive colistin prior to isolation of the colistin-resistant isolate. Genotyping (by both PFGE and by rep-PCR) revealed two closely related clones. The authors concluded that this outbreak was strongly linked to patient-to-patient transmission. However, emergence of resistance due to exposure to colistin was still a postulated mechanism [49].

An additional report from the University of Pittsburgh Medical Center, Pennsylvania, USA, was published in 2011 [56]. Five cases of infection due to colistin-resistant, bla_{KPC}-producing *K. pneumoniae* belonging to the international epidemic clone ST-258 (per multi-locus sequence typing [MLST]) occurred over a 4-month period.

Colistin resistance was attributed partly to selective antimicrobial pressure as well as transmission of already resistant organisms from patient-to-patient. The colistin-resistant isolates were able to persist in the absence of selective pressure when tested *in vitro* [56].

These reports suggest that in order to limit the emergence and spread of resistance, colistin should be used judiciously and dosed appropriately for patients with infections due to XDR-GNB strains that are susceptible to colistin, and that infection control measures are necessary to prevent the spread of XDR-GNB in healthcare facilities [57]. There are several mechanisms of resistance to colistin among XDR-GNB [44, 52, 53, 58, 59]. This topic is reviewed in detail elsewhere, in Chap. 5. Recently, plasmid-mediated outbreaks of genes (e.g., mcr-1) conferring resistance to colistin have been reported from multiple parts of the world, originating mainly from the food industry [60, 61]. However, some resistance mechanisms are chromosomally encoded, subject to induction in the presence of colistin [53]. As reviewed above, clinical studies demonstrated the intuitive and rational correlation between recent colistin exposure and emergence of colistin resistance, leading XDR-GNB strains to become pan-resistant isolates [1]. Therefore, it is important that colistin be prescribed to the appropriate patients, in optimal doses, in order to achieve maximum efficacy with minimum toxicity, while reducing the rate of emerging resistance to colistin among susceptible strains. One theoretical method to reduce the emergence of resistance to colistin is to use it in combination with another synergistic agent or agents [2, 5, 12].

12.4 Combination Therapy

12.4.1 Colistin Use in Combination with Other Synergistic Antibacterial Agents

Theoretical benefits of using colistin in combination with other agents include improved efficacy and preventing the emergence of colistin resistance [62]. Numerous agents have been studied as potential synergistic adjuncts to colistin for the treatment of XDR-GNB including rifampin, extended-spectrum cephalosporins, penicillins combined with beta-lactamase inhibitors, tigecycline, tetracyclines, aminoglycosides, macrolides, fosfomycin, and fluoroquinolones [63–87]. However, carbapenems are the class that has attracted most of the attention with regards to synergism with colistin [12, 88–91]. Although the majority of XDR-GNB isolates causing human infections are resistant to carbapenems, carbapenems are safe beta-lactam agents, and according to several *in-vitro* reports, have strong synergism in combination with colistin, against various XDR-GNB pathogens [92].

A systematic review and meta-analysis recently summarized studies which examined the *in-vitro* interactions of antimicrobial combinations consisting of any carbapenem with colistin or polymyxin B against various GNB [89]. Synergy was tested using various methodologies including time-kill, checkerboard, and E test. The meta-analysis summarized 39 published studies and 15 conference proceedings, reporting on 246 different analyses conducted on 1054 bacterial isolates. In time-kill studies, combination therapy of colistin with a carbapenem demonstrated synergy rates of 77% for *A. baumannii*, 44% for *K. pneumoniae*, and 50% for *P. aeruginosa*. Doripenem showed high synergy rates against all three bacteria. For *A. baumannii*, meropenem was more synergistic than imipenem, whereas for *P. aeruginosa* the opposite was true. Checkerboard and E test studies generally reported lower synergy rates than time-kill studies. The use of combinations led to less *in-vitro* development of resistance [89].

In a multicenter retrospective cohort study, conducted in 3 large Italian teaching hospitals, among 125 patients with BSI caused by KPC-producing *K. pneumoniae* isolates, overall 30-day mortality was 41.6% [43]. However, a significantly higher mortality rate was noted among patients who were treated with colistin monotherapy (54.3%) compared to patients treated with combination therapy (34.1%, $p = 0.02$). Therapy with a combination of colistin, tigecycline, and meropenem was associated with the lowest mortality rate. The authors concluded that in order to improve survival, combination treatment with 2 or more drugs, especially regimens including a carbapenem, may be more clinically effective than monotherapy [43].

In a large prospective randomized, open-label superiority trial conducted at six hospitals in Israel, Greece, and Italy, it was investigated whether combining colistin with meropenem improves clinical outcomes for adults with infections caused by carbapenem-resistant or carbapenemase-producing Gram-negative bacteria [93]. Of the 406 patients enrolled (with bacteraemia, ventilator-associated pneumonia, hospital-acquired pneumonia, or urosepsis), the majority were caused by *A. baumannii* (312/406, 77%). In terms of clinical failure at day 14 following randomization (primary outcome), there were no significant differences between patients assigned to colistin monotherapy (156/198, 79%), versus those randomly assigned to combination therapy with colistin plus meropenem (152/208, 73%). Although combination therapy increased the incidence of diarrhea, it decreased the incidence of mild renal failure (37 [30%] vs. 25 [20%]). This study concluded that combination therapy was not superior to monotherapy, and that the addition of meropenem to colistin did not improve clinical failure in severe *A. baumannii* infections (the trial was under-powered to specifically address other bacteria) [93]. In addition to being open label, another limitation was the extremely high failure rate (almost 75% of patients had clinical failure). Such a high failure rate was similar to what would have been expected if no appropriate therapy had been administered. It is possible that non-infectious causes were the primary drivers of failure [94]. A recent meta-analysis concluded that for MDR and XDR *A. baumannii*, combination therapy of colistin with sulbactam demonstrated superiority in terms of microbiological cure compared to colistin monotherapy [95]. Thus, it remains uncertain if colistin monotherapy is truly equivalent to combination therapy for treatment of *A. baumannii* infection.

To avoid potential biases, the most definitive way to study the efficacy of combination therapy with colistin is to conduct a prospective randomized, double blind trial. Such a study is on-going, supported by the National Institutes of Health, investigating the efficacy of colistin and meropenem versus colistin monotherapy for treatment of XDR-GNB infections. Secondary end-points are emergence of resistance to colistin, and development of nephrotoxicity. This investigation will help to better elucidate whether the addition of meropenem to colistin has any advantageous role in terms of patients' outcomes.

12.4.2 The Efficacy of Inhaled Colistin Alone or in Combination with Parenteral Colistin for Critically Ill Patients with Pneumonia

In the early years of colistin's revival and use among non-cystic fibrosis (CF) critically ill patients, it was often used to treat patients with XDR *A. baumannii* and/or XDR *P. aeruginosa* infections [27]. Both of these non-glucose fermenting GNBs are common causes of hospital acquired pneumonia (HAP) and ventilator associated pneumonia (VAP) in critically ill patients [5]. Because colistin penetrates poorly to infected and relatively avascular lung tissues [96], it was postulated that a combined regimen, using parenteral colistin in conjunction with inhaled colistin, might improve clinical benefit [97, 98]. This assumption relied on multiple investigations in CF patients examining the efficacy and the expected favorable PK/PD properties of this "combined" regimen [99]. Moreover, it was postulated initially that local delivery of inhaled colistin to the lungs could minimize potential renal and neurological toxicities of systemic colistin, although the adverse events of aerosolized colistin were already recognized, i.e. bronchoconstriction, cough, and chest tightness [98].

One of the earliest publications pertaining to aerosolized colistin use in critically ill non-CF patients was published in 2000, from a single ICU from New England Medical Center, Boston, MA, USA [100]. Improvement in three patients with MDR *P. aeruginosa* VAP was reported, following the addition of aerosolized colistin to antipseudomonal systemic therapy. No treatment-limiting adverse effects were noted [100]. A subsequent small study from Singapore reported on the efficacy of nebulized colistin administered to 21 patients with polymyxin-susceptible *A. baumannii* and *P. aeruginosa* pneumonia. None of the patients received IV colistin. The rate of successful clinical response in this cohort was high, i.e. 86% [101].

In 2005 a small retrospective analysis from a Greek ICU was published, which enrolled 8 patients with pneumonia due to *A. baumannii* (*n* = 7) or *P. aeruginosa* (*n* = 1) from 10/2000 to 1/2004 [102]. Aerosolized colistin was administered in a dose of 1.5 to 6 MU/day (45 to 180 mg/day CBA), divided in 3–4 doses. IV colistin was administered concomitantly to all patients. The cure rate was 88%, which was higher than the historical cure rate of 67% for parenteral colistin alone. However, small numbers precluded statistical significance of the finding. No treatment-limiting adverse effects were noted [102].

Two reviews were published in the following year (2006), pertaining to the role of aerosolized colistin according to the data gathered thus far, both determining that additional controlled data is needed in order to allow determination of the incremental benefit of the addition of aerosolized colistin to systemic antibiotics [98, 103].

The same Greek investigators as in the 2005 report published additional data regarding their experience with aerosolized colistin in 2008 [104]. Sixty critically ill patients with a mean APACHE II score of 16.7, received aerosolized colistin (always along with pareneteral colistin) for the treatment of VAP due to *A. baumannii* (n = 37), *P. aeruginosa* (n = 12) and *K. pneumoniae* strains (n = 11). The mean daily dosage of aerosolized colistin was 2.2 ± 0.7 MU (73 ± 23 mg CBA) and the mean duration of administration was for 16.4 days. Bacteriological and clinical response of VAP was observed in 50/60 (83.3%) patients. No adverse effects related to inhaled colistin were recorded. The authors advocated aerosolized colistin as an adjunctive therapy to IV treatment for patients with VAP due to XDR-GNB. The authors also stated, that controlled comparative studies were needed in order to establish the true effectiveness and safety of this therapeutic regimen [104].

Two studies were published in 2010. The first was a retrospective matched case-control study, performed in a single ICU in Heraklion, Crete, Greece, from 1/2005 through 12/2008 [105]. Forty-three patients with VAP due to XDR-GNB (66 *A. baumannii*, 12 *K. pneumoniae*, and 8 *P. aeruginosa*) received aerosolized colistin plus IV colistin and were matched on the basis of age and APACHE II score with 43 control patients who had received IV colistin alone. Demographic characteristics, clinical status, and the relative proportion of XDR pathogens within the group, were similar between the two treatment groups. No significant differences between the groups were observed regarding eradication of pathogens, clinical cure, and mortality. Eight patients (19%) in each treatment group developed reversible renal dysfunction. No aerosolized colistin-related adverse events were recorded. The authors concluded that the addition of aerosolized colistin to IV colistin did not provide additional therapeutic benefit to patients with VAP due to XDR-GNB [105]. The second study was a comparative analysis from a different Greek institution [106]. Seventy-eight patients with VAP received IV plus inhaled colistin, whereas 43 patients received IV colistin alone. Groups were not matched on any specific parameter pertaining to a patient's background condition or characteristics and/or to any indices related to the severity of the acute illness. In addition, the mean daily dosage of IV colistin was higher in the group that received aerosolized colistin (7 ± 2.4 MU [231 ± 79 mg CBA] vs. 6.4 ± 2.3 MU [211 ± 76 mg CBA], respectively), although the difference was not significant (p = 0.13). The outcome was cure for 62/78 (79.5%) patients who received IV and inhaled colistin vs. 26/43 (60.5%) patients who received IV colistin alone (p = 0.025). In multivariable analysis of VAP "cure", the use of inhaled colistin remained an independent significant predictor for cure (OR = 2.53, CI-95% = 1.1–5.8). However, the definition used for "cure" was not optimal or clinically meaningful. Moreover, all-cause in-hospital mortality and all-cause ICU mortality did not differ between the study groups. In the multivariable analysis of mortality, one of the independent predictors was lower dosages of systemic colistin. Thus, because the group that received colistin alone without inhaled colistin received lower doses of IV colistin, analyzing the impact of treatment regimen on clinical outcomes was biased [106].

As of today, no randomized controlled trials (RCT) have been published analyzing the role of inhaled colistin, and the matter is still being debated. It is unclear how often aerosolized colistin is used today. A multi-national survey published in 2013 was answered by 611 departments from 70 countries. Eighty percent of respondents had a "positive opinion" concerning use of nebulized colistin [107]. A recent report suggested that inhaled colistin could induce severe pulmonary eosinophilia [108].

A systematic review published in 2012 analyzed in a meta-regression the overall efficacy of colistin in GNB VAPs, compared to other non-colistin containing appropriate regimens. In the colistin arm, both IV colistin alone and in combination with aerosolized colistin, were included [109]. In the control arm, GNB VAP cases treated appropriately with non-polymyxin regimens were eligible. Six controlled studies met inclusion criteria. Clinical response did not differ significantly between the colistin and control group (OR, 1.14; $p = 0.56$). There was no difference in clinical response after controlling for concomitant antibiotic treatment. Treatment with colistin did not impact hospital mortality (OR, 0.92; $p = 0.78$) or nephrotoxicity (OR, 1.14; $p = 0.7$). Although the authors concluded that their results suggest that colistin may be as safe and as efficacious as standard antibiotics for the treatment of VAP [109], we believe that colistin should be reserved for treating only HAP or VAP caused by XDR-GNB, which are susceptible only to colistin. The role of inhaled colistin as an adjunct to systemic colistin for treatment of pneumonia remains unclear and is reviewed in detail in Chap. 15.

12.5 Colistin Use in Central Nervous System (CNS) Infections

Due to the evolving epidemiology of nosocomial infections over the past 2 decades, and the spread of MDR organisms (MDRO) in healthcare settings, the prevalence of post neurosurgical CNS infections caused by XDR-GNB has risen substantially [110, 111]. The XDR-GNB that are sometimes associated with post-neurosurgical CNS infections are frequently susceptible only to certain aminoglycosides, tigecycline, and colistin [110]. Tigecycline is bacteriostatic, expensive, and its role in CNS infection is undefined [110]. Colistin, which penetrates poorly into the CNS, attracted early attention as an agent that could be administered directly into the CNS, for treatment of XDR-GNB ventriculitis or meningitis [110–115].

In 1999 Spanish investigators reported two cases of catheter-associated ventriculitis caused by XDR *A. baumannii* that were successfully treated with intraventricular colistin (5 mg CBA [0.152 MU] q12 hours) [116]. The intraventricular administration of colistin was coupled with systemic parenteral antibiotics other than colistin [116]. In 2000, an additional report from Argentina reported on a patient treated with intrathecal colistin for XDR *A. baumannii* meningitis [117], with a regimen consisting of 5 mg (0.167 MU) per day of intrathecal colistin on day 1 and 10 mg (0.33 MU) of intrathecal colistin per 24 h for 21 days thereafter. No additional systemic antimicrobial agents were administered, and the patient was cured [117]. In Turkey, intrathecal colistin (5 mg CBA [0.152 MU] per day for 21 days) was successfully used to treat a case of ventriculitis caused by XDR *P. aeruginosa* [118]. In 2004, a case of post-surgical meningitis caused by XDR *A. baumannii* was treated successfully by using 40,000 units (1.3 mg CBA) per day intrathecally of the European formulation of colistin [111]. In a different study, a patient was successfully treated with much higher intrathecal doses, i.e. 125,000 units (4.13 mg CBA) administered twice daily [113]. The dosing of intrathecal colistin remains uncertain as no controlled comparative trial had ever been conducted assessing the 'true' role and efficacy of intrathecal colistin administered to critically ill patients with CNS infections. In a case series publication from Thailand, published in 2010, the clinical and microbiological cure rates reported for 24 patients treated with intrathecal / intraventricular colistin therapy were extremely high, at 83 and 92% respectively [115]. Three patients (13%) developed chemical ventriculitis and one (4%) experienced treatment-associated seizures. In 2013, a group from Athens, Greece, published a case-series analysis [119] of six patients and a literature review [120] pertaining to intraventricular and intrathecal colistin usage for the treatment of MDR and XDR *A. baumannii* ventriculitis and meningitis. A recent additional case-series analysis from Novara, Italy, reported on three patients who were cured with concurrent administration of intraventricular and intravenous administration of colistin [121].

In a recent study from Turkey of XDR *A. baumannii* CNS infections treated with intrathecal colistin, that included 77 patients, the overall mortality rate was 57.1% (44 patients died). The variables associated with increased all-cause mortality during hospitalization included old age ($p = 0.026$) and failure to achieve CSF sterilization ($p = 0.01$) [122].

Although there is literature supporting its use, the intrathecal / intraventricular administration of colistin should only be used when absolutely necessary, for patients with post-neurosurgical meningitis caused by microbiologically documented XDR-GNBs, resistant to all other appropriate agents. Many expert clinicians agree that intrathecal administration must be coupled with administration of parenteral appropriate antibiotics (not necessarily colistin), but this too has never been thoroughly studied or proven [117].

12.6 Colistin Use for Selective Decontamination of Mucosal Surfaces in Critically Ill Patients

Selective digestive tract decontamination (SDD) and selective oropharyngeal decontamination (SOD) are measures that have been studied for their possible role in preventing pneumonia, particularly hospital-acquired (HAP) and ventilator-associated (VAP) pneumonias, and reducing overall mortality rates [123]. Colistin had been used as part of SDD and SOD regimens since its early years of distribution in the 1960s [124]. There were numerous trials pertaining to the usage of colistin as part of SOD and/or SDD regimens over the years [125–140]. However, this practice was eventually abandoned by most countries.

In Holland, this practice was revived and successfully used in several national studies [123, 129]. Today, SDD and SOD administration are common practice for many patients in the ICU. The regimen used in the majority of units consists of 4 days of intravenous cefotaxime and topical application of tobramycin, amphotericin B, and colistin [123, 129]. In a recent study involving 13 ICUs in Holland, a total of 5939 patients were randomized to standard care, standard care plus SOD or standard care plus SDD. The mortality rate associated with standard care was 27.5% at day 28, but the rate was significantly reduced by an estimated 3.5 percentage points with standard care plus SDD and by 2.9 percentage points with standard care plus SOD [129]. The same Dutch group, showed in various analyses, all conducted in Dutch ICUs, that SOD/SDD could actually lead to reductions in MDRO acquisitions [141], despite its huge impact on the human gut resistome [142]. However, Holland is well known for its low rates of MDRO [143]. One of the main concerns in other countries with higher rates of MDRO, is that SOD/SDD could select for increased MDRO isolations and possibly colistin resistance [144]. There is evidence that the use of colistin as SOD [145, 146] or SDD [147], can lead to the emergence of resistance to colistin among common human pathogens.

12.7 Topical Colistin

Topical colistin is rarely used in critically ill patients, apart from those purposes that were reviewed in previous sections of this chapter: i.e. intrathecal, in aerosols, or as part of SDD/SOD regimens. Additional topical colistin preparations available in certain countries outside of the US include cutaneous (e.g. cream, ointment), ophthalmic (drops and ointments), and eardrops.

Colistin bladder instillation, as an alternative way of treating XDR-GNB urinary tract infection (UTI), was used in a study from Rome, Italy [148]. Study rationale was based on previous successful usage of aerosol and intrathecal topical colistin. It was speculated that even high concentrations of colistin instilled directly into biological fluids, would be relatively safe. In this case series, three patients were treated with intravesicular intermittent instillations of colistin for XDR *A. baumannii* UTI [148]. A 100,000 unit (3.3 mg CBA) dose of colistin was delivered through a urinary catheter three times a day for 7 days. One patient experienced suprapubic pain at the end of the instillation [148]. Concurrent systemic colistin was not uniformly prescribed to patients throughout the intravesicular treatment course. Bacterial eradication and clinical cure were achieved for all patients [148]. It should be stressed that when CMS is instilled into the bladder, the time-course for the transformation to colistin is not known, and is probably dependent in part upon the rate of hydrolysis, instillation and/or the washout conditions. Nonetheless, there are reports in which bladder irrigation with polymyxins through a urinary catheter was not as effective inpreventing UTI or eliminating bacteriuria [149, 150]. Continuous bladder irrigation of a colistin mixture through a triple lumen catheter is an additional mode of topical use of colistin that has been used to treat XDR *A. baumannii* UTI [151].

In a study published in 1991 that included 321 patients with chlamydial conjunctivitis, the efficacy of colistin being administered as part of an ophthalmologic ointment containing additional antibiotics (tetracycline and chloramphenicol) was compared to ointments containing various antibiotics (with no colistin) along with dexamethasone [152]. The specified efficacy of topical colistin ointment could not be determined from this report. The overall efficacy was higher for ointments composed of antibiotics combined with dexamethasone [152]. Topical colistin combined with other antimicrobials was also used successfully for treatment of *Nocardia* keratitis [153]. Currently, the role of topical colistin for treatment of ophthalmologic or urinary tract infections remains unclear.

12.8 Colistin Use in Pediatric Patients

Although it is sometimes necessary to use colistin in pediatric patients, evidence-based dosing recommendations in children are currently not available [154, 155]. Daily regimens available in the published literature range from 40,000 U/kg/day (1.3 mg CBA) to 5 mg/kg/day CBA (0.152 MU). The role of a loading dose in the pediatric population has not been established [15].

In 2009, a systematic review was published, evaluating the available clinical evidence regarding the effectiveness and safety of systemic colistin in children without CF [156]. Ten case series and fifteen case reports, including a total of 370 children, were included. However, only 17 of the children were from studies published after 1977. A total of 326 children received colistin for the treatment of infections and 44 for surgical prophylaxis or prophylaxis of infections in burns. Overall, 271 of 311 children included in the identified case series were clinically evaluable. Of these 271 children, 235 (86.7%) were cured of the infection, 10/271 (3.7%) improved, 6/271 (2.2%) deteriorated and 20/271 (7.4%) died. Fourteen (70%) of the 20 deaths were attributed to the infection. Nephrotoxicity occurred in 2.8% of children. No sub-analysis pertaining to the cases treated after the revival of colistin in the 'modern era' was executed. The authors concluded that systemic colistin is an effective and acceptably safe option for the treatment of children without CF who have XDR-GNB infections [156]. However, the fact that the review was based on old data limits generalizability of the findings to current day practice.

In 2012, a multicenter study from Turkey analyzed the efficacy and safety of colistin therapy in pediatric patients with severe infection caused by XDR-GNB from pediatric ICUs (PICU) [154]. There were 87 episodes in 79 pediatric patients from five different PICU. The most commonly isolated microorganism was *A. baumannii* and the most common type of infection was VAP. The mean colistin dose in patients without renal failure was 5.4 ± 0.6 mg/kg/day CBA (0.16 ± 0.02 MU/kg/day), the mean therapy duration was 17.2 ± 8.4 days and the favorable outcome rate was 83.9%. Serious side effects were seen in four patients (4.6%): two patients suffered renal failure and the others had seizures. The authors concluded that colistin was effective in the treatment of severe infections caused by XDR-GNBs in PICU [154]. Unlike adults, no controlled comparative data are yet available pertaining to efficacy and safety of colistin, and as in adults, it currently should be used only for infections caused by XDR-GNB susceptible to colistin and resistant to all other treatment options.

An additional more recent trial from a single PICU at Ankara, Turkey, retrospectively investigated the efficacy of colistin among 29 children who were treated with colistin for 38 courses in calendar year 2011 [157]. VAP was the most common clinical syndrome and *A. baumannii* was the most common causative XDR-GNB pathogen. Two colistin formulations were used: 1) colimycin (Kocak Farma) as used in 21 episodes (median dosage was 5 mg/kg/day [i.e. 0.167 MU/kg/day]), and 2) colomycin (Forest Laboratories) was used in 17 episodes (median dosage was 0.075 MU/kg/day [i.e. 2.25 mg/kg/day]. Good or partial clinical response was evident among 30 (79%) patients, and 8 (21%) had a poor clinical response. No statistically significant differences were noted between the two formulations in terms of the clinical efficacy or the clinical toxicity. Ten patients died. The authors suggested that the use of colistin was well-tolerated and efficacious [157]. In the same year, data from a single NICU at Ankara, Turkey, reported the efficacy of colistin administered to 21 preterm neonates [158]. The median duration and dose of colistin therapy were 9 days and 3 mg/kg/day (0.1 MU/kg/day), respectively. Recovery rate was 81% (17/21), and microbiological clearance was 69% (9/13). The major side effects were acute kidney injury (19%) and electrolyte disturbances (24%), specifically magnesium disturbances. Both acute kidney injury and electrolyte disturbances including hypomagnesemia were reversible. The authors concluded that colistin was efficacious among preterm neonates, though renal function tests and serum electrolytes should be closely monitored [158].

The role of inhaled colistin in critically ill pediatric patients without cystic fibrosis was evaluated [159]. Of three children admitted to a PICU in Athens, Greece, between 2004 and 2009, 2 received inhaled colistin (937,500 U [30.9 mg CBA], diluted in 3 ml of normal saline twice daily) as monotherapy for tracheobronchitis, and 1 as adjunctive therapy for necrotizing pneumonia. All three children recovered from the infections. Also, a gradual reduction, and finally total elimination of the microbial load in bronchial secretions was observed during inhaled colistin treatment courses. No bronchoconstriction or other toxicity related to colistin was observed [159].

The use of aerosolized colistin has been evaluated in neonates as well. A neonatal ICU (NICU) from Ankara, Turkey, reported their experience with aerosolized colistin in two preterm and one term neonate with *A. baumannii* and/or *P. aeruginosa*-related VAP who were unresponsive to other antimicrobial regimens [160]. An aerosolized colistin dose of 5 mg/kg CBA (0.152 MU/kg) was administered every 12 h. VAP was treated for a minimum of 14 days. No adverse events were noted, and all patients clinically improved [160].

12.9 Summary

Due to spread of XDR-GNB organisms and lack of new therapeutic alternatives, colistin use was revived after being abandoned for several decades. The worldwide use of colistin has risen

Table 12.2 Authors' recommendations for colistin dosing

Colistin (mg/kg CBA daily) using ideal body weight		Clcr ≥50 ml/min	Clcr 30–49 ml/min	Clcr 10–29 ml/min	Clcr <10 ml/min or hemodialysis
		Start 8 hours after load	Start 12 hours after load	Start 12 hours after load	Start 24 hours after load
	5 mg/kg loading dose (max 300 mg)	5 mg/kg/day divided every 8 hours	3.5 mg/kg/day divided every 12 hours	2.5 mg/kg/day divided every 12 hours	1.5 mg/kg every 24 hours

CBA Colistin base activity, *Clcr* Creatinine clearance

exponentially in the past 20 years. Although understanding of PK/PD properties as well as efficacy of colistin has improved recently, there still exist multiple knowledge gaps pertaining to its clinical use. Currently, controlled clinical studies are underway, the results of which will hopefully standardize the way that this agent is used in the treatment of critically ill patients.

Colistin is a relatively toxic agent, particularly when it is prescribed in the high doses that are often needed to provide the greatest probability of clinical effectiveness in critically ill septic patients [18, 19]. It is also inferior to safer drugs such as beta-lactam agents [26, 34]. Due to these factors, and concern regarding emergence of resistance to colistin following its use [54], the use of colistin should be restricted to instances when other established and safer alternatives are unavailable. Its use should be reserved for infections caused by XDR-GNB susceptible only to colistin (excluding tigecycline susceptibility). Based on recent data, we recommend giving a loading dose of colistin (5 mg/kg CBA or 0.152 MU/kg up to a maximum of 300 mg of CBA or 9 MU) followed by 5 mg/kg/day CBA or 0.152 MU/kg/day in divided doses every 8 h with adjustment for renal insufficiency (see Table 12.2 and Chap. 15).

The use of concurrent synergistic agents, in order to increase treatment efficacy and curb the emergence and spread of colistin-resistant and pan-resistant GNB isolates, should be further explored. However, unfortunately in clinical scenarios when colistin is prescribed, it is often administered after other treatments have failed and the infection has been active for days [11, 26]. Currently, we recommend for the treatment of systemic infections, combining colistin with other active antimicrobials (e.g. possibly tigecycline or aminoglycosides) or synergistic antimicrobials (e.g. carbapenems). Clinical metrics and prediction scores to facilitate the early and appropriate use of colistin for severely septic patients are needed [33, 161].

References

1. Magiorakos AP, Srinivasan A, Carey RB, Carmeli Y, Falagas ME, Giske CG, Harbarth S, Hindler JF, Kahlmeter G, Olsson-Liljequist B, Paterson DL, Rice LB, Stelling J, Struelens MJ, Vatopoulos A, Weber JT, Monnet DL (2012) Multidrug-resistant, extensively drug-resistant and pandrug-resistant bacteria: an international expert proposal for interim standard definitions for acquired resistance. Clin Microbiol Infect 18(3):268–281. https://doi.org/10.1111/j.1469-0691.2011.03570.x
2. Boucher HW, Talbot GH, Bradley JS, Edwards JE, Gilbert D, Rice LB, Scheld M, Spellberg B, Bartlett J (2009) Bad bugs, no drugs: no ESKAPE! An update from the Infectious Diseases Society of America. Clin Infect Dis 48(1):1–12. https://doi.org/10.1086/595011
3. Falagas ME, Kasiakou SK (2005) Colistin: the revival of polymyxins for the management of multidrug-resistant gram-negative bacterial infections. Clin Infect Dis 40(9):1333–1341. CID35354 [pii] https://doi.org/10.1086/429323
4. Schwaber MJ, Carmeli Y (2008) Carbapenem-resistant enterobacteriaceae: a potential threat. JAMA 300 (24):2911–2913. https://doi.org/10.1001/jama.2008.896 300/24/2911 [pii]
5. Peleg AY, Hooper DC (2010) Hospital-acquired infections due to gram-negative bacteria. N Engl J Med 362(19):1804–1813. https://doi.org/10.1056/NEJMra0904124 362/19/1804 [pii]
6. Paterson DL (2006) Clinical experience with recently approved antibiotics. Curr Opin Pharmacol 6 (5):486–490. S1471-4892(06)00136-6 [pii] https://doi.org/10.1016/j.coph.2006.07.001
7. Peterson LR (2008) A review of tigecycline–the first glycylcycline. Int J Antimicrob Agents

32(Suppl 4):S215–S222. https://doi.org/10.1016/S0924-8579(09)70005-6 [pii]
8. Yahav D, Lador A, Paul M, Leibovici L (2011) Efficacy and safety of tigecycline: a systematic review and meta-analysis. J Antimicrob Chemother 66(9):1963–1971. https://doi.org/10.1093/jac/dkr242 [pii]
9. Cai Y, Wang R, Liang B, Bai N, Liu Y (2011) Systematic review and meta-analysis of the effectiveness and safety of tigecycline for treatment of infectious disease. Antimicrob Agents Chemother 55(3):1162–1172. https://doi.org/10.1128/AAC.01402-10 [pii]
10. FDA (2010) Drug safety communication - increased risk of death with Tygacil (tigecycline) compared to other antibiotics used to treat similar infections. FDA
11. Ku K, Pogue JM, Moshos J, Bheemreddy S, Wang Y, Bhargava A, Campbell M, Khandker N, Lephart PR, Chopra T, Hayakawa K, Martin ET, Abreu-Lanfranco O, Dhar S, Kaye KS, Marchaim D (2012) Retrospective evaluation of colistin versus tigecycline for the treatment of Acinetobacter baumannii and/or carbapenem-resistant Enterobacteriaceae infections. Am J Infect Control 40(10):983–987. https://doi.org/10.1016/j.ajic.2011.12.014 S0196-6553(11)01336-8 [pii]
12. Gilad J, Carmeli Y (2008) Treatment options for multidrug-resistant Acinetobacter species. Drugs 68(2):165–189 doi:6823 [pii]
13. Liscio JL, Mahoney MV, Hirsch EB (2015) Ceftolozane/tazobactam and ceftazidime/avibactam: two novel beta-lactam/beta-lactamase inhibitor combination agents for the treatment of resistant gram-negative bacterial infections. Int J Antimicrob Agents 46(3):266–271. https://doi.org/10.1016/j.ijantimicag.2015.05.003 S0924-8579(15)00203-4 [pii]
14. Falagas ME, Rafailidis PI (2008) Re-emergence of colistin in today's world of multidrug-resistant organisms: personal perspectives. Expert Opin Investig Drugs 17(7):973–981. https://doi.org/10.1517/13543784.17.7.973
15. Garonzik SM, Li J, Thamlikitkul V, Paterson DL, Shoham S, Jacob J, Silveira FP, Forrest A, Nation RL (2011) Population pharmacokinetics of colistin methanesulfonate and formed colistin in critically ill patients from a multicenter study provide dosing suggestions for various categories of patients. Antimicrob Agents Chemother 55(7):3284–3294. https://doi.org/10.1128/AAC.01733-10 [pii]
16. Theuretzbacher U (2014) Product information for parenteral colistin varies substantially across Europe. J Antimicrob Chemother 69(7):1987–92. https://doi.org/10.1093/jac/dku064 [pii]
17. Plachouras D, Karvanen M, Friberg LE, Papadomichelakis E, Antoniadou A, Tsangaris I, Karaiskos I, Poulakou G, Kontopidou F, Armaganidis A, Cars O, Giamarellou H (2009) Population pharmacokinetic analysis of colistin methanesulfonate and colistin after intravenous administration in critically ill patients with infections caused by gram-negative bacteria. Antimicrob Agents Chemother 53 (8):3430–3436. https://doi.org/10.1128/AAC.01361-08 [pii]
18. Pogue JM, Marchaim D, Kaye D, Kaye KS (2011) Revisiting "older" antimicrobials in the era of multidrug resistance. Pharmacotherapy 31(9):912–921. https://doi.org/10.1592/phco.31.9.912
19. Pogue JM, Lee J, Marchaim D, Yee V, Zhao JJ, Chopra T, Lephart P, Kaye KS (2011) Incidence of and risk factors for colistin-associated nephrotoxicity in a large academic health system. Clin Infect Dis 53(9):879–884. https://doi.org/10.1093/cid/cir611 [pii]
20. Nazer LH, Anabtawi N (2017) Optimizing colistin dosing: is a loading dose necessary? Am J Health Syst Pharm 74(1):e9–e16. https://doi.org/10.2146/ajhp150876 74/1/e9 [pii]
21. Vardakas KZ, Rellos K, Triarides NA, Falagas ME (2016) Colistin loading dose: evaluation of the published pharmacokinetic and clinical data. Int J Antimicrob Agents 48(5):475–484. S0924-8579(16)30258-8 [pii] https://doi.org/10.1016/j.ijantimicag.2016.08.009
22. Dalfino L, Puntillo F, Mosca A, Monno R, Spada ML, Coppolecchia S, Miragliotta G, Bruno F, Brienza N (2012) High-dose, extended-interval colistin administration in critically ill patients: is this the right dosing strategy? A preliminary study. Clin Infect Dis 54(12):1720–1726. https://doi.org/10.1093/cid/cis286 [pii]
23. Falagas ME, Rafailidis PI, Ioannidou E, Alexiou VG, Matthaiou DK, Karageorgopoulos DE, Kapaskelis A, Nikita D, Michalopoulos A (2010) Colistin therapy for microbiologically documented multidrug-resistant Gram-negative bacterial infections: a retrospective cohort study of 258 patients. Int J Antimicrob Agents 35(2):194–199. https://doi.org/10.1016/j.ijantimicag.2009.10.005 S0924-8579(09)00469-5 [pii]
24. Zaidi ST, Al Omran S, Al Aithan AS, Al Sultan M (2014) Efficacy and safety of low-dose colistin in the treatment for infections caused by multidrug-resistant gram-negative bacteria. J Clin Pharm Ther. https://doi.org/10.1111/jcpt.12138
25. Nation RL, Li J, Cars O, Couet W, Dudley MN, Kaye KS, Mouton JW, Paterson DL, Tam VH, Theuretzbacher U, Tsuji BT, Turnidge JD (2014) Consistent global approach on reporting of colistin doses to promote safe and effective use. Clin Infect Dis 58(1):139–141. https://doi.org/10.1093/cid/cit680 [pii]
26. Paul M, Bishara J, Levcovich A, Chowers M, Goldberg E, Singer P, Lev S, Leon P, Raskin M, Yahav D, Leibovici L (2010) Effectiveness and safety of colistin: prospective comparative cohort study. J Antimicrob Chemother 65(5):1019–1027. https://doi.org/10.1093/jac/dkq069 [pii]

27. Markou N, Apostolakos H, Koumoudiou C, Athanasiou M, Koutsoukou A, Alamanos I, Gregorakos L (2003) Intravenous colistin in the treatment of sepsis from multiresistant Gram-negative bacilli in critically ill patients. Crit Care 7(5):R78–R83. https://doi.org/10.1186/cc2358
28. Michalopoulos AS, Tsiodras S, Rellos K, Mentzelopoulos S, Falagas ME (2005) Colistin treatment in patients with ICU-acquired infections caused by multiresistant gram-negative bacteria: the renaissance of an old antibiotic. Clin Microbiol Infect 11(2):115–121. CLM1043 [pii] https://doi.org/10.1111/j.1469-0691.2004.01043.x
29. Kallel H, Bahloul M, Hergafi L, Akrout M, Ketata W, Chelly H, Hamida CB, Rekik N, Hammami A, Bouaziz M (2006) Colistin as a salvage therapy for nosocomial infections caused by multidrug-resistant bacteria in the ICU. Int J Antimicrob Agents 28(4):366–369. S0924-8579(06)00290-1 [pii] https://doi.org/10.1016/j.ijantimicag.2006.07.008
30. Yilmaz GR, Bastug AT, But A, Yildiz S, Yetkin MA, Kanyilmaz D, Akinci E, Bodur H (2013) Clinical and microbiological efficacy and toxicity of colistin in patients infected with multidrug-resistant gram-negative pathogens. J Infect Chemother 19(1):57–62. https://doi.org/10.1007/s10156-012-0451-2
31. Reina R, Estenssoro E, Saenz G, Canales HS, Gonzalvo R, Vidal G, Martins G, Das Neves A, Santander O, Ramos C (2005) Safety and efficacy of colistin in Acinetobacter and Pseudomonas infections: a prospective cohort study. Intensive Care Med 31(8):1058–1065. https://doi.org/10.1007/s00134-005-2691-4
32. Hachem RY, Chemaly RF, Ahmar CA, Jiang Y, Boktour MR, Rjaili GA, Bodey GP, Raad II (2007) Colistin is effective in treatment of infections caused by multidrug-resistant Pseudomonas aeruginosa in cancer patients. Antimicrob Agents Chemother 51(6):1905–1911. AAC.01015-06 [pii] https://doi.org/10.1128/AAC.01015-06
33. Martin ET, Tansek R, Collins V, Hayakawa K, Abreu-Lanfranco O, Chopra T, Lephart PR, Pogue JM, Kaye KS, Marchaim D (2013) The carbapenem-resistant Enterobacteriaceae score: a bedside score to rule out infection with carbapenem-resistant Enterobacteriaceae among hospitalized patients. Am J Infect Control 41(2):180–182. https://doi.org/10.1016/j.ajic.2012.02.036 S0196-6553(12)00716-X [pii]
34. Livermore DM, Hill RL, Thomson H, Charlett A, Turton JF, Pike R, Patel BC, Manuel R, Gillespie S, Balakrishnan I, Barrett SP, Cumberland N, Twagira M (2010) Antimicrobial treatment and clinical outcome for infections with carbapenem- and multiply-resistant Acinetobacter baumannii around London. Int J Antimicrob Agents 35(1):19–24. https://doi.org/10.1016/j.ijantimicag.2009.09.014 S0924-8579(09)00437-3 [pii]
35. Yahav D, Farbman L, Leibovici L, Paul M (2012) Colistin: new lessons on an old antibiotic. Clin Microbiol Infect 18(1):18–29. https://doi.org/10.1111/j.1469-0691.2011.03734.x
36. Schwaber MJ, Lev B, Israeli A, Solter E, Smollan G, Rubinovitch B, Shalit I, Carmeli Y (2011) Containment of a country-wide outbreak of carbapenem-resistant Klebsiella pneumoniae in Israeli hospitals via a nationally implemented intervention. Clin Infect Dis 52(7):848–855. https://doi.org/10.1093/cid/cir025 [pii]
37. Cuzon G, Naas T, Truong H, Villegas MV, Wisell KT, Carmeli Y, Gales AC, Venezia SN, Quinn JP, Nordmann P (2010) Worldwide diversity of Klebsiella pneumoniae that produce beta-lactamase blaKPC-2 gene. Emerg Infect Dis 16(9):1349–1356. https://doi.org/10.3201/eid1609.091389
38. Munoz-Price LS, Poirel L, Bonomo RA, Schwaber MJ, Daikos GL, Cormican M, Cornaglia G, Garau J, Gniadkowski M, Hayden MK, Kumarasamy K, Livermore DM, Maya JJ, Nordmann P, Patel JB, Paterson DL, Pitout J, Villegas MV, Wang H, Woodford N, Quinn JP (2013) Clinical epidemiology of the global expansion of Klebsiella pneumoniae carbapenemases. Lancet Infect Dis 13(9):785–796. https://doi.org/10.1016/S1473-3099(13)70190-7 S1473-3099(13)70190-7 [pii]
39. Livermore DM (2005) Tigecycline: what is it, and where should it be used? J Antimicrob Chemother 56(4):611–614. dki291 [pii] https://doi.org/10.1093/jac/dki291
40. Alvarez-Marin R, Lopez-Rojas R, Marquez JA, Gomez MJ, Molina J, Cisneros JM, Ortiz-Leyba C, Aznar J, Garnacho-Montero J, Pachon J (2016) Colistin dosage without loading dose is efficacious when treating Carbapenem-resistant Acinetobacter baumannii ventilator-associated pneumonia caused by strains with high susceptibility to colistin. PLoS One 11(12):e0168468. https://doi.org/10.1371/journal.pone.0168468 PONE-D-16-24297 [pii]
41. Roberts JA, Lipman J (2012) Editorial commentary: closing the loop--a colistin clinical study to confirm dosing recommendations from PK/PD modeling. Clin Infect Dis 54(12):1727–1729. https://doi.org/10.1093/cid/cis311 [pii]
42. Sorli L, Luque S, Segura C, Campillo N, Montero M, Esteve E, Herrera S, Benito N, Alvarez-Lerma F, Grau S, Horcajada JP (2017) Impact of colistin plasma levels on the clinical outcome of patients with infections caused by extremely drug-resistant Pseudomonas aeruginosa. BMC Infect Dis 17(1):11. https://doi.org/10.1186/s12879-016-2117-7 [pii]
43. Tumbarello M, Viale P, Viscoli C, Trecarichi EM, Tumietto F, Marchese A, Spanu T, Ambretti S, Ginocchio F, Cristini F, Losito AR, Tedeschi S, Cauda R, Bassetti M (2012) Predictors of mortality in bloodstream infections caused by Klebsiella pneumoniae carbapenemase-producing K. pneumoniae: importance of combination therapy. Clin Infect Dis 55(7):943–950. https://doi.org/10.1093/cid/cis588 [pii]

44. Li J, Nation RL, Milne RW, Turnidge JD, Coulthard K (2005) Evaluation of colistin as an agent against multi-resistant Gram-negative bacteria. Int J Antimicrob Agents 25(1):11–25. S0924-8579(04)00365-6 [pii] https://doi.org/10.1016/j.ijantimicag.2004.10.001
45. Kalin G, Alp E, Coskun R, Demiraslan H, Gundogan K, Doganay M (2012) Use of high-dose IV and aerosolized colistin for the treatment of multidrug-resistant Acinetobacter baumannii ventilator-associated pneumonia: do we really need this treatment? J Infect Chemother 18(6):872–877. https://doi.org/10.1007/s10156-012-0430-7
46. Benattar YD, Omar M, Zusman O, Yahav D, Zak-Doron Y, Altunin S, Elbaz M, Daitch V, Granot M, Leibovici L, Paul M (2016) The effectiveness and safety of high-dose colistin: prospective cohort study. Clin Infect Dis 63(12):1605–1612. ciw684 [pii] https://doi.org/10.1093/cid/ciw684
47. Breakpoint tables for interpretation of mics and zone diameters (2011) Eucast - European Committee on Antimicrobial Susceptibility Testing
48. Performance standards for antimibrobial susceptibility testing. Nineteenth informational supplement. Approved standard M100-S19 (2009). CLSI, Wayne, PA
49. Marchaim D, Chopra T, Pogue JM, Perez F, Hujer AM, Rudin S, Endimiani A, Navon-Venezia S, Hothi J, Slim J, Blunden C, Shango M, Lephart PR, Salimnia H, Reid D, Moshos J, Hafeez W, Bheemreddy S, Chen TY, Dhar S, Bonomo RA, Kaye KS (2011) Outbreak of colistin-resistant, carbapenem-resistant Klebsiella pneumoniae in metropolitan Detroit, Michigan. Antimicrob Agents Chemother 55(2):593–599. https://doi.org/10.1128/AAC.01020-10 [pii]
50. Mentzelopoulos SD, Pratikaki M, Platsouka E, Kraniotaki H, Zervakis D, Koutsoukou A, Nanas S, Paniara O, Roussos C, Giamarellos-Bourboulis E, Routsi C, Zakynthinos SG (2007) Prolonged use of carbapenems and colistin predisposes to ventilator-associated pneumonia by pandrug-resistant Pseudomonas aeruginosa. Intensive Care Med 33(9):1524–1532. https://doi.org/10.1007/s00134-007-0683-2
51. Antoniadou A, Kontopidou F, Poulakou G, Koratzanis E, Galani I, Papadomichelakis E, Kopterides P, Souli M, Armaganidis A, Giamarellou H (2007) Colistin-resistant isolates of Klebsiella pneumoniae emerging in intensive care unit patients: first report of a multiclonal cluster. J Antimicrob Chemother 59(4):786–790. dkl562 [pii] https://doi.org/10.1093/jac/dkl562
52. Falagas ME, Rafailidis PI, Matthaiou DK (2010) Resistance to polymyxins: mechanisms, frequency and treatment options. Drug Resist Updat 13(4–5):132–138. https://doi.org/10.1016/j.drup.2010.05.002 S1368-7646(10)00031-2 [pii]
53. Adams MD, Nickel GC, Bajaksouzian S, Lavender H, Murthy AR, Jacobs MR, Bonomo RA (2009) Resistance to colistin in Acinetobacter baumannii associated with mutations in the PmrAB two-component system. Antimicrob Agents Chemother 53(9):3628–3634. https://doi.org/10.1128/AAC.00284-09 [pii]
54. Matthaiou DK, Michalopoulos A, Rafailidis PI, Karageorgopoulos DE, Papaioannou V, Ntani G, Samonis G, Falagas ME (2008) Risk factors associated with the isolation of colistin-resistant gram-negative bacteria: a matched case-control study. Crit Care Med 36(3):807–811. https://doi.org/10.1097/CCM.0B013E3181652FAE 00003246-200803000-00021 [pii]
55. Zarkotou O, Pournaras S, Voulgari E, Chrysos G, Prekates A, Voutsinas D, Themeli-Digalaki K, Tsakris A (2010) Risk factors and outcomes associated with acquisition of colistin-resistant KPC-producing Klebsiella pneumoniae: a matched case-control study. J Clin Microbiol 48(6):2271–2274. https://doi.org/10.1128/JCM.02301-09 [pii]
56. Bogdanovich T, Adams-Haduch JM, Tian GB, Nguyen MH, Kwak EJ, Muto CA, Doi Y (2011) Colistin-resistant, Klebsiella pneumoniae carbapenemase (KPC)-producing Klebsiella pneumoniae belonging to the international epidemic clone ST258. Clin Infect Dis 53(4):373–376. https://doi.org/10.1093/cid/cir401 [pii]
57. Bogan C, Marchaim D (2013) The role of antimicrobial stewardship in curbing carbapenem resistance. Future Microbiol 8(8):979–991. https://doi.org/10.2217/fmb.13.73
58. Cai Y, Chai D, Wang R, Liang B, Bai N (2012) Colistin resistance of Acinetobacter baumannii: clinical reports, mechanisms and antimicrobial strategies. J Antimicrob Chemother 67(7):1607–1615. https://doi.org/10.1093/jac/dks084 [pii]
59. Lim LM, Ly N, Anderson D, Yang JC, Macander L, Jarkowski A, 3rd, Forrest A, Bulitta JB, Tsuji BT (2010) Resurgence of colistin: a review of resistance, toxicity, pharmacodynamics, and dosing. Pharmacotherapy 30(12):1279–1291. https://doi.org/10.1592/phco.30.12.1279 [pii]
60. Kawanishi M, Abo H, Ozawa M, Uchiyama M, Shirakawa T, Suzuki S, Shima A, Yamashita A, Sekizuka T, Kato K, Kuroda M, Koike R, Kijima M (2017) Prevalence of colistin resistance gene mcr-1 and absence of mcr-2 in Escherichia coli isolated from healthy food-producing animals in Japan. Antimicrob Agents Chemother 61(1). e02057-16 [pii] https://doi.org/10.1128/AAC.02057-16 [pii]
61. Meinersmann RJ, Ladely SR, Plumblee JR, Cook KL, Thacker E (2017) Prevalence of mcr-1 in the Cecal contents of food animals in the United States. Antimicrob Agents Chemother 61 (2). e02244-16 [pii] https://doi.org/10.1128/AAC.02244-16 [pii]
62. Tascini C, Ferranti S, Messina F, Menichetti F (2000) In vitro and in vivo synergistic activity of colistin, rifampin, and amikacin against a multiresistant

Pseudomonas aeruginosa isolate. Clin Microbiol Infect 6(12):690–691 doi:clm169 [pii]
63. Giamarellos-Bourboulis EJ, Sambatakou H, Galani I, Giamarellou H (2003) In vitro interaction of colistin and rifampin on multidrug-resistant Pseudomonas aeruginosa. J Chemother 15(3):235–238. https://doi.org/10.1179/joc.2003.15.3.235
64. Gunderson BW, Ibrahim KH, Hovde LB, Fromm TL, Reed MD, Rotschafer JC (2003) Synergistic activity of colistin and ceftazidime against multiantibiotic-resistant Pseudomonas aeruginosa in an in vitro pharmacodynamic model. Antimicrob Agents Chemother 47(3):905–909
65. Tascini C, Gemignani G, Ferranti S, Tagliaferri E, Leonildi A, Lucarini A, Menichetti F (2004) Microbiological activity and clinical efficacy of a colistin and rifampin combination in multidrug-resistant Pseudomonas aeruginosa infections. J Chemother 16(3):282–287. https://doi.org/10.1179/joc.2004.16.3.282
66. Petersen PJ, Labthavikul P, Jones CH, Bradford PA (2006) In vitro antibacterial activities of tigecycline in combination with other antimicrobial agents determined by chequerboard and time-kill kinetic analysis. J Antimicrob Chemother 57(3):573–576. dki477 [pii] https://doi.org/10.1093/jac/dki477
67. Timurkaynak F, Can F, Azap OK, Demirbilek M, Arslan H, Karaman SO (2006) In vitro activities of non-traditional antimicrobials alone or in combination against multidrug-resistant strains of Pseudomonas aeruginosa and Acinetobacter baumannii isolated from intensive care units. Int J Antimicrob Agents 27(3):224–228. S0924-8579(05)00331-6 [pii] https://doi.org/10.1016/j.ijantimicag.2005.10.012
68. Zhanel GG, Mayer M, Laing N, Adam HJ (2006) Mutant prevention concentrations of levofloxacin alone and in combination with azithromycin, ceftazidime, colistin (Polymyxin E), meropenem, piperacillin-tazobactam, and tobramycin against Pseudomonas aeruginosa. Antimicrob Agents Chemother 50(6):2228–2230. 50/6/2228 [pii] https://doi.org/10.1128/AAC.01620-05
69. Biancofiore G, Tascini C, Bisa M, Gemignani G, Bindi ML, Leonildi A, Giannotti G, Menichetti F (2007) Colistin, meropenem and rifampin in a combination therapy for multi-drug-resistant Acinetobacter baumannii multifocal infection. A case report. Minerva Anestesiol 73(3):181–185 doi:R02064890 [pii]
70. Chitnis S, Chitnis V, Chitnis DS (2007) In vitro synergistic activity of colistin with aminoglycosides, beta-lactams and rifampin against multidrug-resistant gram-negative bacteria. J Chemother 19(2):226–229. https://doi.org/10.1179/joc.2007.19.2.226
71. Lee NY, Wang CL, Chuang YC, Yu WL, Lee HC, Chang CM, Wang LR, Ko WC (2007) Combination carbapenem-sulbactam therapy for critically ill patients with multidrug-resistant Acinetobacter baumannii bacteremia: four case reports and an in vitro combination synergy study. Pharmacotherapy 27(11):1506–1511. https://doi.org/10.1592/phco.27.11.1506
72. Song JY, Kee SY, Hwang IS, Seo YB, Jeong HW, Kim WJ, Cheong HJ (2007) In vitro activities of carbapenem/sulbactam combination, colistin, colistin/rifampicin combination and tigecycline against carbapenem-resistant Acinetobacter baumannii. J Antimicrob Chemother 60(2):317–322. dkm136 [pii] https://doi.org/10.1093/jac/dkm136
73. Tan TY, Ng LS, Tan E, Huang G (2007) In vitro effect of minocycline and colistin combinations on imipenem-resistant Acinetobacter baumannii clinical isolates. J Antimicrob Chemother 60(2):421–423. dkm178 [pii] https://doi.org/10.1093/jac/dkm178
74. Guzel CB, Gerceker AA (2008) In vitro activities of various antibiotics, alone and in combination with colistin methanesulfonate, against Pseudomonas aeruginosa strains isolated from cystic fibrosis patients. Chemotherapy 54(2):147–151. https://doi.org/10.1159/000119741 [pii]
75. Petrosillo N, Ioannidou E, Falagas ME (2008) Colistin monotherapy vs. combination therapy: evidence from microbiological, animal and clinical studies. Clin Microbiol Infect 14(9):816–827. https://doi.org/10.1111/j.1469-0691.2008.02061.x CLM2061 [pii]
76. Tascini C, Urbani L, Biancofiore G, Rossolini GM, Leonildi A, Gemignani G, Bindi ML, Mugnaioli C, Filipponi F, Menichetti F (2008) Colistin in combination with rifampin and imipenem for treating a blaVIM-1 metallo-beta-lactamase-producing Enterobacter cloacae disseminated infection in a liver transplant patient. Minerva Anestesiol 74(1–2):47–49 doi:R02075090 [pii]
77. Aoki N, Tateda K, Kikuchi Y, Kimura S, Miyazaki C, Ishii Y, Tanabe Y, Gejyo F, Yamaguchi K (2009) Efficacy of colistin combination therapy in a mouse model of pneumonia caused by multidrug-resistant Pseudomonas aeruginosa. J Antimicrob Chemother 63(3):534–542. https://doi.org/10.1093/jac/dkn530 [pii]
78. Maisetta G, Mangoni ML, Esin S, Pichierri G, Capria AL, Brancatisano FL, Di Luca M, Barnini S, Barra D, Campa M, Batoni G (2009) In vitro bactericidal activity of the N-terminal fragment of the frog peptide esculentin-1b (Esc 1-18) in combination with conventional antibiotics against Stenotrophomonas maltophilia. Peptides 30(9):1622–1626. https://doi.org/10.1016/j.peptides.2009.06.004 S0196-9781(09)00222-8 [pii]
79. Principe L, D'Arezzo S, Capone A, Petrosillo N, Visca P (2009) In vitro activity of tigecycline in combination with various antimicrobials against multidrug resistant Acinetobacter baumannii. Ann Clin Microbiol Antimicrob 8:18. https://doi.org/10.1186/1476-0711-8-18 [pii]
80. Song JY, Cheong HJ, Lee J, Sung AK, Kim WJ (2009) Efficacy of monotherapy and combined antibiotic

therapy for carbapenem-resistant Acinetobacter baumannii pneumonia in an immunosuppressed mouse model. Int J Antimicrob Agents 33(1):33–39. https://doi.org/10.1016/j.ijantimicag.2008.07.008 S0924-8579(08)00317-8 [pii]
81. Souli M, Rekatsina PD, Chryssouli Z, Galani I, Giamarellou H, Kanellakopoulou K (2009) Does the activity of the combination of imipenem and colistin in vitro exceed the problem of resistance in metallo-beta-lactamase-producing Klebsiella pneumoniae isolates? Antimicrob Agents Chemother 53(5):2133–2135. https://doi.org/10.1128/AAC.01271-08 AAC.01271-08 [pii]
82. Dizbay M, Tozlu DK, Cirak MY, Isik Y, Ozdemir K, Arman D (2010) In vitro synergistic activity of tigecycline and colistin against XDR-Acinetobacter baumannii. J Antibiot (Tokyo) 63(2):51–53.https://doi.org/10.1038/ja.2009.117 [pii]
83. Rodriguez CH, De Ambrosio A, Bajuk M, Spinozzi M, Nastro M, Bombicino K, Radice M, Gutkind G, Vay C, Famiglietti A (2010) In vitro antimicrobials activity against endemic Acinetobacter baumannii multiresistant clones. J Infect Dev Ctries 4(3):164–167
84. Bercot B, Poirel L, Dortet L, Nordmann P (2011) In vitro evaluation of antibiotic synergy for NDM-1-producing Enterobacteriaceae. J Antimicrob Chemother 66(10):2295–2297. doi:https://doi.org/10.1093/jac/dkr296 [pii]
85. Liang W, Liu XF, Huang J, Zhu DM, Li J, Zhang J (2011) Activities of colistin- and minocycline-based combinations against extensive drug resistant Acinetobacter baumannii isolates from intensive care unit patients. BMC Infect Dis 11:109. https://doi.org/10.1186/1471-2334-11-109 [pii]
86. Pournaras S, Vrioni G, Neou E, Dendrinos J, Dimitroulia E, Poulou A, Tsakris A (2011) Activity of tigecycline alone and in combination with colistin and meropenem against Klebsiella pneumoniae carbapenemase (KPC)-producing Enterobacteriaceae strains by time-kill assay. Int J Antimicrob Agents 37(3):244–247. https://doi.org/10.1016/j.ijantimicag.2010.10.031 S0924-8579(10)00533-9 [pii]
87. Trotman RL, Williamson JC, Shoemaker DM, Salzer WL (2005) Antibiotic dosing in critically ill adult patients receiving continuous renal replacement therapy. Clin Infect Dis 41(8):1159–1166. CID35961 [pii] https://doi.org/10.1086/444500
88. Falagas ME, Lourida P, Poulikakos P, Rafailidis PI, Tansarli GS (2014) Antibiotic treatment of infections due to carbapenem-resistant Enterobacteriaceae: systematic evaluation of the available evidence. Antimicrob Agents Chemother 58(2):654–663. https://doi.org/10.1128/AAC.01222-13 [pii]
89. Zusman O, Avni T, Leibovici L, Adler A, Friberg L, Stergiopoulou T, Carmeli Y, Paul M (2013) Systematic review and meta-analysis of in vitro synergy of polymyxins and carbapenems. Antimicrob Agents Chemother 57(10):5104–5111. https://doi.org/10.1128/AAC.01230-13 [pii]
90. Bergen PJ, Tsuji BT, Bulitta JB, Forrest A, Jacob J, Sidjabat HE, Paterson DL, Nation RL, Li J (2011) Synergistic killing of multidrug-resistant Pseudomonas aeruginosa at multiple inocula by colistin combined with doripenem in an in vitro pharmacokinetic/pharmacodynamic model. Antimicrob Agents Chemother 55(12):5685–5695. https://doi.org/10.1128/AAC.05298-11 [pii]
91. Molina J, Cordero E, Palomino J, Pachon J (2009) [Aminoglycosides and polymyxins]. Enferm Infecc Microbiol Clin 27(3):178–188. https://doi.org/10.1016/j.eimc.2009.02.001 S0213-005X(09)00098-6 [pii]
92. Marchaim D, Perez F, Lee J, Bheemreddy S, Hujer AM, Rudin S, Hayakawa K, Lephart PR, Blunden C, Shango M, Campbell ML, Varkey J, Manickam P, Patel D, Pogue JM, Chopra T, Martin ET, Dhar S, Bonomo RA, Kaye KS (2012) "Swimming in resistance": co-colonization with carbapenem-resistant Enterobacteriaceae and Acinetobacter baumannii or Pseudomonas aeruginosa. Am J Infect Control 40(9):830–835. https://doi.org/10.1016/j.ajic.2011.10.013 S0196-6553(11)01267-3 [pii]
93. Paul M, Daikos GL, Durante-Mangoni E, Yahav D, Carmeli Y, Benattar YD, Skiada A, Andini R, Eliakim-Raz N, Nutman A, Zusman O, Antoniadou A, Pafundi PC, Adler A, Dickstein Y, Pavleas I, Zampino R, Daitch V, Bitterman R, Zayyad H, Koppel F, Levi I, Babich T, Friberg LE, Mouton JW, Theuretzbacher U, Leibovici L (2018) Colistin alone versus colistin plus meropenem for treatment of severe infections caused by carbapenem-resistant Gram-negative bacteria: an open-label, randomised controlled trial. Lancet Infect Dis 18(4):391–400. https://doi.org/10.1016/S1473-3099(18)30099-9
94. Zavascki AP (2018) Colistin versus colistin plus meropenem for severe infections. Lancet Infect Dis 18(5):493–494. https://doi.org/10.1016/S1473-3099(18)30212-3
95. Kengkla K, Kongpakwattana K, Saokaew S, Apisarnthanarak A, Chaiyakunapruk N (2018) Comparative efficacy and safety of treatment options for MDR and XDR Acinetobacter baumannii infections: a systematic review and network meta-analysis. J Antimicrob Chemother 73(1):22–32. https://doi.org/10.1093/jac/dkx368
96. Bergen PJ, Landersdorfer CB, Zhang J, Zhao M, Lee HJ, Nation RL, Li J (2012) Pharmacokinetics and pharmacodynamics of 'old' polymyxins: what is new? Diagn Microbiol Infect Dis 74 (3):213–223. https://doi.org/10.1016/j.diagmicrobio.2012.07.010 S0732-8893(12)00309-4 [pii]
97. Athanassa ZE, Markantonis SL, Fousteri MZ, Myrianthefs PM, Boutzouka EG, Tsakris A, Baltopoulos GJ (2012) Pharmacokinetics of inhaled colistimethate sodium (CMS) in mechanically ventilated critically ill patients. Intensive Care Med 38(11):1779–1786. https://doi.org/10.1007/s00134-012-2628-7

98. Linden PK, Paterson DL (2006) Parenteral and inhaled colistin for treatment of ventilator-associated pneumonia. Clin Infect Dis 43Suppl 2:S89–S94. CID38720 [pii] https://doi.org/10.1086/504485
99. Beringer P (2001) The clinical use of colistin in patients with cystic fibrosis. Curr Opin Pulm Med 7(6):434–440
100. Hamer DH (2000) Treatment of nosocomial pneumonia and tracheobronchitis caused by multidrug-resistant Pseudomonas aeruginosa with aerosolized colistin. Am J Respir Crit Care Med 162(1):328–330. https://doi.org/10.1164/ajrccm.162.1.9910071
101. Kwa AL, Loh C, Low JG, Kurup A, Tam VH (2005) Nebulized colistin in the treatment of pneumonia due to multidrug-resistant Acinetobacter baumannii and Pseudomonas aeruginosa. Clin Infect Dis 41 (5):754–757. CID36576 [pii] https://doi.org/10.1086/432583
102. Michalopoulos A, Kasiakou SK, Mastora Z, Rellos K, Kapaskelis AM, Falagas ME (2005) Aerosolized colistin for the treatment of nosocomial pneumonia due to multidrug-resistant Gram-negative bacteria in patients without cystic fibrosis. Crit Care 9(1):R53–R59. cc3020 [pii] https://doi.org/10.1186/cc3020
103. Falagas ME, Kasiakou SK, Tsiodras S, Michalopoulos A (2006) The use of intravenous and aerosolized polymyxins for the treatment of infections in critically ill patients: a review of the recent literature. Clin Med Res 4(2):138–146. doi: 4/2/138 [pii]
104. Michalopoulos A, Fotakis D, Virtzili S, Vletsas C, Raftopoulou S, Mastora Z, Falagas ME (2008) Aerosolized colistin as adjunctive treatment of ventilator-associated pneumonia due to multidrug-resistant Gram-negative bacteria: a prospective study. Respir Med 102(3):407–412. S0954-6111(07)00433-7 [pii] https://doi.org/10.1016/j.rmed.2007.10.011
105. Kofteridis DP, Alexopoulou C, Valachis A, Maraki S, Dimopoulou D, Georgopoulos D, Samonis G (2010) Aerosolized plus intravenous colistin versus intravenous colistin alone for the treatment of ventilator-associated pneumonia: a matched case-control study. Clin Infect Dis 51(11):1238–1244. https://doi.org/10.1086/657242
106. Korbila IP, Michalopoulos A, Rafailidis PI, Nikita D, Samonis G, Falagas ME (2010) Inhaled colistin as adjunctive therapy to intravenous colistin for the treatment of microbiologically documented ventilator-associated pneumonia: a comparative cohort study. Clin Microbiol Infect 16(8):1230–1236. https://doi.org/10.1111/j.1469-0691.2009.03040.x CLM3040 [pii]
107. Ehrmann S, Roche-Campo F, Sferrazza Papa GF, Isabey D, Brochard L, Apiou-Sbirlea G (2013) Aerosol therapy during mechanical ventilation: an international survey. Intensive Care Med 39(6):1048–1056. https://doi.org/10.1007/s00134-013-2872-5
108. Lepine PA, Dumas A, Boulet LP (2017) Pulmonary eosinophilia from inhaled colistin. Chest 151(1):e1–e3. S0012-3692(16)52646-4 [pii] https://doi.org/10.1016/j.chest.2016.07.001
109. Florescu DF, Qiu F, McCartan MA, Mindru C, Fey PD, Kalil AC (2012) What is the efficacy and safety of colistin for the treatment of ventilator-associated pneumonia? A systematic review and meta-regression. Clin Infect Dis 54(5):670–680.https://doi.org/10.1093/cid/cir934 [pii]
110. Ng J, Gosbell IB, Kelly JA, Boyle MJ, Ferguson JK (2006) Cure of multiresistant Acinetobacter baumannii central nervous system infections with intraventricular or intrathecal colistin: case series and literature review. J Antimicrob Chemother 58(5):1078–1081. dkl347 [pii] https://doi.org/10.1093/jac/dkl347
111. Benifla M, Zucker G, Cohen A, Alkan M (2004) Successful treatment of Acinetobacter meningitis with intrathecal polymyxin E. J Antimicrob Chemother 54 (1):290–292. https://doi.org/10.1093/jac/dkh289 [pii]
112. Kasiakou SK, Rafailidis PI, Liaropoulos K, Falagas ME (2005) Cure of post-traumatic recurrent multiresistant Gram-negative rod meningitis with intraventricular colistin. J Infect 50 (4):348–352. S0163445304000982 [pii] https://doi.org/10.1016/j.jinf.2004.05.008
113. Bukhary Z, Mahmood W, Al-Khani A, Al-Abdely HM (2005) Treatment of nosocomial meningitis due to a multidrug resistant Acinetobacter baumannii with intraventricular colistin. Saudi Med J 26(4):656–658. doi:20040735 [pii]
114. Berlana D, Llop JM, Fort E, Badia MB, Jodar R (2005) Use of colistin in the treatment of multiple-drug-resistant gram-negative infections. Am J Health Syst Pharm 62(1):39–47
115. Khawcharoenporn T, Apisarnthanarak A, Mundy LM (2010) Intrathecal colistin for drug-resistant Acinetobacter baumannii central nervous system infection: a case series and systematic review. Clin Microbiol Infect 16(7):888–894. https://doi.org/10.1111/j.1469-0691.2009.03019.x CLM3019 [pii]
116. Fernandez-Viladrich P, Corbella X, Corral L, Tubau F, Mateu A (1999) Successful treatment of ventriculitis due to carbapenem-resistant Acinetobacter baumannii with intraventricular colistin sulfomethate sodium. Clin Infect Dis 28(4):916–917. https://doi.org/10.1086/517243
117. Vasen W, Desmery P, Ilutovich S, Di Martino A (2000) Intrathecal use of colistin. J Clin Microbiol 38(9):3523
118. Yagmur R, Esen F (2006) Intrathecal colistin for treatment of Pseudomonas aeruginosa ventriculitis: report of a case with successful outcome. Crit Care 10(6):428. cc5088 [pii] https://doi.org/10.1186/cc5088
119. Karaiskos I, Galani L, Baziaka F, Katsouda E, Ioannidis I, Andreou A, Paskalis H, Giamarellou H (2013) Successful treatment of extensively drug-resistant Acinetobacter baumannii ventriculitis and

meningitis with intraventricular colistin after application of a loading dose: a case series. Int J Antimicrob Agents 41(5):480–483. https://doi.org/10.1016/j.ijantimicag.2013.02.010 S0924-8579(13)00070-8 [pii]
120. Karaiskos I, Galani L, Baziaka F, Giamarellou H (2013) Intraventricular and intrathecal colistin as the last therapeutic resort for the treatment of multidrug-resistant and extensively drug-resistant Acinetobacter baumannii ventriculitis and meningitis: a literature review. Int J Antimicrob Agents 41(6):499–508. https://doi.org/10.1016/j.ijantimicag.2013.02.006 S0924-8579(13)00054-X [pii]
121. Bargiacchi O, Rossati A, Car P, Brustia D, Brondolo R, Rosa F, Garavelli PL, De Rosa FG (2014) Intrathecal/intraventricular colistin in external ventricular device-related infections by multi-drug resistant Gram negative bacteria: case reports and review. Infection 42(5):801–809. https://doi.org/10.1007/s15010-014-0618-0
122. Ceylan B, Arslan F, Sipahi OR, Sunbul M, Ormen B, Hakyemez IN, Turunc T, Yildiz Y, Karsen H, Karagoz G, Tekin R, Hizarci B, Turhan V, Senol S, Oztoprak N, Yilmaz M, Ozdemir K, Mermer S, Kokoglu OF, Mert A (2017) Variables determining mortality in patients with Acinetobacter baumannii meningitis/ventriculitis treated with intrathecal colistin. Clin Neurol Neurosurg 153:43–49. S0303-8467(16)30432-2 [pii] https://doi.org/10.1016/j.clineuro.2016.12.006
123. de Jonge E, Schultz MJ, Spanjaard L, Bossuyt PM, Vroom MB, Dankert J, Kesecioglu J (2003) Effects of selective decontamination of digestive tract on mortality and acquisition of resistant bacteria in intensive care: a randomised controlled trial. Lancet 362(9389):1011–1016. S0140-6736(03)14409-1 [pii] https://doi.org/10.1016/S0140-6736(03)14409-1
124. Martin WJ (1964) Newer antimicrobial agents having current or potential clinical application. Med Clin North Am 48:255–292
125. Abele-Horn M, Dauber A, Bauernfeind A, Russwurm W, Seyfarth-Metzger I, Gleich P, Ruckdeschel G (1997) Decrease in nosocomial pneumonia in ventilated patients by selective oropharyngeal decontamination (SOD). Intensive Care Med 23(2):187–195
126. Saidel-Odes L, Polachek H, Peled N, Riesenberg K, Schlaeffer F, Trabelsi Y, Eskira S, Yousef B, Smolykov R, Codish S, Borer A (2012) A randomized, double-blind, placebo-controlled trial of selective digestive decontamination using oral gentamicin and oral polymyxin E for eradication of carbapenem-resistant Klebsiella pneumoniae carriage. Infect Control Hosp Epidemiol 33(1):14–19. https://doi.org/10.1086/663206
127. Agusti C, Pujol M, Argerich MJ, Ayats J, Badia M, Dominguez MA, Corbella X, Ariza J (2002) Short-term effect of the application of selective decontamination of the digestive tract on different body site reservoir ICU patients colonized by multi-resistant Acinetobacter baumannii. J Antimicrob Chemother 49(1):205–208
128. Bergmans DC, Bonten MJ, Gaillard CA, Paling JC, van der Geest S, van Tiel FH, Beysens AJ, de Leeuw PW, Stobberingh EE (2001) Prevention of ventilator-associated pneumonia by oral decontamination: a prospective, randomized, double-blind, placebo-controlled study. Am J Respir Crit Care Med 164(3):382–388. https://doi.org/10.1164/ajrccm.164.3.2005003
129. de Smet AM, Kluytmans JA, Cooper BS, Mascini EM, Benus RF, van der Werf TS, van der Hoeven JG, Pickkers P, Bogaers-Hofman D, van der Meer NJ, Bernards AT, Kuijper EJ, Joore JC, Leverstein-van Hall MA, Bindels AJ, Jansz AR, Wesselink RM, de Jongh BM, Dennesen PJ, van Asselt GJ, te Velde LF, Frenay IH, Kaasjager K, Bosch FH, van Iterson M, Thijsen SF, Kluge GH, Pauw W, de Vries JW, Kaan JA, Arends JP, Aarts LP, Sturm PD, Harinck HI, Voss A, Uijtendaal EV, Blok HE, Thieme Groen ES, Pouw ME, Kalkman CJ, Bonten MJ (2009) Decontamination of the digestive tract and oropharynx in ICU patients. N Engl J Med 360(1):20–31. https://doi.org/10.1056/NEJMoa0800394 360/1/20 [pii]
130. Godard J, Guillaume C, Reverdy ME, Bachmann P, Bui-Xuan B, Nageotte A, Motin J (1990) Intestinal decontamination in a polyvalent ICU. A double-blind study. Intensive Care Med 16(5):307–311
131. Gomez EC, Markowsky SJ, Rotschafer JC (1992) Selective decontamination of the digestive tract in intensive care patients: review and commentary. Ann Pharmacother 26(7–8):963–976
132. Hammond JM, Potgieter PD (1995) Is there a role for selective decontamination of the digestive tract in primarily infected patients in the ICU? Anaesth Intensive Care 23(2):168–174
133. Hammond JM, Potgieter PD (1995) Long-term effects of selective decontamination on antimicrobial resistance. Crit Care Med 23(4):637–645
134. Hammond JM, Potgieter PD, Saunders GL (1994) Selective decontamination of the digestive tract in multiple trauma patients--is there a role? Results of a prospective, double-blind, randomized trial. Crit Care Med 22(1):33–39
135. Hammond JM, Potgieter PD, Saunders GL, Forder AA (1992) Double-blind study of selective decontamination of the digestive tract in intensive care. Lancet 340(8810):5–9. doi:0140-6736(92)92422-C [pii]
136. Hartenauer U, Thulig B, Diemer W, Lawin P, Fegeler W, Kehrel R, Ritzerfeld W (1991) Effect of selective flora suppression on colonization, infection, and mortality in critically ill patients: a one-year, prospective consecutive study. Crit Care Med 19(4):463–473
137. Herruzo-Cabrera R, Garcia Gonzalez JI, Garcia-Magan P, del Rey-Calero J (1994) Nosocomial infection in a neonatal intensive care unit and its prevention with selective intestinal decolonization.

A multivariant evaluation of infection reduction. Eur J Epidemiol 10(5):573–580
138. Melsen WG, de Smet AM, Kluytmans JA, Bonten MJ (2012) Selective decontamination of the oral and digestive tract in surgical versus non-surgical patients in intensive care in a cluster-randomized trial. Br J Surg 99(2):232–237. https://doi.org/10.1002/bjs.7703
139. Nardi G, Valentinis U, Proietti A, De Monte A, Di Silvestre A, Muzzi R, Peressutti R, Troncon MG, Giordano F (1993) Epidemiological impact of prolonged systematic use of topical SDD on bacterial colonization of the tracheobronchial tree and antibiotic resistance. A three year study. Intensive Care Med 19(5):273–278
140. Misset B, Kitzis MD, Conscience G, Goldstein F, Fourrier A, Carlet J (1994) Mechanisms of failure to decontaminate the gut with polymixin E, gentamicin and amphotericin B in patients in intensive care. Eur J Clin Microbiol Infect Dis 13(2):165–170
141. Houben AJ, Oostdijk EA, van der Voort PH, Monen JC, Bonten MJ, van der Bij AK (2014) Selective decontamination of the oropharynx and the digestive tract, and antimicrobial resistance: a 4 year ecological study in 38 intensive care units in the Netherlands. J Antimicrob Chemother 69(3):797–804. https://doi.org/10.1093/jac/dkt416 [pii]
142. Buelow E, Gonzalez TB, Versluis D, Oostdijk EA, Ogilvie LA, van Mourik MS, Oosterink E, van Passel MW, Smidt H, D'Andrea MM, de Been M, Jones BV, Willems RJ, Bonten MJ, van Schaik W (2014) Effects of selective digestive decontamination (SDD) on the gut resistome. J Antimicrob Chemother. dku092 [pii] https://doi.org/10.1093/jac/dku092
143. Kaiser AM, Schultsz C, Kruithof GJ, Debets-Ossenkopp Y, Vandenbroucke-Grauls C (2004) Carriage of resistant microorganisms in repatriates from foreign hospitals to The Netherlands. Clin Microbiol Infect 10(11):972–979. CLM1000 [pii] https://doi.org/10.1111/j.1469-0691.2004.01000.x
144. Bonten MJ, Waterer GW (2014) Antibiotics might do more than cause resistance. Am J Respir Crit Care Med 189(10):1157–1158. https://doi.org/10.1164/rccm.201404-0672ED
145. Halaby T, Al Naiemi N, Kluytmans J, van der Palen J, Vandenbroucke-Grauls CM (2013) Emergence of colistin resistance in Enterobacteriaceae after the introduction of selective digestive tract decontamination in an intensive care unit. Antimicrob Agents Chemother 57(7):3224–3229. https://doi.org/10.1128/AAC.02634-12 [pii]
146. Heininger A, Meyer E, Schwab F, Marschal M, Unertl K, Krueger WA (2006) Effects of long-term routine use of selective digestive decontamination on antimicrobial resistance. Intensive Care Med 32(10):1569–1576. https://doi.org/10.1007/s00134-006-0304-5
147. Oostdijk EA, Smits L, de Smet AM, Leverstein-van Hall MA, Kesecioglu J, Bonten MJ (2013) Colistin resistance in gram-negative bacteria during prophylactic topical colistin use in intensive care units. Intensive Care Med 39(4):653–660. https://doi.org/10.1007/s00134-012-2761-3
148. Giua R, Pedone C, Cortese L, Antonelli Incalzi R (2014) Colistin bladder instillation, an alternative way of treating multi-resistant Acinetobacter urinary tract infection: a case series and review of literature. Infection 42(1):199–202. https://doi.org/10.1007/s15010-013-0507-y
149. Pearman JW, Bailey M, Harper WE (1988) Comparison of the efficacy of "Trisdine" and kanamycin-colistin bladder instillations in reducing bacteriuria during intermittent catheterisation of patients with acute spinal cord trauma. Br J Urol 62(2):140–144
150. Waites KB, Canupp KC, Roper JF, Camp SM, Chen Y (2006) Evaluation of 3 methods of bladder irrigation to treat bacteriuria in persons with neurogenic bladder. J Spinal Cord Med 29(3):217–226
151. Volkow-Fernandez P, Rodriguez CF, Cornejo-Juarez P (2012) Intravesical colistin irrigation to treat multidrug-resistant Acinetobacter baumannii urinary tract infection: a case report. J Med Case Rep 6(1):426. https://doi.org/10.1186/1752-1947-6-426 [pii]
152. Maichuk YF (1991) Dexamethasone in the complex treatment of chlamydial conjunctivitis. Rev Int Trach Pathol Ocul Trop Subtrop Sante Publique 68:83–93
153. Matuska S, Rama P, Cavallero A, Paganoni G, Spinelli A, Brancato R (2006) Nocardia keratitis: a case report. Eur J Ophthalmol 16(1):164–167
154. Paksu MS, Paksu S, Karadag A, Sensoy G, Asilioglu N, Yildizdas D, Akyildiz BN, Kendirli T, Demirkol D, Akgun M, Alp E, Ciftci E, Guney AK, Murat N (2012) Old agent, new experience: colistin use in the paediatric intensive care unit--a multicentre study. Int J Antimicrob Agents 40(2):140–144. https://doi.org/10.1016/j.ijantimicag.2012.04.010 S0924-8579(12)00180-X [pii]
155. Logan LK (2012) Carbapenem-resistant enterobacteriaceae: an emerging problem in children. Clin Infect Dis 55(6):852–859. https://doi.org/10.1093/cid/cis543 [pii]
156. Falagas ME, Vouloumanou EK, Rafailidis PI (2009) Systemic colistin use in children without cystic fibrosis: a systematic review of the literature. Int J Antimicrob Agents 33(6):503 e501–503 e513. https://doi.org/10.1016/j.ijantimicag.2008.10.021 S0924-8579(08)00550-5 [pii]
157. Karbuz A, Ozdemir H, Yaman A, Kocabas BA, Odek C, Guriz H, Aysev AD, Ciftci E, Kendirli T, Ates C, Ince E (2014) The use of colistin in critically ill children in a pediatric intensive care unit. Pediatr Infect Dis J 33(1):e19–e24. https://doi.org/10.1097/INF.0000000000000117 00006454-201401000-00012 [pii]
158. Alan S, Yildiz D, Erdeve O, Cakir U, Kahvecioglu D, Okulu E, Ates C, Atasay B, Arsan S (2014) Efficacy and safety of intravenous colistin in preterm infants

with nosocomial sepsis caused by Acinetobacter baumannii. Am J Perinatol 31(12):1079–1086. https://doi.org/10.1055/s-0034-1371361

159. Falagas ME, Sideri G, Korbila IP, Vouloumanou EK, Papadatos JH, Kafetzis DA (2010) Inhaled colistin for the treatment of tracheobronchitis and pneumonia in critically ill children without cystic fibrosis. Pediatr Pulmonol 45(11):1135–1140. https://doi.org/10.1002/ppul.21302
160. Celik IH, Oguz SS, Demirel G, Erdeve O, Dilmen U (2012) Outcome of ventilator-associated pneumonia due to multidrug-resistant Acinetobacter baumannii and Pseudomonas aeruginosa treated with aerosolized colistin in neonates: a retrospective chart review. Eur J Pediatr 171(2):311–316. https://doi.org/10.1007/s00431-011-1537-z
161. Falagas ME, Rafailidis PI (2007) When to include polymyxins in the empirical antibiotic regimen in critically ill patients with fever? A decision analysis approach. Shock 27(6):605–609. https://doi.org/10.1097/01.shk.0000246899.73315.cb 00024382-200706000-00003 [pii]

Clinical Use of Colistin in Biofilm-Associated Infections

13

Jaime Lora-Tamayo, Oscar Murillo, and Javier Ariza

Abstract

Biofilm is an adaptive bacterial strategy whereby microorganisms become encased in a complex glycoproteic matrix. The low concentration of oxygen and nutrients in this environment leads to heterogeneous phenotypic changes in the bacteria, with antimicrobial tolerance being of paramount importance. As with other antibiotics, the activity of colistin is impaired by biofilm-embedded bacteria. Therefore, the recommendation for administering high doses in combination with a second drug, indicated for planktonic infections, remains valid in this setting. Notably, colistin has activity against metabolically inactive biofilm-embedded cells located in the inner layers of the biofilm structure. This is opposite and complementary to the activity of other antimicrobials that are able to kill metabolically active cells in the outer layers of the biofilm. Several experimental models havc shown a higher activity of colistin when used in combination with other agents, and have reported that this can avoid the emergence of colistin-resistant subpopulations. Most experience of colistin in biofilm-associated infections comes from patients with cystic fibrosis, where the use of nebulized colistin allows high concentrations to reach the site of the infection. However, limited clinical experience is available in other scenarios, such as osteoarticular infections or device-related central nervous system infections caused by multi-drug resistant microorganisms. In the latter scenario, the use of intraventricular or intrathecal colistin also permits high local concentrations and good clinical results. Overall, the efficacy of intravenous colistin seems to be poor, but its association with a second antimicrobial significantly increases the response rate. Given its activity against inner bioflm-embedded cells, its possible role in combination with other antibiotics, beyond last-line therapy situations, should be further explored.

Keywords

Polymyxin · Biofilm · Cystic fibrosis · Prosthetic joint infection · Implant-associated infection

J. Lora-Tamayo (✉)
Department of Internal Medicine, Hospital Universitario 12 de Octubre, Madrid, Spain
e-mail: jaime@lora-tamayo.es

O. Murillo · J. Ariza
Department of Infectious Diseases, Hospital Universitario de Bellvitge, Barcelona, Spain

The presence of bacterial biofilms in nature, industry, and pathological processes in the human body has attracted increasing interest in recent

J. Li et al. (eds.), *Polymyxin Antibiotics: From Laboratory Bench to Bedside*, Advances in Experimental Medicine and Biology 1145, https://doi.org/10.1007/978-3-030-16373-0_13

years [1, 2]. The biofilm crucially conditions the bacterial susceptibility to disinfecting substances and antimicrobial molecules, including polymyxins, and has led to a paradigm change in particular clinical scenarios [1, 3, 4]. In the current setting of increasing antimicrobial resistance [5], polymyxins are a last-line therapy, also for biofilm-associated infections. In this chapter, we review the most important features of the biofilm structure, and focus on the activity of polymyxins against biofilm-embedded bacteria. Furthermore, we will analyze the use of high doses and combination therapy in the management of biofilm-associated infections, before outlining the different clinical applications of polymyxins in this scenario.

13.1 The Biofilm Paradigm, The Clinical Problem

The bacterial biofilm is a universal and sophisticated adaptive mechanism of bacterial survival, defined as a structured bacterial population embedded in a self-produced glycoproteic three-dimensional matrix. The formation of a biofilm starts with bacteria attaching to a surface [6], which is typically inert and belongs to a foreign body such as a pacemaker or prosthetic joint, but it may also be the surface of organic tissue such as occurs in the bronchial tree in cystic fibrosis or in sequestrated bone in chronic osteomyelitis [1, 3].

The initial reversible adhesion to the surface is sensed by the bacteria, which induces the expression of several genes that allow a more sustained attachment and the excretion of a polymeric matrix composed of glycoproteins, polysaccharides, and ribonucleic acids [7, 8]. Consequently, cells become encased by a slime-like substance within which the concentration of nutrients and oxygen dramatically reduces. In this particular environment, bacterial cells undergo phenotypic changes and significantly reduce their metabolism: in short, they consume less energy and decrease the rate of replication [1, 9].

Far from being a passive adaptive form, the biofilm structure is a complex and dynamic 3-dimensional matrix. Maturation of the biofilm leads to inner channels being formed that allow media and nutrients to be circulated [6, 10]. When the biofilm achieves a critical size, the outer layers may then detach from the structure, which allows the cells encased within to be released and to recover their planktonic properties. Subsequently, the cells are able to attach to new surfaces and to restart the process. The detachment of the outer layers may occur due to the physical conditions under which the biofilm develops, or may be due to the excretion of digestive enzymes that disrupt the extracellular matrix and release the bacteria [1, 6].

The production of these enzymes is just one example of the bacterial specialization and coordination that occurs throughout the biofilm because of both the local concentration of nutrients and the biochemical system of communication and signaling known as *quorum sensing* [6, 9, 11]. Indeed, when the number of bacteria excreting a particular compound (signal) reaches a critical concentration threshold, new gene expression is triggered in distant cells, which leads to a heterogeneous phenotypic pattern of bacteria within the biofilm and to the presence of specialized subpopulations [7].

In the outer biofilm layers where the concentrations of oxygen and nutrients are higher, bacteria are metabolically more active, whereas the rate of replication is much lower in the deeper layers [3]. Intracellular bacteria may also be found in the biofilm structure [12], as well as specialized surviving forms such as small colony variants [13]. Indeed, the existence of bacteria at various metabolic stages, with specific abilities and phenotypes, is believed to increase the chances of survival when faced with a particular threat.

Importantly, biofilms are known to be tolerant to antimicrobials, and so can survive when exposed to biocidal substances or antibiotics at clinically achievable concentrations. The reasons for this are beyond the classical mechanisms of resistance and can be summarized as follows [1, 3, 4, 14]:

(i) Impaired diffusion of molecules in the biofilm. While most antibiotics are able to diffuse within the glycoproteic matrix, the transition of some may be impaired in the case of voluminous or polymeric molecules [1]. In addition, the presence of extracellular hydrolytic enzymes may inactivate the antibiotic before it reaches bacterial cells [4, 15], or the antibiotic molecule may be held by physical forces, as in the case of positively charged aminoglycosides that are held in the negatively charged biofilm [3, 16].

(ii) Biofilm-embedded bacteria become intrinsically less susceptible to most antibiotics. This is mainly due to the metabolic changes that bacteria undergo when exposed to low nutrient and oxygen concentrations. Antibiotics with activity that is highly dependent on the rate of bacterial growth, for example β-lactams, are particularly affected by the resulting low replication rates of adherent bacteria. This has also been observed in planktonic bacteria which, when exposed to high bacterial density and low nutrient concentrations, enter a phase of stationary growth that makes them tolerant to antimicrobials [3].

(iii) Biofilm-embedded bacteria may express very different phenotypes according to the local environmental conditions and the *quorum sensing*. Thus, they can differentiate to subpopulations that may be particularly resistant to external chemical or physical threats; these constitute the so-called *persisters* [13, 17]. Also, some bacteria found in biofilm-associated infections are known to be intracellular [12]. These phagocytosis-surviving microorganisms may become infection reservoirs because they are less exposed to antibiotics that are unable to either penetrate the eucaryotic cell or reach specific intracellular compartments [18].

(iv) Finally, both the humoral and cellular immune responses have proved to be ineffective for clearing biofilm-associated infections, but instead contribute to the chronic inflammation and damage observed in the surrounding tissues.

Biofilm-embedded bacteria may also express antimicrobial resistance due to conventional mechanisms such as modification of the antibiotic target or cell permeability, the use of efflux-pumps, or the expression of hydrolysing enzymes. Moreover, horizontal gene transmission is increased in biofilms, thus raising the likelihood of resistance developing [4, 19]. In addition, although the rate of cell replication is significantly decreased for biofilm-embedded cells, some bacteria may increase their mutation frequency, especially when it is not normally very high in the planktonic state [20, 21].

13.2 Activity of Colistin in Biofilms

Polymyxin activity on the biofilm of gram-negative microorganisms has been demonstrated in several *in vitro* and *in vivo* models. Most studies have addressed the activity of colistin on biofilms associated with *Pseudomonas aeruginosa* [8, 14, 22–29] because of its prominence in lung infections in patients with cystic fibrosis.

Colistin has a characteristic but different behavior against *P. aeruginosa* biofilms when compared with other antibiotics [8, 14], being dependent on the 3-dimensional structure of the biofilm [30]. As previously discussed, the biofilm contains many different phenotypes of the same bacteria. Indeed, *P. aeruginosa* biofilms grown *in vitro* develop a characteristic 3-dimensional structure that looks like a mushroom. Bacterial cells contained in the outer part of the structure (the cap of the mushroom) are larger, show mobility, and have more active metabolism when compared with cells located in the deeper layers of the inner structure (the stalk of the mushroom) [7, 8]. Other families of antibiotics, such as the aminoglycosides or fluoroquinolones, are able to kill bacteria located in the cap of the mushroom, but are inactive against the less metabolically active bacteria within the stalk [14]. In contrast, various studies have observed that colistin behaves in an opposite manner: it is able to kill the cells within the stalk of the mushroom structure, but has no activity against the bacterial cells in the outer layers of the cap [7, 8, 14].

As in the case of planktonic bacteria, the mechanism of tolerance to colistin of these metabolically active cells in the outer layers of biofilm is due to modifications in the membrane lipids of the bacterial cells. This depends on the synthesis of 4-amino-4-deoxyarabinoside (4A4D), which binds to the lipid A of the lipopolysaccharide of the membrane and reduces its negative charge. As a result, the affinity of the positively charged colistin is significantly decreased (Fig. 13.1) [8, 26, 31]. Among other regulatory systems, the polymyxin resistance *pmr* operon induces the synthesis of 4A4D in response to various stimuli, including the presence of sub-inhibitory concentrations of colistin or low concentrations of magnesium or calcium [8, 32]. Pamp *et al* demonstrated that the activity of *pmr* was energy-dependent, thus explaining the heterogeneous distribution of colistin tolerance within the biofilm [14].

In addition, the regulatory system PhoPQ influences the synthesis of molecules that modify the electrical charge of the membrane [27, 31, 32]. In *P. aeruginosa*, the activator BrlR induces the expression of several types of efflux pumps such as MexAB-OprM and MexEF-OprN [27]. Chambers and Sauer observed that BrlR down-regulates the PhoPQ system in *P. aeruginosa* biofilms and that higher susceptibility to colistin could compensate for the tolerance of those biofilms to quinolone or aminoglycoside antibiotics [27].

Regarding the particular bactericidal activity shown by colistin against less metabolically active cells located in the stalk part of the mushroom, it is important to note that colistin does not

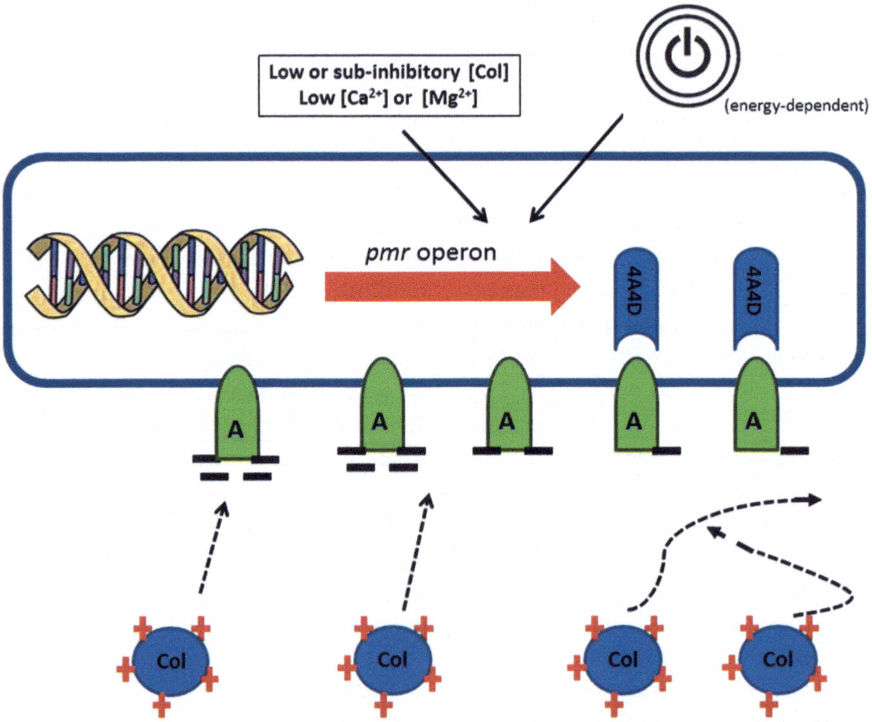

Fig. 13.1 Loss of activity of colistin (Col) in *Pseudomonas aeruginosa* caused by a modification in the net electrical charge of the bacterial outer membrane. This is an energy-dependent mechanism of resistance regulated by the operon pmr. Various stimuli, such as sub-inhibitory concentrations of colistin or low concentrations of magnesium and calcium, may lead to the synthesis of 4-amino-4-deoxyarabinoside (4A4D), which binds to the lipid A of the lipopolysaccharide of the membrane (A), thus reducing its negative charge [7, 14, 27, 31]

seem to be dependent on oxidative stress when targeting *P. aeruginosa*. While the cumulative production of hydroxyl radicals is a common mechanism of killing in metabolically active cells that are exposed to most antibiotic classes, irrespective of their cellular target (e.g., β-lactams, aminoglycosides, fluoroquinolones, and glycopeptides), this is not the case for cationic peptides such as colistin and other substances that modify the membrane permeability. This behaviour may explain the specific activity of colistin against these less metabolically active cells [25].

Several *in vitro* models have shown a decrease in the number of viable biofilm-embedded *P. aeruginosa* cells when exposed to colistin [8, 14, 24, 28, 29]. However, most of these studies have used high concentrations of colistin (10–25 mg/L), which are too optimistic from a clinical perspective, especially in the setting of a biofilm-associated infection. The classic microbiological concepts of minimal inhibitory and bactericidal concentrations (MIC and MBC, respectively) are helpful to predict the activity of antibiotics in planktonic infection, but may not be as useful for biofilm-associated infections [33]. Minimal biofilm inhibitory and eradicative concentrations (MBIC and MBEC, respectively) more accurately reflect the activity of antimicrobials when tested for biofilms [33]. As previously discussed, virtually all antimicrobials are less active against biofilm-embedded bacteria than against their planktonic counterparts [1, 9]. The degree of tolerance against a particular antimicrobial will depend on the specific microorganism and on the maturity and characteristics of the biofilm [2, 22, 34].

To some degree, this is also the case for polymyxins against gram-negative bacilli. In an *in vitro* model of *P. aeruginosa* biofilm, Hengzhuang *et al* showed that colistin had an MBIC of 8 mg/L or 16 mg/L for young or old biofilms, respectively; these required concentrations that were 4–8 times higher than the MIC of 2 mg/L [22]. In this study, the MBEC of a young biofilm was 128 mg/L, which was confirmed by an *in vivo* model of biofilm-associated lung-infection in mice [23]. It is unlikely that intravenous administration of colistimethate sodium (CMS, colistin's prodrug) could provide the required plasma colistin concentrations [35–37].

Therefore, the concerns that exist for achieving sufficiently high colistin concentrations for the treatment of planktonic infections may be extended to biofilm-associated infections. Of course, sub-inhibitory concentrations of colistin may be associated with inadequate therapeutic efficacy [36]. What is more, colistin heteroresistance has been described for several strains of *Acinetobacter* spp. [38], *Klebsiella* spp. [39], and *Pseudomonas aeruginosa* [40]. Selection and amplification of resistant subpopulations after exposure to sub-inhibitory concentrations of colistin is a potential danger that must be considered. Indeed, a recent study using an *in vitro* PK/PD dynamic model of *P. aeruginosa* biofilm found that colistin monotherapy at clinically relevant concentrations initially had no bactericidal activity, which was followed by regrowth and the emergence of colistin-resistance [29].

Based on this PK/PD problem, and supported by experimental models, current recommendations suggest high doses of CMS be administered in combination with a second antimicrobial [37, 41]. The rationale for this is based on the potential for subpopulation synergy: that is, each drug would target the subpopulation that the other drug is not able to kill. In addition, a mechanistic synergy has been proposed based on colistin's mechanism of action [42, 43]. As a cationic peptide, colistin targets the bacterial external membrane and enhances its own uptake, together with that of other molecules, which may favor the penetration of other antibiotics in the bacterial cell [41–43]. With this combination strategy, clinically achievable lower doses of colistin become efficacious and heteroresistant colistin strains might not develop.

In the setting of biofilm infections by gram-negative bacilli, the rationale for colistin in combination with another antibiotic is even greater: as mentioned, colistin is active against the less active bacteria located in the inner biofilm layers, in contrast with other antibiotics. Therefore, it may have a relevant, distinctive and complementary role in the treatment of biofilm infections caused by gram-negative bacilli. Indeed, the combination of colistin and a second antimicrobial, such as ciprofloxacin or tobramycin, has been shown to be more efficacious than the use of each antibiotic alone, presumably due to the synergistic activity against the whole bacterial population of the biofilm [14, 24].

This may also imply that colistin could be useful not only as a last-line therapy against biofilm-associated multidrug-resistant bacterial infection but also in other settings with poor prognosis, such as prosthetic joint infection caused by fluoroquinolone-resistant gram-negative bacilli [44]. Indeed, in a recent study of foreign-body infection caused by extended-spectrum beta-lactamase-producing *E. coli* in guinea pigs, a higher activity of colistin was demonstrated when combined with gentamycin, fosfomycin, or tigecycline [45].

In the previously mentioned *in vitro* dynamic biofilm model study [29], additivity, synergy, and avoidance of colistin-resistance was observed when colistin was combined with doripenem at clinically relevant concentrations. Interestingly, this was observed not only for carbapenem-susceptible strains of *P. aeruginosa* but also for carbapenem-resistant strains (including two different mechanisms of resistance). Thus, it is suggested that modifications in the bacterial membrane induced by colistin could overcome, at least partially, the mechanisms of resistance to doripenem.

In summary, the rationale for administering high-dose CMS in combination with a second drug remains valid in the setting of biofilm-associated infections in which the overall activity of colistin is typically decreased, as occurs with other antimicrobials. Both *in vitro* and *in vivo* experimental models of biofilm infection support the administration of colistin in combination with other antimicrobials, as each agent targets different sites within the biofilm structure. Finally, there is some evidence to suggest that the modifications in cell permeability caused by colistin may enhance the activity of the second drug against biofilm-embedded bacteria.

13.3 Clinical Experience of Colistin and Biofilm-Associated Infections

Biofilm growth may occur in human infections with or without the presence of foreign bodies. In the former, such as intravascular catheter or pacemaker infections, device removal should be performed whenever possible to improve the cure rate [46]. However, special difficulties exist when seeking to cure those biofilm-related infections that involve non-debrided human tissues or retained medical devices, as can occur in cystic fibrosis and prosthetic joint infection.

The presence of multi-drug resistant (MDR) gram-negative bacilli in the context of a limited therapeutic repertoire of active antimicrobials only adds more complexity to the treatment of biofilm-related infections. Consequently, colistin has mainly been used in the last-line therapy of these difficult-to-treat infections; however, according to the potential benefits of colistin in the setting of bacterial biofilms noted in Sect. 2, this antibiotic might be a suitable alternative to conventional agents against infections caused by other susceptible gram-negative bacilli.

To date, limited clinical experience exists for the use of colistin in the treatment of biofilm-related infections. Apart from aerosolized and intravenous administration, colistin has been administered locally, using the intraventricular route or in cement spacers for central nervous system (CNS) or prosthetic joint infections, respectively. However, neither the optimal dosage of colistin nor the comparative efficacy between colistin alone or in combination have been assessed in this setting. Thus, we review the clinical experience of the use of colistin for the treatment of severe biofilm-related infections in cystic fibrosis and non-cystic fibrosis bronchiectasis, prosthetic joint infection, and CNS device-related infections.

13.3.1 Cystic Fibrosis and Non-cystic Fibrosis Bronchiectasis

13.3.1.1 Cystic Fibrosis

To date, the vast majority of experience with colistin in biofilm-associated scenarios has been accumulated in the context of cystic fibrosis. This congenital disorder produces mutations in the cystic fibrosis transmembrane conductance regulator gene that cause the chloride channel to malfunction. Consequently, the disease affects several human systems, including the lungs, the respiratory airways, the pancreas, and the small intestine, with clinical manifestations dependent on the system affected. In the lungs and respiratory tract, the malfunction produces decreased paraciliary fluid and clearance of microorganisms, which leads to bronchial obstruction, superinfection, inflammation, bronchiectasis, and a loss of respiratory function [47–50].

The life expectancy of patients with cystic fibrosis has improved significantly over recent years, probably because of more aggressive antimicrobial therapy, both in the treatment of infections and as maintenance therapy [48, 51]. Chronic *P. aeruginosa* lung infection is the main cause of morbidity and mortality in cystic fibrosis; indeed, 30% of infants aged 2–5 years and 80% of adults are colonized [52]. Several adaptive mechanisms of *P. aeruginosa* have been related to its ability to survive for long periods in the lungs of patients with cystic fibrosis. Probably the most relevant mechanism is the mucoid-biofilm mode of growth, which allows *P. aeruginosa* to tolerate the immune system, antibiotic therapy, and an anaerobic environment [47].

Colistin is widely used for the treatment of infections by *P. aeruginosa* in patients with cystic fibrosis, both as first line therapy and as a salvage treatment against MDR strains [53, 54]. To date, much of the clinical experience refers to the use of nebulized colistin in intermittently colonized or chronically infected patients [53, 54], with minimal information regarding its intravenous use [55, 56]. However, *in vitro* and experimental studies suggest that both the upper and the lower respiratory airways are infected by *P. aeruginosa*, thus supporting combination therapy with inhaled and intravenous antibiotics, especially in acute exacerbations [47].

The administration of nebulized colistin allows low serum levels and high lung concentrations to be achieved (at least 10 times greater than the MIC value), thus leading to higher efficacy and less drug-related toxicity [57–59]. Furthermore, the conversion of CMS to colistin is higher with nebulized than intravenous administration, probably because there is no renal clearance of CMS in the bronchi, which allows a higher proportion of the prodrug to be hydrolysed to colistin [59]. Thus, compared with intravenous administration, it has been reported that nebulized administration may lead to higher bronchial colistin levels than the MBIC of colistin reported with *P. aeruginosa* biofilm [22, 23]. Currently, a dose of two million IU of CMS every 8–12 h is recommended; although some bronchoconstriction may occur, this dose is usually well tolerated [54, 60]. Recently, significant efforts have been made to improve the aerosolized delivery to achieve superior drug distribution along the airways and to increase medication compliance, as reviewed by Heijerman *et al* [54]. Thanks to modern portable devices, inhalation of colistin as a dry powder is possible over just 2–3 min [61].

In previous studies, the efficacy of aerosolized colistin in combination with oral ciprofloxacin was evaluated and was found to postpone *P. aeruginosa* infection significantly and to maintain pulmonary function [62, 63]. In addition, other studies have revealed similar efficacy between nebulized colistin and tobramycin, especially in decreasing the *P. aeruginosa* sputum density [64, 65]. Of interest, it seems that the emergence of drug-resistant *P. aeruginosa* strains when using aerosolized colistin is less common than occurs when using intravenous colistin for other infections. This is probably related to the higher drug levels achieved with the aerosol route. Furthermore, the emergence of resistance is also less common when colistin is compared with other nebulized drugs [51, 66].

Regarding the intravenous administration of colistin [55, 56, 66], its safety profile and optimal dosage remain unclear [67]. Previous studies have evaluated the efficacy of colistin, mainly in combination therapy, in the treatment of acute pulmonary exacerbations of cystic fibrosis; in one study, a greater improvement in pulmonary function was reported [56]. Overall, results that are more consistent are needed to evaluate the current role of intravenous colistin for the treatment of pulmonary infections due to gram-negative bacilli in patients with cystic fibrosis. Moreover, the emergence of MDR gram-negative bacilli poses a greater challenge for clinicians, and there is a need to improve our knowledge of the efficacy of colistin in this setting.

13.3.1.2 Non-cystic Fibrosis Bronchiectasis

In comparison with the clinical experience in managing patients with cystic fibrosis, there is much less knowledge of the use of colistin for the treatment of infections in patients with non-cystic fibrosis bronchiectasis. Where research is present, this is limited to the use of inhaled colistin in a limited number of studies. A detailed review of these results is beyond the scope of the present chapter, and we direct readers to a recently published systematic review for further detail [68]. In summary, inhaled colistin has some proven benefits, such as a greater reduction in sputum *P. aeruginosa* load [69]; however, further studies are needed to demonstrate its benefit in the long-term eradication of *P. aeruginosa* or in ameliorating the number of acute pulmonary exacerbations.

13.3.2 Prosthetic Joint and Other Osteoarticular Device-Related Infections

Orthopedic devices (joint prostheses or osteosynthesis hardware) are widely used in current clinical practice to improve the quality of life of patients [70]. However, infection of the devices raises serious concerns, not least because the resulting biofilm-related infections are difficult to treat [46]. The treatment of orthopedic device-related infections must include appropriate and prolonged antibiotic therapy, usually administered at high doses and combined with adequate surgical intervention [70–72].

Depending on the specific type of prosthetic joint infection, management may include debridement and implant retention, replacement with a new prosthesis, or definitive removal of the joint prosthesis [70–72]. However, maintaining the prosthesis poses a major challenge when trying to cure the infection. Concerning infections of osteosynthesis hardware, it seems that internal fixation also shares similarities with prosthetic joint infection. In general, fixation-device related infections are more frequently managed by device removal when compared with prosthetic joint infection.

In this difficult clinical scenario, the most appropriate antibiotic therapy for infections caused by MDR gram-negative bacilli remains a matter of great concern. Again, colistin has been used as a last-line therapy, but its efficacy and the potential benefits of combination therapy with other drugs have yet to be properly evaluated.

Older reports found that the diffusion of colistin into bone was poor [73]; therefore, it was exclusively administered in local beads and cement spacers in the past [74, 75]. This use has progressively been abandoned because there is insufficient knowledge regarding the most appropriate concentration and elution of colistin needed for such cements [76]. While some authorities have discouraged the use of cement spacers in the presence of MDR microorganisms [71], our opinion and that of others argue that the use of colistin-loaded cement spacers might be useful for the treatment of some cases of prosthetic joint infection by MDR gram-negative bacilli [77]. Further studies should explore the potential benefits of administering colistin locally for the treatment of device-related infections.

Clinical experience with intravenous CMS in this field is limited and mainly based on its efficacy as a last-line therapy in difficult-to-treat bone and joint infections caused by MDR gram-negative bacilli [77–80]. Neither the optimal dosage nor the optimal pharmacodynamic parameters of colistin are known for treating these infections.

A recent study was conducted by Valour *et al* with 19 patients suffering from bone and joint infections caused by MDR and extensively-drug resistant (XDR) gram-negative bacilli [78]. In that study, 12 cases were associated with an orthopedic device, and colistin alone was used as salvage therapy in 90% of cases. The authors reported clinical remission in 74% of cases (median follow-up, 28 weeks); however, the outcome of orthopedic-device related infections was clearly worse, leading to a treatment failure of 42%.

Over the last 10 years, we have accumulated data on 22 cases of osteoarticular infection caused by XDR *P. aeruginosa* (Ribera *et al*, communication in the ICAAC, Washington D.C., 2014). In 15 cases (68%), an orthopedic device was involved (8 prosthetic joints). While the combination of colistin and a β-lactam achieved a cure rate of 80%, colistin monotherapy (57% cases) led to poorer results (cure rate, 29%). Of interest, when faced with device-related infections, the combination of colistin with a β-lactam was better when the latter was administered as a continuous infusion (cure rate, 83%) as compared with intermittent boluses (cure rate, 67%).

Overall, colistin seems to offer clinical efficacy against orthopedic device-related infection by MDR or XDR gram-negative bacilli, especially when used in combination with other antimicrobials. In the case of *P. aeruginosa* with full resistance or intermediate susceptibility to β-lactams, the administration of colistin in combination with a β-lactam can improve the outcomes of these infections. Further studies are needed to confirm these results and to explore alternative therapeutic combinations with colistin (i.e., fosfomycin, tigecycline).

Finally, the preclinical and clinical studies that highlight the potential activity of colistin against biofilm-associated infection suggest that the efficacy of colistin needs to be evaluated as a first line therapy in a wider range of orthopedic device-related infections. These include those caused by not only MDR and XDR microorganisms but also less resistant gram-negative bacilli. The poor outcomes associated with prosthetic joint infection by ciprofloxacin-resistant gram-negative bacilli, which is usually treated with β-lactam monotherapy, has been reported [44]. Moreover, the best therapy for the treatment of infections caused by extended-spectrum beta-lactamase or carbapenemase producing *Enterobacteriaceae* needs to be defined.

13.3.3 Central Nervous System Device-Related Infections

Colistin may be the only therapeutic option for CNS infections caused by MDR gram-negative bacilli such as *A. baumannii*, MDR *P. aeruginosa,* or carbapenemase-producing *Klebsiella pneumoniae*. These typically occur in a nosocomial setting in patients with brain damage, as well as those with external ventricular drainage (EVD) or ventriculo-peritoneal shunt (VPS) devices [81]. Patients are usually critically ill, with the infection representing a life-threatening complication that requires optimal antimicrobial therapy. Although the foreign body should be removed to cure the infection [46, 82], the patient frequently needs a replacement CSF diversion to be placed at the same time as the infected material is removed. Therefore, the sterility of the CSF during this procedure is of paramount importance.

In addition to the presence of a foreign-body and bacterial biofilm, the blood-brain barrier (BBB) may significantly impair the diffusion of colistin; thus, excessively low concentrations may occur at the site of infection when the drug is administered intravenously. Experimental and clinical data suggest that only 5% of plasma colistin is able to diffuse into the CNS [83–86]. Diffusion through the BBB does not seem to be affected by efflux pumping by P-glycoproteins, but depends on the permeability of the endothelial tight junctions, which may be increased by inflammatory cytokines [84]. Indeed, the percentage of colistin able to reach the cerebrospinal fluid (CSF) is higher during meningeal inflammation; however, the absolute concentration is usually less than 0.5 mg/L, which is less than that required for most gram-negative bacilli [86, 87].

To achieve higher concentrations of colistin at the infection site, clinicians directly administer CMS into the ventricular or meningeal space [81]. CMS has been observed to convert to colistin in CSF [88] and the available data suggests that repeated doses of intraventricular CMS do not lead to accumulation [84, 85]. In addition to its bactericidal activity, the characteristic anti-endotoxin effect of polymyxins could have a potential favorable effect when treating meningitis. Indeed, the affinity of colistin for the lipopolysaccharide molecule may critically reduce the inflammatory response in the meningeal space [81]. The administration of CMS may be via externalized VPS or EVD ports under extremely sterile conditions: first, 5 mL of CSF must be gently removed to prevent an increase of intracranial pressure; then, the CMS dose is diluted in 3 mL of saline and administered as a bolus over 1–2 min followed by a 2-mL saline flush. Provided the intracranial pressure does not raise too much, the drainage must be closed for at least 60 min to avoid excessive clearance of CMS/colistin [88]. Alternatively, in patients without a VPS or EVD, an Ommaya device or a lumbar drainage may be implanted, although intraventricular administration seems to provide superior diffusion in the CNS than the intrathecal route [81].

Depending on the case, CNS device-related infection may involve the ventricles or the meninges differently. Here, the concomitant administration of intravenous antibiotics with locally administered colistin is desirable, especially when meningitis is present. When MDR gram-negative bacilli are responsible for such infections, intravenous colistin has also been used in some cases [89–92]. While awaiting further studies with greater consistency, some research has found that higher concentrations of colistin are achieved by combined intravenous-intraventricular administration [86], and others have reported the success of this combined approach [90–93]. Given these factors, we consider that the concomitant administration of colistin via the intraventricular route, together with a second intravenous antimicrobial agent, is appropriate for the treatment of severe CNS device-related infection caused by MDR gram-negative bacilli.

Overall, clinical experience with colistin in this setting is mainly based on case reports or small case series, with wide variability in dosing, administration routes (intraventricular versus intrathecal), concomitant intravenous antimicrobials, and the presence or absence of foreign devices. Of note, the greatest experience involves infections caused by *A. baumannii* and, to a lesser degree, *P. aeruginosa*; little information exists regarding other MDR gram-negative bacilli such as the carbapenemase-producing Enterobacteriaceae. Each of these microorganisms may present with different virulence, but it is beyond the scope of this chapter to discuss this in detail. However, regardless of their MDR status, infections by bacteria such as *P. aeruginosa* and *K. pneumoniae* typically cause clinical presentations that are more aggressive, are difficult to treat, and have worse prognoses.

To date, the dose of intrathecal or intraventricular CMS has not been standardized, with doses ranging from 20,000 IU twice daily to 500,000 IU once daily [89, 94–96]. Imberti *et al* studied the pharmacokinetics of colistin in CSF after various doses of intrathecal CMS in 9 patients [88]. Doses of 60,000 IU/d gave C_{max} values of 7–22.1 mg/L and C_{trough} values $\geq$2 mg/L. The authors concluded that a dose of 125,000 IU/d, as recommended in the Infectious Diseases Society of America guidelines, was probably appropriate based on the notable inter-patient variability observed [82, 88]. In a recent review of more than 100 cases of CNS infection by MDR gram-negative bacilli, Bargiacchi *et al* found no differences in the clinical and microbiological cure rates among patients receiving either $\geq$125,000 IU/d or < 125,000 IU/d [95].

Clinical and microbiological cure with the use of intrathecal or intraventricular colistin is reported to be high [89, 94–96]. Karaiskos *et al* recently performed a literature review of 83 episodes of CNS infection by MDR *A. baumannii* treated with locally administered CMS (either in monotherapy or with other systemic antimicrobials). A foreign body was present in 63% of cases, and the cure rate was 89%. Toxicity was observed in 11%, which was mainly due to reversible chemical ventriculitis or meningitis, although there were some cases that involved seizures too [94].

The duration of intrathecal or intraventricular treatment is also highly variable, ranging from 2 to 56 days [94–96]. In the report by Karaiskos, the median time needed to sterilize the CSF was 4 days [94], while Bargiacchi reported that treatments shorter than 7 days in their review had a significantly higher failure rate than longer treatments [95].

In summary, in patients with CNS infection by MDR-gram-negative bacilli, intrathecal or intraventricular administration of CMS is recommended at doses of 125,000 IU per day over at least 7 days. While the eventual foreign body (e.g., the EVD or VPS) will probably need to be removed, locally administered colistin appears to be helpful in sterilizing the CSF before implanting new foreign material. Supported by the current knowledge suggesting colistin heteroresistance and the potential for synergistic interactions, it is also recommended to administer colistin in combination with an intravenous antimicrobial.

13.4 Conclusions

Antimicrobial therapy must be optimized in the case of difficult-to-treat biofilm-associated infections. In many cases, colistin represents an effective last-line therapeutic option because of the increasing incidence of MDR microorganisms. Given that the targets of colistin in the biofilm are different and complementary to those of other antimicrobials, it continues to be recommended that high doses of colistin are appropriate in combination with a second antimicrobial in this setting. The local administration of colistin at the infection site, either nebulized for cystic fibrosis or intraventricular for CNS infections, increases local antibiotic concentrations and improves clinical results. Intravenous administration of colistin seems to be less effective although use in combination with a second antimicrobial significantly increases the response rate. The possible role of colistin in combination with other antibiotics, beyond last-line therapy, should be further explored.

Acknowledgments We thank Michael Maudsley for reviewing the English manuscript.

References

1. Costerton JW, Stewart PS, Greenberg EP (1999) Bacterial biofilms: a common cause of persistent infections. Science 284:1318–1322
2. Coenye T, Nelis H (2010) *In vitro* and *in vivo* model systems to study microbial biofilm formation. J Microbiol Methods 83:89–105
3. Fux CA, Costerton JW, Stewart PS, Stoodley P (2005) Survival strategies of infectious biofilms. Trends Microbiol 13:34–40
4. Høiby N, Bjarnsholt T, Givkov M, Molin S, Ciofu O (2010) Antibiotic resistance of bacterial biofilms. Int J Antimicrob Agents 35:322–332
5. Boucher HW, Talbot GH, Bradley JS, Edwards JE, Gilbert D, Rice LB et al (2009) Bad bugs, no drugs: no ESKAPE! An update from the Infectious Diseases Society of America. Clin Infect Dis 48:1–12
6. Høiby N, Johansen HK, Moser C, Song Z, Ciofu O, Kharazmi A (2001) *Pseudomonas aeruginosa* and the in vitro and in vivo biofilm mode of growth. Microbes Infect 3:23–35
7. Klausen M, Aaes-Jørgensen MS, Tolker-Nielsen T (2003) Involvement of bacterial migration in the development of complex multicellular structures in *Pseudomonas aeruginosa* biofilms. Mol Microbiol 50:61–68
8. Haagensen JAJ, Klausen M, Ernst RK, Si M, Folkesson A, Tolker-Nielsen T et al (2007) Differentiation and distribution of colistin- and sodium dodecyl sulfate-tolerant cells in *Pseudomonas aeruginosa* biofilms. J Bacteriol 189:28–37
9. Stewart PS, Costerton JW (2001) Antibiotic resistance of bacterial biofilms. Lancet 358:135–138
10. Bardouniotis E, Ceri H, Olson ME (2003) Biofilm formation and biocide susceptibility testing of *Mycobacterium fortuitum* and *Mycobacterium marinum*. Curr Microbiol 46(1):28–32
11. Horswill AR, Stoodley P, Stewart PS, Parsek MR (2007) The effect of the chemical, biological, and physical environment on quorum sensing in structured microbial communities. Anal Bioanal Chem 387:371–380
12. Murillo O, Pachón ME, Euba G, Verdaguer R, Carreras M, Cabellos C et al (2009) Intracellular antimicrobial activity appearing as a relevant factor in antibiotic efficacy against an experimental foreign body infection caused by *Staphylococcus aureus*. J Antimicrob Chemother 64:1062–1066

13. Sendi P, Rohrbach M, Graber P, Frei R, Ochsner PE, Zimmerli W (2006) *Staphylococcus aureus* small colony variants in prosthetic joint infection. Clin Infect Dis 43:961–967
14. Pamp SJ, Gjermansen M, Johansen HK, Tolker-Nielsen T (2008) Tolerance to the antimicrobial peptide colistin in *Pseudomonas aeruginosa* biofilms is linked to the metabolically active cells, and depends on the *pmr* and *mexAB-oprM* genes. Mol Microbiol 68:223–240
15. Anderl JN, Franklin MJ, Stewart PS (2000) Role of antibiotic penetration limitation in *Klebsiella pneumoniae* biofilm resistance to ampicillin and ciprofloxacin. Antimicrob Agents Chemother 44:1818–1824
16. Gilbert DN, Legget JE (2010) Aminoglycosides. In: Mandell GL, Bennett JE, Dolin R (eds) Principles and practice of infectious diseases. Churchill Livingston and Elsevier, Philadelphia, pp 359–384
17. Proctor RA, Peters G (1998) Small colony variants in staphylococcal infections: diagnostic and therapeutic implications. Clin Infect Dis 27:419–423
18. Marin M, Raoult D (1997) Intracellular organisms. Int J Antimicrob Agents 9:61–70
19. Molin S, Tolker-Nielsen T (2003) Gene transfer occurs with enhanced efficiency in biofilms and induces enhanced stabilisation of the biofilm structure. Curr Opin Biotechnol 14:255–261
20. Driffield K, Miller K, Bostock JM, O'Neill AJ, Chopra I (2008) Increased mutability of *Pseudomonas aeruginosa* in biofilms. J Antimicrob Chemother 61:1053–1056
21. García-Castillo M, del Campo R, Baquero F, Morosini MI, Turrientes MC, Zamora J et al (2011) Stationary biofilm growth normalizes mutation frequencies and mutant prevention concentrations in *Pseudomonas aeruginosa* from cystic fibrosis patients. Clin Microbiol Infect 17:704–711
22. Hengzhuang W, Wu H, Ciofu O, Song Z, Høiby N (2011) Pharmacokinetics/pharmacodynamics of colistin and imipenem on mucoid and nonmucoid *Pseudomonas aeruginosa* biofilms. Antimicrob Agents Chemother 55:4469–4474
23. Hengzhuang W, Wu H, Ciofu O, Song Z, Høiby N (2012) In vivo pharmacokinetics/pharmacodynamics of colistin and imipenem in Pseudomonas aeruginosa biofilm infection. Antimicrob Agents Chemother 56:2683–2690
24. Herrmann G, Yang L, Wu H, Song Z, Wang H, Høiby N et al (2010) Colistin-tobramycin combinations are superior to monotherapy concerning the killing of biofilm *Pseudomonas aeruginosa*. J Infect Dis 202:1585–1592
25. Brochmann RP, Toft A, Ciofu O, Briales A, Kolpen M, Hempel C et al (2014) Bactericidal effect of colistin on planktonic *Pseudomonas aeruginosa* is independent of hydroxyl radical formation. Int J Antimicrob Agents 43:140–147
26. Chiang WC, Pamp SJ, Nilsson M, Givskov M, Tolker-Nielsen T (2012) The metabolically active subpopulation in *Pseudomonas aeruginosa* biofilms survives exposure to membrane-targeting antimicrobials via distinct molecular mechanisms. FEMS Immunol Med Microbiol 65:245–256
27. Chambers JR, Sauer K (2013) The MerR-like regulator BrlR impairs *Pseudomonas aeruginosa* biofilm tolerance to colistin by repressing PhoPQ. J Bacteriol 195:4678–4688
28. Cai Y, Li R, Lianq B, Bai N, Liu Y, Wang R (2010) *In vitro* antimicrobial activity and mutant prevention concentration of colistin against *Acinetobacter baumannii*. Antimicrob Agents Chemother 54:3998–3999
29. Lora-Tamayo J, Murillo O, Bergen PJ, Nation RL, Poudyal A, Luo X et al (2014) Activity of colistin combined with doripenem at clinically relevant concentrations against multidrug-resistant *Pseudomonas aeruginosa*in an *in vitro* dymanic biofilm model. J Antimicrob Chemother 69(9):2434–2342
30. Folkesson A, Haagensen JAJ, Zampaloni C, Sternberg C, Molin S (2008) Biofilm induced tolerance towards antimicrobial peptides. PLoS One 3:e1891
31. Moskowitz SM, Ernst RK, Miller SI (2004) PmrAB, a two-component regulatory system of *Pseudomonas aeruginosa* that modulates resistance to cationic antimicrobial peptides and addition of aminoarabinose to lipid A. J Bacteriol 186:575–579
32. Cummins J, Reen FJ, Baysse C, Mooij MJ, O'Gara F (2009) Subinhibitory conecentrations of the cationic antimicrobial peptide colistin induce the pseudomonas quinolone signal in *Pseudomonas aeruginosa*. Microbiology 155:2826–2837
33. Ceri H, Olson ME, Stremick C, Read RR, Morck D, Buret A (1999) The Calgary Biofilm Device: new technology for rapid determination of antibiotic susceptibilities of bacterial biofilms. J Clin Microbiol 37:1771–1776
34. Tanaka G, Shigeta M, Komatsuzawa H, Sugai M, Suginaka H, Usui T (1999) Effect of the growth rate of *Pseudomonas aeruginosa* biofilms on the susceptibility to antimicrobial agents: β-lactams and fluoroquinolones. Chemotherapy 45:28–36
35. Garonzik SM, Li J, Thamlikitkul V, Paterson DL, Shoham S, Jacob J et al (2011) Population pharmacokinetics of colistin methanesulfonate and formed colistin in critically ill patients from a multicenter study provide dosing suggestion for various categories of patients. Antimicrob Agents Chemother 55:3284–3294
36. Plachouras D, Karvanen M, Friberg LE, Papadomichelakis E, Antoniadou A, Tsangaris I et al (2009) Population pharmacokinetic analysis of colistin methanesulfonate and colistin after intravenous administration in critically ill patients with infection caused by Gramnegative bacteria. Antimicrob Agents Chemother 53:3430–3436
37. Nation RL, Li J (2009) Colistin in the 21st century. Curr Opin Infect Dis 22:535–543
38. Li J, Rayner CR, Nation RL, Owen RJ, Spelman D, Tan KE, Liolios L (2006) Heteroresistance to colistin in multidrug-resistant *Acinetobacter baumannii*. Antimicrob Agents Chemother 50:2946–2950

39. Poudyal A, Howden BP, Bell JM, Gao W, Owen RJ et al (2008) *In vitro* pharmacodynamics of colistin against multidrug-resistant *Klebsiella pneumoniae*. J Antimicrob Chemother 62:1311–1318
40. Bergen PJ, Bulitta JB, Forrest A, Tsuji BT, Li J, Nation RL (2010) Pharmacokinetic/phamracodynamic investigation of colistin against *Pseudomonas aeruginosa* using an *in vitro* model. Antimicrob Agents Chemother 54:3783–3789
41. Bergen PJ, Forrest A, Bulitta JB, Tsuji BT, Sidjabat HE, Paterson DL et al (2011) Clinically relevant plasma concentrations of colistin in combination with imipenem enhance pharmacodynamics activity against multidrug-resistant *Pseudomonas aeruginosa* at multiple inocula. Antimicrob Agents Chemother 55:5134–5142
42. Hancock RE, Wong PG (1984) Compounds which increase the permeability of the Pseudomonas aeruginosa outer membrane. Antimicrob Agents Chemother 26:48–52
43. Hancock RE (1997) Peptide antibiotics. Lancet 349:418–422
44. Rodríguez-Pardo D, Pigrau C, Lora-Tamayo J, Soriano A, del Toro MD, Cobo J et al (2014) Gram-negative prosthetic joint infection: outcome of a debridement, antibiotics and implant retention approach. A large multicenter study. Clin Microbiol Infect 20(11):O911–O919
45. Corvec S, Furustrand U, Betrisey B, Borens O, Trampuz A (2013) Activities of fosfomycin, tigecyclin, colistin, and gentamycinm against extended-spectrum-β-lactamase-producing *Escherichia coli* in a foreign body infection model. Antimicrob Agents Chemother 57:1421–1427
46. Darouiche RO (2004) Treatment of infections associated with surgical implants. N Engl J Med 350:1422–1429
47. Høiby N (2011) Recent advances in the treatment of Pseudomonas aeruginosa infections in cystic fibrosis. BMC Med 9:32
48. Cantón R, Cobos N, de Gracia J, Baquero F, Honorato J, Gartner S et al (2005) Antimicrobial therapy for pulmonary pathogenic colonisation and infection by *Pseudomonas aeruginosa* in cystic fibrosis patients. Clin Microbiol Infect 11:690–703
49. O'Sullivan BP, Freedman SD (2009) Cystic fibrosis. Lancet 373:1891–1904
50. Maíz L, Girón RM, Olveira C, Quintana E, Lamas A, Pastor D et al (2013) Inhaled antibiotics for the treatment of chronic bronchopulmonary *Pseudomonas aeruginosa* infection in cystic fibrosis: systematic review of randomised controlled trials. Expert Opin Pharmacother 14:1135–1149
51. Valenza G, Radike K, Schoen C, Horn S, Oesterlein A, Frosch M et al (2010) Resistance to tobramycin and colistin in isolates of *Pseudomonas aeruginosa* from chronically colonized patients with cystic fibrosis under antimicrobial treatment. Scand J Infect Dis 42:885–889
52. Pitt TL, Sparrow M, Warner M, Stefanidou M (2003) Survey of resistance of *Pseudomonas aeruginosa* from UK patients with cystic fibrosis to six commonly prescribed antimicrobial agents. Thorax 58:794–796
53. Döring G, Conway SP, HGM H, Hodson M, Høiby N, Smyth A et al (2000) Antibiotic therapy against *Pseudomonas aeruginosa* in cystic fibrosis: a European consensus. Eur Respir J 16:749–767
54. Heijerman H, Westerman E, Conway S, Touw D, Döring G (2009) Inhaled medication and inhalation devices for lung disease in patients with cystic fibrosis: a European consensus. J Cyst Fibros 8:295–315
55. Conway SP, Pond MN, Watson A, Etherington C, Robey HL, Goldman MH (1997) Intravenous colistin sulphomethate in acute respiratory exacerbations in adult patients with cystic fibrosis. Thorax 52:987–993
56. Ledson MJ, Gallagher MJ, Cowperthwaite C, Convery RP, Walshaw MJ (1998) Four years' experience of intravenous colomycin in an adult cystic fibrosis unit. Eur Respir J 12:592–594
57. Ratjen F, Rietschel E, Kasel D, Schwiertz R, Starke K, Beier H et al (2006) Pharmacokinetics of inhaled colistin in patients with cystic fibrosis. J Antimicrob Chemother 57:306–311
58. Yapa WS, Li J, Porter CJ, Nation RL, Patel K, McIntosh MP (2013) Population pharmacokinetics of colistin methanesulfonate in rats: achieving sustained lung concentrations of colistin for targeting respiratory infections. Antimicrob Agents Chemother 57:5087–5095
59. Yapa WS, Li J, Patel K, Wilson JW, Dooley MJ, George J, Clark D et al (2014) Pulmonary and systemic pharmacokinetics of inhaled and intravenous colistin methanesulfonate in cystic fibrosis patients: targeting advantage of inhalational administration. Antimicrob Agents Chemother 58:2570–2579
60. Alothman GA, Ho B, Alsaadi MM, Ho SL, O'Drowsky L, Louca E, Coates AL (2005) Bronchial constriction and inhaled colistin in cystic fibrosis. Chest 127:522–529
61. Westerman EM, De Boer AH, Le Brun PP, Touw DJ, Roldaan AC, Frijlink HW et al (2007) Dry powder inhalation of colistin in cystic fibrosis patients: a single dose pilot study. J Cyst Fibros 6:284–292
62. Frederiksen B, Koch C, Høiby N (1997) Antibiotic treatment of initial colonization with *Pseudomonas aeruginosa* postpones chronic infection and prevents deterioration of pulmonary function in cystic fibrosis. Pediatr Pulmonol 23:330–335
63. Valerius NH, Koch C, Høiby N (1991) Prevention of chronic *Pseudomonas aeruginosa* colonization in cystic fibrosis by early treatment. Lancet 338:725–726
64. Hodson ME, Gallagher CG, Govan JRW (2002) A randomised clinical trial of nebulised tobramycin or colistin in cystic fibrosis. Eur Respir J 20:658–664
65. Adeboyeku D, Scott S, Hodson ME (2006) Open follow-up study of tobramycin nebuliser solution and colistin in patients with cystic fibrosis. J Cyst Fibros 5:261–263

66. Beringer P (2001) The clinical use of colistin in patients with cystic fibrosis. Curr Opin Pulm Med 7:434–440
67. Li J, Coulthard K, Milne R, Nation RL, Conway S, Peckham D et al (2003) Steady-state pharmacokinetics of intravenous colistin methanesulphonate in patients with cystic fibrosis. J Antimicrob Chemother 52:987–992
68. Brodt AM, Stovold E, Zhang L (2014) Inhaled antibiotics for stable non-cystic fibrosis bronchiectasis: a systematic review. Eur Respir J 44:382–393
69. Haworth CS, Foweraker JE, Wilkinson P, Kenyon RF, Bilton D (2014) Inhaled colistin in patients with bronchiectasis and chronic *Pseudomonas aeruginosa* infection. Am J Respir Crit Care Med 189:975–982
70. Del Pozo JL, Patel R (2009) Infection associated with prosthetic joints. N Engl J Med 361:787–794
71. Zimmerli W, Trampuz A, Ochsner PE (2004) Prosthetic-joint infections. N Engl J Med 351:1645–1654
72. Cobo J, Del Pozo JL (2011) Prosthetic joint infection: diagnosis and management. Expert Rev Anti-Infect Ther 9:787–802
73. Falagas ME, Kasiakou SK (2005) Colistin: the revival of polymyxins for the management of multidrug-resistant gram-negative bacterial infections. Clin Infect Dis 40:1333–1341
74. Rosenthal AL, Rovell JM, Girard AE (1976) Polyacrylic bone cement containing erythromycin and colistin. I. In vitro bacteriological activity and diffusion properties of erythromycin, colistin and erythromycin/colistin combination. J Int Med Res 4:296–304
75. Murray WR (1984) Use of antibiotic-containing bone cement. Clin Orthop Relat Res 190:89–95
76. Waterman P, Barber M, Weintrob AC, VanBrakle R, Howard R, Kozar MP et al (2012) The elution of colistimethate sodium from polymethylmethacrylate and calcium phosphate cement beads. Am J Orthop 41:256–259
77. Papagelopoulos PJ, Mavrogenis AF, Giannitsioti E, Kikilas A, Kanellakopoulou K, Soucacos PN (2007) Management of a multidrug-resistant Pseudomonas aeruginosa infected total knee arthroplasty using colistin. A case report and review of the literature. J Arthroplast 22:457–463
78. Valour F, Dutronc H, Dinh A, Cazorla C, Pavèse P, Lesens O et al (2013) Difficult-to-treat Gram-negative bone and joint infections: efficacy and safety of prolonged intravenous colistin. Int J Antimicrob Agents 41:197–199
79. Kasiakou SK, Fragoulis K, Tzagarakis G, Mistidis P, Kapaskelis A, Falagas ME (2005) Cure of multidrug-resistant Acinetobacter baumannii fixation device-related orthopedic infections in two patients with intravenous colistin. Microb Drug Resist 11(3):287–289
80. de Sanctis J, Teixeira L, van Duin D, Odio C, Hall G, Tomford JW et al (2014) Complex prosthetic joint infections due to carbapenemase-producing Klebsiella pneumoniae: a unique challenge in the era of untreatable infections. Int J Infect Dis 25:73–78
81. Imberti I, Iotti GA, Regazzi M (2014) Intraventricular or intrathecal colistin for the treatment of central nervous system infections caused by multidrug-resistant Gram-negative bacteria. Expert Rev Anti-Infect Ther 12:471–478
82. Tunkel AR, Harman BJ, Kaplan SL, Kaufman BA, Roos KL, Scheld WM et al (2004) Practice guidelines for the management of bacterial meningitis. Clin Infect Dis 39:1267–1284
83. Lin L, Li J, Nation RL, Nicolazzo JA (2009) Brain penetration of colistin in mice assessed by a novel high-performance liquid chromatographic technique. Antimicrob Agents Chemother 53:4247–4251
84. Jin L, Li J, Nation RL, Nicolazzo JA (2011) Impact of P-glycoprotein inhibition and lipopolysaccharide administration on blood-brain barrier transport of colistin in mice. Antimicrob Agents Chemother 55:502–507
85. Markantonis SL, Markou N, Fousteri M, Sakellaridis N, Karatzas S, Alamanos I et al (2009) Penetration of colistin into cerebrospinal fluid. Antimicrob Agents Chemother 53:4907–4910
86. Ziaka M, Markantonis SL, Fousteri M, Zygoulis P, Panidis D, Karvouniaris M et al (2013) Combined intravenous and intraventricular administration of colistin methanesulfonate in critically ill patients with central nervous system infection. Antimicrob Agents Chemother 57:1938–1940
87. Antachopoulos C, Karvanen M, Iosifidis E, Jansson B, Plachouras D, Cars O et al (2010) Serum and cerebrospinal fluid levels of colistin in pediatric patients. Antimicrob Agents Chemother 54:3985–3987
88. Imberti R, Cusato M, Accetta G, Marinò V, Procaccio F, Del Gaudio A et al (2012) Pharmacokinetics of colistin in cerebrospinal fluid after intraventricular administration of colistin methanesulfonate. Antimicrob Agents Chemother 56:4416–4421
89. Falagas ME, Bliziotis IA, Tam VH (2007) Intraventricular or intrathecal use of polymyxins in patients with Gram-negative meningitis: a systematic review of the available evidence. Int J Antimicrob Agents 29:9–25
90. Kasiakou SK, Rafailidis PI, Liaropoulos K, Falagas ME (2005) Cure of post-traumatic recurrent multiresistant gram-negative rod meningitis with intraventricular colistin. J Infect 50:348–352
91. Karagoz G, Kadanali A, Dede B, Sahin OT, Comoglu S, Altug SB et al (2014) Extensively drug-resistant Pseudomonas aeruginosa ventriculitis and meningitis treated with intrathecalcolistin. Int J Antimicrob Agents 43:93–94
92. Fernández-Viladrich P, Corbella X, Corral L, Tubau F, Mateu A (1999) Successful treatment of ventriculitis due to carbapenem-resistant *Acinetobacter baumannii* with intraventricular colistin sulfomethate sodium. Clin Infect Dis 28:916–917
93. Jiménez-Mejías ME, Pichardo-Guerrero C, Márquez-Rivas FJ, Martín-Lozano D, Prados T, Pachón J (2002)

Cerebrospinal fluid penetration and pharmacokinetic/pharmacodynamic parameters of intravenously administered colistin in a case of multidrug-resistant *Acinetobacter baumannii* meningitis. Eur J Clin Microbiol Infect Dis 21:212–214
94. Karaiskos I, Galani L, Baziaka F, Giamarellou H (2013) Intraventricular and intratechal colistin as the last therapeutic resort for the treatment of multidrug-resistant and extensively drug-resistant *Acinetobacter baumannii* ventriculitis and meningitis: a literature review. Int J Antimicrob Agents 41:499–508
95. Bargiacchi O, Rossati A, Car P, Brustia D, Brondolo R, Rosa F et al (2014) Intrathecal/intraventricular colistin in external ventricular device-related infections by multi-drug resistant Gram negative bacteria: case reports and review. Infection 42(5):801–809
96. Khawcharoenporn T, Apisarnthanarak A, Mundy LM (2010) Intrathecal colistin for drug-resistant *Acinetobacter baumannii* central nervous system infection: a case series and systematic reviews. Clin Microbiol Infect 16:888–894

Clinical Use of Polymyxin B

14

Maria Helena Rigatto, Diego R. Falci,
and Alexandre P. Zavascki

Abstract

Polymyxin B is another clinically available polymyxin that has re-emerged in clinical practice to treat infections caused by multi-drug (MDR) or extensively-drug-resistant (XDR) Gram-negative bacteria (GNB). Its chemical structure is very similar to the structure of polymyxin E (colistin). However, since the latter is administered as a prodrug, there are major pharmacokinetic differences between both polymyxins that may potentially determine different clinical and microbiological outcomes. Studies addressing clinical or microbiological outcomes in patients treated with polymyxin B for MDR or XDR GNB are reviewed in this chapter.

Keywords

Polymyxin B · Clinical outcome · Dosing · Mortality

M. H. Rigatto
Infectious Diseases Service, Hospital São Lucas da Pontifícia Universidade Católica do Rio Grande do Sul, Porto Alegre, Brazil

Medical School, Pontifícia Universidade Católica do Rio Grande do Sul, Porto Alegre, Brazil

D. R. Falci
Post-Graduate Program in Health and Human Development, Universidade La Salle, Canoas, Brazil

Infectious Diseases Service, Hospital de Clínicas de Porto Alegre, Porto Alegre, Brazil

A. P. Zavascki (✉)
Infectious Diseases Service, Hospital de Clínicas de Porto Alegre, Porto Alegre, Brazil

Department of Internal Medicine, School of Medicine, Universidade Federal do Rio Grande do Sul, Porto Alegre, Brazil
e-mail: azavascki@hcpa.edu.br

14.1 Introduction

Similar to polymyxin E (colistin), polymyxin B is another clinically available polymyxin that re-emerged in clinical practice to treat infections caused by multi-drug (MDR) or extensively-drug-resistant (XDR) Gram-negative bacteria (GNB) [1–4]. Its chemical structure is very similar to the structure of polymyxin E (colistin) as discussed in Chap. 3. However, since the latter is administered as a pro-drug, there are major pharmacokinetic differences between polymyxins (see Chap. 15) that may potentially determine different clinical and microbiological outcomes, although it has not been clearly demonstrated so far by well-designed prospective studies.

In contrast to colistin, commercially available as colistin methanesulphonate (CMS), which has a worldwide distribution for clinical utilization, polymyxin B use is much more restricted, being only available in a few countries [5, 6]. Polymyxin B is most commonly used in Brazil, Malaysia, Singapore, New York State in USA and some

J. Li et al. (eds.), *Polymyxin Antibiotics: From Laboratory Bench to Bedside*, Advances in Experimental Medicine and Biology 1145, https://doi.org/10.1007/978-3-030-16373-0_14

regions in India [5–7]. It should be noted that there are many countries where there is no information regarding which polymyxin, if there is one, is commercially available for clinical use. Thus, it is possible that polymyxin B is available in other countries.

As a consequence of its more restricted utilization, there are fewer studies assessing polymyxin B against GNB, particularly XDR GNB. Many of clinical findings have been extrapolated from CMS/colistin studies considering that both have the same mechanism of action. Although it may be suitable for *in vitro* or even experimental studies which evaluate colistin and polymyxin B, the direct extrapolation of all clinical data may be done with some caution, considering that, as stated above, there are major PK differences between the two commercially available polymyxins, and clinical studies actually present results of the administration of CMS to the patients, not of colistin [5, 8, 9]. The complex and inter-related PK of CMS and colistin may have some impact in determining distinct clinical outcomes from patients treated with polymyxin B, including toxicity.

The most relevant clinical use of polymyxin B is for the treatment of infections caused by XDR GNB through its intravenous (IV) administration [2–4], and this chapter will mainly focus on this use. The chapter is divided by major sites of infections where these bacteria are often involved as etiological agents.

Polymyxin B can also be administered by intramuscular (IM), intraventricular or intrathecal routes for the treatment of systemic infections and ventriculitis/meningitis, respectively. The intramuscular (IM) route is rarely used nowadays. It was used when polymyxin B was one of the few antimicrobial agents against GNB infections. IM injection causes local pain and it is no longer appropriate for the current indications of polymyxins B, i.e. severe systemic infections caused by XDR GNB. Additionally, such infections may require high polymyxin B doses that may be inappropriate for IM injection. On the other hand, intrathecal administration is still used as an adjuvant therapy for central nervous system infections, particularly, meningitis, although there is much more experience and many more publications with CMS administered through this route [10]. Nonetheless, intraventricular and intrathecal administration of polymyxin B will be reviewed in this chapter.

Another important use of polymyxins B is through aerosolized form for the treatment of respiratory tract infections by XDR GNB, usually as an adjuvant to systemic therapy [11], and this route will be reviewed here. The use of aerosolized polymyxins in special populations such as cystic fibrosis patients will be reviewed in Chap. 13.

Finally, topical polymyxin B has been largely used as otologic and ophthalmic solutions, and these uses will be just briefly reviewed here, since it is not specific for difficult to treat infections such as those caused by XDR GNB. Another topical use of polymyxins has been for selective digestive tract decontamination and this will also be briefly presented in this Chapter. Polymyxin B has also been used as an anti-endotoxin drug and this will be reviewed in Chap. 19.

14.2 Evaluation of Polymyxin B Use for the Treatment of Gram-Negative Bacteria

Beyond the fact that there are fewer studies addressing IV polymyxin B in the treatment of systemic infections by XDR GNB, many other factors impose some difficulties for an appropriate systematic evaluation of clinical use of this polymyxin. There is a great heterogeneity of study designs that precludes a direct comparison among studies. Most of the studies involved patients with many kinds of infections, and a few are restricted to one infection site, for example, pneumonia or bloodstream infections [12–15]. Dosage regimes are not always described, nor is the time between the onset of infection and the beginning of therapy, making difficult a correct evaluation of the efficacy of this drug, since adequate dosage and early initiation of therapy are both critical for favorable outcomes in severe infections. It is also important to note that in some studies the dose is described in international units (IU), while in others it is described in mg. As discussed in Chap. 3, 1 mg is equivalent to approximately 10,000 IU of polymyxins B.

Another problem is the fact that in many reports there is no distinction between each polymyxin used because polymyxins were evaluated as a class [16–18]. The use of combination therapy with distinct antibiotics in many studies also hampers comparison between them. Finally, the heterogeneity of patients studied, although most were critically ill patients, impairs a clear comparison among studies.

In the next sections, we will summarize the major findings of the published studies addressing polymyxin B in the most frequent sites of infections, with particular emphasis on pneumonia, urinary tract and bloodstream infections. Authors' comments on specific syndromes are presented at the end of each section.

14.2.1 Pneumonia and Other Respiratory Tract Infections

One of the most common indications for polymyxin B is for the treatment of hospital-acquired pneumonia, especially ventilator-associated pneumonia [19, 20]. The main reason is that pneumonia is the leading infection among hospitalized patients, which in turn is most frequently caused by Gram-negative bacteria, notably *P. aeruginosa*, *A. baumannii* and species of Enterobacteriaceae [19]. Most of the reported experience with pneumonia relates to infections caused by the first two pathogens, which are frequently the most common etiologic agents of pneumonia. Moreover, carbapenem-resistant infections were initially almost exclusive of *P. aeruginosa* infections by a combination of multiple resistance mechanisms [21]. The emergence of major acquired carbapenemases in a worldwide scale first occurred in both *P. aeruginosa* and *A. baumannii*, and it also explains why most experience is with these organisms. Finally, multi-drug and extensively-drug resistance initially emerged most frequently in the intensive care unit (ICU) setting where antimicrobial use is common and most severely ill patients are found; thus, ventilator-associated pneumonia (VAP) caused by MDR or XDR *P. aeruginosa* or *A. baumannii* are among the most common reported infections for which polymyxin B has been prescribed.

The first experience with polymyxin B for the treatment of MDR pneumonia was published in 2003 by Ouderkirk et al. [22]. It was a retrospective study carried out in a tertiary-care hospital from Manhattan with patients who received polymyxin B from October 1999 to September 2000. It included 60 patients of whom 39 (65%) had lung infections [22]. Most infections (77%) were caused by *A. baumannii* although it was not stratified by site of infection. The results of this first report were not disappointing since the overall mortality (not specified by site of infection) was 20% and the incidence of renal failure was only 14%, although it was probably underestimated by the definition utilized in this study, which was a doubling of serum creatinine to a value of ≥2.0 mg/dL [22].

In 2004, Sobieszczyk et al. reported the first study evaluating the use of polymyxin B only in respiratory tract infections [15]. This was also a retrospective study conducted at a tertiary-care hospital from Manhattan, including 25 patients from January 2000 to June 2003 [15]. There were a total of 29 episodes of which in six only aerosolized polymyxin B was administered. Most infections were caused by *A. baumannii* (55%) and *P. aeruginosa* (41%) and all patients received another antimicrobial against GNB, mostly (65%) either imipenem or meropenem. End-of-treatment mortality based on each course of polymyxin B was 21%, including episodes with aerosolized use of the drug, while overall discharge mortality was 48% [15]. Nephrotoxicity, defined as the doubling of serum creatinine during therapy, was also low at 10%.

Following this report, a Brazilian study, performed at a university-affiliated hospital in São Paulo, evaluated 74 patients with *P. aeruginosa* pneumonia treated with polymyxin B from January 1997 to 31 December 2004 [12]. Overall in-hospital mortality rate was 74.3%. Of all patients, 35 (47.3%) had a favorable clinical response, which was defined as an improvement of signs and symptoms at the end of the therapy. These patients were compared to patients with unfavorable response. The factors related to unfavorable response in the multivariate analysis were a higher APACHE II score, septic shock, acute respiratory distress syndrome and use of sedatives [12].

Another study conducted in Brazil compared polymyxins with ampicillin-sulbactam for the treatment of *A. baumannii* infections [17]. This study included 28 patients with pneumonia treated with polymyxins. Unfortunately, both polymyxins were evaluated and the number of patients treated with each of them was not described, nor were the outcomes stratified by site of infection.

Other studies assessed the use of polymyxin B in patients with infections at many sites, of which pneumonia was the most common site, ranging from 16.3 up to 75.7% [23–28]. The overall mortality rates found in these studies ranged from 28% to 61% [23–28]; however, the specific outcomes of patients with pneumonia were not shown in any of them.

Only two other studies, both prospective, have specifically addressed the use of polymyxin B in patients with VAP or ventilator-associated tracheobronchitis (VAT) [29, 14]. The first evaluated the use of polymyxin B monotherapy in 29 patients presenting VAP caused by carbapenem-resistant *P. aeruginosa* in a teaching hospital from São Paulo, Brazil, during the period of January 2004 to December 2006 [29]. Overall in-hospital mortality was 72.4% (21/29) and infection-related mortality (within 14 days of the diagnosis) was 51.7% (15/29) [29]. Dosages of polymyxin B were not described.

The largest study specifically assessing the efficacy of polymyxin B in patients with VAP or VAT by *P. aeruginosa* or *A. baumannii* was conducted in a Brazilian tertiary-care teaching hospital including patients from February 2009 to December 2010 [14]. This prospective cohort study evaluated the 30-day mortality of 67 episodes of VAP or VAT in which 45 of them the treatment was IV polymyxin B. The crude 30-day mortality was 53%, which was higher than the mortality of patients treated with other antimicrobials (27%; 6 of 22 patients). After adjusting for some covariates a ≥100% increase in baseline serum creatinine value, length of hospital stay and APACHE II score in a Cox-regression model, the use of polymyxin B in the treatment of VAP or VAT was associated with increased mortality rate (adjusted Hazard Ratio of 3.9, 95% Confidence Interval, 1.41–10.76, P = 0.009) [14]. This was the first study suggesting that polymyxin B might be inferior to other antibiotics in the treatment of VAP and VAT [14]. The median (interquartile range) total daily dose of polymyxin B was 150 mg (150–200 mg; 1 mg = 10,000 IU), administered in divided doses every 12 h. This was possibly the first study in which dosages were not adjusted for renal dysfunction, as current recommendation following more recent pharmacokinetic studies [30, 31]. Although the 30-day mortality was high and higher than the comparator group, the length of mechanical ventilation after treatment initiation and the rates of superinfection were very similar between groups [14]. However, one of the most remarkable findings of this study was the very low bacterial eradication from tracheal aspirates, both in patients treated with polymyxin B and comparators (see Table 14.1) [14].

14.2.1.1 Bacterial Clearance

Bacterial clearance from respiratory tract has neither been assessed in all studies nor exclusively in patients with pneumonia. Nonetheless, it is possible to conclude that bacterial eradication from respiratory tract secretions is low, particularly, for *P. aeruginosa* isolates [14]. However, in the most recent prospective study, eradication rates were not different between patients treated with polymyxin B and comparators, suggesting that many factors associated with the pathogen and the host may be more important than the drug administered [14].

Despite this, there is still no conclusive study indicating that eradication of the pathogen is associated with improved outcomes in patients treated with polymyxin B. Nevertheless, it might be expected that it may influence recurrence rates and it may be important for infection control practices.

Table 14.1 Studies assessing bacterial eradication in patients treated with polymyxin B

Reference	Year	N	Polymyxin B dose Recommended Actually received[a]	Combination therapy	Bacteria	Microbiological clearance
[22]	1999–2000	60 patients:	96% of patients received 1.5–2.5 mg/kg/day	Ampicillin-sulbactam 52%, Imipenem 52%, Aminoglycosides 47%, Cephalosporins 30%, Fluoroquinolones 25, Other extended-spectrum penicillins15%	Not specific from respiratory material	Evaluated in 41 of 50 with multiresistant *Acinetobacter* or *Pseudomonas* bacteria:
		50 with bacterial recovery				
		39 (65%) with lung infection	Mean daily dose 1,100,000 IU (range 120,000–2,250,000)		*A. baumannii* 46 (77%)	36 (81%) bacterial eradication
					P. aeruginosa 2 (3%)	5 (12%) bacterial persistence (4 from pulmonary site)
					Both 2 (3%)	
[15]	2000–2003	29 episodes:	6 patients received only aerolized polymyxin B (most frequent 2.5 mg/kg/day, divided in 4 doses)	All patients received combination therapy:	*A. baumannii* 16(55%)	Evaluated in 22 of 29:
		All respiratory infections	21 patients received IV dose: 2.5–3 mg/kg/day (adjusted to renal function in 31% of them)	Imipenem or meropenem 65%, amikacin 28%, tobramycin 10%, cefepime 10%, quinolone 7%, ampicillin–sulbactam 10%, aztreonam 3%	*P. aeruginosa* 12(41%)	–9 (41%) bacterial eradication
			2 aerosolized + IV		*Alcaligenes xylosoxidans* 1 (3%)	
			Actual dose: Not described			
[14]	2009–2010	45 patients with VAP or VAT	2.5 mg/kg/day every 12 h	29 (64.4); antibiotics not described	*P. aeruginosa* 16 (35.6%)	18 (41.9%) bacterial erradication
			Median daily dose: 150 mg (25th–75th = 150–200 mg)		*A. baumannii* 27 (60%)	23 (65.7%) *A. baumannii* of 35 isolates were erradicated
					Both 2 (4.4%)	3 (11.1%) of 27 *P. aeruginosa* were erradicated[b]

VAP Ventilator-associated pneumonia, *VAT* Ventilator-associated tracheobronchitis
[a]1 mg of polymyxin = 10,000 IU
[b]In this study, prospective daily collections of tracheal aspirates were performed

14.2.1.2 Inhalatory Therapy

The use of aerosolized polymyxin B has been advocated to be useful in the treatment of respiratory tract infections either alone or, more commonly, as an adjuvant therapy to parenteral therapy [11]. There have been some concerns regarding the penetration of IV polymyxins into epithelial lining fluid (ELF) that relied mostly on empirical observation than on solid pharmacokinetic studies. Indeed, there is no study assessing the concentration of polymyxin B in ELF in humans. There are animal studies indicating that high and sustained concentrations of colistin may be achieved in ELF after IV administration of CMS owing to the slow conversion of CMS to colistin in the lungs [32], although the inhalatory administration has been shown to provide increased drug concentrations in the airways [33, 34]. However, these data cannot be extrapolated to polymyxin B which is administered as the active compound. A single animal study has evaluated the ELF concentrations of polymyxin B during the first 6 h after the intravenous administration of 3 mg/kg of the drug in mice [35]. Of the four polymyxin B components evaluated, only polymyxin B1 could be detected for up to 4 h, while polymyxin B2, B3 and isoleucine-B1 for only 2 h [35]. The area under the curve 0–6 h of polymyxin B1 was approximately 54% of that found in serum, while these values ranged from 68% to 103% for the other polymyxin B components [35]. Although the bacteriological and clinical responses of aqueous polymyxin B in this study were inferior to those found for liposomal polymyxin B, attributed to a higher concentration in ELF for the latter formulation, the concentration achieved in ELF in this experimental model could not be considered negligible. It is clear that studies assessing pulmonary concentrations of polymyxin B in humans are urgently required.

Regardless of the pending issue of polymyxin B ELF penetration, the inhalatory use of this antimicrobial is an attractive option as an adjuvant to parenteral therapy. Nevertheless, in contrast to CMS/colistin, the reported experience with aerosolized polymyxin B is scarce. Despite the lack of clinical comparative studies, polymyxins B has been associated with higher rates of bronchoconstriction episodes, the most frequent adverse effect of inhalatory polymyxins, which may contribute to the lower experience compared with CMS/colistin [11].

The use of polymyxin B as aerosol for treatment of MDR GNB respiratory tract infections was assessed in a retrospective study from 2000 to 2003 [15]. Twenty-nine courses of polymyxin B were included, 6 of these were only by the inhalation route with no parenteral polymyxin B and 2 by combined aerosol and IV polymyxin B. No difference in mortality or in favorable clinical outcome was noted when comparing patients who received only IV or aerosolized polymyxin B therapy [15]. Of course, the small number of patients precludes any definitive conclusion.

The use of inhaled polymyxin B was also tested as salvage therapy for pneumonia and initial treatment for tracheobronchites by MDR GNB infections, mostly by *P. aeruginosa* [36]. Fourteen patients had pneumonia with failure of treatment with IV polymyxin B, defined as persistence of radiologic images and bacterial recovery from tracheal secretion. Inhaled polymyxin B was added to therapy in those patients. Five patients had tracheobronchitis and received inhaled polymyxin B alone. Inhalation was done for an average of 14 days with 50 mg twice a day in a 5-mL solution of distilled water, administered after 30 min of beta-2 agonist inhalation [36]. Cure was achieved in ten patients (53%), improvement in eight (42%) and only one patient was considered as a failure. The main adverse event was cough and bronchospasm that occurred in 4 patients, but resolved after reduction of polymyxin B dose [36].

As stated above, most clinical experience with aerosolized therapy is with CMS and although there is still no prospective study demonstrating a clear benefit of this strategy in clinical outcomes, there are many studies pointing towards advantages in bacterial eradication from the respiratory tract [37–39]. Additionally, a recent large retrospective study showed significantly higher clinical cure rates in patients with VAP treated with adjuvant inhalation of CMS plus IV CMS in comparison with patients treated with IV CMS alone (69.2% vs. 54.8%, respectively, $P = 0.03$) [40]. Furthermore, a shorter length of mechanical ventilation after pneumonia was also found in patients treated with inhalatory CMS [40].

Despite the fact that no comparative study with polymyxin B has been performed, there is preliminary evidence with CMS/colistin that encourages the use of inhalatory therapy of polymyxins, particularly if bacterial eradication from respiratory secretions is a target aim. The rationale of increased alveolar levels of the drug without increasing systemic toxicity also supports the use of this strategy.

Comments The mortality rate of patients treated with polymyxin B for pneumonia broadly ranged in published reports, but it is usually above 40%.

This high mortality certainly reflects in some degree the severity of baseline disease since most of them were critically ill patients. Variations in total dose administered can also have influenced results. The dose recommended and applied in these few studies mostly ranged from 1.5 to 3.0 mg/kg/day, a variation of up to 100%. Additionally, except in one, in all studies the dose of polymyxin B was adjusted (lowered) in the presence of decreased creatinine clearance, which is not in accordance with most recent PK findings [30, 31]. Finally, the time to start polymyxin B, although not described in some studies, was likely delayed, as a consequence of many factors that retard the initiation of this drug. For example, most patients, particularly in centers where XDR organisms are not highly prevalent, usually receive a polymyxin agent only when the bacteria is identified and susceptibility tests are finished, which may require from 48 h, at best, to 96 h. This certainly retards polymyxins initiation potentially affecting outcomes.

Unfortunately, there is still no large study assessing prognostic factors associated with polymyxin B therapy, especially dosage regimes or use of combination therapy that could lead to improved clinical and microbiological responses. There are also no data regarding the relation of bacteria eradication with clinical outcomes, including recurrence rates. This information will be useful since eradication from the respiratory tract of GNB is very low, particularly in patients under mechanical ventilation, and strategies that can increase microbiological clearance rates, such as inhalatory therapy, may be more widely recommended, if such eradication affects clinical outcomes.

14.2.2 Bloodstream Infections

Bloodstream infections (BSIs) are among the major health-care associated infections and they are associated with high mortality rates [41]. Many studies have reported higher mortality rates when the causative agent is a MDR GNB, when compared to episodes caused by susceptible organisms [42–44]. Carbapenem-resistant *P. aeruginosa*, *A. baumannii* and Enterobacteriaceae (notably, *K. pneumoniae* and *Enterobacter* spp.) are currently common or even endemic in many parts of the world and BSIs caused by these organisms have been occurring with increased frequency. Polymyxins are often the unique therapeutic option for BSIs caused by these resistant bacteria. As occurred with other infections, there is more reported experience with CMS than with polymyxin B.

A retrospective cohort study including patients from 2003 to 2009 enrolled 276 patients who received IV polymyxin B during ≥72 h in a Brazilian hospital [25]. In a subgroup analysis, the authors evaluated the outcome of 53 (19.2%) patients who presented BSI, caused either by *P. aeruginosa* in 32 (60.4%) or *A. baumannii* in 21 (39.6%) patients. The in-hospital mortality of patients was 34% (18 out of 53). In this study, as discussed below, use of polymyxin B doses ≥200 mg/day was associated with lower mortality [25].

The first study specifically assessing the efficacy of polymyxin B in the treatment of BSIs caused by *P. aeruginosa* retrospectively evaluated 133 patients from 2004 to 2009 [13]. Forty-five (33.8%) patients were treated with polymyxin B and 88 (66.2%) with other antimicrobials (comparators), of which most were beta-lactams. The mean average daily dose of polymyxin B was 141 ± 54 mg. Overall in-hospital mortality was 41.4% (55/133), and it was significantly higher in patients who received polymyxin B (66.7%) compared to those who were treated with other antimicrobials (28.4%), $p \leq 0.001$ [13]. This difference remained significant after adjusting for confounding variables such as Pitt bacteremia score, mechanical ventilation at the onset of infection and primary bloodstream infection, showing an approximately twofold increased risk of mortality of patients treated with polymyxin B [13].

A recent study carried out in Singapore from January 2006 to December 2010 evaluated 186 patients with XDR and non-XDR *A. baumannii*

BSIs and found 30-day mortality rates of 64.5% and 58.1%, respectively, p = 0.366 [45]. Unfortunately, there is no description of how many patients were treated with polymyxin B; however, it was reported that survivors at 30 days had received higher polymyxin B daily doses than non-survivors (median of 840,000 IU, 25th and 75th percentile of 200,000 and 2,000,000 IU vs. 700,000 IU, 160,000 and 1,500,000 IU, respectively) [45].

Comments Only a single study suggests that polymyxin B may be related to worse outcomes in patients with BSIs by susceptible organisms receiving other antimicrobial classes [13]. However, it is clear the scarcity of studies addressing polymyxin B in such infections necessitates additional studies to confirm such a finding. Meanwhile, the main principles of BSI treatment must be followed when using polymyxin B, i.e. early initiation of therapy, appropriate dosage, controlling of source of infection when applicable, among others.

Unfortunately, there is no clear description of time to initiation of polymyxin B in most of the studies (not only with BSIs). Polymyxin B was only used when XDR bacteria had been identified in many of them, with empirical use adopted only in institutions where XDR organisms were highly endemic. This usually results in a delayed initiation that likely adversely affects patients' outcomes. Additionally, doses may be considered low in most studies taking into account the polymyxin B PK, and it was almost always lowered in the presence of decreased creatinine clearance, certainly leading to suboptimal plasma concentrations of the drug. PK/PD data support the use of high dose of polymyxin B, and although only demonstrated in two studies, survival has been associated with the use of high doses of polymyxin B [25, 45]. Thus, beyond the general measures recommended for the management of BSIs, it is critical that appropriate doses of polymyxin B are prescribed (see section below).

14.2.3 Urinary Tract Infections

Urinary tract infections by MDR GNB are becoming more frequent especially after the emergence of carbapenemase-producing Enterobacteriaceae isolates, notably *K. pneumoniae* carbapenemase (KPC) and New Delhi Metallo-beta-lactamase (NDM) [46, 47]. Currently, with the worldwide dissemination of carbapenem-resistant Enterobacteriaceae isolates, urinary tract infections are among the main infections that required treatment with polymyxins.

Although polymyxin B is active against most of these isolates, the concentration of this drug in the urine is usually very low, around 1–4% of the administered IV dose [30, 31]. Nonetheless, the few studies reporting the use of polymyxin B for the treatment of UTI have suggested that it is an efficacious agent, although further properly designed studies are required to deeply evaluate the efficacy of this antibiotic in the treatment of UTI. The therapeutic success despite the low concentration of the drug may be explained by possible adequate concentrations of the drug in the bladder tissue.

Most of the case series report UTIs treated with polymyxin B, but a few have specifically assessed this site of infection. One of them was a retrospective cohort study performed in New York, USA, between January 2005 and June 2010 [48]. It included 87 patients with positive urine culture for carbapenem-resistant *K. pneumoniae* treated for at least 72 h with polymyxin B, an aminoglycoside or tigecycline. Patients were excluded if there was no follow up culture, if the bacteria were not susceptible *in vitro* to the antibiotic studied or if more than one active agent was used. The microbiologic clearance rate for the 25 patients treated with polymyxin B was 64% [48]. The median daily dose was 2.25 mg/kg (range, 1.1–3.3 mg/kg/day; 1 mg = 10,000 IU). Although the bacterial eradication was lower than that reached with an aminoglycoside (88%, p = 0.02), it may be considered satisfactory taking into account the low concentration of polymyxin B in urine. Additionally, eradication was clearly higher than tigecycline (43%) and untreated patients (36%). There was neither difference in BSIs nor mortality rates among treatment groups [48]. It should also be noted that in this study, the index culture colony count of >10^5 CFU/mL was found in 88% of patients treated with polymyxin B (22 of 25) and in 73% (30 of 41) of patients treated with aminoglycosides, a fact that may contribute to higher bacterial clearance rates obtained in the latter group.

Another positive experience with polymyxin B in the treatment of UTI was from a retrospective study with solid organ transplant patients [26]. It was conducted at a tertiary-hospital in Brazil from January 2001 to December 2007 and included 92 patients treated with either polymyxin B (90) or CMS (2 patients) [26]. UTI was the most common type of infection, affecting 38 patients. There was no definition of high or low UTI among these patients. The median daily dose of polymyxin B was 1,000,000 IU. Of the 24 patients from whom a follow-up urine culture had been collected, all presented microbiological cure, defined as the clearance in subsequent cultures of the pathogen initially isolated [26]. This finding is particularly relevant considering such a vulnerable patient population.

In a tertiary-hospital in New York, another retrospective study assessed 40 patients from January 2007 to August 2011 who were treated with polymyxin B monotherapy for carbapenem-resistant *K. pneumoniae* infections [24]. They found a clinical cure rate in 10 (83.3%) of 12 patients treated for UTIs [24]. Seventeen (53%) of 32 patients who had a follow-up culture presented documented microbiological clearance. Unfortunately, there was no description of the infection sites from which the cultures were collected and it is not possible to know how many microbiological cures were from UTIs.

Another very small study showed bacterial clearance in four of four patients with MDR Gram-negative UTI with IV polymyxin B therapy [49].

Comments As can be seen in Chap. 15, urinary concentrations of polymyxin B are low, while those of colistin are relatively high, owing to the conversion of CMS to colistin in the bladder. For this reason, it has been speculated that CMS should be preferred instead of polymyxin B in the treatment of UTI [5]. However, it is important to differentiate between high and low UTI, since this theoretical advantage may be applied only to the latter, because high concentrations of polymyxin B have been found in kidney parenchyma [50, 51]. Additionally, as demonstrated above, clinical cure and bacterial eradication have been demonstrated in patients treated with polymyxin B, which may be explained by possible adequate concentration of the drug in bladder tissue. Thus, this possible superiority must be further evaluated by clinical studies.

A clearer advantage would be regarding bacterial eradication, since bacteria attached to the bladder mucosa may play an important role in recurrence of the disease [52]. Thus, a high concentration of the antimicrobial in urine could theoretically increase the efficacy for bacterial eradication by a topical action.

Finally, intravesical instillation of polymyxin B and CMS has been reported in a few studies [53, 54]. Doses and mode of administration must be further evaluated; nonetheless, it may potentially be an interesting alternative for patients with uncomplicated low UTI in order to avoid systemic adverse effects.

14.2.4 Central Nervous System Infections

The development of invasive procedures and devices in the central nervous system (CNS) along with the increasing number of immunocompromised individuals have been determining an increasing incidence in GNB infections in this site, such as meningitis and/or ventriculitis [10, 55]. The emergence of CNS infections by MDR and XDR GNB is of special concern because few antibiotics are available to treat these infections, and polymyxins have poor CNS penetration following intravenous administration, although recent studies have only evaluated CMS/colistin [56, 57]. Nonetheless, since neither CMS nor colistin have been found in adequate concentrations in CNS after intravenous dosing of CMS, it could be expected that polymyxin B penetration must also be low. Intraventricular or intrathecal use is a strategy adopted in this situation in order to reach higher CNS concentration of the drug [10].

A systematic review evaluated 64 episodes of proven GNB meningitis which were treated with intratventricular or intrathecal polymyxins [10]. In most cases intrathecal polymyxins were used after failure to respond to IV regimens. Intrathecal polymyxin B was mostly used before 1974, whereas CMS was the drug most commonly evaluated in more recent studies [10]. The dose of polymyxin B applied was 50,000 IU/day in most cases and treatment length varied from 1 to 9 weeks [10]. Overall cure rate was 80% (51/64). From the 40 episodes treated with polymyxin B, 4 were considered to have failure to polymyxin B therapy, 4 had an intermediate outcome (in most cases intermediate outcome was considered when there was improvement in clinical symptoms, but polymyxin B had to be stopped for adverse effects) and 32 were cured. The main signs of toxicity were characterized as meningeal irritation and were related to the use of high daily doses of polymyxins (100,000–200,000 IU in adults and 20,000 IU in infants weighting 3 kg). Fortunately, all cases of toxicity were reversible with the suspension of the drug [10].

Comments There is no doubt that CNS infections by XDR GNB are among the major therapeutic challenges. There is no study defining an unequivocal benefit of adjuvant intrathecal or intraventricular administration neither of polymyxin B nor of CMS. However, considering the low penetration of the drugs in cerebral spinal fluid following intravenous administration, there is strong rationale for adopting direct administration to the infection site, at least in those cases in which initial response is poor. For infections of the CNS parenchyma, the rationale for using this strategy is less clear. While PK and clinical studies are not available to define the best dosage regime for the treatment of meningitis or ventriculitis by XDR GNB, it is recommended the highest IV dosage regime must be prescribed with or without intraventricular or intrathecal instillation of polymyxin B in the dosages recommended in IDSA guidelines that are 50,000 IU daily in adults and 20,000 IU daily in children [58].

14.2.5 Other Infections

Other important and common infectious syndromes are intra-abdominal and skin and soft tissue infections, including surgical wound infections. As occurs with other types of infections, there is an increasing incidence of XDR GNB causing intra-abdominal infections, usually as a complication of a surgical procedure, and cellulitis, often as a complication of surgical wound infections or in patients with ischemic or diabetic ulcers, or other kinds of soft tissue injury [23–28].

Although such infections have been reported with variable rates in studies assessing IV polymyxin B for infections caused by MDR or XDR GNB, there is no reported experience with polymyxin B in such specific syndromes. However, as occurs in other situations, the management of such infections must follow the general principles of infections caused by susceptible organisms. Appropriate surgical intervention should be indicated whenever a persistent focus or necrotized tissue is present. For intra-abdominal infections, the highest dose possible should be prescribed, because these infections are usually severe complications of a subjacent disease, and frequently present as intra-abdominal (including retroperitoneal) abscesses or peritonitis (authors' personal experience).

Cellulitis or other soft tissue infections in the authors' experience commonly present in patients with underlying comorbidities and some kind of skin injury. Superinfections by XDR GNB in patients undergoing treatment for erysipelas and cellulitis caused by usual pathogens, such as *Streptococcus pyogenes* and *Staphylococcus aureus,* have also been observed. Clinical worsening in patients with an initial favorable response to treatment for these pathogens may warrant attention for the possibility of XDR GNB infection, notably in patients with prolonged hospitalization and previous broad spectrum antibiotic exposure.

For skin and soft tissue infections, the polymyxin B dose can be decided according to the severity of presentation. However, the authors usually prescribe a dose not lower than 2.5 mg/kg/day for moderate to severe infections, and we discourage the use of 1.5 mg/kg/day which can be found in the product label (see dosage recommendations below).

14.3 Topical Use of Polymyxins

14.3.1 Ophthalmic and Otological Use

Polymyxin B has been used for a long time as a topical agent in the treatment of conjunctivitis and otitis externa. This use is not specific against XDR GNB, but for its general GNB spectrum. Polymyxin B as eye drops in combination with other antibiotics against Gram-positive organisms has been used for conjunctivitis since the 1980s [59]. It has been shown to be as effective as other more recent drugs, with lower treatment costs [60]. For acute otitis externa, ear drops containing a combination of polymyxin B, neomycin and hydrocortisone have been found to be an effective treatment of this condition [61, 62]. As for conjunctivitis, the combination of polymyxin B with an anti-Gram-positive agent has been used for otitis treatment as a topical agent since the 1980s [63].

14.3.2 Selective Digestive Decontamination

Selective digestive decontamination (SDD) is defined as the prophylactic application of topical, non-absorbable antimicrobials in the oropharynx and stomach with the objective of eliminating potentially pathogenic bacteria without affecting anaerobic microbiota [64]. SDD and selective oropharyngeal decontamination (SOD) have been shown to reduce the incidence of hospital-acquired infection, particularly, ventilator-associated pneumonia, and it possibly has a positive impact in decreasing overall mortality in intensive care units [65]. Because of its broad spectrum against potentially pathogenic GNB, polymyxin B is one of the most commonly used antibiotics typically in combination with other agents.

Despite the unequivocal evidence-based benefits of SDD and a recent meta-analyses that could not demonstrate any relation between the use of SDD or SOD and the development of overall antimicrobial resistance in pathogens in patients in the intensive care unit [64], there have been alarming reports of emergence of resistance to polymyxins after implementation of SDD with these drugs [66–68]. Of great concern is the fact that such resistance has emerged in carbapenemase-producing GNB for which polymyxins are the last resort therapy [66–68]. Thus, considering that the impact of SDD on intensive care unit antimicrobial resistance rates is still understudied [64], and the emergence of resistance to polymyxins in XDR GNB may have catastrophic consequences, the use of both polymyxins should be reconsidered in SDD regimes.

14.3.3 Other Topical Forms of Administration

Successful treatment with polymyxin B was achieved in a case of peritoneal dialysis related peritonitis by a carbapenemase-producing *K. pneumoniae* isolate which had failed treatment with amikacin and meropenem [69]. Intraperitoneal topical use of polymyxin B was also described in a case report of peritoneal dialysis related peritonitis by an XDR *A. baumannii* in addition to intravenous and intraperitoneal ampicilin-sulbactam, without the need for catheter removal [70].

Polymyxin B was also used topically, actually sprayed into the posterior pharynx and tracheal tube, for preventing pulmonary infections in critically ill patients [71]. The administration of polymyxin B (2.5 mg/kg/day – 6 divided doses for 2 months) to 355 patients was compared to a placebo group of 337 patients. Although incidences of *P. aeruginosa* pneumonia were low, these rates were significantly lower in the polymyxin B arm (0.80%) than in the placebo arm (4.6%). No improvement in overall mortality was found [71].

14.4 Toxicity

Toxicity of polymyxins is a major concern in clinical practice. High rates of nephrotoxicity have been reported along with neurologic and other less common adverse events. Although it has been shown that adverse effects are not as frequent as previously reported and that most are reversible, toxic effects are still the major limiting factor in the use of both polymyxins. This topic is fully reviewed in Chap. 17. Thus, in this section, just a brief summary of the main studies involving polymyxin B in the modern literature will be presented.

14.4.1 Nephrotoxicity

The defining criteria for nephrotoxicity greatly varies among studies, mostly because definitions of renal toxicity have broadly ranged among them. The absence of standardized criteria for nephrotoxicity in older studies impairs their direct comparison. Only recently, a small number of studies evaluated acute kidney injury (AKI) based on RIFLE criteria or AKIN criteria [72–75], which have been used as standard criteria for AKI definitions [76]. Their major findings are summarized in Table 14.2.

Although still not properly evaluated, AKI is generally reversible. It seems to be dose dependent and avoidance of other nephrotoxic drugs has been recommended whenever possible. The authors' opinion is that proper dose administration should not be regarded as a secondary plan for preventing AKI. Doses should be administered according to the patients' needs and any AKI event should be posteriorly handled. Further details on nephrotoxicity and its risk factors are discussed in Chap. 17.

14.4.2 Neurotoxicity

The relation of polymyxins and neurotoxicity is not fully understood and is mainly based on case reports. There are not many reports of neurotoxicity in the recent literature, but specifically with polymyxin B, it has been described in six patients, and clinically manifested as paresthesias (3), altered mental status (1), seizures (1) and neuromuscular weakness (1) [77, 15, 78].

Reports of respiratory arrest related to polymyxin B infusion have been described in the 1960s and were attributed to neuromuscular blockade [79, 80]. More recently two other cases of respiratory arrest were reported [81]. In one of them a 48 year-old man receiving oxygen by nasal cannula with oxygen saturation of 100% developed apnea and became unresponsive after 1 h of 125 mg (1.6 mg/kg) of polymyxin B infusion. There was a rapid recovery after supporting management [81]. In another case, a 58 year-old man received IV polymyxin B at a loading dose of 200 mg (2.9 mg/kg) IV followed by 80 mg (1.1 mg/kg) IV [81]. Three hours after the fourth infusion the patient suddenly developed acute respiratory distress also requiring endotracheal intubation. Other explanations for the event were excluded through multiple exams. The day after, the tube was removed and the patient described he felt unable to breath or move his arms [81]. Later on, in the same hospital stay, the patient's clinical state worsened and polymyxin B was reinitiated under ICU monitoring. Two hours after commencing the infusion of polymyxin B, the patient developed a witnessed respiratory arrest requiring emergency airway management followed again by extubation without sequelae on the following day. Applying Naranjo adverse drug reaction scale [81, 82] the first episode would be reported as possible and the second as probable related event.

In the authors' own experience, paresthesia is a relatively common side effect. It is most frequently observed during polymyxin B infusion, although not exclusively. More details of neurotoxicity of polymyxins are found in Chap. 17.

14.4.3 Cutaneous Hyperpigmentation

Skin hyperpigmentation has been occasionally described in clinical practice although very few case reports have documented these findings.

Table 14.2 Studies assessing the incidence of acute kidney injury (AKI) in patient treated with polymyxin B using standardized criteria

		Polymyxin B			
Reference	Year/n of patients	Protocol dose	Actual dose	Therapy duration	AKI Development by RIFLE or AKIN criteria
[73]	2010/73	2.5–3 mg/kg/day	Median daily dose 1.8 mg/kg (interquartile range, 1.3–2.4)	Median = 11, interquartile range = 5–16 days	44 (60%) developed AKI by **RIFLE** criteria:
		Adjusted for renal function not fully described: if creatinine clearance <80 ml/min: 1.5 mg/kg/day daily or every other day	Median total dose 18 mg/kg (interquartile range, 8–32)		None met criteria for loss or end stage disease.
					The number of patients classified as risk, injury or failure was not specified
[72]	2008–2009/67	1.5–2.5 mg/kg/day	AKI patients: 2.1 mg/kg/day	Mean 11.7 ± 11.4 days	28 (41.8%) developed AKI by **RIFLE** criteria:
		Continuous infusion in 24 h	Non-AKI patients: 1.7 mg/kg/day		R: 13.4%
					I: 19.4%
					F: 9.0%
					L: 0%
					E: 0%
[7]	2009–2010/32	1.2–2.5 mg/kg/day administered every 8 h	AKI patients: 10.67 ± 3.71 x 10^5 IU/day	AKI patients: 7.69 ± 5.27 days	6 (18.7%) developed AKI by **RIFLE** criteria:
		Dose adjustment according to creatinine clearance (not described)	Non-AKI patients: 12.92 ± 5.1 x 10^5 IU/day	Non-AKI patients: 6.33 ± 3.26 days	R: 15.6%
					I: 3.1%
					F: 0%
					L: 0%
					E: 0%

(continued)

Table 14.2 (continued)

		Polymyxin B			
[75]	2009–2011/96	2.5 mg/kg/day every 12 h	Median = 200 mg/day; interquartile range, 100–200 mg	Median time to develop AKI = 6.5 days (interquartile range, 5.5–11.5)	20 (20.8%) developed AKI by AKIN criteria:
		Not adjusted for renal dysfunction			Stage 1: 11.5%
					Stage 2: 8.3%
					Stage 3: 1%
[74]	2006–2011/104	Not described	Mean 1.5 ± 0.5 mg/kg/day by actual body weight and 1.8 ± 0.6 mg/kg/day by ideal body weight	Mean 11.7 ± 7.2 days	24 (23.1%)
					R: 4.8%
					I: 6.7%
					F: 11.5%
					L: 0%
					E: 0%

RIFLE Risk, Injury, Failure, Loss, End-stage renal disease [76]
AKIN Acute Kidney Injury Network [103]

Two case reports described darkness of face, ears, neck and upper chest after infusion of 14 and 30 days of polymyxin B use [83]. Both patients received tigecycline along with polymyxin B which is an important confusing factor, since skin hyperpigmentation has been described in minocycline derivatives. As described previously, three solid organ transplant patients developed grey discoloration of skin during polymyxin use [84].

14.4.4 Polymyxin B in Pregnancy

Data regarding possible embryotoxic effect of polymyxin B are very limited. A case series of seven pregnant patients exposed to polymyxin B during the first trimester did not result in congenital abnormalities in any of the cases [85]. A large case-control study including 22,843 infants with congenital abnormalities matched to 38,151 infants in the control group showed that 6 infants in the case group had mothers that received polymyxin B treatment during gestation compared to 13 of the control group, resulting in no statistically significant association [86, 87]. Recently a case report of a pregnant patient treated for 7 days with 1,000,000 IU of polymyxin B per day in the third month of gestation resulted in the delivery of a healthy child [87]. Despite the scarce data on this issue, there is no current evidence for embryotoxic effect of polymyxin B.

14.5 Recommendations for Clinical Use of Polymyxin B

There are some practical issues regarding the clinical use of polymyxin B in patients with infections by MDR or XDR GNB. Dose regimes, adjustment for renal function and the use of combination therapy are some of these topics and we summarize some authors' recommendations on the use of polymyxin B based on published evidence and on their own experience.

14.5.1 Dose

The historically recommended doses of polymyxin B and found in the product label ranged from 1.5 to 3.0 mg/kg/day, either divided in two doses or by continuous infusion. However, there has been no specific recommendation for each dose chosen in each specific situation, and this choice has been made in a totally empirical, almost intuitive, manner. It was a common practice to use lower dosages of the drug because of toxicity concerns. As the PK/PD of polymyxins have become clearer, the toxicity has been shown to be acceptable and manageable, along with recent PK knowledge on polymyxin B [30] and some preliminary demonstration that higher doses have been associated with improved outcomes [25], there seems to be a transition in prescribing practices, and many have been using higher dosage regimes to treat XDR GNB infections.

It has been demonstrated that free (unbound) area under the concentration–time curve (*f*AUC)/minimal inhibitory concentration (MIC) is the PK/PD index that best predicts polymyxin activity against GNB [88–90], indicating that an adequate time-averaged exposure to polymyxins is necessary to optimize their bactericidal properties [91]. Thus, the higher is the MIC of infecting bacteria, the higher the *f*AUC required to attain the target relation.

The MICs of XDR GNB for polymyxin B have a broad variation within the susceptibility range (breakpoint of 2 mg/L) from concentrations as low as $\leq$0.125 mg/L up to 2 mg/L [92]. Considering a recent population PK study with Monte Carlo simulation a dose of 2.5–3.0 mg/kg/day divided in two administrations has been recommended for all severe infections with GNB with MIC of 0.5–2 mg/L [9]. It must be noted that it is very likely that a reasonable proportion of patients infected with a GNB with an MIC = 2 mg/L will not achieve an appropriate PK/PD target even receiving these 'maximal' doses of polymyxin B [30]. Additionally, in all severe infections, especially in patients with septic shock and/or in patients infected by GNB with MIC = 1 or 2 mg/L, a loading dose is strongly recommended to achieve steady-state levels more promptly.

Lower doses may be prescribed for less severe infections with organisms with MICs of 0.25, 0.125 mg/L or even below these values. A dose of 2.0 mg/kg/day may also be adequate for infection by GNB with MIC = 0.5 mg/L causing mild infections when a good clinical outcome may be expected even without achieving the optimal PK/PD target of the drug. Considering the PK profile of polymyxin B and its therapeutic efficacy, along with the potential for emergence of resistance during treatment, the authors strongly discourage the use of 1.5 mg/kg/day in any situation.

The clinical impact of dose on mortality was evaluated in a retrospective cohort study conducted from 2003–2009 including 276 patients that received at least 72 h of polymyxin B [25]. Overall in-hospital mortality was 60.5% (167/276). Risk factors for mortality in the multivariate model were severe sepsis or septic shock at the onset of infection, mechanical ventilation, higher Charlson comorbidity score, and older age. The only protective factor for mortality was a dose ≥200 mg/day (adjusted Odds Ratio = 0.35, 95% Confidence Interval = 0.17–0.71, p = 0.007). In subgroup analysis of patients with microbiologically confirmed infections (212 patients) and BSIs (53 patients), doses ≥200 mg/day remained as an independent protective factor in the multivariate model [25]. Noteworthy, the benefit of these higher doses in reducing mortality was observed even in the presence of renal dysfunction during therapy, which was more commonly observed in patients receiving higher doses [25].

The ideal administration interval of polymyxin B has also been a matter of debate. Only a small retrospective study conducted from 2001 to 2004 compared continuous versus intermittent infusion of polymyxin B regarding efficacy and toxicity [93]. Fourteen patients received a 24-hour continuous infusion and 13 patients intermittent infusion (every 12 h). Both groups of patients received low total daily doses and these were even lower in the group of continuous infusion compared to intermittent infusion; 68.6 ± 31.4 vs 96.5 ± 19.6 mg/day, respectively (p = 0.09). However, doses were adjusted for renal impairment and after excluding those patients, the mean dose was comparable between groups. No difference in efficacy was found, evaluated through resolution of signs and symptoms at the end of therapy and mortality during therapy [93]. Toxicity, defined as 50% increase in serum creatinine or 50% clearance reduction, was similar between groups [93]. Nevertheless, the study included few patients, a fact that precludes any definitive conclusion. We recommend administration every 12 h in a 1-hour infusion, because there is a preclinical rational for such posology. First, recent PK studies in rats have shown that the renal toxicity seems to be related to the continued exposure to the drug rather than the peak concentrations [50, 51]; additionally, static PD studies showed a concentration-dependent activity of polymyxin B [94], although still requiring confirmation of its potential clinical impact; and finally, because of the practical difficulties in administering a continuous infusion drug.

14.5.2 Renal Impairment and Hemodialysis

Renal adjustment of polymyxin B dose has been done for many years and it has been reported in most published studies, usually following the recommendation of the first review article on polymyxins published in the resurgence era of these drugs [95]. However, although necessary for CMS, recent evaluation of polymyxin B PK has shown that its total body clearance is independent of creatinine clearance [30, 31]. This indicates that dose reduction in renal impairment is inappropriate for polymyxin B and will probably lead to reduced serum concentration of the drug [30, 8].

There are no data of polymyxin B in patients receiving intermittent hemodialysis and only one recent study using appropriate methodology for measuring polymyxin B concentrations has been published so far in two patients receiving continuous venovenous hemodialysis [96]. On day 8 and day 10 of therapy, antibiotic concentration was quantified at 8 different time points, in plasma and dialysate. Only 12.2% and 5.62% of the dose was recovered in each patient in the dialysate after a 12-hour interval [96]. Although a higher proportion of polymyxin B was recovered in dialysate compared to the urinary recovery of <1–4% found in previous studies [30, 31], it still represents a small proportion of the total dose. This single study combined with the knowledge that polymyxin B is not cleared by the kidneys preliminarily suggests that polymyxin B dose should not be reduced in patients receiving continuous venovenous hemodialysis. In contrast, even an incremental dose of around 10% may be considered [9]. Additional studies are still required to evaluate polymyxin B in distinct modalities of dialysis.

14.5.3 Combination Therapy

There is no definitive clinical evidence that combining another antimicrobial with polymyxins improves clinical and microbiological outcomes of patients with MDR or XDR GNB infections [97]. Nonetheless, it has been the current practice considering preclinical data supporting the use of combination therapy and some preliminary evidence from cohort studies suggesting that it may be superior to monotherapy with polymyxins in patients with carbapenem-resistant Enterobacteriaceae, although this therapeutic strategy still lacks more robust clinical evidence to support it [98–101, 97]. Notably, most of this benefit has been attributed to the addition of a carbapenem drug in schemes with polymyxins in patients with low level resistance to these former agents [98–101]. No cohort study has shown similar results with other GNB such as *A. baumannii* and *P. aeruginosa* [97] and combination therapy with carbapenems against such GNB is more complicated considering that low level resistance in these pathogens is much less common than in Enterobacteriaceae isolates [97]. In addition, it is important to highlight that all studies evaluating combination therapy have analyzed CMS/colistin.

The rationale for combination therapy is fully discussed in Chap. 16 and it is beyond the scope of this chapter to review all the aspects of this strategy. The general lines for clinical choice of the combining agents should follow an evaluation of the previous clinical experience with the drug, the MICs of the organisms for the specific antimicrobials along with their PK and issues regarding toxicities, especially, renal toxicity. Thus, most suitable candidates are the carbapenems, tigecycline (not for *P. aeruginosa*), ampicillin-sulbactam (only for *A. baumannii*), aztreonam (for metallo-beta-lactamase-only-producing GNB) and fosfomycin (where a parenteral formulation is available). Rifampicin has been an attractive option, but it showed no benefit in a recent randomized clinical trial against *A. baumannii* [102]. Finally, although aminoglycosides show *in vitro* susceptibility in some MDR and XDR isolates, their use should be done with extreme caution, since the high doses required to achieve reasonable concentrations against high-MIC pathogens substantially increase renal toxicity, with no clear therapeutic benefit [97].

References

1. Kwa A, Kasiakou SK, Tam VH, Falagas ME (2007) Polymyxin B: similarities to and differences from colistin (polymyxin E). Expert Rev Anti-Infect Ther 5(5):811–821. https://doi.org/10.1586/14787210.5.5.811
2. Landman D, Georgescu C, Martin DA, Quale J (2008) Polymyxins revisited. Clin Microbiol Rev 21(3):449–465. https://doi.org/10.1128/cmr.00006-08
3. Yuan Z, Tam VH (2008) Polymyxin B: a new strategy for multidrug-resistant Gram-negative organisms. Expert Opin Investig Drugs 17(5):661–668. https://doi.org/10.1517/13543784.17.5.661
4. Zavascki AP, Goldani LZ, Li J, Nation RL (2007) Polymyxin B for the treatment of multidrug-resistant pathogens: a critical review. J Antimicrob Chemother 60(6):1206–1215. https://doi.org/10.1093/jac/dkm357
5. Nation RL, Velkov T, Li J (2014) Colistin and polymyxin B: peas in a pod, or chalk and cheese? Clin

Infect Dis 59(1):88–94. https://doi.org/10.1093/cid/ciu213

6. Wertheim H, Van Nguyen K, Hara GL, Gelband H, Laxminarayan R, Mouton J et al (2013) Global survey of polymyxin use: a call for international guidelines. J Glob Antimicrob Resis 1(3):131–134. https://doi.org/10.1016/j.jgar.2013.03.012
7. Nandha R, Sekhri K, Mandal AK (2013) To study the clinical efficacy and nephrotoxicity along with the risk factors for acute kidney injury associated with parenteral polymyxin B. Indian J Crit Care Med 17(5):283–287. https://doi.org/10.4103/0972-5229.120319
8. Zavascki AP (2011) Dose adjustment of polymyxins for renal insufficiency. Antimicrob Agents Chemother 55(10):4940. https://doi.org/10.1128/aac.05080-11
9. Zavascki AP (2014) Polymyxins for the treatment of extensively-drug-resistant Gram-negative bacteria: from pharmacokinetics to bedside. Expert Rev Anti-Infect Ther 12(5):531–533. https://doi.org/10.1586/14787210.2014.902307
10. Falagas ME, Bliziotis IA, Tam VH (2007) Intraventricular or intrathecal use of polymyxins in patients with Gram-negative meningitis: a systematic review of the available evidence. Int J Antimicrob Agents 29(1):9–25. https://doi.org/10.1016/j.ijantimicag.2006.08.024
11. Falagas ME, Kasiakou SK, Tsiodras S, Michalopoulos A (2006) The use of intravenous and aerosolized polymyxins for the treatment of infections in critically ill patients: a review of the recent literature. Clin Med Res 4(2):138–146
12. Furtado GH, d'Azevedo PA, Santos AF, Gales AC, Pignatari AC, Medeiros EA (2007) Intravenous polymyxin B for the treatment of nosocomial pneumonia caused by multidrug-resistant Pseudomonas aeruginosa. Int J Antimicrob Agents 30(4):315–319. https://doi.org/10.1016/j.ijantimicag.2007.05.017
13. Kvitko CH, Rigatto MH, Moro AL, Zavascki AP (2011) Polymyxin B versus other antimicrobials for the treatment of pseudomonas aeruginosa bacteraemia. J Antimicrob Chemother 66(1):175–179. https://doi.org/10.1093/jac/dkq390
14. Rigatto MH, Ribeiro VB, Konzen D, Zavascki AP (2013) Comparison of polymyxin B with other antimicrobials in the treatment of ventilator-associated pneumonia and tracheobronchitis caused by Pseudomonas aeruginosa or Acinetobacter baumannii. Infection 41(2):321–328. https://doi.org/10.1007/s15010-012-0349-z
15. Sobieszczyk ME, Furuya EY, Hay CM, Pancholi P, Della-Latta P, Hammer SM et al (2004) Combination therapy with polymyxin B for the treatment of multidrug-resistant Gram-negative respiratory tract infections. J Antimicrob Chemother 54(2):566–569. https://doi.org/10.1093/jac/dkh369
16. Esterly JS, Griffith M, Qi C, Malczynski M, Postelnick MJ, Scheetz MH (2011) Impact of carbapenem resistance and receipt of active antimicrobial therapy on clinical outcomes of Acinetobacter baumannii bloodstream infections. Antimicrob Agents Chemother 55(10):4844–4849. https://doi.org/10.1128/aac.01728-10
17. Oliveira MS, Prado GV, Costa SF, Grinbaum RS, Levin AS (2008) Ampicillin/sulbactam compared with polymyxins for the treatment of infections caused by carbapenem-resistant Acinetobacter spp. J Antimicrob Chemother 61(6):1369–1375. https://doi.org/10.1093/jac/dkn128
18. Sekhri K, Nandha R, Mandal A, Bhasin D, Singh H (2013) Parenteral polymyxins: assessing efficacy and safety in critically ill patients with renal dysfunction. Indian J Pharmacol 45(6):608–611. https://doi.org/10.4103/0253-7613.121373
19. Barbier F, Andremont A, Wolff M, Bouadma L (2013) Hospital-acquired pneumonia and ventilator-associated pneumonia: recent advances in epidemiology and management. Curr Opin Pulm Med 19(3):216–228. https://doi.org/10.1097/MCP.0b013e32835f27be
20. Florescu DF, Qiu F, McCartan MA, Mindru C, Fey PD, Kalil AC (2012) What is the efficacy and safety of colistin for the treatment of ventilator-associated pneumonia? A systematic review and meta-regression. Clin Infect Dis 54(5):670–680. https://doi.org/10.1093/cid/cir934
21. Zavascki AP, Carvalhaes CG, Picao RC, Gales AC (2010) Multidrug-resistant Pseudomonas aeruginosa and Acinetobacter baumannii: resistance mechanisms and implications for therapy. Expert Rev Anti-Infect Ther 8(1):71–93. https://doi.org/10.1586/eri.09.108
22. Ouderkirk JP, Nord JA, Turett GS, Kislak JW (2003) Polymyxin B nephrotoxicity and efficacy against nosocomial infections caused by multiresistant gram-negative bacteria. Antimicrob Agents Chemother 47(8):2659–2662
23. Crusio R, Rao S, Changawala N, Paul V, Tiu C, van Ginkel J et al (2014) Epidemiology and outcome of infections with carbapenem-resistant Gram-negative bacteria treated with polymyxin B-based combination therapy. Scand J Infect Dis 46(1):1–8. https://doi.org/10.3109/00365548.2013.844350
24. Dubrovskaya Y, Chen TY, Scipione MR, Esaian D, Phillips MS, Papadopoulos J et al (2013) Risk factors for treatment failure of polymyxin B monotherapy for carbapenem-resistant Klebsiella pneumoniae infections. Antimicrob Agents Chemother 57(11):5394–5397. https://doi.org/10.1128/aac.00510-13
25. Elias LS, Konzen D, Krebs JM, Zavascki AP (2010) The impact of polymyxin B dosage on in-hospital mortality of patients treated with this antibiotic. J Antimicrob Chemother 65(10):2231–2237. https://doi.org/10.1093/jac/dkq285
26. Mostardeiro MM, Pereira CA, Marra AR, Pestana JO, Camargo LF (2013) Nephrotoxicity and efficacy assessment of polymyxin use in 92 transplant patients. Antimicrob Agents Chemother 57(3):1442–1446. https://doi.org/10.1128/aac.01329-12

27. Oliveira MS, Prado GV, Costa SF, Grinbaum RS, Levin AS (2009) Polymyxin B and colistimethate are comparable as to efficacy and renal toxicity. Diagn Microbiol Infect Dis 65(4):431–434. https://doi.org/10.1016/j.diagmicrobio.2009.07.018
28. Ramasubban S, Majumdar A, Das PS (2008) Safety and efficacy of polymyxin B in multidrug resistant Gram-negative severe sepsis and septic shock. Indian J Crit Care Med 12(4):153–157. https://doi.org/10.4103/0972-5229.45074
29. Furtado GH, Gales AC, Perdiz LB, Santos AF, Medeiros EA (2011) Prevalence and clinical outcomes of episodes of ventilator-associated pneumonia caused by SPM-1-producing and non-producing imipenem-resistant Pseudomonas aeruginosa. Rev Soc Bras Med Trop 44(5):604–606
30. Sandri AM, Landersdorfer CB, Jacob J, Boniatti MM, Dalarosa MG, Falci DR et al (2013) Population pharmacokinetics of intravenous polymyxin B in critically ill patients: implications for selection of dosage regimens. Clin Infect Dis 57(4):524–531. https://doi.org/10.1093/cid/cit334
31. Zavascki AP, Goldani LZ, Cao G, Superti SV, Lutz L, Barth AL et al (2008) Pharmacokinetics of intravenous polymyxin B in critically ill patients. Clin Infect Dis 47(10):1298–1304. https://doi.org/10.1086/592577
32. WSY S, Li J, Porter CJ, Nation RL, Patel K, McIntosh MP (2013) Population pharmacokinetics of colistin methanesulfonate in rats: achieving sustained lung concentrations of colistin for targeting respiratory infections. Antimicrob Agents Chemother 57(10):5087–5095. https://doi.org/10.1128/aac.01127-13
33. Gontijo AV, Gregoire N, Lamarche I, Gobin P, Couet W, Marchand S (2014) Biopharmaceutical characterization of nebulized antimicrobial agents in rats. 2. Colistin. Antimicrob Agents Chemother 58(7):3950–3956. https://doi.org/10.1128/aac.02819-14
34. Yapa SWS, Li J, Patel K, Wilson JW, Dooley MJ, George J et al (2014) Pulmonary and systemic pharmacokinetics of inhaled and intravenous colistin methanesulfonate in cystic fibrosis patients: targeting advantage of inhalational administration. Antimicrob Agents Chemother 58(5):2570–2579. https://doi.org/10.1128/aac.01705-13
35. He J, Abdelraouf K, Ledesma KR, Chow DS, Tam VH (2013) Pharmacokinetics and efficacy of liposomal polymyxin B in a murine pneumonia model. Int J Antimicrob Agents 42(6):559–564. https://doi.org/10.1016/j.ijantimicag.2013.07.009
36. Pereira GH, Muller PR, Levin AS (2007) Salvage treatment of pneumonia and initial treatment of tracheobronchitis caused by multidrug-resistant Gram-negative bacilli with inhaled polymyxin B. Diagn Microbiol Infect Dis 58(2):235–240. https://doi.org/10.1016/j.diagmicrobio.2007.01.008
37. Kofteridis DP, Alexopoulou C, Valachis A, Maraki S, Dimopoulou D, Georgopoulos D et al (2010) Aerosolized plus intravenous colistin versus intravenous colistin alone for the treatment of ventilator-associated pneumonia: a matched case-control study. Clin Infect Dis 51(11):1238–1244. https://doi.org/10.1086/657242
38. Kuo SC, Lee YT, Yang SP, Chen CP, Chen TL, Hsieh SL et al (2012) Eradication of multidrug-resistant Acinetobacter baumannii from the respiratory tract with inhaled colistin methanesulfonate: a matched case-control study. Clin Microbiol Infect 18(9):870–876. https://doi.org/10.1111/j.1469-0691.2011.03682.x
39. Michalopoulos A, Fotakis D, Virtzili S, Vletsas C, Raftopoulou S, Mastora Z et al (2008) Aerosolized colistin as adjunctive treatment of ventilator-associated pneumonia due to multidrug-resistant Gram-negative bacteria: a prospective study. Respir Med 102(3):407–412. https://doi.org/10.1016/j.rmed.2007.10.011
40. Tumbarello M, De Pascale G, Trecarichi EM, De Martino S, Bello G, Maviglia R et al (2013) Effect of aerosolized colistin as adjunctive treatment on the outcomes of microbiologically documented ventilator-associated pneumonia caused by colistin-only susceptible gram-negative bacteria. Chest 144(6):1768–1775. https://doi.org/10.1378/chest.13-1018
41. Goto M, Al-Hasan MN (2013) Overall burden of bloodstream infection and nosocomial bloodstream infection in North America and Europe. Clin Microbiol Infect 19(6):501–509. https://doi.org/10.1111/1469-0691.12195
42. Tam VH, Rogers CA, Chang KT, Weston JS, Caeiro JP, Garey KW (2010) Impact of multidrug-resistant Pseudomonas aeruginosa bacteremia on patient outcomes. Antimicrob Agents Chemother 54(9):3717–3722. https://doi.org/10.1128/aac.00207-10
43. Tumbarello M, Repetto E, Trecarichi EM, Bernardini C, De Pascale G, Parisini A et al (2011) Multidrug-resistant Pseudomonas aeruginosa bloodstream infections: risk factors and mortality. Epidemiol Infect 139(11):1740–1749. https://doi.org/10.1017/s0950268810003055
44. Viale P, Giannella M, Lewis R, Trecarichi EM, Petrosillo N, Tumbarello M (2013) Predictors of mortality in multidrug-resistant Klebsiella pneumoniae bloodstream infections. Expert Rev Anti-Infect Ther 11(10):1053–1063. https://doi.org/10.1586/14787210.2013.836057
45. Ng TM, Teng CB, Lye DC, Apisarnthanarak A (2014) A multicenter case-case control study for risk factors and outcomes of extensively drug-resistant Acinetobacter baumannii bacteremia. Infect Control Hosp Epidemiol 35(1):49–55. https://doi.org/10.1086/674387
46. Dortet L, Poirel L, Nordmann P (2014) Worldwide dissemination of the NDM-type carbapenemases in gram-negative bacteria. Biomed Res Int 2014:249856. https://doi.org/10.1155/2014/249856
47. Munoz-Price LS, Poirel L, Bonomo RA, Schwaber MJ, Daikos GL, Cormican M et al (2013)

Clinical epidemiology of the global expansion of Klebsiella pneumoniae carbapenemases. Lancet Infect Dis 13(9):785–796. https://doi.org/10.1016/s1473-3099(13)70190-7
48. Satlin MJ, Kubin CJ, Blumenthal JS, Cohen AB, Furuya EY, Wilson SJ et al (2011) Comparative effectiveness of aminoglycosides, polymyxin B, and tigecycline for clearance of carbapenem-resistant Klebsiella pneumoniae from urine. Antimicrob Agents Chemother 55(12):5893–5899. https://doi.org/10.1128/aac.00387-11
49. Pastewski AA, Caruso P, Parris AR, Dizon R, Kopec R, Sharma S et al (2008) Parenteral polymyxin B use in patients with multidrug-resistant gram-negative bacteremia and urinary tract infections: a retrospective case series. Ann Pharmacother 42(9):1177–1187. https://doi.org/10.1345/aph.1K346
50. Abdelraouf K, Braggs KH, Yin T, Truong LD, Hu M, Tam VH (2012) Characterization of polymyxin B-induced nephrotoxicity: implications for dosing regimen design. Antimicrob Agents Chemother 56(9):4625–4629. https://doi.org/10.1128/aac.00280-12
51. Abdelraouf K, He J, Ledesma KR, Hu M, Tam VH (2012) Pharmacokinetics and renal disposition of polymyxin B in an animal model. Antimicrob Agents Chemother 56(11):5724–5727. https://doi.org/10.1128/aac.01333-12
52. Barber AE, Norton JP, Spivak AM, Mulvey MA (2013) Urinary tract infections: current and emerging management strategies. Clin Infect Dis 57(5):719–724. https://doi.org/10.1093/cid/cit284
53. Giua R, Pedone C, Cortese L, Antonelli Incalzi R (2014) Colistin bladder instillation, an alternative way of treating multi-resistant Acinetobacter urinary tract infection: a case series and review of literature. Infection 42(1):199–202. https://doi.org/10.1007/s15010-013-0507-y
54. Suzuki Y, Toyota S, Katoh S, Fukushi Y, Orikasa S, Majima K (1989) Clinical effects of intravesical instillation of polymyxin B in patients with acute cystitis. Hinyokika Kiyo Acta Urologica Japonica 35(4):593–596
55. Kim BN, Peleg AY, Lodise TP, Lipman J, Li J, Nation R et al (2009) Management of meningitis due to antibiotic-resistant Acinetobacter species. Lancet Infect Dis 9(4):245–255. https://doi.org/10.1016/s1473-3099(09)70055-6
56. Jin L, Li J, Nation RL, Nicolazzo JA (2009) Brain penetration of colistin in mice assessed by a novel high-performance liquid chromatographic technique. Antimicrob Agents Chemother 53(10):4247–4251. https://doi.org/10.1128/aac.00485-09
57. Jin L, Li J, Nation RL, Nicolazzo JA (2011) Impact of p-glycoprotein inhibition and lipopolysaccharide administration on blood-brain barrier transport of colistin in mice. Antimicrob Agents Chemother 55(2):502–507. https://doi.org/10.1128/aac.01273-10
58. Tunkel AR, Hartman BJ, Kaplan SL, Kaufman BA, Roos KL, Scheld WM et al (2004) Practice guidelines for the management of bacterial meningitis. Clin Infect Dis 39(9):1267–1284. https://doi.org/10.1086/425368
59. Gigliotti F, Hendley JO, Morgan J, Michaels R, Dickens M, Lohr J (1984) Efficacy of topical antibiotic therapy in acute conjunctivitis in children. J Pediatr 104(4):623–626
60. Williams L, Malhotra Y, Murante B, Laverty S, Cook S, Topa D et al (2013) A single-blinded randomized clinical trial comparing polymyxin B-trimethoprim and moxifloxacin for treatment of acute conjunctivitis in children. J Pediatr 162(4):857–861. https://doi.org/10.1016/j.jpeds.2012.09.013
61. Drehobl M, Guerrero JL, Lacarte PR, Goldstein G, Mata FS, Luber S (2008) Comparison of efficacy and safety of ciprofloxacin otic solution 0.2% versus polymyxin B-neomycin-hydrocortisone in the treatment of acute diffuse otitis externa. Curr Med Res Opin 24(12):3531–3542. https://doi.org/10.1185/03007990802583845
62. Rahman A, Rizwan S, Waycaster C, Wall GM (2007) Pooled analysis of two clinical trials comparing the clinical outcomes of topical ciprofloxacin/dexamethasone otic suspension and polymyxin B/neomycin/hydrocortisone otic suspension for the treatment of acute otitis externa in adults and children. Clin Ther 29(9):1950–1956. https://doi.org/10.1016/j.clinthera.2007.09.005
63. Gyde MC (1981) A double-blind comparative study of trimethoprim-polymyxin B versus trimethoprim-sulfacetamide-polymyxin B otic solutions in the treatment of otorrhea. J Laryngol Otol 95(3):251–259
64. Daneman N, Sarwar S, Fowler RA, Cuthbertson BH (2013) Effect of selective decontamination on antimicrobial resistance in intensive care units: a systematic review and meta-analysis. Lancet Infect Dis 13(4):328–341. https://doi.org/10.1016/s1473-3099(12)70322-5
65. de Smet AM, Kluytmans JA, Cooper BS, Mascini EM, Benus RF, van der Werf TS et al (2009) Decontamination of the digestive tract and oropharynx in ICU patients. N Engl J Med 360(1):20–31. https://doi.org/10.1056/NEJMoa0800394
66. Halaby T, Al Naiemi N, Kluytmans J, van der Palen J, Vandenbroucke-Grauls CM (2013) Emergence of colistin resistance in Enterobacteriaceae after the introduction of selective digestive tract decontamination in an intensive care unit. Antimicrob Agents Chemother 57(7):3224–3229. https://doi.org/10.1128/aac.02634-12
67. Lubbert C, Faucheux S, Becker-Rux D, Laudi S, Durrbeck A, Busch T et al (2013) Rapid emergence of secondary resistance to gentamicin and colistin following selective digestive decontamination in patients with KPC-2-producing Klebsiella pneumoniae: a single-centre experience. Int J Antimicrob Agents 42(6):565–570. https://doi.org/10.1016/j.ijantimicag.2013.08.008

68. Oostdijk EA, Smits L, de Smet AM, Leverstein-van Hall MA, Kesecioglu J, Bonten MJ (2013) Colistin resistance in gram-negative bacteria during prophylactic topical colistin use in intensive care units. Intensive Care Med 39(4):653–660. https://doi.org/10.1007/s00134-012-2761-3
69. Parchuri S, Mohan S, Cunha BA (2005) Extended spectrum beta-lactamase-producing Klebsiella pneumoniae chronic ambulatory peritoneal dialysis peritonitis treated successfully with polymyxin B. Heart Lung 34(5):360–363. https://doi.org/10.1016/j.hrtlng.2004.11.001
70. Fitzpatrick MA, Esterly JS, Postelnick MJ, Sutton SH (2012) Successful treatment of extensively drug-resistant Acinetobacter baumannii peritoneal dialysis peritonitis with intraperitoneal polymyxin B and ampicillin-sulbactam. Ann Pharmacother 46(7–8):e17. https://doi.org/10.1345/aph.1R086
71. Klick JM, du Moulin GC, Hedley-Whyte J, Teres D, Bushnell LS, Feingold DS (1975) Prevention of gram-negative bacillary pneumonia using polymyxin aerosol as prophylaxis. II. Effect on the incidence of pneumonia in seriously ill patients. J Clin Invest 55(3):514–519. https://doi.org/10.1172/jci107957
72. Akajagbor DS, Wilson SL, Shere-Wolfe KD, Dakum P, Charurat ME, Gilliam BL (2013) Higher incidence of acute kidney injury with intravenous colistimethate sodium compared with polymyxin B in critically ill patients at a tertiary care medical center. Clin Infect Dis 57(9):1300–1303. https://doi.org/10.1093/cid/cit453
73. Kubin CJ, Ellman TM, Phadke V, Haynes LJ, Calfee DP, Yin MT (2012) Incidence and predictors of acute kidney injury associated with intravenous polymyxin B therapy. J Infect 65(1):80–87. https://doi.org/10.1016/j.jinf.2012.01.015
74. Phe K, Lee Y, McDaneld PM, Prasad N, Yin T, Figueroa DA et al (2014) In vitro assessment and multicenter cohort study of comparative nephrotoxicity rates associated with colistimethate versus polymyxin B therapy. Antimicrob Agents Chemother 58(5):2740–2746. https://doi.org/10.1128/aac.02476-13
75. Tuon FF, Rigatto MH, Lopes CK, Kamei LK, Rocha JL, Zavascki AP (2014) Risk factors for acute kidney injury in patients treated with polymyxin B or colistin methanesulfonate sodium. Int J Antimicrob Agents 43(4):349–352. https://doi.org/10.1016/j.ijantimicag.2013.12.002
76. Palevsky PM, Liu KD, Brophy PD, Chawla LS, Parikh CR, Thakar CV et al (2013) KDOQI US commentary on the 2012 KDIGO clinical practice guideline for acute kidney injury. Am J Kidney Dis 61(5):649–672. https://doi.org/10.1053/j.ajkd.2013.02.349
77. Holloway KP, Rouphael NG, Wells JB, King MD, Blumberg HM (2006) Polymyxin B and doxycycline use in patients with multidrug-resistant Acinetobacter baumannii infections in the intensive care unit. Ann Pharmacother 40(11):1939–1945. https://doi.org/10.1345/aph.1H353
78. Weinstein L, Doan TL, Smith MA (2009) Neurotoxicity in patients treated with intravenous polymyxin B: two case reports. Am J Health Syst Pharm 66(4):345–347. https://doi.org/10.2146/ajhp080065
79. Lindesmith LA, Baines RD Jr, Bigelow DB, Petty TL (1968) Reversible respiratory paralysis associated with polymyxin therapy. Ann Intern Med 68(2):318–327
80. Pohlmann G (1966) Respiratory arrest associated with intravenous administration of polymyxin B sulfate. JAMA 196(2):181–183
81. Wunsch H, Moitra VK, Patel M, Dzierba AL (2012) Polymyxin use associated with respiratory arrest. Chest 141(2):515–517. https://doi.org/10.1378/chest.11-1483
82. Naranjo CA, Busto U, Sellers EM, Sandor P, Ruiz I, Roberts EA et al (1981) A method for estimating the probability of adverse drug reactions. Clin Pharmacol Ther 30(2):239–245
83. Knueppel RC, Rahimian J (2007) Diffuse cutaneous hyperpigmentation due to tigecycline or polymyxin B. Clin Infect Dis 45(1):136–138. https://doi.org/10.1086/518706
84. Bergamasco MD, Barroso Barbosa M, de Oliveira Garcia D, Cipullo R, Moreira JC, Baia C et al (2012) Infection with Klebsiella pneumoniae carbapenemase (KPC)-producing K. pneumoniae in solid organ transplantation. Transpl Infect Dis 14(2):198–205. https://doi.org/10.1111/j.1399-3062.2011.00688.x
85. Heinonen OP, Slone D, Shapiro S (1977) Birth defects and drugs in pregnancy. Publishing Sciences Group, Littleton
86. Czeizel AE (2001) The ethical issues of the post-marketing surveillance of drug teratogenicity in Hungary. Pharmacoepidemiol Drug Saf 10(7):635–639. https://doi.org/10.1002/pds.635
87. Kazy Z, Puho E, Czeizel AE (2005) Parenteral polymyxin B treatment during pregnancy. Reprod Toxicol (Elmsford, NY) 20(2):181–182. https://doi.org/10.1016/j.reprotox.2005.03.004
88. Bergen PJ, Bulitta JB, Forrest A, Tsuji BT, Li J, Nation RL (2010) Pharmacokinetic/pharmacodynamic investigation of colistin against Pseudomonas aeruginosa using an in vitro model. Antimicrob Agents Chemother 54(9):3783–3789. https://doi.org/10.1128/aac.00903-09
89. Cheah SE, Wang J, Nguyen VT, Turnidge JD, Li J, Nation RL (2015) New pharmacokinetic/pharmacodynamic studies of systemically administered colistin against Pseudomonas aeruginosa and Acinetobacter baumannii in mouse thigh and lung infection models: smaller response in lung infection. J Antimicrob Chemother 70(12):3291–3297. https://doi.org/10.1093/jac/dkv267
90. Landersdorfer CB, Wang J, Wirth V, Chen K, Kaye KS, Tsuji BT, Li J, Nation RL (2018) Pharmacokinetics/pharmacodynamics of systemi-

cally administered polymyxin B against Klebsiella pneumoniae in mouse thigh and lung infection models. J Antimicrob Chemother 73(2):462–468. https://doi.org/10.1093/jac/dkx409
91. Bergen PJ, Landersdorfer CB, Zhang J, Zhao M, Lee HJ, Nation RL et al (2012) Pharmacokinetics and pharmacodynamics of 'old' polymyxins: what is new? Diagn Microbiol Infect Dis 74(3):213–223. https://doi.org/10.1016/j.diagmicrobio.2012.07.010
92. Gales AC, Jones RN, Sader HS (2011) Contemporary activity of colistin and polymyxin B against a worldwide collection of Gram-negative pathogens: results from the SENTRY Antimicrobial Surveillance Program (2006-09). J Antimicrob Chemother 66(9):2070–2074. https://doi.org/10.1093/jac/dkr239
93. Teng CB, Koh PT, Lye DC, Ang BS (2008) Continuous versus intermittent infusion of polymyxin B in the treatment of infections caused by multidrug-resistant Gram-negative bacteria. Int J Antimicrob Agents 31(1):80–82. https://doi.org/10.1016/j.ijantimicag.2007.08.004
94. Tam VH, Schilling AN, Vo G, Kabbara S, Kwa AL, Wiederhold NP et al (2005) Pharmacodynamics of polymyxin B against Pseudomonas aeruginosa. Antimicrob Agents Chemother 49(9):3624–3630. https://doi.org/10.1128/aac.49.9.3624-3630.2005
95. Evans ME, Feola DJ, Rapp RP (1999) Polymyxin B sulfate and colistin: old antibiotics for emerging multiresistant gram-negative bacteria. Ann Pharmacother 33(9):960–967
96. Sandri AM, Landersdorfer CB, Jacob J, Boniatti MM, Dalarosa MG, Falci DR et al (2013) Pharmacokinetics of polymyxin B in patients on continuous venovenous haemodialysis. J Antimicrob Chemother 68(3):674–677. https://doi.org/10.1093/jac/dks437
97. Zavascki AP, Bulitta JB, Landersdorfer CB (2013) Combination therapy for carbapenem-resistant Gram-negative bacteria. Expert Rev Anti-Infect Ther 11(12):1333–1353. https://doi.org/10.1586/14787210.2013.845523
98. Daikos GL, Tsaousi S, Tzouvelekis LS, Anyfantis I, Psichogiou M, Argyropoulou A et al (2014) Carbapenemase-producing Klebsiella pneumoniae bloodstream infections: lowering mortality by antibiotic combination schemes and the role of carbapenems. Antimicrob Agents Chemother 58(4):2322–2328. https://doi.org/10.1128/aac.02166-13
99. Qureshi ZA, Paterson DL, Potoski BA, Kilayko MC, Sandovsky G, Sordillo E et al (2012) Treatment outcome of bacteremia due to KPC-producing Klebsiella pneumoniae: superiority of combination antimicrobial regimens. Antimicrob Agents Chemother 56(4):2108–2113. https://doi.org/10.1128/aac.06268-11
100. Tumbarello M, Viale P, Viscoli C, Trecarichi EM, Tumietto F, Marchese A et al (2012) Predictors of mortality in bloodstream infections caused by Klebsiella pneumoniae carbapenemase-producing K. pneumoniae: importance of combination therapy. Clin Infect Dis 55(7):943–950. https://doi.org/10.1093/cid/cis588
101. Zarkotou O, Pournaras S, Tselioti P, Dragoumanos V, Pitiriga V, Ranellou K et al (2011) Predictors of mortality in patients with bloodstream infections caused by KPC-producing Klebsiella pneumoniae and impact of appropriate antimicrobial treatment. Clin Microbiol Infect 17(12):1798–1803. https://doi.org/10.1111/j.1469-0691.2011.03514.x
102. Durante-Mangoni E, Signoriello G, Andini R, Mattei A, De Cristoforo M, Murino P et al (2013) Colistin and rifampicin compared with colistin alone for the treatment of serious infections due to extensively drug-resistant Acinetobacter baumannii: a multicenter, randomized clinical trial. Clin Infect Dis 57(3):349–358. https://doi.org/10.1093/cid/cit253
103. Mehta RL, Kellum JA, Shah SV, Molitoris BA, Ronco C, Warnock DG et al (2007) Acute Kidney Injury Network: report of an initiative to improve outcomes in acute kidney injury. Crit Care (London, UK) 11(2):R31. https://doi.org/10.1186/cc5713

15 Clinical Pharmacokinetics, Pharmacodynamics and Toxicodynamics of Polymyxins: Implications for Therapeutic Use

Roger L. Nation and Alan Forrest

Abstract

The availability of sensitive, accurate and specific analytical methods for the measurement of polymyxins in biological fluids has enabled an understanding of the pharmacokinetics of these important antibiotics in healthy humans and patients. Colistin is administered as its inactive prodrug colistin methanesulfonate (CMS) and has especially complex pharmacokinetics. CMS undergoes conversion *in vivo* to the active entity colistin, but the rate of conversion varies from brand to brand and possibly from batch to batch. The extent of conversion is generally quite low and depends on the relative magnitudes of the conversion clearance and other clearance pathways for CMS of which renal excretion is a major component. Formed colistin in the systemic circulation undergoes very extensive tubular reabsorption; the same mechanism operates for polymyxin B which is administered in its active form. The extensive renal tubular reabsorption undoubtedly contributes to the propensity for the polymyxins to cause nephrotoxicity. While there are some aspects of pharmacokinetic behaviour that are similar between the two clinically used polymyxins, there are also substantial differences. In this chapter, the pharmacokinetics of colistin, administered as CMS, and polymyxin B are reviewed, and the therapeutic implications are discussed.

Keywords

Colistimethate · Colistin · Polymyxin B · Pharmacokinetics in humans · Comparison of colistin and polymyxin B pharmacokinetics · Pharmacodynamics and toxicodynamics · Clinical implications

This article is dedicated to the memory of Alan Forrest, a friend of many and an inspiring researcher and teacher.

R. L. Nation (✉)
Drug Delivery, Disposition and Dynamics, Monash Institute of Pharmaceutical Sciences, Monash University, Melbourne, VIC, Australia
e-mail: roger.nation@monash.edu

A. Forrest (deceased)
Eshelman School of Pharmacy, University of North Carolina at Chapel Hill, Chapel Hill, NC, USA

As discussed in Chap. 3, the polymyxins were discovered in the 1940s. While the polymyxin family of antibiotics comprises several members, only colistin (also known as polymyxin E) and polymyxin B were developed for clinical use. These two polymyxins differ from each other by just one amino acid in the heptapeptide ring (Fig. 15.1) and they possess very similar *in vitro* antibacterial activity [1–3]. Since they are products of fermentation, they are multicomponent mixtures with the major respective components being colistin A and B, and polymyxin B1 and

J. Li et al. (eds.), *Polymyxin Antibiotics: From Laboratory Bench to Bedside*, Advances in Experimental Medicine and Biology 1145, https://doi.org/10.1007/978-3-030-16373-0_15

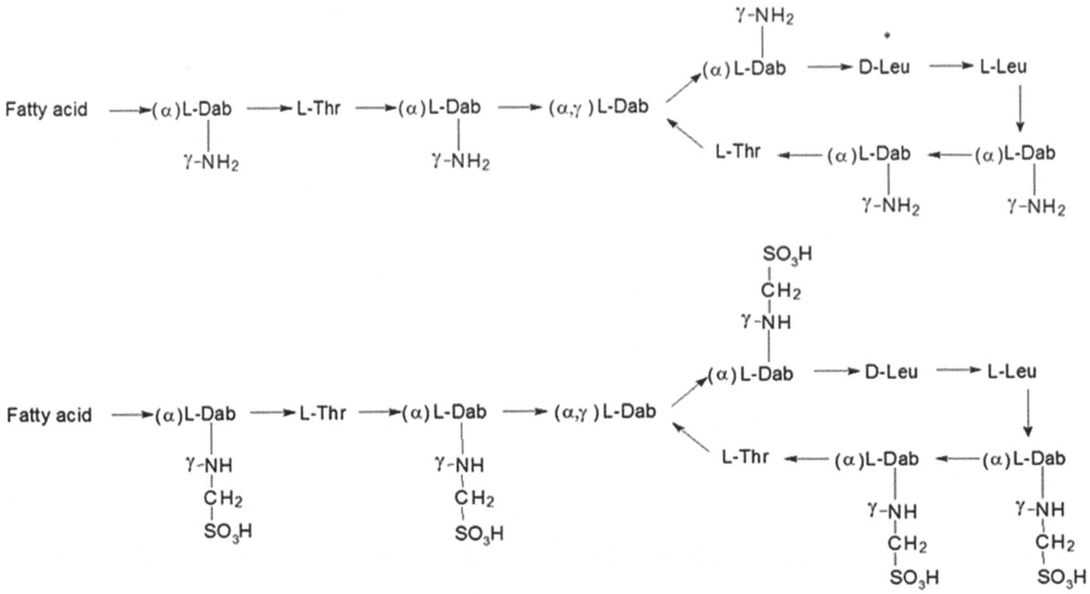

Fig. 15.1 Structures of colistin A and B, and polymyxin B1 and B2 are shown in the upper panel. In polymyxin B, D-Phe (phenylalanine) replaces the D-Leu (leucine) marked with the asterisk. The lower panel depicts structures of colistin methanesulfonate (CMS) A and B. All five primary amines of colistin are shown as being methanesulfonated, but it should be recognized that CMS is a complex mixture containing numerous partially methanesulfonated derivatives [1]. In addition, recently it has been shown that some primary amines may not be derivatized while others may have two methanesulfonate groups attached [128]. Fatty acid: 6-methyloctanoic acid for colistin A and polymyxin B1; and 6-methylheptanoic acid for colistin B and polymyxin B2. Thr: threonine; Dab: α,γ-diaminobutyric acid. α and γ indicate the respective –NH_2 involved in the peptide linkage

B2. Notwithstanding the similarities mentioned above, colistin and polymyxin B differ substantially in regard to the chemical form that is contained within the parenteral and inhalational products that are administered to patients [1, 4]. Polymyxin B is administered parenterally as its sulfate salt; that is, patients directly receive the active antibacterial entity. In contrast, colistin is administered as the sodium salt of colistin methanesulfonate (CMS, also known as colistimethate) (Fig. 15.1). It is now well established that CMS is an inactive prodrug [5, 6], and it requires conversion *in vivo* to colistin to unmask antibacterial activity [5, 7–9]. However, CMS is an extremely complex mixture of up to ~30 methanesulfonated derivatives for each colistin component, and the composition of CMS pharmaceutical products may vary from brand-to-brand and even from batch-to-batch [10]. The numerous chemical pathways for the demethanesulfonation process of the multiple CMS components is likely to affect the overall time-course for *in vivo* generation of colistin. Indeed, in a study conducted in rats the time-course of plasma colistin concentrations differed across four brands of parenteral CMS [10]. As discussed in this chapter, the difference in the chemical forms of colistin and polymyxin B as administered to patients has a major effect on the pharmacokinetic profiles of the active forms of the two polymyxins, with significant clinical pharmacological implications [1].

The pharmacokinetic studies conducted on the polymyxins during the twentieth century relied on microbiological assays for quantification of the drugs in biological fluids; unfortunately, such methods are still used in some more recent studies. As discussed in Chap. 6, these methods have general limitations in regard to assay performance criteria such as selectivity (interference by coadministered antibiotics), sensitivity (inability to quantify low concentrations), accuracy and reproducibility. Microbiological assays are especially problematic for quantification of 'colistin' concentrations in biological fluids following administration of CMS [8]. These assays are incapable of differentiating the colistin present in a biological sample at the time of its collection from a patient and the colistin formed *in vitro* by hydrolysis of CMS during the microbiological assay. The resultant measured concentration of 'colistin' can substantially over-estimate the true concentration present in the patient, especially in the early hours after administration of a dose of CMS when its concentration vastly exceeds that of colistin (see below). Prior to 2015, the pharmacokinetic information provided in all product labels (summary of product characteristics (SPCs)) for CMS parenteral products around the world was still based upon the erroneous plasma concentration *versus* time profiles that were generated with microbiological assays decades ago [8, 11, 12]. Fortunately, in December 2014 the European Commission accepted recommendations of the European Medicines Agency (EMA) to modernize the SPCs in use across Europe [13]. It is regrettable that such a process has not yet occurred in the USA and many other parts of the world. In the present chapter, in general only pharmacokinetic data from studies in which appropriate sample handling and chromatographic methods [8] (see also Chap. 6) were used will be reviewed. Studies that do not satisfy these criteria will only be mentioned to highlight the importance of appropriate sample handling and bioanalytical procedures.

In the present chapter, the pharmacokinetic, and where possible the pharmacodynamic and toxicodynamic, data for CMS/colistin and polymyxin B will be reviewed in turn. For each of the polymyxins, initially data obtained following intravenous dosing will be considered as this is the most common mode of administration and the route for which most information exists. Where data are available, other routes, such as nebulization delivery to the airways and intrathecal dosing, will also be considered. As discussed in Chap. 7 in relation to disposition in animals and below in regard to humans, there are major and clinically important differences in the pharmacokinetics of CMS and colistin. However, there is no substantial difference in the pharmacokinetics of the major components of the polymyxins (colistin A and B; polymyxin B1 and B2) [14–18]. Therefore the pharmacokinetics of the individual components will not be considered in this review.

15.1 Colistin Methanesulfonate and Formed Colistin

Doses of CMS below are expressed in terms of both milligrams of colistin base activity (CBA) and number of international units (IU). The conversion factor between these two conventions is: 1 million IU is equivalent to ~33 mg CBA [19].

15.1.1 Intravenous Administration

Healthy Volunteers With a new drug, typically the disposition in healthy volunteers is elucidated early in the clinical development of the drug, and this provides baseline information for comparison with data obtained later in various patients groups. There have been only three published reports in which chromatographic methods were used to define the disposition of CMS and colistin in healthy volunteers [18, 20, 21]. The first two of these studies were reported in 2011 while the third study was reported in 2018, approximately 50–60 years after colistin, administered as its inactive prodrug CMS, began to be used in patients.

In the first of these studies [18], the disposition of CMS and formed colistin was examined in 12 young (mean ± SD; 29.5 ± 5.5 years) healthy male volunteers (creatinine clearance 121 ± 18 mL/min) in France. The intravenous dose of CMS was 1 million IU (~33 mg CBA) and this was administered as a 1-h infusion. CMS was predominantly cleared by excretion into urine; the typical values of the total and renal clearances of CMS were 148 and 103 mL/min, respectively; the latter was similar to the glomerular filtration rate (GFR) estimated from the creatinine clearance values. As had been reported from an earlier study in patients with cystic fibrosis [22], the formation of colistin *in vivo* was observed in healthy human volunteers [18]. Maximum plasma concentrations of formed colistin were achieved at ~2 to 3 h after commencement of the CMS infusion. The terminal half-life of colistin (~3 h) was longer than that of the prodrug (~2 h) indicating that the disposition of formed colistin was rate-limited by its own elimination, not its conversion from CMS; [18] the same finding had been reported from an earlier study in rats [7]. Population pharmacokinetic analysis (structural model had two compartments for CMS and one compartment for colistin) was performed, but subject factors (e.g. renal function) influencing the disposition of CMS and colistin were not identified. This is not surprising given the small sample size and the fact that all subjects were healthy young males. Both CMS and formed colistin were recovered in urine; however, it was recognized based upon earlier studies in rats [7, 14], that much of the recovered colistin was formed from excreted CMS that underwent spontaneous hydrolysis in tubular and/or bladder urine. After correcting for this post-excretion conversion of CMS to colistin, the renal clearance of the latter was only 1.9 mL/min. While plasma protein binding of colistin was not examined in this study [18], subsequent studies have shown that colistin is approximately 50% bound in the plasma of humans [23]. Thus, the renal clearance value of colistin (1.9 mL/min) [18] is much less than the product of the unbound fraction in plasma and GFR indicating extensive renal tubular reabsorption of colistin that is formed from CMS prior to renal excretion. Very extensive renal tubular reabsorption of colistin following its direct administration to rats had been observed previously [14]. Based upon a urinary recovery of CMS and colistin (the latter mainly formed within the urinary tract) of approximately 70% of the administered dose, the authors speculated that ~30% of the prodrug was converted to colistin within the body prior to the renal excretion events [18]. This assumes that the only clearance pathways for CMS are conversion to colistin within the body and renal excretion. However, this approach almost certainly over-estimates the percentage conversion of the prodrug within the body, prior to renal excretion events. In an earlier study in rats in which the area under the plasma concentration-time curve of formed colistin after intravenous administration of CMS [7] was compared with the dose-normalized area after direct administration of colistin [14] it was found that only ~7% of a dose of CMS was converted to colistin within the body; [7] other similar studies in rats concluded that ~2.5–12% of a CMS dose was converted systemically to colistin [24–26]. These figures for percentage systemic conversion (~2.5–12%) [7, 24–26] are substantially lower than what would be predicted (~40%) based upon the urinary recovery (as CMS and colistin) of ~60% of the dose in rats intravenously administered CMS [7]. Thus, the percentage of a CMS dose converted systemically to colistin in the study in healthy young human volunteers was very likely lower than the 30% proposed by the authors [18]. Clearly, CMS is an extremely inefficient prodrug in both rats and humans with good kidney function.

In the second study in healthy human volunteers [20], the disposition of CMS and formed colistin was investigated in 15 male Japanese subjects (age 26.3 ± 6.7 years; creatinine clearance 125 ± 29 mL/min/1.73 m^2). CMS was administered as a 0.5-h intravenous infusion of a single dose (2.5 mg/kg CBA (~75,000 IU per kg)); after a 7-day washout, 14 of the subjects received the same dose twice daily for 2.5 days. Non-compartmental pharmacokinetic analysis was used for the plasma and urinary data from both the single- and multiple-dose phases of the study [20]. It is not clear why the authors chose to extrapolate the area under the plasma concentration-time curve (AUC) to infinite time after both single and repeat dosing, because for a drug exhibiting linear pharmacokinetics the AUC to infinite time after the first (single) dose should be the same as the AUC across a dosage interval at steady state [27]. It is also not understood why the accumulation (steady-state) ratios for CMS and formed colistin were determined as the ratio of the respective AUC over a 12-h dosing interval on day 3 to the AUC to infinite time after the single dose; [20] both AUCs should have been determined over a 12-h interval following the respective doses [27]. In the Japanese subjects, ~40% of the dose of CMS was excreted in urine as CMS plus colistin [20]. The authors made the same assumption as that discussed for the study conducted in France [18] and therefore concluded that ~60% of each dose of CMS was converted to colistin in the body. As discussed above, that almost certainly over-estimates the actual extent of systemic conversion.

More recently, the pharmacokinetics of CMS and formed colistin in healthy Chinese subjects after single and multiple intravenous doses of a new product of CMS developed in China have been reported [21]. A total of 12 subjects (6 female and 6 male; age 25.6 ± 3.2 years; creatinine clearance 129 ± 9.8 mL/min) received nominally 2.5 mg CBA per kg as a single dose infused over 1 h; the same dose was administered twice daily for 7 days to another group of 12 subjects, again with an equal number of women and men (age 25.4 ± 3.2 years; creatinine clearance 133 ± 13.8 mL/min). Because the maximum dose to be administered at any time was 150 mg CBA, the average (range) actual CMS doses in the single- and multiple-dose studies were 2.36 (2.19–2.50) mg CBA per kg and 2.35 (2.01–2.50) mg CBA per kg, respectively. The plasma and urinary data on CMS and colistin were subjected to non-compartmental pharmacokinetic analysis. As for the study in Japanese subjects [20], AUC from zero time to infinity was determined for both the single- and multiple-dose studies in the Chinese subjects [21]. In the latter study, the accumulation ratio was determined as the ratio of the AUC over 12 h on day 7 of the multiple-dose study to the AUC over 12 h after the single dose in the other group of subjects. As occurred in the other two studies in healthy subjects [18, 20], the fraction of CMS converted to colistin in the body in the Chinese subjects was estimated as one minus the fraction of CMS in the body that was excreted in urine. Thus, the resulting estimate of ~37% for the fractional conversion of CMS to colistin in the Chinese subjects [21] very likely is an over-estimate of the actual fraction of each CMS dose converted systemically to colistin. Indeed, in discussing their results, the authors suggested that the fractional conversion of CMS to colistin may increase by three to fivefold in renally impaired patients [21], which is clearly not possible if the fractional conversion in healthy subjects (*i.e.* with normal kidney function) is actually ~37%.

Across all three studies in healthy subjects [18, 20, 21] there were a number of similarities [21]. In all studies: (a) the maximum plasma concentration of CMS occurred at the end of the infusion of the prodrug while that of formed colistin occurred 1–3 h later; (b) the maximum plasma concentration of CMS was substantially higher than that of colistin; (c) the terminal half-life of formed colistin was longer than that of its prodrug, indicating that the disposition of colistin was rate-limited by its own elimination; (d) the renal clearance of CMS (84–103 mL/min) was substantially greater than that of colistin (1.9–10.5 mL/min); (e) given that colistin is ~50% unbound in human plasma [23] the very low renal clearance of colistin in all three studies was consistent with extensive net renal tubular reabsorption; and, (f) the conversion of CMS to colistin in the body was far from complete, and the fraction converted was very likely over-estimated in all three studies.

The major difference across the three studies in healthy volunteers is the dose-normalized plasma concentration of formed colistin following a single dose. In the studies in Japanese [20] and Chinese [21] subjects, the CMS dose administered was approximately 150 mg CBA, and the average maximum plasma colistin concentration was 0.69 mg/L and 2.55 mg/L, respectively, representing ~3.7-fold difference across these two studies [21]. In the study conducted in France [18], the single dose administered was only 33 mg CBA and the average maximum plasma colistin concentration was 0.83 mg/L [21]. When normalized to a dose of 150 mg CBA, the maximum plasma colistin concentration predicted from the study in France would be ~3.8 mg/L. Thus, across the three studies the range of plasma colistin concentrations (0.69 mg/L to ~3.8 mg/L) differs by ~5.5-fold. The possible reasons for the wide range of plasma colistin concentrations achieved from the same dose of CMS are: (a) brand-to-brand or batch-to-batch variation in the composition of the CMS products administered; [10] (b) differing extents of unrecognized conversion of CMS to colistin during sample collection, processing and analysis; [5, 6, 28] and, (c) inter-ethnic differences.

Patients with Cystic Fibrosis Reed et al. [29] were the first to report a study of the pharmacokinetics of 'colistin' in patients with cystic fibrosis in which an HPLC method was used for quantification of concentrations in biological fluids. The report did not indicate that blood samples had been collected and plasma harvested and stored in such a way as to minimize *ex vivo* conversion of CMS to colistin [30, 31]. In addition, unfortunately their analytical method involved heating at 54 °C for 2 h during preparation of the samples for HPLC analysis, conditions under which a substantial portion of the CMS contained within samples would undergo conversion to colistin [28]. This likely accounts for the very high plasma concentrations of 'colistin' and the very low values of clearance and volume of distribution that they reported, relative to later studies [22, 32]. Thus, the pharmacokinetic parameters reported from the study by Reed et al. [29] may not be reliable. This highlights the importance of ensuring that sample handling procedures and analytical methods do not lead to *in vitro* conversion of CMS to colistin.

Li et al. conducted a study in which the pharmacokinetics of CMS and formed colistin were determined across a dosage interval at steady state [22]. The study subjects were 12 patients (6 female; age 21.7 ± 6.9 years (mean ± SD); body weight 56 ± 9 kg) with cystic fibrosis, none of whom had renal impairment. According to the clinical protocol at the study site, patients weighing more than 50 kg received 2 million IU of CMS (~66 mg CBA) whereas those less than 50 kg were administered 1 million IU (~33 mg CBA) intravenously every 8 h, with each dose infused over 15–60 min. Plasma concentrations of CMS and formed colistin were quantified by two previously validated HPLC assays [33, 34]. Thus, this study was the first to demonstrate the *in vivo* conversion of CMS to colistin in humans and report the time-course of both species in plasma [22]. The total body clearance, volume of distribution and half-life of CMS were 2.01 ± 0.46 mL/min/kg, 340 ± 95 mL/kg and 124 ± 52 min, respectively. Colistin had a significantly longer mean half-life of 251 ± 79 min; the half-life of formed colistin was longer than that of CMS in each of the 12 patients. Apparent clearance of formed colistin was not reported in this study. It was noted that the protocol-driven dosage regimens employed resulted in plasma colistin concentrations across a dosage interval (maximum and minimum concentration ranges of 1.2–3.1 mg/L and 0.14–1.3 mg/L, respectively) that were low based upon emerging pharmacodynamic data at the time [35]. On that basis the authors suggested that dose-ranging studies to examine higher daily doses should be considered [22]. The dosage regimens used in this study (3–6 million IU per day, equivalent to ~100–200 mg CBA per day) [22] would now be regarded as low for patients without renal impairment.

More recently, the pharmacokinetics of intravenously administered CMS and the colistin formed from it were determined in cystic fibrosis patients as part of a study by Yapa et al. [32] to elucidate the pulmonary and systemic pharmacokinetics of inhaled and intravenous CMS. One of the study arms involved administration of a single intravenous dose of CMS (150 mg CBA, equivalent to ~4.5 million IU) infused over 45 min. The six study subjects were male patients with cystic fibrosis (age range 20–35 years; body weight 56–85 kg; creatinine clearance 103–148 mL/min/1.73m^2). The mean ± SD values for clearance, volume of distribution and terminal half-life of the administered CMS were 5.96 ± 1.07 L/h, 16.9 ± 4.68 L and 2.66 ± 0.60 h, respectively. When clearance and volume of distribution are normalized for body weight they are in good agreement with the values reported several years earlier by Li et al. [22], and the CMS half-life values reported from both studies are also similar. Across the 6 subjects, the plasma concentrations of formed colistin increased slowly to maxima of 0.40–0.77 mg/L within ~5 h after commencement of the CMS infusion. The time of the maximum colistin concentration was later than had been observed in healthy volunteers [18] and the previous study in patients with cystic fibrosis [22]; this may be the result of brand-to-brand variability in the complex composition of CMS products with impact on the rate of *in vivo* conversion to colistin, as has been observed in rats [10]. In the study of Yapa et al. [32], the post-maximal colistin concentrations declined with a longer half-life than that of CMS in each subject (mean ± SD 7.34 ± 1.41 h). Of the intravenous dose of CMS, 40.0% ± 18.7% was recovered as CMS and colistin in urine collected over 24 h, with approximately half of the recovered CMS dose (19.5% ± 8.79%) in the form of colistin. Based on evidence from other studies [7, 18, 28], most of the colistin recovered in urine would have been formed within the urinary tract. An important finding of the study of Yapa et al. was that negligible concentrations of colistin were measured in sputum samples collected across the 12 h sampling period [32]. While care is needed in the interpretation of this finding from a single-dose study, it may have implications for the ability to achieve colistin concentrations in lung fluids sufficiently high to elicit an adequate antibacterial effect in the many hours after initiation of therapy; any binding of colistin to mucin [36] or other components within the lung must also be considered. This study also investigated the potential targeting advantage that may be achieved by inhalational administration [32], and this will be discussed below.

In summary, the two evaluable studies in patients with cystic fibrosis [22, 32] revealed pharmacokinetics of CMS and formed colistin that were remarkably consistent with each other and also with the overall disposition profile that has been observed in healthy volunteers [18]. Notable features were: a total clearance of CMS similar to creatinine clearance and dominated by renal excretion of CMS with subsequent ongoing formation of colistin within the urinary tract; relatively slow and variable formation of colistin from CMS; a terminal half-life of CMS of approximately 2–3 h while that of formed colistin was 1.5–2.5 times longer. All of these studies had relatively small numbers of subjects with quite homogeneous demographic and clinical presentations. Thus, while the studies provide essential information on the overall disposition profiles of CMS and formed colistin in healthy volunteers and patients with cystic fibrosis, they were not designed to explore the full spectrum of patient characteristics (e.g. renal function) that may influence the exposure to colistin from a given dosage regimen of CMS.

Critically-Ill Patients Li et al. [37] were the first to report plasma concentrations of CMS and formed colistin in a critically-ill patient. The patient had developed multiple organ failure and was receiving continuous venovenous hemodiafiltration. The product information for CMS provided no guidance on dosage selection for such a patient and so the patient was administered 2.5 mg CBA (~76,000 IU) per kg every 48 h; this regimen had been proposed in a review on antibiotic dosing in patients receiving continuous renal replacement therapy [38], although there was no supporting data for the colistin regimen proposed. The case report of Li et al. [37] demonstrated that both CMS and colistin were cleared by the extracorporeal system and that plasma concentrations of colistin were substantially lower than 1 mg/L (the minimum inhibitory concentration (MIC) of the infecting organism) for ~85% of the 48-h dosage interval. Unfortunately, the patient succumbed. This report highlighted the urgent need for pharmacokinetic information to guide dosage regimens of CMS in various categories of critically-ill patients.

Subsequently, Makou et al. [39] reported plasma concentrations of formed colistin (CMS was not quantified) across a CMS dosage interval in 14 critically-ill patients (1 female, age range 18–84 years; body weight 60–85 kg; creatinine clearance 46–200 mL/min) who were receiving 3 million IU of CMS (~100 mg CBA) every 8 h. The half-life of formed colistin was 7.4 ± 1.7 h (mean ± SD). The maximum plasma colistin concentration within the dose interval ranged from 1.15 to 5.14 mg/L (2.93 ± 1.24 mg/L) and the corresponding range for the minimum concentration was 0.35–1.70 mg/L (1.03 ± 0.44 mg/L); 8 of the 14 patients had minimum plasma colistin concentrations less than 1 mg/L. A similar study was conducted by Imberti et al. [40] in 13 critically-ill patients (3 female, age range 20–70 years; body weight 55–110 kg; creatinine clearance 96–215 mL/min) who were receiving 2 million IU of CMS (~66 mg CBA) every 8 h. Blood samples were collected across a dosage interval at least 2 days after commencement of the regimen, and plasma was analysed for colistin concentration. The terminal half-life of colistin was 5.9 ± 2.6 h (mean ± SD) and maximum and minimum plasma colistin concentrations were 2.21 ± 1.08 mg/L and 1.03 ± 0.69 mg/L, respectively. The authors of both reports expressed concern about the relatively low plasma colistin concentrations achieved in their patients, all of whom had creatinine clearance greater than ~50 mL/min [39, 40]. Because of the small sample sizes in these studies, it was not possible to identify patient factors influencing exposure to colistin. A study of similar size (15 critically-ill patients) was reported by Karnik et al. [41]. The investigators collected blood samples after the first dose and across a dosage interval on the fourth day of therapy with CMS. The resultant plasma colistin concentration *versus* time profiles were highly unusual. The profiles for the two doses were very similar with little evidence of accumulation that has been observed in other studies [15, 31, 42]. The reported maximal plasma colistin concentrations were as high as 22–23 mg/L and occurred at the end of the 30 min infusion of CMS, while the median half-life of colistin after the first (2.7 h, range 1.1–4.6 h) and day 3–4 dose (3.3 h, 1.2–5.4 h) [41] was very short in comparison with other studies in critically-ill patients [15, 31, 39, 42]. It seems likely that these findings were the result at least in part of *ex vivo* conversion of CMS to colistin artificially elevating the measured plasma colistin concentrations at early time points after administration of a dose.

An important report by Plachouras et al. [15] identified a major problem that may arise if CMS regimens are commenced without administration of a loading dose. The study involved 18 critically-ill patients (6 female, age range 40–83 years; body weight 65–110 kg; creatinine clearance 41–126 mL/min). The CMS regimens (either 3 million IU (~100 mg CBA) or 2 million IU (~66 mg CBA) every 8 h) for these patients were commenced without administration of a loading dose. The result was a very gradual increase in plasma concentrations of formed colistin over several hours after the initial dose. Indeed, the population estimate of the terminal half-life of formed colistin in the critically-ill patients was ~14 h; thus, in the absence of a loading dose steady-state plasma concentrations of colistin would not be achieved before ~2 days of therapy. This raised concern about the possible negative impact on antibacterial effect in the early hours and days after initiation of a regimen. In a subsequent study from the same group [42], the pharmacokinetics of CMS and formed colistin were evaluated following administration of a loading dose of 6 million IU (~200 mg CBA) in 10 critically-ill patients (4 female, age range 32–88 years; body weight 60–140 kg; creatinine clearance 25–192 mL/min). As expected, the plasma concentrations of formed colistin across the first 8 h after this loading dose were higher than those observed in the earlier study by this group [15]. However, even with a loading dose of 6 million IU, only three of the ten patients had achieved a plasma colistin concentration of 2 mg/L by 8 h. A later study by this group examined the administration of a CMS loading dose of 9 million IU (~300 mg CBA) in 19 critically-ill patients (8 female, age range 18–86 years, body weight 50–120 kg, creatinine clearance 29–220 mL/min) [43]. The average maximum plasma concentration of formed colistin after the loading dose was 2.65 mg/L, with a time to maximum concentration of 8 h. It should be noted, however, that a wide variation was observed in the maximum plasma colistin concentration (0.95–5.1 mg/L) after the loading dose of CMS.

Based on these [15, 42, 43] and other [31, 44] studies CMS regimens should commence with a loading dose [13], but even when this is done there is still a delay of several hours in achievement of plasma concentrations of colistin greater than 2 mg/L. It is evident that the enormously complex chemistry of CMS [Chap. 3, 1] has the potential to lead to variation across brands and even across batches in the composition of partially methanesulfonated derivatives of colistin. This in turn may impact the rate of *in vivo* conversion of CMS to colistin, as has been observed in rats administered different brands of parenteral CMS [10]. The same situation may apply to CMS in patients, and indeed may account for the wide inter-patient variability in the plasma concentrations of formed colistin observed both after a loading dose and at steady state [31, 43–45]. The need for a loading dose in critically-ill patients would be greatest under circumstances where the rate of *in vivo* conversion to colistin is slow. Unfortunately, there is no *a priori* mechanism for determining the likely rate of *in vivo* conversion among different brands and batches of parenteral CMS.

The three aforementioned studies reported by the same group [15, 42, 43], involved collection of plasma samples after the first dose and again after the fourth to eighth dose administered every 8 h. The data from all three studies were pooled and subjected to population pharmacokinetic analysis [43]. The authors needed to invoke a different availability to account for the somewhat higher concentrations of the A and B forms of colistimethate and colistin measured in the third study. Patients with creatinine clearance >80 mL/min had a reduced ability to achieve maximum plasma colistin concentrations above 2 mg/L at steady state despite the administration of a maintenance dose of 9 million IU (300 mg CBA) per day. Only 4 of 12 patients with creatinine clearance >80 mL/min achieved maximum plasma concentrations of 2 mg/L at steady state, and the proportion of patients may have been even lower if an average steady-state plasma concentration ($C_{ss,avg}$) of colistin of 2 mg/L had been considered. This confirmed an earlier report which first indicated the difficulty of achieving such a colistin concentration in patients with good renal function [31].

Up until the end of 2016, the largest population pharmacokinetic study of CMS and formed colistin in critically-ill patients was that reported by Garonzik et al. [31]. The study involved 105 patients (37 female, age range 19–92 years; body weight 30–106 kg) with a wide range of renal function. The report [31] presented an interim analysis while the study proceeded to eventually recruit 215 patients; the population pharmacokinetic analysis on the full cohort of patients was reported in 2017 [46] and the results are discussed below.

Of the 105 patients in the interim population pharmacokinetic analysis [31], 89 were not receiving renal replacement therapy (creatinine clearance range 3–169 mL/min/1.73 m^2) while 16 patients were recipients of such support (12 patients receiving intermittent hemodialysis and 4 continuous renal replacement therapy [3 continuous veno-venous hemodialysis, 1 continuous veno-venous hemofiltration]). The CMS dosage regimen for each patient was at the discretion of the respective treating physician (range of maintenance doses 75–410 mg CBA per day, equivalent to ~2.3 to 12.4 million IU per day). Blood samples for quantification of plasma CMS and colistin concentrations were collected across an inter-dosing interval on the third or fourth day of the CMS regimen. The plasma CMS and colistin concentrations achieved varied greatly across the patients (Fig. 15.2); there was an approximate 20-fold range in the $C_{ss,avg}$ of the active antibacterial colistin (0.48–9.38 mg/L). Preliminary analysis of the data suggested that, in addition to the daily dose of CMS, renal function was an important contributor to the wide range of plasma colistin concentrations (Fig. 15.3). The data in this figure also indicate that in patients with creatinine clearance greater than ~80 mL/min/1.73 m^2, administration of a daily dose of CMS at or in the vicinity of the upper limit of the recommended dose range (300 mg CBA (~9 million IU) per day) [8, 13, 47] was not able to reliably generate a plasma colistin $C_{ss,avg}$ of 2 mg/L. This concentration may be regarded as a reasonable target, based upon: [47] translation of current evidence from pharmacokinetic/pharmacodynamic studies of colistin against *Pseudomonas aeruginosa* and *Acinetobacter baumannii* in the mouse thigh infection model; [23] and, as discussed in more detail below, on the basis of pharmacokinetic/toxicodynamic analyses in patients which indicate that the risk of nephrotoxicity substantially increases above a plasma colistin $C_{ss,avg}$ of ~2–3 mg/L [48, 49]. Unfortunately, in patients with creatinine clearance >80 mL/min it may not be possible to simply increase the daily dose of CMS to compensate for the low conversion to colistin because of the increased risk of nephrotoxicity, which is the

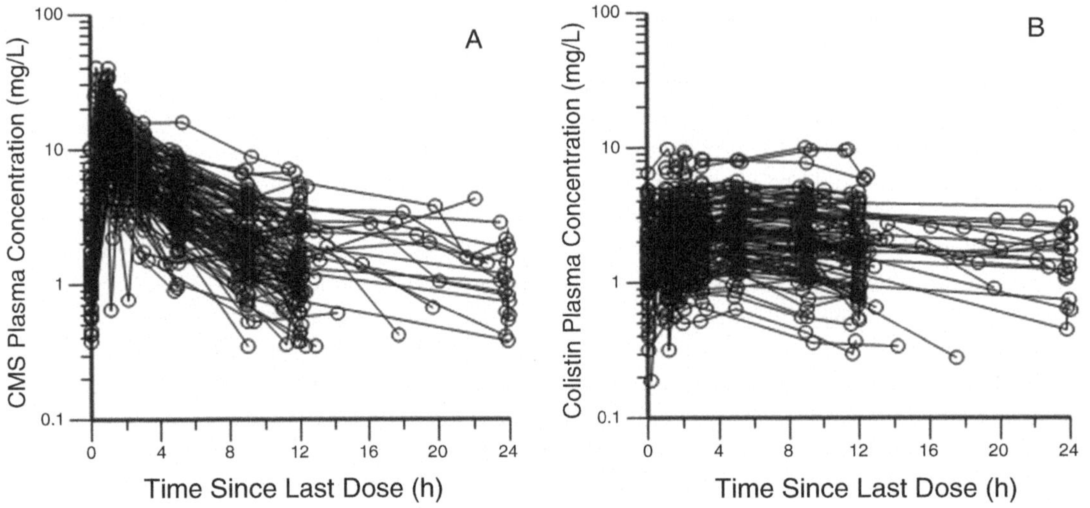

Fig. 15.2 Steady-state plasma concentration *versus* time profiles of the prodrug CMS (Panel A) and formed colistin (Panel B) in 105 critically-ill patients (89 not on renal replacement, 12 on intermittent hemodialsys and 4 on continuous renal replacement therapy) [31]. Physician-selected CMS dosage intervals ranged from 8 to 24 h and hence the inter-dosing blood sampling interval spanned the same range. (Reproduced with permission from Garonzik et al. [31] Copyright © American Society for Microbiology, Antimicrobial Agents and Chemotherapy 55 (7):3284–3294. doi:https://doi.org/10.1128/AAC.01733-10)

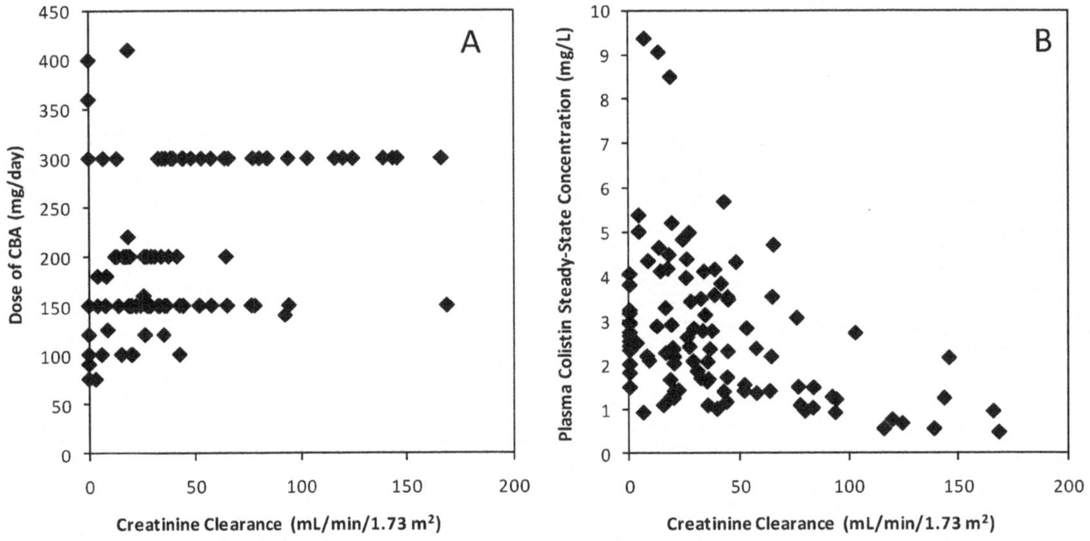

Fig. 15.3 Relationship of physician-selected daily dose of CMS (expressed as colistin base activity (CBA)) (Panel A) and the resultant steady-state plasma colistin concentration (Panel B) with creatinine clearance in 105 critically-ill patients [31]. (Reproduced with permission from Garonzik et al. [31] Copyright © American Society for Microbiology, Antimicrobial Agents and Chemotherapy 55 (7):3284–3294. doi:https://doi.org/10.1128/AAC.01733-10)

major dose-limiting adverse effect [8, 50]. It should be noted that studies in the mouse lung infection model revealed that these infections were more resilient than those in the mouse thigh; [23] accordingly a plasma colistin $C_{ss,avg}$ of 2 mg/L may be insufficient for such infections. Thus, combination therapy should be carefully considered in patients with relatively good renal function, especially if the MIC for the infecting organism is towards the upper end of the current breakpoint range and/or the patient is being treated for a respiratory tract infection [31, 47]. As discussed in Chap. 11, uncertainty still surrounds the effectiveness of colistin combination therapy compared with colistin monotherapy.

In the study by Garonzik et al. [31], the influence of renal function on the overall disposition of CMS and formed colistin was demonstrated *via* population pharmacokinetic analysis (structural model with two compartments for CMS and one compartment for colistin). Creatinine clearance was an important covariate on the total clearance of both CMS and colistin. It is easy to understand the dependence of total clearance of CMS on creatinine clearance because in animals [7] and humans [18] with good renal function, the prodrug CMS is mainly cleared by renal excretion; and only a relatively small fraction of a dose is converted to colistin [7, 18]. As a result, it is not unexpected that the total clearance of CMS declines with creatinine clearance. Animal studies involving direct administration of colistin have revealed that it is cleared predominantly by non-renal mechanisms, with only a very small fraction of the administered dose recovered in urine in unchanged form; [14] this is very similar to the disposition of polymyxin B in animals [17] and patients [16, 51]. Thus, it may seem surprising that renal function would influence the plasma $C_{ss,avg}$ of formed colistin achieved from a given daily dose of CMS. This relationship occurs because of the relatively complex overall disposition of CMS and formed colistin (Fig. 15.4, left panel). As discussed above, in subjects with good kidney function only a small fraction of each dose of CMS is converted to colistin. Because CMS is cleared predominantly by renal excretion, as renal function declines a progressively larger fraction of each dose of CMS is converted to colistin. Thus, the apparent clearance of colistin (i.e. the actual clearance divided by the fraction of the CMS dose converted to colistin in a given patient) declines with reduction in kidney function [31]. Accordingly, creatinine clearance was the patient factor incorporated into the algorithm developed by the authors to calculate the CMS daily maintenance dose needed to generate a desired target plasma concentration of formed colistin in a patient not receiving renal replacement therapy. It should be noted, however, that at a given creatinine clearance there was a very large degree of inter-patient variability (up to ~tenfold) in the apparent clearance of colistin and consequently in the CMS dosage requirements to achieve a desired steady-state plasma colistin concentration, reflected by the data in Fig. 15.3. The variability probably arises because of the complexity of the chemical composition, and the inefficiency, of CMS as a prodrug (i.e. substantially less than 100% conversion *in vivo*). These factors impact the fractional conversion to colistin at a given creatinine clearance [1]. This is consistent with a study in rats which demonstrated that the time-course of plasma concentration of formed colistin varied across four different brands of parenteral CMS [10]. The inter-patient variability in the plasma colistin concentration achieved at a given creatinine clearance and daily dose of CMS complicates the use of CMS, particularly since colistin has a low therapeutic index [1].

Of the 89 patients not on renal replacement in the above-mentioned report [31], all but two had been prescribed a CMS maintenance dose of 300 mg CBA (~9 million IU) per day or less, and 48% had a rise in serum creatinine of >50% from baseline; this is in keeping with nephrotoxicity rates of up to ~60% in other studies [50, 52–55]. In population pharmacokinetic/toxicodynamic analyses based upon 153 patients not receiving renal replacement at the start of CMS therapy (from the full cohort of 215 patients mentioned above), the time-course of changes in renal function after commencing CMS have been examined throughout the entire course of treatment [49]. Possible relationships with a range of factors,

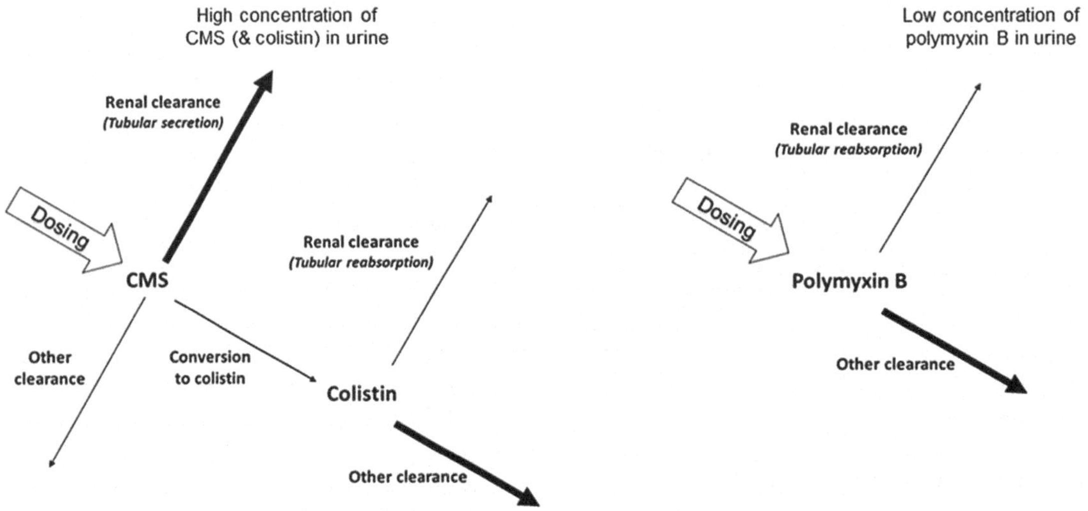

Fig. 15.4 Overview of the pharmacokinetic pathways for colistin methanesulfonate (CMS) and colistin (left panel) and polymyxin B (right panel). The thickness of the arrows indicates the relative magnitude of the respective clearance pathways when kidney function is normal. CMS includes fully and all partially methanesulfonated derivatives of colistin. After administration of CMS, extensive renal excretion of the prodrug occurs with some of the excreted CMS converting to colistin within the urinary tract. In addition to renal clearance of CMS and conversion to colistin, one or more 'other clearance' pathways must exist for the prodrug, although the mechanisms involved are not known [1]. (Reproduced with permission from Nation et al. [1])

including pharmacokinetic exposure to CMS and colistin have been sought. As a component of these analyses, risk factors for a colistin-associated decline in kidney function have been identified. A creatinine clearance >80 mL/min and a plasma colistin $C_{ss,avg}$ > 1.9–2.3 mg/L were identified as independent risk factors; the presence of both factors was at least additive [49]. In a study involving 102 patients receiving CMS, the pre-dose plasma colistin concentration on day 3 of CMS therapy was a predictor of acute kidney injury on day 7 and at end of treatment (the only two time-points considered). When the plasma colistin concentration on day 3 was evaluated as a categorical variable, the breakpoints that better predicted acute kidney injury were 3.33 mg/L on day 7 and 2.42 mg/L at end of treatment [48]. The end-of-treatment breakpoint of 2.42 mg/L was subsequently validated in a small study [56].

Sixteen of the critically-ill patients in the report of Garonzik et al. [31] were recipients of renal support at the time of initiating the CMS regimen (12 intermittent hemodialysis and 4 continuous renal replacement). The results from these patients revealed the substantial impact of these modalities on the plasma colistin concentration achieved from a given daily dose of CMS [31], in accord with other reports [37, 57–67]. The reason why renal replacement has such a large impact is easily understood when one considers that the rate of extracorporeal removal of a drug (or prodrug) is equal to the product of the intrinsic dialysis clearance (this is determined by the physicochemical properties of the drug (or prodrug) and the membrane) and the plasma concentration of the compound being delivered to the extracorporeal circuit. Not surprisingly given the similar molecular size of CMS and colistin, the intrinsic dialysis clearances of the two are of similar magnitude [37, 60, 63, 66]. However, the plasma concentration of CMS is substantially higher than that of colistin for a significant portion of a dosage interval (Fig. 15.2). Thus, much of the CMS/colistin that is removed by these renal replacement modalities is in the form of the prodrug, before it has had an opportunity to be converted *in vivo* to the active form, colistin. An additional reason for the impact of renal replacement relates to the respective handling of formed colistin in a functioning kidney *versus* that occurring in a renal replacement cartridge. In the kidney, an extremely large fraction of colistin filtered at glomeruli undergoes carrier-mediated tubular reabsorption [14], whereas in an extracorporeal cartridge no such reabsorprtion mechanism is in operation for colistin that diffuses into dialysate. The clinical implication is that dosage regimens for such patients must be carefully tailored to the circumstances. By way of population pharmacokinetic modeling, Garonzik et al. [31] proposed daily doses of CMS to achieve a target plasma colistin concentration in patients receiving intermittent hemodialysis. The dosage regimen algorithm included administration of a supplemental dose of CMS at the completion of dialysis to replace CMS and colistin lost during the session. These authors also provided an algorithm for the daily dose of CMS to achieve a desired plasma colistin concentration in patients receiving continuous renal replacement therapy [31]. In contrast to the efficient extracorporeal clearance of CMS and colistin via intermittent hemodialysis and continuous renal replacement [31], continuous ambulatory peritoneal dialysis has been reported to contribute little to the overall clearance of CMS and colistin; consequently it has been suggested that daily doses of CMS should not be increased to accommodate any drug loss via this renal support modality [68].

Following on from the interim analysis and dosage suggestions reported by Garonzik et al. [31], the results of the population pharmacokinetic analysis on the complete cohort of patients have been recently reported [46]. That analysis included data from a total of 215 patients, 29 of whom were receiving renal replacement therapy on the pharmacokinetic study day. Briefly, the characteristics for the 214 patients whose data were fully evaluable were: age range 19–101 years; body weight 30–122 kg; Appache II score 4–43; creatinine clearance 0–236 mL/min; 29 were receiving renal replacement therapy (16 intermittent hemodialysis, 4 sustained low-efficiency dialysis, 9 continuous renal replacement therapy)). Protein binding was conducted on plasma collected from 66 of the critically-ill patients in the study by Garonzik et al. and Nation et al. [31, 46]. The median and mean unbound fraction in plasma was approximately 0.5. The population pharmacokinetic analysis on the full cohort [46] confirmed the finding from Garonzik et al. [31] that the apparent clearance of formed colistin was related to the renal function of the patient. The impact of that relationship is shown in Fig. 15.5 which portrays the daily dose of CBA needed to achieve each 1 mg/L of plasma colistin $C_{ss,avg}$. The extensive inter-individual variability that had been observed by Garonzik et al. [31] is clearly evident in both panels of Fig. 15.5. The relationship in the right-hand panel of Fig. 15.5 formed the basis of a renal function-based algorithm to determine the daily dose needed to achieve a desired plasma colistin $C_{ss,avg}$. In developing the algorithm, the investigators needed to minimize the proportion of patients who would be likely to achieve plasma colistin concentrations that may increase the risk of colistin-associated nephrotoxicity [46]. Also reported from the population pharmacokinetic analysis were suggested daily doses for patients receiving renal replacement therapy; those suggestions included the need for any supplemental dosing after intermittent forms of dialysis and the timing of CMS dosing relative to the dialysis session [46]. All of the daily dose suggestions updated those that had been reported by Garonzik et al. [31] The ability of the updated daily dose suggestions to achieve a target plasma colistin $C_{ss,avg} \geq 2$ mg/L were evaluated for various creatinine clearance clusters, for patients not in receipt of renal support. A plasma colistin $C_{ss,avg}$ of 2 mg/L was chosen for the reasons discussed above and also because the MIC of the infecting pathogen may not be known at the time CMS treatment is initiated. While it was possible to achieve target attainment rates for a plasma colistin $C_{ss,avg}$ of ≥2 mg/L in 80–90% of patients with creatinine clearance <80 mL/min, the attainment rate was only ~40% in patients with higher creatinine clearance. This highlighted the potential importance of considering active combination therapy, especially in patients with good renal function. Target attainment rates for a plasma colistin $C_{ss,avg}$ of ≥2 mg/L with the proposed daily dose suggestions for patients receiving renal support were 85–89% [46].

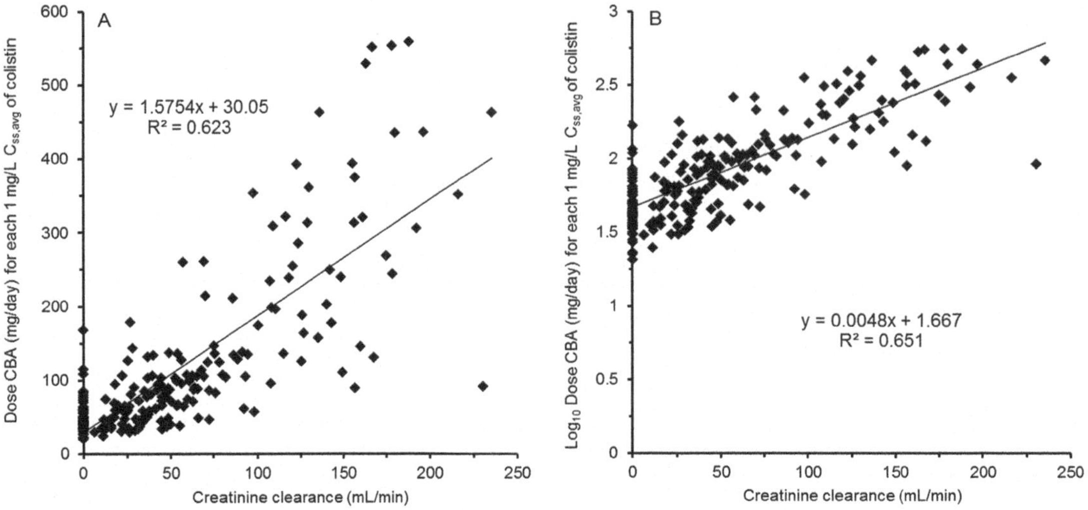

Fig. 15.5 Relationship between the dose of colistin base activity (CBA) per day needed for each 1 mg/L of plasma colistin $C_{ss,avg}$ and creatinine clearance. Panels A and B show the relationship in linear-linear and log-linear forms, respectively. (Reproduced with permission from Nation et al. [46])

In summary, the study by Garonzik et al. [31] and Nation et al. [46] lead to the proposing of the first scientifically-based dosage regimens for various categories of critically-ill patients. The authors reported an algorithm for determination of a loading dose of CMS, based upon body weight being a covariate on the volume of the central compartment of CMS. Others have questioned whether it may be more appropriate to use a non-weight based loading dose [69]. Following on from the work of Garonzik et al. [31], Nation et al. [46] proposed daily maintenance doses of CMS for patients receiving or not receiving renal support.

Results from the final population pharmacokinetic analysis on the data from 162 patients not in receipt of renal replacement therapy have been used to evaluate the recently updated daily dose guidelines approved by the European Medicines Agency (EMA) [13] and the US Food and Drug Administration (FDA) [70], as reported [47]. The EMA-approved daily doses had greatly superior attainment rates for plasma colistin $C_{ss,avg} \geq 0.5$, ≥ 1, ≥ 2 and ≥ 4 mg/L than did the dose suggestions approved by the FDA [47]. The EMA-approved dosing suggestions had been informed by a thorough review of recent pharmacological data [19], especially key studies discussed above [31, 42], while those approved by the FDA had not.

Little is known about the pharmacokinetics of CMS and colistin in pediatric patients. An investigation involving three patients aged 1.5 months (this patient was also studied during two other courses of CMS at 2.5 and 5.5 months of age), 5.5 years and 14 years (1 female; body weight range 6.2–40 kg) has been reported [71]. The CMS dosage regimens across the five courses were 0.06, 0.13, 0.2, 0.2 and 0.225 million IU/kg/day, corresponding to ~2.0, 4.3, 6.6, 6.6 and 7.4 mg/kg/day of CBA. The daily dose was divided and administered as 20-min infusions at 8 h intervals. After administration of at least 4 doses, blood samples were collected immediately before and 30 min after the end of a CMS infusion; cerebrospinal fluid samples were collected concomitantly (results discussed in the following paragraph). The authors noted that despite the CMS daily doses being in the range of those currently recommended and even higher, the serum colistin concentrations were greater than 2 mg/L in only the 14-year old patient (mean pre-dose and post-dose concentrations in this patient of 2.29 and 2.20 mg/L). Consequently, a key observation from this study was that CMS daily doses higher than previously recommended may be needed for pediatric patients to treat bloodstream infections caused by Gram-negative bacteria, especially if the causative organism exhibits borderline susceptibility to colistin [71]. The observation of the likely need for higher than traditionally used body-weight-based doses of CMS in neonates and in children is supported by results of other studies [72, 73]. Unfortunately, plasma colistin concentrations reported from another study as occurring in seven pediatric patients [74] were almost certainly spuriously high measurements due to ongoing conversion of CMS to colistin after samples were collected [75, 76]. Clearly, substantially more reliable information on the pharmacokinetics of colistin in pediatric patients is required to guide therapy.

There are very sparse data on the distribution of colistin into extravascular sites where infection may exist. The available data suggest that following intravenous administration of CMS, the concentrations of colistin in cerebrospinal fluid (CSF) [71, 77, 78] are only ~5% of concurrent plasma concentrations; very limited data from the study in pediatric patients discussed above suggest increased penetration in the presence of meningitis [71]. Similarly, the available data indicate that following intravenous administration of CMS, only relatively low concentrations of colistin are achieved in sputum of patients with cystic fibrosis [32] and in bronchoalveolar lavage (BAL) fluid from critically-ill patients [40, 79]. Care is required when interpreting the BAL concentrations from one of the latter studies [40] because the limit of quantification of the assay (0.05 mg/L) was for the analysed matrix (BAL) and there was ~100-fold dilution of the epithelial lining fluid (ELF) in performing the lavage procedure. Microdialysis studies in pigs have revealed ELF concentrations of colistin substantially lower than concomitant unbound concentrations in plasma [80]. Binding of colistin to mucin (and/or other components) in the airways must be considered as this may serve to decrease antibacterial activity [36]. As noted above, mouse lung infections were shown to be substantially more difficult to treat with colistin administered systemically than were thigh infections in the same species [23]. Thus, direct administration to these sites may be advantageous. Pharmacokinetics following intrathecal/intraventricular and inhalation administration of CMS are discussed below.

Burn Patients Severely burned patients are very vulnerable to nosocomial infections with Gram-negative bacteria. Because burn injury can lead to many changes in physiological status, including blood flows, fluid distribution and glomerular filtration rate [81], it is important to understand the impact of such changes on the disposition of antibiotics. Lee et al. conducted a population pharmacokinetic study in 50 patients who had burns to 4–85% (mean ± SD; 50.5 ± 21.8%) of total body surface area [82]. The patients (11 female; age range 26–80 years; body weight 50–98 kg; creatinine clearance 23–309 mL/min, with 17 patients receiving continuous renal replacement therapy of unspecified type; 18 patients with edema) were administered 150 mg of CBA (~4.5 million IU) infused over 30 min every 12 h. Blood samples were collected before and across the 8 h following a CMS infusion, at least 3 days after the first dose of CMS. The plasma concentration *versus* time data for colistin (CMS was not quantified) were subjected to population pharmacokinetic analysis. The structural model involved one compartment for colistin; in the absence of plasma concentrations for the prodrug, a single compartment was assumed for CMS. The terminal half-life of colistin for the typical burn patient was 6.6 h, which is somewhat shorter than that observed in critically-ill patients [15, 31]. The identified covariates in the final model for burn patients were the presence of edema and, as in the study by Garonzik et al. in critically-ill patients [31], creatinine clearance. It was estimated that the fractional conversion of CMS to colistin in an anephric burn patient was approximately five times greater than in a patient with a creatinine clearance of ~120 mL/min, consistent with expectations (Fig. 15.4, left panel). The final population pharmacokinetic model was used in simulations for 1000 virtual burn patients to estimate steady-state values for the area under the plasma colistin concentration-time curve across a day at steady state for a dosage regimen of 150 mg CBA as a 30-min infusion every 12 h. Based upon the results of the simulation presented graphically in Figure 5 of the paper [82], this dosage regimen would achieve a plasma colistin $C_{ss,avg}$ in a typical burn patient with a creatinine clearance <70 mL/min of ~1.5 mg/L and only ~1 mg/L in a patient with a creatinine clearance of ≥70 mL/min. The authors suggested that these relatively low concentrations may prompt the need to consider escalating doses above the currently approved dose that was used in this study. That patients with burns may require higher doses of CMS was supported by a recent brief report describing data from eight patients (median (IQR) age, weight, body mass index and serum creatinine concentration were 62 years (49–69 years), 69 kg (65–75 kg), 23.5 kg/m^2 (21.8–26.4 kg/m^2) and 0.75 mg/dL (0.68–1.36 mg/dL)) [83]. The patients received a median CMS dose of 2.25 million IU (2.00–3.00 million IU) (median ~74 mg CBA (IQR ~66–100 mg CBA)) every 12 h, but the plasma colistin concentrations were below the limit of quantification of the assay (~0.1–0.2 mg/L). The possible need for high doses in burn patients was exemplified in a case report describing a patient with burns who required a daily dose substantially higher than the currently approved dose range [84]. It is to be expected that renal function would be an important determinant of CMS dose requirements and that patients with augmented renal function in particular would be candidates for high doses. Data from two burn patients during continuous venovenous hemofiltration suggested efficient extracorporeal clearance of colistin [85], similar to findings in critically-ill patients as discussed above. However, it appears that CMS, rather than colistin (sulfate), was used to establish the calibration curve for colistin in this study.

Future investigations in burn patients should aim to provide additional information, including relationships of bacteriological and clinical outcomes and of nephrotoxicity with plasma colistin concentrations.

15.1.2 Direct Administration to the Lungs and Central Nervous System

One of the most common reasons for administration of CMS/colistin is the treatment of pulmonary infections such as ventilator-associated pneumonia in an intensive care setting. While the need to treat Gram-negative infections within the central nervous system (CNS) is less common, the critical and sensitive nature of this site poses special challenges. The common goal in treating these infections is the delivery of adequate colistin concentrations to the respective infection site, without exposing that site or systemic regions (e.g. kidneys) to concentrations likely to cause toxicity. As noted above, it is evident from the data available that intravenous administration of CMS results in relatively low colistin concentrations in CSF and pulmonary fluids. In contrast, direct administration to these sites leads to a very substantial targeting advantage, with significantly higher concentrations of colistin in the region to which CMS is delivered relative to the colistin concentrations reaching the systemic circulation.

A pulmonary targeting advantage has been demonstrated in six cystic fibrosis patients where relative sputum *versus* plasma exposures (area under the respective colistin concentration – time curve) were determined after intravenous and nebulization administration of CMS on different occasions [32]. The systemic availability of CMS was low (<10%) following nebulization of 2 and 4 million IU (~66 mg and 130 mg of CBA, respectively) and the plasma colistin concentrations were below the limit of quantification of the assay (0.125 mg/L). It was not possible in the study to perform bronchoalvealoar lavage and it remains to be determined whether the findings for sputum are representative of those for ELF. In this context, the relative concentrations of colistin in ELF and serum have been determined after inhalation of CMS (1 million IU (~33 mg CBA) every 8 h) in 20 mechanically ventilated critically-ill patients [86]. Median (25–75% interquartile range) colistin concentrations in ELF were 6.7 (4.8–10.1) mg/L, 3.9 (2.5–6.0) mg/L and 2.0

(1.0–3.8) mg/L at 1, 4 and 8 h post-nebulization, respectively. These concentrations were ~five-fold higher than those achieved concurrently in serum. The relative concentrations in the two biological fluids were not determined after intravenous administration and therefore it is not possible from this study to determine the targeting advantage of inhalational delivery. A recent study in adult critically-ill patients has demonstrated the ability of nebulized CMS to achieve concentrations of colistin in ELF much higher than concomitant concentrations in plasma, and substantially higher than the ELF colistin concentrations achieved with intravenous administration of CMS [79]. Similarly, a brief report has described six ventilated neonates (median gestational age 39 weeks (range, 32–39 weeks)) with ventilator-associated pneumonia who received a single dose of nebulized CMS (not clear if the dose was 120,000 IU/kg or 120,000 IU) [87]. Tracheal aspirate concentrations of colistin were very much higher than those in plasma. The results from these studies in cystic fibrosis [32] and critically-ill [79, 86–88] patients, and in a large animal model [89], suggest that inhalational delivery of CMS (perhaps together with intravenous administration) may be advantageous for the treatment of lung infections. This would be expected to allow attainment of higher colistin concentrations in lung fluid/tissue and at the same time lower systemic exposure to spare the kidneys. Recent clinical studies and systematic reviews and meta analyses reported in Chap. 11 and elsewhere [90–96] are suggestive of a beneficial effect of nebulized CMS either together with intravenous CMS or alone (including non-inferiority to intravenous CMS but with lower incidence of nephrotoxicity). A recent report of a relatively small retrospective clinical study suggested lack of benefit of inhaled CMS added to intravenous CMS versus intravenous CMS, although the study may have been confounded by a very high proportion of patients in both groups who were co-administered at least one other antibiotic [97]. Larger well-controlled studies are required.

Assuming no or negligible publication bias, there is a growing body of evidence that intrathecal or intraventricular administration of CMS (often in combination with intravenous therapy) is a generally safe and effective treatment for CNS infections (ventriculitis/meningitis) caused by Gram-negative bacteria [98–104]. From a relatively low dose of CMS administered *via* these routes, it is possible to achieve CSF colistin concentrations substantially higher than can be achieved by intravenous administration of a very much larger dose [71, 77, 78]. Interestingly, there is little information on the plasma colistin concentrations achieved from the relatively low doses of CMS administered by the intrathecal or intraventricular routes, but it would be expected that the systemic exposure would be low. It is not surprising that colistin-associated nephrotoxicity is an extremely rare occurrence with intrathecal or intraventricular administration [98, 100]. Importantly, administration of CMS *via* these routes gives rise to concentrations of colistin in CSF that could never be safely achieved with intravenous administration alone.

15.2 Polymyxin B

At the outset, it is important to recognize that there has been only one report of a pharmacokinetic/pharmacodynamic study on systemically administered polymyxin B (against *Klebsiella pneumoniae*) in an animal infection model [105], and there has not been a prospective pharmacokinetic/toxicodynamic study in patients to identify the association between plasma polymyxin B concentration and risk of nephrotoxicity. However, recently Lakota et al. proposed a tentative plasma polymyxin B concentration range of 2–4 mg/L [106] based upon animal pharmacokinetic/pharmacodynamic data for colistin against *P. aeruginosa* and *A. baumannii* [23] and polymyxin B against *K. pneumoniae* [105] to define the lower end of the range, and a pharmacokinetic/toxicodynamic meta-analysis of retrospective nephrotoxicity data gleaned from the literature to define the upper end of the range [106].

Studies on the disposition of intravenously administered polymyxin B in healthy humans with analysis of concentrations in plasma by chromatographic methods have not been performed. There is also a lack of information on pharmacokinetics of polymyxin B after administration to patients *via* routes other than intravenous (e.g. inhalation, intrathecal, intraventricular). Thus, the focus of the review below will be the pharmacokinetics of polymyxin B following intravenous administration in patients, most of whom were critically ill. The review will only include studies in which investigators used chromatographic methods for the analysis of drug concentration in biological fluids.

15.2.1 Intravenous Administration

The first reports on the disposition of polymyxin B in the modern era were in 2008 [16, 107]. The study by Kwa et al. [107] was conducted in nine adult patients (one female; age range 16–70 years; body weight 46–80 kg), all of whom were described as being among the general patient population and having normal renal function; patients with cystic fibrosis were excluded. The dosage regimens were at the discretion of the attending physicians and ranged from 0.75 to 2 million units (~75–200 mg of polymyxin B base) per day; each dose was infused over 2–6 h. Sparse blood sampling was performed (19 samples from the 9 patients) and serum was analysed for the concentration of one of the components (polymyxin B1) of the material administered. The resulting data were subjected to population pharmacokinetic analysis (one-compartment structural model) that yielded a typical half-life and clearance of 13.1 h and 2.2 L/h, respectively. The latter value may have over-estimated the clearance because only polymyxin B1 was quantified.

Zavascki et al. [16] studied the pharmacokinetics of polymyxin B in eight critically-ill patients (four female; age range 42–86 years; body weight 50–80 kg; creatinine clearance <10 to 246 mL/min). The physician-selected dosage regimens of polymyxin B ranged from 0.5 mg/kg (~5000 units/kg) every 48 h to 1.25 mg/kg (~12,500 units/kg) every 12 h. At the time, it was the practice at the clinical site to use lower daily doses in patients with poorer kidney function; three patients with the lowest creatinine clearances (<10, <10 and 26 mL/min) were dosed every 48 h and the remainder were dosed every 12 h. Each dose was infused over 60 min, and after at least 2 days of therapy seven blood samples were collected from each patient across a dosage interval including samples at the end of the 1-h infusion and at 0.5, 1 and 2 h thereafter. The two patients with creatinine clearance <10 mL/min were receiving intermittent hemodialysis; they were studied on a non-dialysis day. Importantly, for four patients it was possible to quantitatively collect urine over the dosage interval. The samples of plasma and urine were analysed for the concentration of polymyxin B with the resultant data subjected to non-compartmental pharmacokinetic analysis. Peak plasma concentration at the end of the short-term infusion ranged from 2.38 to 13.9 mg/L. There were several key findings from the pharmacokinetic analysis. First, after the infusion ceased the plasma polymyxin B concentrations declined in a multi-exponential manner indicating that more than one pharmacokinetic compartment was involved in the disposition of polymyxin B. Second, notwithstanding the diverse renal functions in the patients, there was remarkably little inter-individual variability in the total body clearance (range, 0.27–0.81 mL/min/kg) and volume of distribution (71–194 mL/kg) of polymyxin B. Third, urinary recovery of unchanged drug was extremely low, with each of the 4 patients for whom data were available excreting <1% of the administered dose in urine. The same very low urinary recovery (<1%) has been observed in rats for polymyxin B [17] and colistin [14], each after administration of the respective sulfate salt. Fourth, consideration of the relative magnitudes of the calculated glomerular filtration clearance of polymyxin B and the ultimate overall renal clearance indicated that very extensive renal tubular reabsorption must have been occurring in the patients. Again, a similar conclusion was reached for the renal handling of colistin in rats [14]. Fifth, and arguably most important from a clinical pharmacokinetic perspective, even though the renal clearance of polymyxin B varied ~12-fold among patients, consistent with the wide range of creatinine clearance values, there was very little variability (~threefold range) in the total body clearance of polymyxin B. This arose because renal clearance was only a very small component (<1%) of the total body clearance. The study by Zavascki et al. [16] was the first to imply that dose adjustment based upon renal function (as may be appropriate for CMS) may not be required for polymyxin B, a proposition supported by the pharmacokinetic data from a single case report 3 years later involving a patient with renal insufficiency [108].

Subsequently, Sandri and coworkers reported on a larger study on the pharmacokinetics of polymyxin B in critically-ill patients [51]. That study involved 24 patients, including 2 receiving continuous venovenous hemodialysis (CVVHD) for whom greater detail was reported separately [109]. The 24 patients were typical of the intensive care setting (11 female; age range 21–87 years; total body weight 41–250 kg; lean body weight 29–99 kg; creatinine clearance in the 22 non-CVVHD patients 10–143 mL/min). One patient was extremely obese (250 kg); the next heaviest patient was 110 kg. The study design was similar to that described in the earlier report from the same group as discussed above [16]; in the study of Sandri et al. eight blood samples were collected across the dosage interval from each patient, with even more intensive sampling shortly after the end of the polymyxin B infusion [51]. The plasma polymyxin B concentration – time profiles resulting from the physician-selected dosage regimens are shown in Fig. 15.6. These data were well described in a population pharmacokinetic analysis (two-compartment structural model). The concentration-time profiles from all patients (i.e. on and not on renal replacement therapy) were successfully described simultaneously with one set of population pharmacokinetic parameter estimates because only a small amount was removed by CVVHD (12.2% and 5.62% for the two patients). The apparent small effect of extracorporeal clearance on total body clearance and hence dosage requirements of polymyxin B contrasts with the situation for CMS/colistin [31, 37, 57, 59, 62]. This difference may be explained by the fact that in the latter case a substantial amount of the CMS/colistin removed by the cartridge is in the form of the prodrug which circulates in plasma at concentrations that are generally higher than those of colistin across the early stage of a dosage interval. Sandri et al., the authors of the population pharmacokinetic study on polymyxin B, reported that the median unbound fraction in plasma was 0.42 [51]. Urinary excretion data for polymyxin B were available for 17 non-CVVHD patients. The median percentage of the polymyxin B dose excreted in urine was 4.04% (range 0.98–17.4%). Thus, renal clearance was only a very small component of the total body clearance. While renal

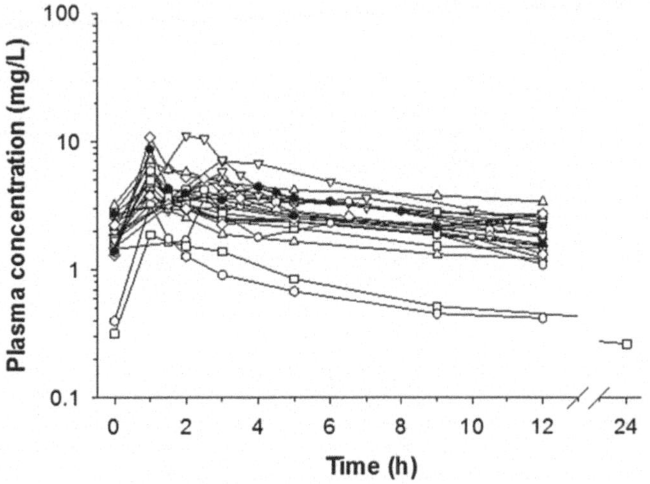

Fig. 15.6 Plasma concentration – time profiles of polymyxin B in 24 critically-ill patients. Concentrations from the two patients receiving continuous venovenous hemodialysis [109] are shown by filled symbols [51]. (Reproduced with permission from Sandri et al. [51])

clearance of polymyxin B is a minor elimination pathway, it is important to understand the renal handling mechanisms because of the very high probability of a link with the propensity to cause nephrotoxicity. Application of physiological concepts to the polymyxin B renal clearance data revealed extensive renal tubular reabsorption [51], consistent with previous findings for polymyxin B in critically-ill patients [16].

In the study reported by Sandri and coworkers, the population estimate of polymyxin B clearance not adjusted for body weight was 1.87 L/h and that scaled by body weight was 0.0276 L/h/kg [51]. In view of the minor contribution of renal clearance to total clearance, it was not surprising that the population pharmacokinetic analysis did not identify creatinine clearance as a covariate of the total body clearance of polymyxin B (Fig. 15.7). The use of unscaled (i.e. absolute) clearance (left panel of the figure) resulted in the obese patient being an outlier. In the population pharmacokinetic analysis, linear scaling of polymyxin B clearance by total body weight reduced the unexplained between subject variability. Allometric scaling of polymyxin B clearance by total body weight, and linear and allometric scaling by lean body weight were also considered, but they resulted in only marginal improvement. To simplify application in the clinic, the results from the model with linear scaling by total body weight were adopted. After linear scaling of clearance by total body weight, the parameter estimates for the two patients receiving CVVHD, including the patient with 250-kg total body weight, were within the range of estimates from the other patients (Fig. 15.7, right panel). It is evident that neither the unscaled nor the scaled polymyxin B clearance values were related to creatinine clearance. Notwithstanding the very wide range of total body weights and creatinine clearance values there was only approximately threefold variation in the total body weight scaled clearance values of polymyxin B (Fig. 15.7, right panel). Accordingly, the population pharmacokinetic model yielded a between subject variability in clearance (coefficient of variation 32.4%) that was remarkably low in this diverse group of critically-ill patients [51]. The authors went on to perform Monte Carlo simulations for a number of clinically relevant dosage regimens scaled by total body weight.

In 2017, Thamlikitkul et al. reported pharmacokinetic data on polymyxin B in 19 patients (creatinine clearance range 15–110 mL/min) [110]. At least 48 h after initiating polymyxin B, four blood samples were collected across a dosage interval. The resulting plasma

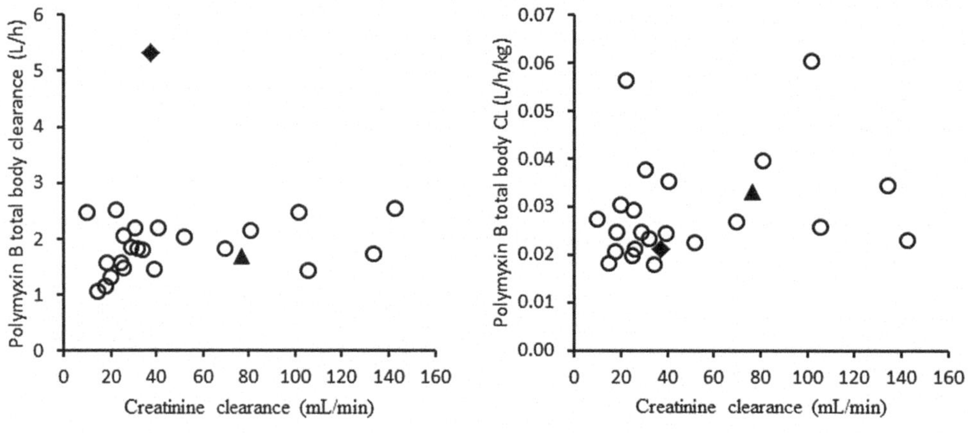

Fig. 15.7 Individual polymyxin B clearance estimates *versus* creatinine clearance. Polymyxin B clearance was either unscaled (L/h, left panel) or scaled by total body weight (L/h/kg, right panel). Open circles represent patients not receiving continuous venovenous hemodialysis (CVVHD), the filled diamond represents the CVVHD patient who weighed 250 kg, and the filled triangle the lean CVVHD patient [51]. (Reproduced with permission from Sandri et al. [51])

concentration-time data were subjected to two forms of analysis: a standard two-stage approach which involved a one-compartment model being used to describe the data; and a maximum *a posteriori* Bayesian approach (which used the results of Sandri et al. [51] as Bayes priors) resulting in a two-compartment model providing the best fit to the data. The primary subsequent analysis involved comparison of the dose-normalized plasma exposure to polymyxin B in 5 patients with creatinine clearance ≥80 mL/min with that of the 14 patients with creatinine clearance <80 mL/min; there was no difference in exposure ($P > 0.4$) [110], consistent with results reported previously [16, 51, 108].

In a paper published in 2018 Manchandani et al. reported a population pharmacokinetic study in 35 patients (12 female; age range 25–89 years; body weight 36–112 kg; creatinine clearance 15–175 mL/min) [111]. Each polymyxin B dose was infused over 1–3 h, and four blood samples were collected over one dosage interval at steady state (prior to the dose, at 1–2 h and 8–12 h after the end of the infusion, and prior to the next dose). Population pharmacokinetic analysis was conducted and a one-compartment model was chosen as the best-fit model. The population estimate of polymyxin B clearance was 2.5 L/h, a value similar to those reported from earlier studies [16, 51, 107, 110], and the inter-patient variability was similar to that reported by Sandri et al. [51]. In the study reported by Manchandani and coworkers, aside from one patient who presented as an outlier, the polymyxin B clearance estimates for the remaining 34 patients were within an approximate four to fivefold range across the creatinine clearance spectrum represented in this patient population. While creatinine clearance was found to be a statistically significant covariate of polymyxin B clearance, the magnitude was regarded as clinically insignificant [111]. Volume of distribution of polymyxin B was poorly predicted by actual body weight, but given the small number of samples collected from each patient, the timing of the samples and the resultant application of a one-compartment model it is possible that the volume estimates were subject to bias.

Kubin et al. [112] and Miglis et al. [113] published companion papers in 2018 reporting the population pharmacokinetics of polymyxin B in patients, other than those with cystic fibrosis. Both studies involved sparse blood sample collection (samples collected primarily for therapeutic drug monitoring (TDM)) with an average of three blood samples per patient. The study by Kubin and coworkers included 43 patients while that of Miglis et al. included data for 52 patients, but 43 of those patients were from the study of Kubin et al. (*i.e.* 83% of the patients reported in the second study were also included in the first study). Interestingly, even though the data from the 43 patients in the study of Kubin et al. were used to develop the base model described in Miglis et al., the former study found that a one-compartment model satisfactorily described the data while a two-compartment model proved superior in the study by Miglis et al. [112, 113]. A finding in common across both reports was that actual body weight was not a covariate of polymyxin B clearance; the commonality of findings is not surprising given the very large degree of overlap in the patient populations across the two studies. The mean population estimates of polymyxin B clearance reported by Kubin et al. (2.37 L/h) and Miglis et al. (2.63 L/h) were similar to those reported previously [51, 110, 111]. However, it was noticeable that the inter-patient variability in polymyxin B clearance was substantially greater than in the earlier studies; even after disregarding an outlier, the report by Miglis et al. suggests there was >tenfold inter-patient variability in the clearance of polymyxin B [113]. It is possible that this difference across studies arose because of the sparse sampling used by Miglis et al.; the resultant inability to accurately define area under the plasma concentration-time curve very likely resulted in biased estimates of clearance [112]. The sparse blood sampling approach was also used by the same group to undertake a pilot pharmacokinetic study in nine adult patients with cystic fibrosis [114]. In that study, patients had a median of two blood samples collected during polymyxin B therapy for the purpose of TDM. The authors commented that while the pharmacokinetic parameters in the

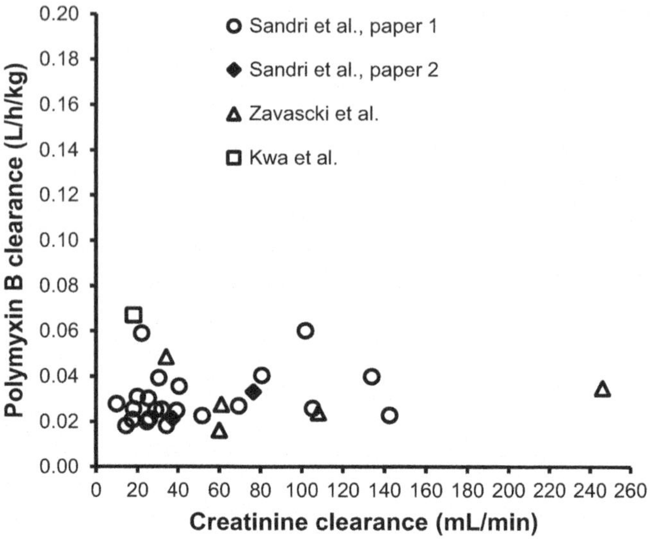

Fig. 15.8 Individual polymyxin B clearance estimates *versus* creatinine clearance; data pooled from four reports. (The figure was prepared using data from the studies of Sandri et al., paper 1 [51], Sandri et al., paper 2 [109], Zavascki et al. [16] and Kwa et al. [108])

patients with cystic fibrosis were similar to those without this condition, substantial care is required in regard to interpreting this and other findings because of the small number of patients and the sparse blood sampling [114].

It is instructive to consider the relationship between polymyxin B clearance and creatinine clearance using data pooled from studies reviewed above (Fig. 15.8) [16, 51, 108, 109]. It is evident from these studies and more recent investigations [110, 111, 113] that polymyxin B total body clearance is not influenced by renal function to a clinically important extent. It is also evident from Fig. 15.8 that polymyxin B clearance is subject to a remarkably low degree of inter-individual variability across the wide range of creatinine clearance values examined. These two characteristics contrast sharply with those for the apparent clearance of formed colistin after administration of CMS, as discussed above. The clinically insignificant effect of kidney function on the total clearance of polymyxin B is behaviour expected of a drug that is excreted in urine to only a minor extent; see Fig. 15.4 (right panel) for a diagrammatic representation of the overall disposition of polymyxin B. Therefore, based on currently available data, in contrast to information provided in package inserts for polymyxin B, daily doses should not be based on renal function. A reduction in daily dose in a renally-impaired patient will potentially have negative impact on microbiological and clinical responses [51, 108, 115].

15.3 Commentary and Conclusions

From the foregoing, it is clear that while very significant progress has been made in understanding key features of the clinical pharmacology of colistin and polymyxin B, there is still much to be learned. Large clinical population pharmacokinetic, pharmacodynamic and toxicodynamic studies have been performed on colistin, but there are much sparser data for polymyxin B and large-scale studies are urgently needed. There is a dearth of pharmacokinetic information for both polymyxins in important groups, including obese patients, those with burns, and neonates and children. There is also need to delineate the role of inhaled polymyxin in treatment of pneumonia, and of when and how to implement polymyxin combination regimens [12].

Even at this stage, it is very evident that while both clinically used polymyxins share similar characteristics in some areas (e.g. similar *in vitro* antibacterial activity against important species of Gram-negative bacteria (Chap. 3) [2, 3], propensity to cause damage to renal tubular cells (Chap.

18) [52, 116]), they differ substantially in some important aspects of their clinical pharmacology. The difference in *in vivo* behaviour arises because colistin is administered as an inactive prodrug (CMS) that is predominantly renally cleared before significant conversion can occur, while polymyxin B is administered directly as its sulfate salt (Fig. 15.4).

The potential clinical pharmacological consequences of the different disposition profiles have been discussed previously [1] and therefore they will be presented only briefly here. Firstly, the generally slow, incomplete and variable (due to batch-to-batch variability in CMS) *in vivo* conversion to colistin often impedes the ability to achieve desired plasma concentrations of colistin in the first several hours of therapy, even with a loading dose [10, 31, 42, 43]. This may have a negative impact on prognosis given the link between delayed initiation of appropriate antibiotic therapy and patient outcome [117, 118]. As noted earlier, the need for a loading dose would be greater if administering a parenteral product of CMS that is subject to very slow *in vivo* conversion to colistin; unfortunately, the rate of conversion of the product available in a given hospital is generally not known by the clinicians caring for a patient. Because polymyxin B is not administered as a prodrug it is possible to more rapidly attain plasma concentrations that are likely to be effective [51]. Second, in patients with creatinine clearance greater than ~80 mL/min, the avid renal clearance of CMS competes very successfully with the pathway for conversion to colistin (Fig. 15.4), and therefore it is not possible to reliably achieve plasma colistin concentrations of ≥2 mg/L in such patients, even with daily doses at the upper limit of the current prescribing information (Fig. 15.3) [31, 43, 46, 47]. This is not the case with polymyxin B [16, 51]. Third, because the apparent clearance of colistin is related to creatinine clearance, the daily maintenance dose of CMS may require adjustment based upon renal function [31, 43, 46, 47]. Based upon current evidence, this is not appropriate for polymyxin B because its total body clearance is not dependent on creatinine clearance to a clinically significant degree [16, 51, 108, 110, 111]. Fourth, based upon studies with intensive blood sampling across a dosage interval to define the area under the plasma concentration *versus* time curve, there is substantially greater inter-individual variability in the apparent clearance of colistin at a given creatinine clearance (up to ~tenfold variability) [31, 46] than there is in the clearance of polymyxin B across a very wide range of creatinine clearance values (only ~threefold variability in the clearance of polymyxin B) [51]. This point together with that immediately above means that it is substantially easier to predict a daily maintenance dose to achieve a desired steady-state plasma concentration for polymyxin B than it is for colistin. Fifth, both colistin and polymyxin B have low therapeutic indices and as such TDM is likely to be beneficial [12]. Some laboratories are already providing a TDM service to assist in optimization of dosage regimens [37, 84, 119, 120]. There is an even stronger case for TDM (and ideally adaptive feedback control (AFC) [121]) with CMS/colistin because of the very substantial variability in the overall pharmacokinetic profile. However, that variability will render difficult the successful implementation of AFC for colistin. In addition, TDM is inherently more difficult for CMS/colistin, because of the critical need to ensure that all sample collection, handling and assay procedures do not lead to ongoing *in vitro* conversion of CMS to colistin [30]. Thus, application of TDM and AFC is expected to be more straightforward for polymyxin B than for CMS/colistin; indeed, early work towards the application of AFC for polymyxin B has already occurred [106]. Sixth, CMS/colistin and polymyxin B differ in the concentrations of the active antibacterial that can be achieved in urine. This occurs because CMS is extensively excreted into urine and then it undergoes partial conversion to colistin within the urinary tract as a result of instability of CMS in an aqueous environment; [7, 18, 122] in contrast, polymyxin B undergoes minimal excretion into urine [16, 51] (Fig. 15.4).

Thus, polymyxin B would appear to have superior pharmacokinetic characteristics for infections where it is important to rapidly and reliably achieve and then maintain a desired concentration in plasma. Interestingly, a meta-analysis of five recent clinical studies [52, 53, 123–125] which compared relative rates of nephrotoxicity associated with CMS/colistin *versus* polymyxin B revealed that the risk was lower for polymyxin B [126], probably related to the administration of colistin as a prodrug [127]. CMS may be the preferred polymyxin for the treatment of urinary tract infections [1]. The relative safety and effectiveness of CMS and polymyxin B delivered directly to sites such as the lungs and CSF needs to be evaluated. CMS may be less irritating to membranes lining the lungs and to the meninges, and there may be ongoing conversion to colistin from the prodrug 'trapped' in these regions due to its slow absorption into the systemic circulation and eventual delivery to the kidneys. It would be advantageous for clinicians in all parts of the world to have access to parenteral products of both CMS and polymyxin B, so that they can choose between the two in particular circumstances [12]. Head-to-head comparisons of the two polymyxins against various types of infections are needed, and such studies should evaluate both efficacy and toxicity endpoints. A compilation of high priority aspects requiring attention to optimize the clinical use of the polymyxins has been formulated [12]. It is hoped that progress on addressing these areas will be made with good speed. In the meantime, it is pleasing that international consensus guidelines for the optimal use of the polymyxins have been published [129]. The information used to formulate the dosing regimens of colistin and polymyxin B recommended in those guidelines was from key studies reviewed in the present chapter.

References

1. Nation RL, Velkov T, Li J (2014) Colistin and polymyxin B: peas in a pod, or chalk and cheese? Clin Infect Dis 59:88–94. https://doi.org/10.1093/cid/ciu213
2. Gales AC, Jones RN, Sader HS (2011) Contemporary activity of colistin and polymyxin B against a worldwide collection of Gram-negative pathogens: results from the SENTRY Antimicrobial Surveillance Program (2006–09). J Antimicrob Chemother 66(9):2070–2074. https://doi.org/10.1093/jac/dkr239
3. Sader HS, Rhomberg PR, Farrell DJ, Jones RN (2015) Differences in potency and categorical agreement between colistin and polymyxin B when testing 15,377 clinical strains collected worldwide. Diagn Microbiol Infect Dis 83(4):379–381. https://doi.org/10.1016/j.diagmicrobio.2015.08.013
4. Kwa A, Kasiakou SK, Tam VH, Falagas ME (2007) Polymyxin B: similarities to and differences from colistin (polymyxin E). Expert Rev Anti-Infect Ther 5(5):811–821. https://doi.org/10.1586/14787210.5.5.811
5. Barnett M, Bushby SR, Wilkinson S (1964) Sodium sulphomethyl derivatives of polymyxins. Br J Pharmacol Chemother 23:552–574
6. Bergen PJ, Li J, Rayner CR, Nation RL (2006) Colistin methanesulfonate is an inactive prodrug of colistin against Pseudomonas aeruginosa. Antimicrob Agents Chemother 50(6):1953–1958. https://doi.org/10.1128/AAC.00035-06
7. Li J, Milne RW, Nation RL, Turnidge JD, Smeaton TC, Coulthard K (2004) Pharmacokinetics of colistin methanesulphonate and colistin in rats following an intravenous dose of colistin methanesulphonate. J Antimicrob Chemother 53(5):837–840. https://doi.org/10.1093/jac/dkh167
8. Li J, Nation RL, Turnidge JD, Milne RW, Coulthard K, Rayner CR, Paterson DL (2006) Colistin: the re-emerging antibiotic for multidrug-resistant Gram-negative bacterial infections. Lancet Infect Dis 6(9):589–601. https://doi.org/10.1016/S1473-3099(06)70580-1
9. Gregoire N, Aranzana-Climent V, Magreault S, Marchand S, Couet W (2017) Clinical pharmacokinetics and pharmacodynamics of colistin. Clin Pharmacokinet 56:1441–1460. https://doi.org/10.1007/s40262-017-0561-1
10. He H, Li JC, Nation RL, Jacob J, Chen G, Lee HJ, Tsuji BT, Thompson PE, Roberts K, Velkov T, Li J (2013) Pharmacokinetics of four different brands of colistimethate and formed colistin in rats. J Antimicrob Chemother 68(10):2311–2317. https://doi.org/10.1093/jac/dkt207
11. Theuretzbacher U (2014) Product information for parenteral colistin varies substantially across Europe. J Antimicrob Chemother 69:1987–1992. https://doi.org/10.1093/jac/dku064
12. Nation RL, Li J, Cars O, Couet W, Dudley MN, Kaye KS, Mouton JW, Paterson DL, Tam VH, Theuretzbacher U, Tsuji BT, Turnidge JD (2015) Framework for optimisation of the clinical use of colistin and polymyxin B: the Prato polymyxin consensus. Lancet Infect Dis 15(2):225–234. https://doi.org/10.1016/s1473-3099(14)70850-3
13. European Medicines Agency (2014) Polymyxin-containing medicines. Polymyxin Article 31 referral – annex III amendments to relevant sections of the summary of product characteristics and the package leaflets. http://www.ema.europa.eu/docs/en_GB/document_library/Referrals_document/

Polymyxin_31/WC500176332.pdf. Last accessed 16 Jun 2019
14. Li J, Milne RW, Nation RL, Turnidge JD, Smeaton TC, Coulthard K (2003) Use of high-performance liquid chromatography to study the pharmacokinetics of colistin sulfate in rats following intravenous administration. Antimicrob Agents Chemother 47(5):1766–1770
15. Plachouras D, Karvanen M, Friberg LE, Papadomichelakis E, Antoniadou A, Tsangaris I, Karaiskos I, Poulakou G, Kontopidou F, Armaganidis A, Cars O, Giamarellou H (2009) Population pharmacokinetic analysis of colistin methanesulfonate and colistin after intravenous administration in critically ill patients with infections caused by Gram-negative bacteria. Antimicrob Agents Chemother 53(8):3430–3436. https://doi.org/10.1128/AAC.01361-08
16. Zavascki AP, Goldani LZ, Cao G, Superti SV, Lutz L, Barth AL, Ramos F, Boniatti MM, Nation RL, Li J (2008) Pharmacokinetics of intravenous polymyxin B in critically ill patients. Clin Infect Dis 47(10):1298–1304. https://doi.org/10.1086/592577
17. Abdelraouf K, He J, Ledesma KR, Hu M, Tam VH (2012) Pharmacokinetics and renal disposition of polymyxin B in an animal model. Antimicrob Agents Chemother 56(11):5724–5727. https://doi.org/10.1128/AAC.01333-12
18. Couet W, Gregoire N, Gobin P, Saulnier PJ, Frasca D, Marchand S, Mimoz O (2011) Pharmacokinetics of colistin and colistimethate sodium after a single 80-mg intravenous dose of CMS in young healthy volunteers. Clin Pharmacol Ther 89(6):875–879. https://doi.org/10.1038/clpt.2011.48
19. European Medicines Agency (2014) Assessment report on polymyxin-based products. Referral under Article 31 of Directive 2001/83/EC. http://www.ema.europa.eu/docs/en_GB/document_library/Referrals_document/Polymyxin_31/WC500179664.pdf. Last accessed 16 Jun 2019
20. Mizuyachi K, Hara K, Wakamatsu A, Nohda S, Hirama T (2011) Safety and pharmacokinetic evaluation of intravenous colistin methanesulfonate sodium in Japanese healthy male subjects. Curr Med Res Opin 27(12):2261–2270. https://doi.org/10.1185/03007995.2011.626557
21. Zhao M, Wu XJ, Fan YX, Zhang YY, Guo BN, Yu JC, Cao GY, Chen YC, Wu JF, Shi YG, Li J, Zhang J (2018) Pharmacokinetics of colistin methanesulfonate (CMS) in healthy Chinese subjects after single and multiple intravenous doses. Int J Antimicrob Agents 51(5):714–720. https://doi.org/10.1016/j.ijantimicag.2017.12.025
22. Li J, Coulthard K, Milne R, Nation RL, Conway S, Peckham D, Etherington C, Turnidge J (2003) Steady-state pharmacokinetics of intravenous colistin methanesulphonate in patients with cystic fibrosis. J Antimicrob Chemother 52(6):987–992. https://doi.org/10.1093/jac/dkg468
23. Cheah SE, Wang J, Nguyen VT, Turnidge JD, Li J, Nation RL (2015) New pharmacokinetic/pharmacodynamic studies of systemically administered colistin against Pseudomonas aeruginosa and Acinetobacter baumannii in mouse thigh and lung infection models: smaller response in lung infection. J Antimicrob Chemother 70(12):3291–3297. https://doi.org/10.1093/jac/dkv267
24. Marchand S, Gobin P, Brillault J, Baptista S, Adier C, Olivier JC, Mimoz O, Couet W (2010) Aerosol therapy with colistin methanesulfonate: a biopharmaceutical issue illustrated in rats. Antimicrob Agents Chemother 54(9):3702–3707. https://doi.org/10.1128/AAC.00411-10
25. Marchand S, Lamarche I, Gobin P, Couet W (2010) Dose-ranging pharmacokinetics of colistin methanesulphonate (CMS) and colistin in rats following single intravenous CMS doses. J Antimicrob Chemother 65(8):1753–1758. https://doi.org/10.1093/jac/dkq183
26. Yapa SWS, Li J, Porter CJ, Nation RL, Patel K, McIntosh MP (2013) Population pharmacokinetics of colistin methanesulfonate in rats: achieving sustained lung concentrations of colistin for targeting respiratory infections. Antimicrob Agents Chemother 57(10):5087–5095. https://doi.org/10.1128/AAC.01127-13
27. Gibaldi M, Perrier D (1982) Pharmacokinetics, 2nd edn. Marcel Dekker, New York
28. Li J, Milne RW, Nation RL, Turnidge JD, Coulthard K (2003) Stability of colistin and colistin methanesulfonate in aqueous media and plasma as determined by high-performance liquid chromatography. Antimicrob Agents Chemother 47(4):1364–1370
29. Reed MD, Stern RC, O'Riordan MA, Blumer JL (2001) The pharmacokinetics of colistin in patients with cystic fibrosis. J Clin Pharmacol 41(6):645–654
30. Dudhani RV, Nation RL, Li J (2010) Evaluating the stability of colistin and colistin methanesulphonate in human plasma under different conditions of storage. J Antimicrob Chemother 65(7):1412–1415. https://doi.org/10.1093/jac/dkq134
31. Garonzik SM, Li J, Thamlikitkul V, Paterson DL, Shoham S, Jacob J, Silveira FP, Forrest A, Nation RL (2011) Population pharmacokinetics of colistin methanesulfonate and formed colistin in critically ill patients from a multicenter study provide dosing suggestions for various categories of patients. Antimicrob Agents Chemother 55(7):3284–3294. https://doi.org/10.1128/AAC.01733-10
32. Yapa SWS, Li J, Patel K, Wilson JW, Dooley MJ, George J, Clark D, Poole S, Williams E, Porter CJ, Nation RL, McIntosh MP (2014) Pulmonary and systemic pharmacokinetics of inhaled and intravenous colistin methanesulfonate in cystic fibrosis patients: targeting advantage of inhalational administration. Antimicrob Agents Chemother 58(5):2570–2579. https://doi.org/10.1128/AAC.01705-13
33. Li J, Milne RW, Nation RL, Turnidge JD, Coulthard K, Johnson DW (2001) A simple method for the

assay of colistin in human plasma, using pre-column derivatization with 9-fluorenylmethyl chloroformate in solid-phase extraction cartridges and reversed-phase high-performance liquid chromatography. J Chromatogr B Biomed Sci Appl 761(2):167–175
34. Li J, Milne RW, Nation RL, Turnidge JD, Coulthard K, Valentine J (2002) Simple method for assaying colistin methanesulfonate in plasma and urine using high-performance liquid chromatography. Antimicrob Agents Chemother 46(10):3304–3307
35. Li J, Turnidge J, Milne R, Nation RL, Coulthard K (2001) In vitro pharmacodynamic properties of colistin and colistin methanesulfonate against Pseudomonas aeruginosa isolates from patients with cystic fibrosis. Antimicrob Agents Chemother 45(3):781–785. https://doi.org/10.1128/AAC.45.3.781-785.2001
36. Huang JX, Blaskovich MA, Pelingon R, Ramu S, Kavanagh A, Elliott AG, Butler MS, Montgomery AB, Cooper MA (2015) Mucin binding reduces colistin antimicrobial activity. Antimicrob Agents Chemother 59(10):5925–5931. https://doi.org/10.1128/aac.00808-15
37. Li J, Rayner CR, Nation RL, Deans R, Boots R, Widdecombe N, Douglas A, Lipman J (2005) Pharmacokinetics of colistin methanesulfonate and colistin in a critically ill patient receiving continuous venovenous hemodiafiltration. Antimicrob Agents Chemother 49(11):4814–4815. https://doi.org/10.1128/AAC.49.11.4814-4815.2005
38. Trotman RL, Williamson JC, Shoemaker DM, Salzer WL (2005) Antibiotic dosing in critically ill adult patients receiving continuous renal replacement therapy. Clin Infect Dis 41(8):1159–1166. https://doi.org/10.1086/444500
39. Markou N, Markantonis SL, Dimitrakis E, Panidis D, Boutzouka E, Karatzas S, Rafailidis P, Apostolakos H, Baltopoulos G (2008) Colistin serum concentrations after intravenous administration in critically ill patients with serious multidrug-resistant, Gram-negative bacilli infections: a prospective, open-label, uncontrolled study. Clin Ther 30(1):143–151. https://doi.org/10.1016/j.clinthera.2008.01.015
40. Imberti R, Cusato M, Villani P, Carnevale L, Iotti GA, Langer M, Regazzi M (2010) Steady-state pharmacokinetics and BAL concentration of colistin in critically ill patients after IV colistin methanesulfonate administration. Chest 138(6):1333–1339. https://doi.org/10.1378/chest.10-0463
41. Karnik ND, Sridharan K, Jadhav SP, Kadam PP, Naidu RK, Namjoshi RD, Gupta V, Gore MS, Surase PV, Mehta PR, Gogtay JA, Thatte UM, Gogtay NJ (2013) Pharmacokinetics of colistin in critically ill patients with multidrug-resistant Gram-negative bacilli infection. Eur J Clin Pharmacol 69(7):1429–1436. https://doi.org/10.1007/s00228-013-1493-9
42. Mohamed AF, Karaiskos I, Plachouras D, Karvanen M, Pontikis K, Jansson B, Papadomichelakis E, Antoniadou A, Giamarellou H, Armaganidis A, Cars O, Friberg LE (2012) Application of a loading dose of colistin methanesulfonate in critically ill patients: population pharmacokinetics, protein binding, and prediction of bacterial kill. Antimicrob Agents Chemother 56(8):4241–4249. https://doi.org/10.1128/AAC.06426-11
43. Karaiskos I, Friberg LE, Pontikis K, Ioannidis K, Tsagkari V, Galani L, Kostakou E, Baziaka F, Paskalis C, Koutsoukou A, Giamarellou H (2015) Colistin population pharmacokinetics after application of a loading dose of 9 MU colistin methanesulfonate in critically ill patients. Antimicrob Agents Chemother 59(12):7240–7248. https://doi.org/10.1128/aac.00554-15
44. Gregoire N, Mimoz O, Megarbane B, Comets E, Chatelier D, Lasocki S, Gauzit R, Balayn D, Gobin P, Marchand S, Couet W (2014) New colistin population pharmacokinetic data in critically ill patients suggesting an alternative loading dose rational. Antimicrob Agents Chemother 58:7324–7330. https://doi.org/10.1128/aac.03508-14
45. Gautam V, Shafiq N, Mouton JW, Malhotra S, Kaur S, Ray P (2018) Pharmacokinetics of colistin in patients with multidrug-resistant Gram-negative infections: a pilot study. Indian J Med Res 147(4):407–412. https://doi.org/10.4103/ijmr.IJMR_1464_16
46. Nation RL, Garonzik SM, Thamlikitkul V, Giamarellos-Bourboulis EJ, Forrest A, Paterson DL, Li J, Silveira FP (2017) Dosing guidance for intravenous colistin in critically ill patients. Clin Infect Dis 64(5):565–571. https://doi.org/10.1093/cid/ciw839
47. Nation RL, Garonzik SM, Li J, Thamlikitkul V, Giamarellos-Bourboulis EJ, Paterson DL, Turnidge JD, Forrest A, Silveira FP (2016) Updated US and European dose recommendations for intravenous colistin: how do they perform? Clin Infect Dis 62(5):552–558. https://doi.org/10.1093/cid/civ964
48. Sorli L, Luque S, Grau S, Berenguer N, Segura C, Montero MM, Alvarez-Lerma F, Knobel H, Benito N, Horcajada JP (2013) Trough colistin plasma level is an independent risk factor for nephrotoxicity: a prospective observational cohort study. BMC Infect Dis 13:380. https://doi.org/10.1186/1471-2334-13-380
49. Forrest A, Garonzik SM, Thamlikitkul V, Giamarellos-Bourboulis E, Paterson DL, Li J, Silveira FP, Nation RL (2017) Pharmacokinetic/toxicodynamic analysis of colistin-associated acute kidney injury in critically ill patients. Antimicrob Agents Chemother 61:e01367–e01317
50. Pogue JM, Lee J, Marchaim D, Yee V, Zhao JJ, Chopra T, Lephart P, Kaye KS (2011) Incidence of and risk factors for colistin-associated nephrotoxicity in a large academic health system. Clin Infect Dis 53(9):879–884. https://doi.org/10.1093/cid/cir611
51. Sandri AM, Landersdorfer CB, Jacob J, Boniatti MM, Dalarosa MG, Falci DR, Behle TF, Bordinhao RC, Wang J, Forrest A, Nation RL, Li J, Zavascki AP (2013) Population pharmacokinetics of intravenous polymyxin B in critically ill patients: implications

for selection of dosage regimens. Clin Infect Dis 57(4):524–531. https://doi.org/10.1093/cid/cit334
52. Phe K, Lee Y, McDaneld PM, Prasad N, Yin T, Figueroa DA, Musick WL, Cottreau JM, Hu M, Tam VH (2014) In vitro assessment and multicenter cohort study of comparative nephrotoxicity rates associated with colistimethate versus polymyxin B therapy. Antimicrob Agents Chemother 58(5):2740–2746. https://doi.org/10.1128/AAC.02476-13
53. Akajagbor DS, Wilson SL, Shere-Wolfe KD, Dakum P, Charurat ME, Gilliam BL (2013) Higher incidence of acute kidney injury with intravenous colistimethate sodium compared with polymyxin B in critically ill patients at a tertiary care medical center. Clin Infect Dis 57(9):1300–1303. https://doi.org/10.1093/cid/cit453
54. Hartzell JD, Neff R, Ake J, Howard R, Olson S, Paolino K, Vishnepolsky M, Weintrob A, Wortmann G (2009) Nephrotoxicity associated with intravenous colistin (colistimethate sodium) treatment at a tertiary care medical center. Clin Infect Dis 48(12):1724–1728. https://doi.org/10.1086/599225
55. Deryke CA, Crawford AJ, Uddin N, Wallace MR (2010) Colistin dosing and nephrotoxicity in a large community teaching hospital. Antimicrob Agents Chemother 54(10):4503–4505. https://doi.org/10.1128/AAC.01707-09
56. Horcajada JP, Sorli L, Luque S, Benito N, Segura C, Campillo N, Montero M, Esteve E, Mirelis B, Pomar V, Cuquet J, Marti C, Garro P, Grau S (2016) Validation of a colistin plasma concentration breakpoint as a predictor of nephrotoxicity in patients treated with colistin methanesulfonate. Int J Antimicrob Agents 48(6):725–727. https://doi.org/10.1016/j.ijantimicag.2016.08.020
57. Karvanen M, Plachouras D, Friberg LE, Paramythiotou E, Papadomichelakis E, Karaiskos I, Tsangaris I, Armaganidis A, Cars O, Giamarellou H (2013) Colistin methanesulfonate and colistin pharmacokinetics in critically ill patients receiving continuous venovenous hemodiafiltration. Antimicrob Agents Chemother 57(1):668–671. https://doi.org/10.1128/AAC.00985-12
58. Luque S, Sorli L, Li J, Collado S, Barbosa F, Berenguer N, Horcajada JP, Grau S (2014) Effective removal of colistin methanesulphonate and formed colistin during intermittent haemodialysis in a patient infected by polymyxin-only-susceptible Pseudomonas aeruginosa. J Chemother 26(2):122–124. https://doi.org/10.1179/1973947813Y.0000000104
59. Markou N, Fousteri M, Markantonis SL, Zidianakis B, Hroni D, Boutzouka E, Baltopoulos G (2012) Colistin pharmacokinetics in intensive care unit patients on continuous venovenous haemodiafiltration: an observational study. J Antimicrob Chemother 67(10):2459–2462. https://doi.org/10.1093/jac/dks257
60. Marchand S, Frat JP, Petitpas F, Lemaitre F, Gobin P, Robert R, Mimoz O, Couet W (2010) Removal of colistin during intermittent haemodialysis in two critically ill patients. J Antimicrob Chemother 65(8):1836–1837. https://doi.org/10.1093/jac/dkq185
61. Strunk AK, Schmidt JJ, Baroke E, Bode-Boger SM, Martens-Lobenhoffer J, Welte T, Kielstein JT (2014) Single- and multiple-dose pharmacokinetics and total removal of colistin in a patient with acute kidney injury undergoing extended daily dialysis. J Antimicrob Chemother 69:2008–2010. https://doi.org/10.1093/jac/dku075
62. Leporati M, Bua RO, Mariano F, Carignano P, Stella M, Biancone L, Vincenti M (2014) Determination by LC-MS/MS of colistins A and B in plasma and ultrafiltrate from critically ill patients undergoing continuous venovenous hemodiafiltration. Ther Drug Monit 36(2):182–191. https://doi.org/10.1097/FTD.0b013e3182a8997c
63. Jacobs M, Gregoire N, Megarbane B, Gobin P, Balayn D, Marchand S, Mimoz O, Couet W (2016) Population pharmacokinetics of colistin methanesulphonate (CMS) and colistin in critically ill patients with acute renal failure requiring intermittent haemodialysis. Antimicrob Agents Chemother 60:1788–1793. https://doi.org/10.1128/aac.01868-15
64. Karaiskos I, Friberg LE, Galani L, Ioannidis K, Katsouda E, Athanassa Z, Paskalis H, Giamarellou H (2016) Challenge for higher colistin dosage in critically ill patients receiving continuous venovenous haemodiafiltration. Int J Antimicrob Agents 48(3):337–341. https://doi.org/10.1016/j.ijantimicag.2016.06.008
65. Mariano F, Leporati M, Carignano P, Stella M, Vincenti M, Biancone L (2015) Efficient removal of colistin A and B in critically ill patients undergoing CVVHDF and sorbent technologies. J Nephrol 28(5):623–631. https://doi.org/10.1007/s40620-014-0143-3
66. Jitmuang A, Nation RL, Koomanachai P, Chen G, Lee HJ, Wasuwattakul S, Sritippayawan S, Li J, Thamlikitkul V, Landersdorfer CB (2015) Extracorporeal clearance of colistin methanesulphonate and formed colistin in end-stage renal disease patients receiving intermittent haemodialysis: implications for dosing. J Antimicrob Chemother 70(6):1804–1811. https://doi.org/10.1093/jac/dkv031
67. Menna P, Salvatorelli E, Mattei A, Cappiello D, Minotti G, Carassiti M (2018) Modified colistin regimen for critically ill patients with acute renal impairment and continuous renal replacement therapy. Chemotherapy 63(1):35–38. https://doi.org/10.1159/000484974
68. Koomanachai P, Landersdorfer CB, Chen G, Lee HJ, Jitmuang A, Wasuwattakul S, Sritippayawan S, Li J, Nation RL, Thamlikitkul V (2014) Pharmacokinetics of colistin methanesulfonate and formed colistin in end-stage renal disease patients receiving continuous ambulatory peritoneal dialysis. Antimicrob Agents

Chemother 58(1):440–446. https://doi.org/10.1128/AAC.01741-13

69. Mohamed AF, Cars O, Friberg LE (2014) A pharmacokinetic/pharmacodynamic model developed for the effect of colistin on Pseudomonas aeruginosa in vitro with evaluation of population pharmacokinetic variability on simulated bacterial killing. J Antimicrob Chemother 69(5):1350–1361. https://doi.org/10.1093/jac/dkt520
70. Food and Drug Administration (2013) Approved drug products. Label and approval history for Coly-Mycin M, NDA 050108. http://www.accessdata.fda.gov/drugsatfda_docs/label/2013/050108s030lbl.pdf. Last accessed 16 Jun 2019
71. Antachopoulos C, Karvanen M, Iosifidis E, Jansson B, Plachouras D, Cars O, Roilides E (2010) Serum and cerebrospinal fluid levels of colistin in pediatric patients. Antimicrob Agents Chemother 54(9):3985–3987. https://doi.org/10.1128/AAC.01799-09
72. Nakwan N, Usaha S, Chokephaibulkit K, Villani P, Regazzi M, Imberti R (2016) Pharmacokinetics of colistin following a single dose of intravenous colistimethate sodium in critically ill neonates. Pediatr Infect Dis J 35(11):1211–1214. https://doi.org/10.1097/inf.0000000000001263
73. Magreault S, Petyt C, Senneville E, Fron D, Nectoux E, Loiez C, Marchand S, Gregoire N, Couet W (2018) Pharmacokinetics of colistin in a 8-year-old child with acute bone infection. Clin Microbiol Infect 24(9):1025–1026. https://doi.org/10.1016/j.cmi.2018.04.021
74. Mesini A, Loy A, Gattorno M, Moscatelli A, Bandettini R, Faraci M, Cangemi G, Castagnola E (2018) Colistin area under the time-concentration in children treated with intravenous loading dose and maintenance therapy. Clin Infect Dis 66(5):808–809. https://doi.org/10.1093/cid/cix757
75. Magreault S, Gregoire N, Marchand S, Couet W (2018) Colistin pharmacokinetics in pediatrics. Clin Infect Dis 66(5):809. https://doi.org/10.1093/cid/cix1046
76. Nation RL (2018) Dose suggestions for intravenous colistin in pediatric patients: caution required. Clin Infect Dis 66(5):810–811. https://doi.org/10.1093/cid/cix1048
77. Markantonis SL, Markou N, Fousteri M, Sakellaridis N, Karatzas S, Alamanos I, Dimopoulou E, Baltopoulos G (2009) Penetration of colistin into cerebrospinal fluid. Antimicrob Agents Chemother 53(11):4907–4910. https://doi.org/10.1128/AAC.00345-09
78. Ziaka M, Markantonis SL, Fousteri M, Zygoulis P, Panidis D, Karvouniaris M, Makris D, Zakynthinos E (2013) Combined intravenous and intraventricular administration of colistin methanesulfonate in critically ill patients with central nervous system infection. Antimicrob Agents Chemother 57(4):1938–1940. https://doi.org/10.1128/AAC.01461-12
79. Boisson M, Jacobs M, Gregoire N, Gobin P, Marchand S, Couet W, Mimoz O (2014) Comparison of intra-pulmonary and systemic pharmacokinetics of colistinmethansulphonate (CMS) and colistin after aerosol delivery and intravenous administration of CMS in critically ill patients. Antimicrob Agents Chemother 58:7331–7339. https://doi.org/10.1128/aac.03510-14
80. Rottboll LA, Friis C (2016) Penetration of antimicrobials to pulmonary epithelial lining fluid and muscle and impact of drug physicochemical properties determined by microdialysis. J Pharmacol Toxicol Methods 78:58–65. https://doi.org/10.1016/j.vascn.2015.11.007
81. Weinbren MJ (1999) Pharmacokinetics of antibiotics in burn patients. J Antimicrob Chemother 44(3):319–327
82. Lee J, Han S, Jeon S, Hong T, Song W, Woo H, Yim DS (2013) Population pharmacokinetic analysis of colistin in burn patients. Antimicrob Agents Chemother 57(5):2141–2146. https://doi.org/10.1128/AAC.00271-13
83. Corcione S, Baietto L, Malvasio V, Stella M, Di Perri G, D'Avolio A, De Rosa FG (2017) Pharmacokinetics of colistin methanesulfonate (CMS) in burn patients. J Antimicrob Chemother 72(1):319–321. https://doi.org/10.1093/jac/dkw361
84. Bode-Boger SM, Schopp B, Troger U, Martens-Lobenhoffer J, Kalousis K, Mailander P (2013) Intravenous colistin in a patient with serious burns and borderline syndrome: the benefits of therapeutic drug monitoring. Int J Antimicrob Agents 42(4):357–360. https://doi.org/10.1016/j.ijantimicag.2013.06.009
85. Akers KS, Rowan MP, Niece KL, Stewart IJ, Mende K, Cota JM, Murray CK, Chung KK (2015) Colistin pharmacokinetics in burn patients during continuous venovenous hemofiltration. Antimicrob Agents Chemother 59(1):46–52. https://doi.org/10.1128/aac.03783-14
86. Athanassa ZE, Markantonis SL, Fousteri MZ, Myrianthefs PM, Boutzouka EG, Tsakris A, Baltopoulos GJ (2012) Pharmacokinetics of inhaled colistimethate sodium (CMS) in mechanically ventilated critically ill patients. Intensive Care Med 38(11):1779–1786. https://doi.org/10.1007/s00134-012-2628-7
87. Nakwan N, Lertpichaluk P, Chokephaibulkit K, Villani P, Regazzi M, Imberti R (2015) Pulmonary and systemic pharmacokinetics of colistin following a single dose of nebulized colistimethate in mechanically ventilated neonates. Pediatr Infect Dis J 34(9):961–963. https://doi.org/10.1097/inf.0000000000000775
88. Boisson M, Gregoire N, Cormier M, Gobin P, Marchand S, Couet W, Mimoz O (2017) Pharmacokinetics of nebulized colistin methanesulfonate in critically ill patients. J Antimicrob Chemother 72(9):2607–2612. https://doi.org/10.1093/jac/dkx167

89. Landersdorfer CB, Nguyen TH, Lieu LT, Nguyen G, Bischof RJ, Meeusen EN, Li J, Nation RL, McIntosh MP (2017) Substantial targeting advantage achieved by pulmonary administration of colistin methanesulfonate in a large-animal model. Antimicrob Agents Chemother 61(1). https://doi.org/10.1128/aac.01934-16
90. Gu WJ, Wang F, Tang L, Bakker J, Liu JC (2014) Colistin for the treatment of ventilator-associated pneumonia caused by multidrug-resistant Gram-negative bacteria: a systematic review and meta-analysis. Int J Antimicrob Agents 44:477–485. https://doi.org/10.1016/j.ijantimicag.2014.07.004
91. Valachis A, Samonis G, Kofteridis DP (2015) The role of aerosolized colistin in the treatment of ventilator-associated pneumonia: a systematic review and metaanalysis. Crit Care Med 43(3):527–533. https://doi.org/10.1097/ccm.0000000000000771
92. Liu D, Zhang J, Liu HX, Zhu YG, Qu JM (2015) Intravenous combined with aerosolised polymyxin versus intravenous polymyxin alone in the treatment of pneumonia caused by multidrug-resistant pathogens: a systematic review and meta-analysis. Int J Antimicrob Agents 46(6):603–609. https://doi.org/10.1016/j.ijantimicag.2015.09.011
93. Abdellatif S, Trifi A, Daly F, Mahjoub K, Nasri R, Ben Lakhal S (2016) Efficacy and toxicity of aerosolised colistin in ventilator-associated pneumonia: a prospective, randomised trial. Ann Intensive Care 6(1):26. https://doi.org/10.1186/s13613-016-0127-7
94. Jang JY, Kwon HY, Choi EH, Lee WY, Shim H, Bae KS (2017) Efficacy and toxicity of high-dose nebulized colistin for critically ill surgical patients with ventilator-associated pneumonia caused by multidrug-resistant Acinetobacter baumannii. J Crit Care 40:251–256. https://doi.org/10.1016/j.jcrc.2017.04.004
95. Jung SY, Lee SH, Lee SY, Yang S, Noh H, Chung EK, Lee JI (2017) Antimicrobials for the treatment of drug-resistant Acinetobacter baumannii pneumonia in critically ill patients: a systemic review and Bayesian network meta-analysis. Crit Care 21(1):319. https://doi.org/10.1186/s13054-017-1916-6
96. Vardakas KZ, Voulgaris GL, Samonis G, Falagas ME (2018) Inhaled colistin monotherapy for respiratory tract infections in adults without cystic fibrosis: a systematic review and meta-analysis. Int J Antimicrob Agents 51(1):1–9. https://doi.org/10.1016/j.ijantimicag.2017.05.016
97. Demirdal T, Sari US, Nemli SA (2016) Is inhaled colistin beneficial in ventilator associated pneumonia or nosocomial pneumonia caused by Acinetobacter baumannii? Ann Clin Microbiol Antimicrob 15(1):11. https://doi.org/10.1186/s12941-016-0123-7
98. Bargiacchi O, Rossati A, Car P, Brustia D, Brondolo R, Rosa F, Garavelli PL, De Rosa FG (2014) Intrathecal/intraventricular colistin in external ventricular device-related infections by multi-drug resistant Gram negative bacteria: case reports and review. Infection 42:801–809. https://doi.org/10.1007/s15010-014-0618-0
99. Imberti R, Cusato M, Accetta G, Marino V, Procaccio F, Del Gaudio A, Iotti GA, Regazzi M (2012) Pharmacokinetics of colistin in cerebrospinal fluid after intraventricular administration of colistin methanesulfonate. Antimicrob Agents Chemother 56(8):4416–4421. https://doi.org/10.1128/AAC.00231-12
100. Karaiskos I, Galani L, Baziaka F, Giamarellou H (2013) Intraventricular and intrathecal colistin as the last therapeutic resort for the treatment of multidrug-resistant and extensively drug-resistant Acinetobacter baumannii ventriculitis and meningitis: a literature review. Int J Antimicrob Agents 41(6):499–508. https://doi.org/10.1016/j.ijantimicag.2013.02.006
101. Karaiskos I, Galani L, Baziaka F, Katsouda E, Ioannidis I, Andreou A, Paskalis H, Giamarellou H (2013) Successful treatment of extensively drug-resistant Acinetobacter baumannii ventriculitis and meningitis with intraventricular colistin after application of a loading dose: a case series. Int J Antimicrob Agents 41(5):480–483. https://doi.org/10.1016/j.ijantimicag.2013.02.010
102. De Bonis P, Lofrese G, Scoppettuolo G, Spanu T, Cultrera R, Labonia M, Cavallo MA, Mangiola A, Anile C, Pompucci A (2015) Intraventricular versus intravenous colistin for the treatment of extensively drug resistant Acinetobacter baumannii meningitis. Eur J Neurol 23:68–75. https://doi.org/10.1111/ene.12789
103. Fotakopoulos G, Makris D, Chatzi M, Tsimitrea E, Zakynthinos E, Fountas K (2016) Outcomes in meningitis/ventriculitis treated with intravenous or intraventricular plus intravenous colistin. Acta Neurochir 158(3):603–610. https://doi.org/10.1007/s00701-016-2702-y
104. Inamasu J, Ishikawa K, Oheda M, Nakae S, Hirose Y, Yoshida S (2016) Intrathecal administration of colistin for meningitis due to New Delhi metallo-beta-lactamase 1(NDM-1)-producing Klebsiella pneumoniae. J Infect Chemother 22(3):184–186. https://doi.org/10.1016/j.jiac.2015.10.007
105. Landersdorfer CB, Wang J, Wirth V, Chen K, Kaye KS, Tsuji BT, Li J, Nation RL (2018) Pharmacokinetics/pharmacodynamics of systemically administered polymyxin B against Klebsiella pneumoniae in mouse thigh and lung infection models. J Antimicrob Chemother 73(2):462–468. https://doi.org/10.1093/jac/dkx409
106. Lakota EA, Landersdorfer CB, Nation RL, Li J, Kaye KS, Rao GG, Forrest A (2018) Personalizing polymyxin B dosing using an adaptive feedback control algorithm. Antimicrob Agents Chemother 62(7). https://doi.org/10.1128/aac.00483-18
107. Kwa AL, Lim TP, Low JG, Hou J, Kurup A, Prince RA, Tam VH (2008) Pharmacokinetics of polymyxin B1 in patients with multidrug-resistant Gram-negative bacterial infections. Diagn Microbiol

Infect Dis 60(2):163–167. https://doi.org/10.1016/j.diagmicrobio.2007.08.008
108. Kwa AL, Abdelraouf K, Low JG, Tam VH (2011) Pharmacokinetics of polymyxin B in a patient with renal insufficiency: a case report. Clin Infect Dis 52(10):1280–1281. https://doi.org/10.1093/cid/cir137
109. Sandri AM, Landersdorfer CB, Jacob J, Boniatti MM, Dalarosa MG, Falci DR, Behle TF, Saitovitch D, Wang J, Forrest A, Nation RL, Zavascki AP, Li J (2013) Pharmacokinetics of polymyxin B in patients on continuous venovenous haemodialysis. J Antimicrob Chemother 68(3):674–677. https://doi.org/10.1093/jac/dks437
110. Thamlikitkul V, Dubrovskaya Y, Manchandani P, Ngamprasertchai T, Boonyasiri A, Babic JT, Tam VH (2017) Dosing and pharmacokinetics of polymyxin B in patients with renal insufficiency. Antimicrob Agents Chemother 61(1). https://doi.org/10.1128/aac.01337-16
111. Manchandani P, Thamlikitkul V, Dubrovskaya Y, Babic JT, Lye DC, Lee LS, Tam VH (2018) Population pharmacokinetics of polymyxin B. Clin Pharmacol Ther 104(3):534–538. https://doi.org/10.1002/cpt.981
112. Kubin CJ, Nelson BC, Miglis C, Scheetz MH, Rhodes NJ, Avedissian SN, Cremers S, Yin MT (2018) Population pharmacokinetics of intravenous polymyxin B from clinical samples. Antimicrob Agents Chemother 62(3):e01493–e01417. https://doi.org/10.1128/aac.01493-17. [pii]
113. Miglis C, Rhodes NJ, Avedissian SN, Kubin CJ, Yin MT, Nelson BC, Pai MP, Scheetz MH (2018) Population pharmacokinetics of polymyxin B in acutely ill adult patients. Antimicrob Agents Chemother 62(3):e01475–e01417. https://doi.org/10.1128/aac.01475-17. [pii]
114. Avedissian S, Miglis C, Kubin CJ, Rhodes NJ, Yin MT, Cremers S, Prickett M, Scheetz MH (2018) Polymyxin B pharmacokinetics in adult cystic fibrosis patients. Pharmacotherapy 38:730–738. https://doi.org/10.1002/phar.2129
115. Nelson BC, Eiras DP, Gomez-Simmonds A, Loo AS, Satlin MJ, Jenkins SG, Whittier S, Calfee DP, Furuya EY, Kubin CJ (2015) Clinical outcomes associated with polymyxin B dose in patients with bloodstream infections due to carbapenem-resistant Gram-negative rods. Antimicrob Agents Chemother 59(11):7000–7006. https://doi.org/10.1128/aac.00844-15
116. Roberts KD, Azad MA, Wang J, Horne AS, Thompson PE, Nation RL, Velkov T, Li J (2015) Antimicrobial activity and toxicity of the major lipopeptide components of polymyxin B and colistin: last-line antibiotics against multidrug-resistant Gram-negative bacteria. ACS Infect Dis 1(11):568–575. https://doi.org/10.1021/acsinfecdis.5b00085
117. Kumar A, Roberts D, Wood KE, Light B, Parrillo JE, Sharma S, Suppes R, Feinstein D, Zanotti S, Taiberg L, Gurka D, Kumar A, Cheang M (2006) Duration of hypotension before initiation of effective antimicrobial therapy is the critical determinant of survival in human septic shock. Crit Care Med 34(6):1589–1596. https://doi.org/10.1097/01.ccm.0000217961.75225.e9
118. Luna CM, Aruj P, Niederman MS, Garzon J, Violi D, Prignoni A, Rios F, Baquero S, Gando S (2006) Appropriateness and delay to initiate therapy in ventilator-associated pneumonia. Eur Respir J 27(1):158–164. https://doi.org/10.1183/09031936.06.00049105
119. Spapen HD, Honore PM, Gregoire N, Gobin P, de Regt J, Martens GA, Pierard D, Couet W (2011) Convulsions and apnoea in a patient infected with New Delhi metallo-beta-lactamase-1 Escherichia coli treated with colistin. J Infect 63(6):468–470. https://doi.org/10.1016/j.jinf.2011.07.008
120. Kim EJ, Oh J, Lee K, Yu KS, Chung JY, Hwang JH, Nam EY, Kim HS, Kim M, Park JS, Song KH, Kim ES, Song J, Kim HB (2018) A short communication: pharmacokinetic characteristics and limited sampling strategies for therapeutic drug monitoring of colistin in patients with multidrug-resistant Gram-negative bacterial infections. Ther Drug Monit 41:102–106. https://doi.org/10.1097/ftd.0000000000000572
121. Jelliffe RW, Schumitzky A, Van Guilder M, Liu M, Hu L, Maire P, Gomis P, Barbaut X, Tahani B (1993) Individualizing drug dosage regimens: roles of population pharmacokinetic and dynamic models, Bayesian fitting, and adaptive control. Ther Drug Monit 15(5):380–393
122. Luque S, Escano C, Sorli L, Li J, Campillo N, Horcajada JP, Salas E, Grau S (2017) Urinary concentrations of colistimethate and formed colistin after intravenous administration in patients with multidrug-resistant Gram-negative bacterial infections. Antimicrob Agents Chemother 61(8):e02595-16. https://doi.org/10.1128/aac.02595-16. [pii]
123. Oliveira MS, Prado GV, Costa SF, Grinbaum RS, Levin AS (2009) Polymyxin B and colistimethate are comparable as to efficacy and renal toxicity. Diagn Microbiol Infect Dis 65(4):431–434. https://doi.org/10.1016/j.diagmicrobio.2009.07.018
124. Tuon FF, Rigatto MH, Lopes CK, Kamei LK, Rocha JL, Zavascki AP (2014) Risk factors for acute kidney injury in patients treated with polymyxin B or colistin methanesulfonate sodium. Int J Antimicrob Agents 43(4):349–352. https://doi.org/10.1016/j.ijantimicag.2013.12.002
125. Rigatto MH, Oliveira MS, Perdigao-Neto LV, Levin AS, Carrilho CM, Tanita MT, Tuon FF, Cardoso DE, Lopes NT, Falci DR, Zavascki AP (2016) Multicenter prospective cohort study of renal failure in patients treated with colistin versus polymyxin B. Antimicrob Agents Chemother 60(4):2443–2449. https://doi.org/10.1128/aac.02634-15
126. Vardakas KZ, Falagas ME (2017) Colistin versus polymyxin B for the treatment of patients with multidrug-resistant Gram-negative infec-

tions: a systematic review and meta-analysis. Int J Antimicrob Agents 49(2):233–238. https://doi.org/10.1016/j.ijantimicag.2016.07.023
127. Zavascki AP, Nation RL (2017) Nephrotoxicity of polymyxins: is there any difference between colistimethate and polymyxin B? Antimicrob Agents Chemother 61(3):e02319–e02316. https://doi.org/10.1128/aac.02319-16. [pii]
128. Kenyon R (2015) Structural evidence for a new understanding of the nature of CMS. Paper presented at the 2nd international conference on polymyxins, 22–24 September, San Diego, USA
129. Tsuji BT, Pogue JM, Zavascki AP, Paul M, Daikos GL, Forrest A, Giacobbe DR, Viscoli C, Giamarellou H, Karaiskos I, Kaye D, Mouton JW, Tam VH, Thamlikitkul V, Wunderink RG, Li J, Nation RL, Kaye KS (2019) International consensus guidelines for the optimal use of the polymyxins: endorsed by the American College of Clinical Pharmacy (ACCP), European Society of Clinical Microbiology and Infectious Diseases (ESCMID), Infectious Diseases Society of America (IDSA), International Society for Anti-infective Pharmacology (ISAP), Society of Critical Care Medicine (SCCM), and Society of Infectious Diseases Pharmacists (SIDP). Pharmacotherapy 39(1):10–39

Rational Combinations of Polymyxins with Other Antibiotics

16

Phillip J. Bergen, Nicholas M. Smith, Tyler B. Bedard, Zackery P. Bulman, Raymond Cha, and Brian T. Tsuji

Abstract

Combinations of antimicrobial agents are often used in the management of infectious diseases. Antimicrobial agents used as part of combination therapy are often selected empirically. As regrowth and the emergence of polymyxin (either colistin or polymyxin B) resistance has been observed with polymyxin monotherapy, polymyxin combination therapy has been suggested as a possible means by which to increase antimicrobial activity and reduce the development of resistance. This chapter provides an overview of preclinical and clinical investigations of CMS/colistin and polymyxin B combination therapy. *In vitro* data and animal model data suggests a potential clinical benefit with many drug combinations containing clinically achievable concentrations of polymyxins, even when resistance to one or more of the drugs in combination is present and including antibiotics normally inactive against Gram-negative organisms. The growing body of data on the emergence of polymyxin resistance with monotherapy lends theoretical support to a benefit with combination therapy. Benefits include enhanced bacterial killing and a suppression of polymyxin resistant subpopulations. However, the complexity of the critically ill patient population, and high rates of treatment failure and death irrespective of infection-related outcome make demonstrating a potential benefit for polymyxin combinations extremely challenging. Polymyxin combination therapy in the clinic remains a heavily debated and controversial topic. When combinations are selected, optimizing the dosage regimens for the polymyxin and the combinatorial agent is critical to ensure that the benefits outweigh the risk of the development of toxicity. Importantly, patient characteristics, pharmacokinetics, the site of infection, pathogen and resistance mechanism must be taken into account to define optimal and rational polymyxin combination regimens in the clinic.

Keywords

Polymyxin · Antibiotic combination therapy · Synergy · Pharmacodynamics · Pharmacokinetics

P. J. Bergen (✉)
Centre for Medicine Use and Safety, Monash University, Parkville Campus, Melbourne, VIC, Australia
e-mail: Phillip.Bergen@monash.edu

N. M. Smith · T. B. Bedard · R. Cha · B. T. Tsuji
Laboratory for Antimicrobial Pharmacodynamics, School of Pharmacy and Pharmaceutical Sciences, University at Buffalo, SUNY, Buffalo, NY, USA

Z. P. Bulman
University of Illinois Chicago, College of Pharmacy, Chicago, IL, USA

J. Li et al. (eds.), *Polymyxin Antibiotics: From Laboratory Bench to Bedside*, Advances in Experimental Medicine and Biology 1145, https://doi.org/10.1007/978-3-030-16373-0_16

16.1 Introduction

Combinations of antimicrobial agents have been used in the management of infectious diseases since the 1940s [160]. Reasons for the use of antimicrobial combinations include prevention of resistance selection during treatment, decreased dose-related toxicity as a result of reduced dosage, broadening of spectrum in polymicrobial infections, and 'synergy' [183]. However, it remains controversial whether combination therapy, given empirically or as definitive treatment for many infection types, is warranted. There are also potential disadvantages with combination therapy including a greater risk of drug toxicity, increased cost, and superinfection with even more resistant bacteria [119]. Clinicians often resort to antibiotic combinations as a consequence of limited therapeutic options in the hope of improving the activity of available agents. In clinical practice, antimicrobial agents used as part of combination therapy are often selected empirically by clinicians, mainly by trial and error or based on personal experience. This approach is poorly guided and may not be optimal for patient care.

Polymyxin (colistin [administered as colistin methanesulphonate; CMS] or polymyxin B) combination therapy is increasingly used clinically [10, 11, 30, 51, 62, 78, 120, 139, 141, 142, 162]. However, systematic investigations of such combinations are a relatively recent phenomenon. As outlined in Chap. 15, the emerging pharmacodynamic (PD) and pharmacokinetic (PK) data on CMS/colistin and polymyxin B suggest that caution is required with monotherapy. Given this situation, polymyxin combination therapy has been suggested as a possible means by which to increase antimicrobial activity and reduce the development of resistance [63, 72, 99, 151].

The growing body of data on the emergence of polymyxin resistance with monotherapy lends theoretical support to a benefit with combination therapy. As discussed in Chap. 8, a consistent finding of both *in vitro* and *in vivo* studies is regrowth with colistin or polymyxin B monotherapy, even with concentrations far exceeding those which can be safely achieved clinically [12, 13, 16, 25–27, 39, 67, 88, 89, 93, 103, 104, 128, 147, 165, 173, 174, 201, 208]. Amplification of colistin-resistant subpopulations in heteroresistant isolates, i.e. isolates that are susceptible to polymyxins based upon their minimum inhibitory concentrations (MICs) but which contain resistant subpopulations, has been shown to contribute to the observed regrowth following polymyxin monotherapy [13, 14, 16, 24, 45, 50, 84, 89, 103, 123, 147, 173, 174, 188]. Studies undertaken in *in vitro* pharmacokinetic/pharmacodynamic (PK/PD) models simulating clinically achievable unbound plasma concentration-time profiles of colistin or polymyxin B in critically ill patients demonstrated early regrowth of heteroresistant strains of *Pseudomonas aeruginosa* [16, 103, 173], *Klebsiella pneumoniae* [45, 208] and *Acinetobacter baumannii* [84, 89], with population analysis profiles (PAPs) revealing substantial increases in the proportion of polymyxin-resistant subpopulations; PAPs after 72 h (colistin) or 96 h (polymyxin B) were substantially different from the PAPs prior to polymyxin therapy and those for the growth controls. Similar increases in the proportion of colistin-resistant bacteria with monotherapy have been observed in other *in vitro* studies (both static and dynamic time-kill infection models) [1, 13, 14, 24, 123, 128, 143, 147, 174] and, for *A. baumannii*, murine thigh and lung infection models [50]; many of these studies include polymyxin concentrations well above the MIC of the organism. These observations suggest that the susceptible bacterial populations were selectively eradicated, resulting in unopposed growth of resistant subpopulations (such as LPS-deficient *A. baumannii* [114]; discussed in detail in Chap. 5) and consequently the emergence of resistance over time. Heteroresistance notwithstanding, adaptive resistance (see Chap. 5) may also contribute to regrowth as evidenced by reversion to the susceptible state following serial passaging on drug-free plates of one of three isolates in the study by Tam et al. [173]. Finally, a recent study demonstrated that amino acid alterations in two-component systems such as PmrAB, PhoPQ and ParRS involved in polymyxin resistance (due to modifications of lipopolysaccharides in the Gram-negative cell wall) occur rapidly *in vitro* in the presence of colistin within the period of selection of single-step mutants [32]. This suggests polymyxin treatment may provoke genetic mutations related to resistance as a mutagen within a short period in addition to the selection of pre-existing resistant subpopulations. Such observations highlight the importance of polymyxin combinations to minimize the emergence of polymyxin resistance.

In addition to a reduction in the emergence of polymyxin resistance, combination therapy has the potential to increase bacterial killing via 'synergy'. Two mechanisms have been proposed whereby polymyxin combinations may provide an enhanced PD effect. As regrowth with polymyxin monotherapy is due, at least in part, to amplification of pre-existing polymyxin-resistant subpopulations in heteroresistant strains, it has been suggested that polymyxin combinations may give rise to so-called subpopulation synergy, the process whereby one drug kills the resistant subpopulation(s) of the other drug, and *vice versa* (Fig. 16.1) [23]. Additionally mechanistic synergy, whereby two drugs acting on different cellular pathways increase the rate or extent of killing of the other drug, has been suggested as a mechanism by which polymyxin combinations may lead to an enhanced antimicrobial effect (Fig. 16.1) [23]. The ability of colistin to increase the permeability of the outer membrane of many Gram-negative bacteria (Chap. 4) represents one possible mechanism for mechanistic synergy, potentially allowing better access of other antimicrobial agents to their target sites within the pathogen and thereby improving activity. Potential examples of each type of synergy are discussed subsequently in the PK/PD time-kill studies section. Mechanisms of subpopulation and mechanistic synergy are not mutually exclusive and both may operate simultaneously.

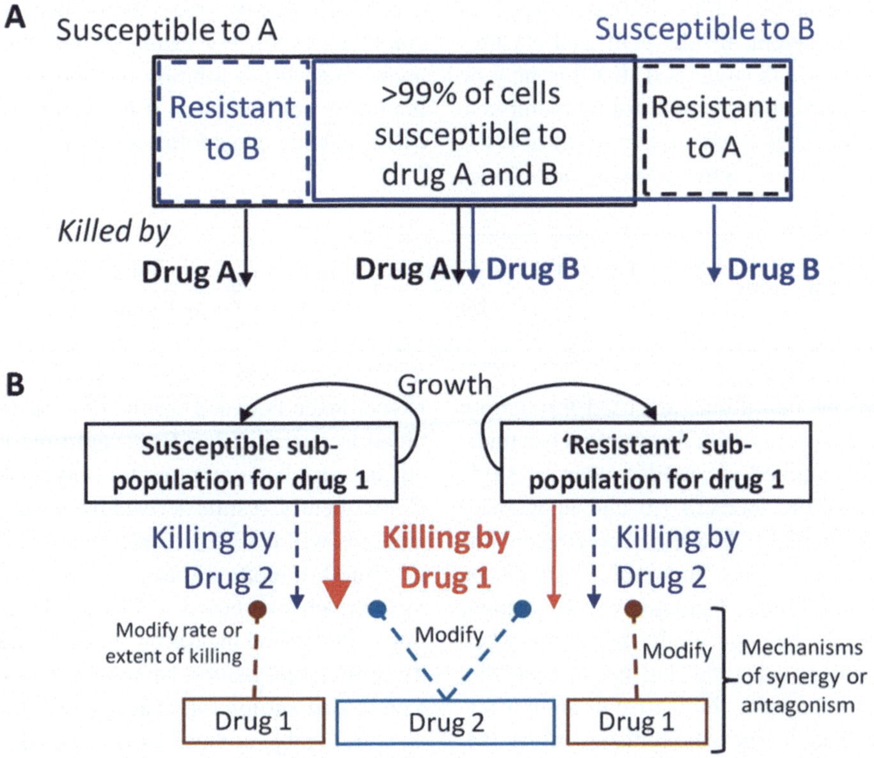

Fig. 16.1 Schematic representations for subpopulation synergy (Panel A) and mechanistic synergy (Panel B). In subpopulation synergy, drug A kills the resistant subpopulations of drug B, and *vice versa*. In mechanistic synergy for drugs acting on different cellular pathways, drug A increases the rate or extent of killing by drug B, and *vice versa*. (Figure adapted from Bulitta et al. [23], with permission)

An important observation of some recent studies which investigated colistin susceptibility has been the substantially increased susceptibility of colistin-resistant isolates of several Gram-negative species to many antibiotics, including some normally considered inactive against Gram-negative organisms (e.g., rifampicin, macrolides, glycopeptides and daptomycin) [25, 61, 66, 86, 92, 109, 187, 190]. For example, Li et al. [92] examined the antibiograms of paired colistin-susceptible and -resistant strains of multidrug-resistant (MDR) *A. baumannii* against a broad range of antibiotics. In that study, the MICs of most colistin-resistant strains were substantially lower against a number of antibiotic classes typically used against Gram-negative organisms than their colistin-susceptible counterparts (e.g. >16 times lower in some cases against the penicillin class and carbapenems). Additionally, the colistin-resistant strains had substantially increased susceptibility to many antibiotics that are typically inactive against Gram-negative bacteria (e.g., rifampicin, fusidic acid, and erythromycin). The authors suggested that this may be due to substantial changes in the outer membrane of *A. baumannii* which occur as a result of resistance to colistin, thereby allowing antibiotics such as rifampicin and the lipopeptides, macrolides and streptogramins greater access to their target sites. This unexpected finding further emphasises the need for rational, systematic examination of polymyxin combination therapy. This chapter will provide an overview of both preclinical and clinical investigations of CMS/colistin and polymyxin B combination therapy.

16.2 Preclinical Studies of CMS/Colistin or Polymyxin B Combination Therapy

16.2.1 *In Vitro* Studies

In vitro studies examining combination therapy most commonly define the pharmacodynamic (PD) interaction of the agents in terms of additivity, synergy, indifference or antagonism, with the method used to determine such interactions dependent upon the experimental system employed [144]. For example, with the checkerboard microbroth dilution method the fractional inhibitory concentration (FIC) index is used. The FIC is calculated as follows [144]:

$$\text{FIC index} = \frac{\text{MIC of drug A in combination}}{\text{MIC of drug A alone}} + \frac{\text{MIC of drug B in combination}}{\text{MIC of drug B alone}}$$

Though various definitions are used throughout the literature, synergy with this method has traditionally been defined as an FIC index of ≤0.5, additivity as an FIC index of 1.0, and antagonism as an FIC index of 2.0. However, more recent criteria suggest that an FIC index of >4 should be applied to definitions of antagonism to account for inherent imprecision of the technique when twofold dilutions are used and because an FIC index of 2.0 is probably indicative of an indifferent, rather than a true antagonistic, effect [6]. Though widely used, the checkerboard method is less discriminatory than other more sophisticated *in vitro* methods (e.g., static or PK/PD time-kill models; discussed below) for assessing the interactions of antimicrobial agents [28, 126, 194]. Discordance between results derived from combination testing using Etest and time-kill methods has also been reported for polymyxins [175]. Consequently, results derived from FIC and Etest methods will not be discussed here.

Time-kill methods have important advantages over the checkerboard technique. Primarily, the time-kill method measures the bactericidal activity of the combination being tested and provides a picture of antimicrobial action over time (based on serial viable counts); in contrast, the checkerboard technique provides only inhibitory data and is usually examined at a single time point (after 16–24 h of incubation) [144]. Time-kill models can be subdivided into static and PK/PD models. In static time-kill models, with the exception of a small degree of loss in drug activity due to bacterial metabolism or inactivation, bacteria are exposed to static (fixed) concentrations of an antibacterial agent over a defined period of time. PK/PD models essentially fall into one of two categories: one-compartment (1-CM) or two-compartment (2-CM) models [65, 186]. In these models, the test organism is presented with a dynamic concentration of drug designed to mimic *in vivo* PK. PK/PD models typically consist of a central reservoir containing the organism, a diluent reservoir and a waste reservoir. Drug is added to the central reservoir to achieve the desired peak concentration and the elimination profile is mimicked by addition of sterile, drug-free media to the central reservoir and removal of an equal volume of drug-containing media into the waste reservoir; various adaptations of this standard model are available to simultaneously mimic the *in vivo* PK of two or more drugs with differing half-lives [21]. Though 1-CM are most common, the 2-CM hollow-fibre infection model (HFIM) – which prevents bacterial elimination by physically separating bacteria from the central reservoir – is now considered gold standard for detailed examination of the effects of different regimens and PK on the time-course of bacterial killing and emergence of resistance [22].

For both static and PK/PD time-kill methods synergy has traditionally been defined as a 100-fold increase in killing at 24 h (as measured by colony counts; i.e. a $\geq$2-$\log_{10}$ lower CFU/mL) with the combination relative to its most active component (Fig. 16.2) [144]; antagonism is defined as a 100-fold decrease (i.e. a $\geq$2-$\log_{10}$ higher CFU/mL) in killing at 24 h with the combination compared with the most active single drug alone. While a strict application of these definitions requires that at least one of the drugs being tested produces no significant inhibition or killing alone, there are no established criteria with which to evaluate interactions when using two or more drugs, each of which has significant activity alone [144]. Consequently, these definitions are commonly applied in practice even when more than one drug displays significant bacterial killing. Variations on, and additions to, these definitions abound in the literature however, complicating comparisons of effect between studies. A typical example is that synergy is sometimes reported as described above, with the qualification that the number of surviving organisms in the presence of the combination must be $\geq$2-$\log_{10}$ CFU/mL below the starting inoculum [53, 61, 134, 149, 167]. In this way, an interaction described as synergistic by the former definition may not be synergistic by the latter. These defini-

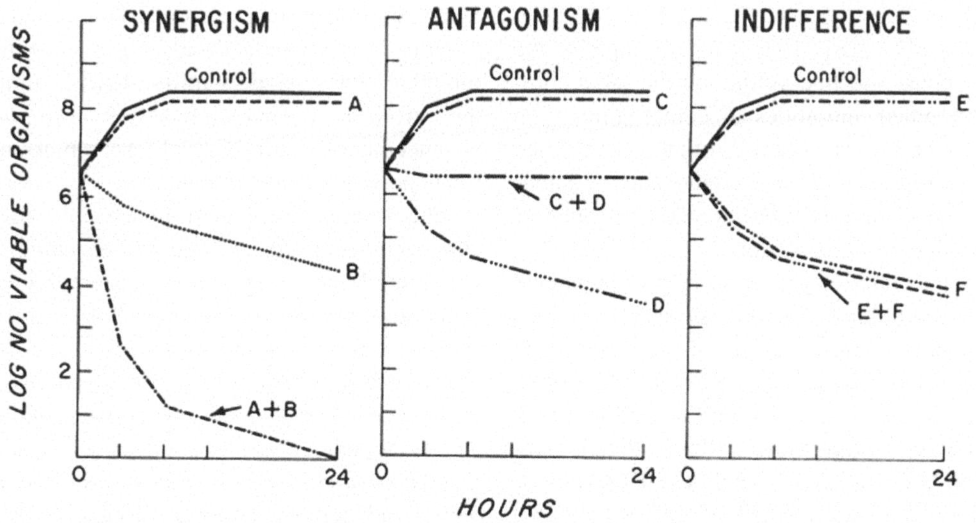

Fig. 16.2 Effects of antimicrobial combinations as measured with the time-kill method. A + B, synergism; C + D, antagonism; E + F, indifference. (Figure adapted from Pillai et al. [144], with permission)

tions are also commonly applied at times other than 24 h.

Numerous *in vitro* studies have used the static or PK/PD time-kill method to examine polymyxin combination therapy, with the majority of studies utilising CMS or colistin (sulphate). However, as discussed in Chap. 3, CMS is an inactive prodrug of colistin and undergoes conversion to colistin in aqueous media [15, 91]. Administration of CMS will therefore result in a variable formation of active colistin over time, making the administering CMS in these *in vitro* systems inappropriate. Unfortunately, as for animal studies discussed above, it is not always possible to ascertain whether the 'colistin' administered was colistin (sulphate) or CMS (sodium). Antimicrobial agents combined with polymyxins in time-kill models include both agents with and without usual activity against Gram-negative pathogens. Studies have included polymyxins combined with rifampicin [8, 9, 19, 60, 82, 84, 94, 124, 177, 179, 180], carbapenems [8, 13, 16, 34, 36, 39, 45, 60, 82, 83, 87, 89, 96, 100, 102, 103, 111, 127, 134, 135, 137, 149, 161, 167, 168, 176–178, 180, 184], tigecycline [4, 18, 19, 27, 37, 40, 44, 60, 68, 80, 116, 122, 137, 148, 177], ampicillin/sulbactam [89, 180], ceftazidime [67], ciprofloxacin [8, 67], aminoglycosides [8, 40, 131, 152, 171], glycopeptides [18, 60, 66, 140, 187, 190], fosfomycin [5, 40, 46, 80, 87, 166, 177, 188, 201, 208] and others [1, 34, 39, 61, 97, 127, 129, 143, 153, 155, 175, 177, 187, 193, 201]; rifampicin, the carbapenems and tigecycline are the most commonly studied antibiotics in combination with colistin. The most common organisms studied are *P. aeruginosa*, *A. baumannii* and *K. pneumoniae*, and these will be the primary focus of the remainder of this section.

Despite a relatively large number of published studies examining polymyxin (primarily colistin) combination therapy there are a number of deficiencies with much of the existing information in addition to the lack of certainty around the form of 'colistin' administered; these deficiencies apply to both static and dynamic (PK/PD) models. Firstly, the vast majority of studies employ a single, generally lower inoculum (~10^5–10^6 CFU/mL). However, as the antibacterial activity of both colistin and polymyxin B is subject to an inoculum effect [24, 173], and as high bacterial densities can be found in some infections [107, 169], it is important to examine the antibacterial activity of combination therapy at multiple inocula. Second, many studies present antibiotic concentrations as multiples of the MIC with little reference to, or discussion of, the clinical relevance of the actual concentrations used. Further to this, many authors judge the 'success' of a particular combination only by whether synergy was attained rather than examining the overall antimicrobial activity of the combination. However, a combination that attains synergy may still achieve poor overall antimicrobial activity and may even be less active overall than another combination considered antagonistic. Third, consideration of polymyin heteroresistance and the effect of combinations on the development of polymyxin resistance have only been examined in a small number of recent studies [1, 5, 13, 16, 27, 39, 45, 68, 84, 89, 100, 102, 103, 143, 201, 208]. As discussed above heteroresistance is known to contribute to regrowth observed following colistin or polymyxin B monotherapy, although its clinical significance is unclear. Given the status of the polymyxins as agents of last resort and reports of increasing polymyxin resistance, it is crucial to systematically examine the effect of combination therapy on the emergence of polymyxin resistance, including on heteroresistant strains, in order to design optimal dosage regimens. Finally, remarkably few studies utilise PK/PD models, the introduction of which has been an important advancement in antimicrobial research, to investigate polymyxins in combination.

The next two sections of this chapter will discuss significant recent static and dynamic (PK/PD) time-kills investigations with polymyxins (colistin or polymyxin B) and will focus primarily on studies involving *P. aeruginosa*, *A. baumannii* and *K. pneumoniae*. Although polymyxins have been reported to be synergistic against a variety of pathogenic fungi including a variety of *Candida*, *Aspergillus* and other species in combination with echinocandins, azoles and amphotericin B [2, 105, 117, 133, 158, 205, 206], these studies will not be discussed here.

16.2.2 Static Time-Kill Studies

Pseudomonas aeruginosa Few studies have examined polymyxin combinations against *P. aeruginosa* using either static or dynamic (the latter discussed below) time-kill models. Two studies by Pankuch et al. combined colistin with either meropenem [134] or doripenem [135] against clinical isolates of *P. aeruginosa*; the proportion of MDR strains was not stated. Sub-MIC concentrations of colistin (0.12–1 mg/L) and meropenem (0.06–8 mg/L) were synergistic against 13 (25.5%) of 51 isolates (all isolates colistin-susceptible; 6 [11.8%] isolates meropenem-resistant) at 24 h, whereas colistin (0.12–16 mg/L) and doripenem (0.03–128 mg/L) demonstrated synergy against 19 (76.0%) of 25 isolates (1 [4%] colistin-resistant isolate; 14 [56%] isolates doripenem-resistant). Urban et al. examined antibiotic combinations using polymyxin B, doripenem, and rifampicin against five MDR isolates of *P. aeruginosa* (one *K. pneumoniae* carbapenemase [KPC]-producing and four non-metallo-β-lactamase [MBL] or KPC-β-lactamase producing) [184]. All isolates were carbapenem-resistant and one polymyxin resistant, and antibiotics were used at a concentration of 0.25× MIC. As monotherapy, none of the tested antibiotics was bactericidal (defined as a ≥3-$\log_{10}$ CFU/mL decrease in 24 h). Triple therapy with the combination of polymyxin B, doripenem and rifampicin was most effective, with bactericidal activity achieved against all isolates at 24 h. Combinations utilising only two antibiotics were less effective, with polymyxin B plus doripenem or rifampicin bactericidal against only one isolate. Despite examining combination therapy 'synergy' was not directly examined in this investigation.

Bergen et al. systematically investigated bacterial killing and resistance emergence with colistin alone and in combination with imipenem against *P. aeruginosa* [13]. Conducted over 48 h this study included five clinical isolates and an ATCC reference strain representing a mixture of colistin and imipenem susceptible and resistant strains, colistin heteroresistant and non-heteroresistant strains, and MDR and non-MDR strains; one isolate contained IMP- and CTX-M-type β-lactamases. Importantly, of the static time-kill studies discussed in this chapter only this study examined the effect of combinations at multiple inocula (~10^6 and ~10^8 CFU/mL); it was also the first study to specifically incorporate colistin-heteroresistant strains and investigate the emergence of polymyxin resistance with polymyxin combination therapy. In combination experiments both antibiotics were studied at concentrations of 0.5×, 4× and 16× MIC for susceptible isolates and 1, 4 and 32 mg/L for colistin and 1, 8 and 32 mg/L for imipenem for resistant isolates; the majority of concentrations for colistin and all concentrations for imipenem can be considered clinically achievable. In total nine colistin/imipenem combinations were examined for each isolate at each inoculum. Regrowth of all isolates was observed with colistin monotherapy even with colistin concentrations well above those which can be safely achieved clinically. The addition of imipenem to colistin at both inocula generally resulted in substantial improvements in bacterial killing over equivalent monotherapy against MDR *P. aeruginosa* isolates resistant to either antibiotic. The improvements in activity against these isolates were observed across the 48-h duration and with all colistin concentrations at the low inoculum, and 4× and 16× MIC (or 4 and 32 mg/L) colistin at the high inoculum. Notably, the total reductions in $\log_{10}$ CFU/mL achieved with combinations containing lower colistin concentrations (0.5× and 4× MIC or 1 and 4 mg/L) were on many occasions similar in magnitude to the reductions achieved with combinations containing 16× MIC colistin, particularly at the 10^6 inoculum. Benefits in overall antibacterial activity for this combination were less pronounced against the three isolates susceptible to both antibiotics, although substantial improvements in initial kill (i.e., up to 6 h) were present. As for the emergence of colistin resistance, colistin monotherapy against the five colistin-susceptible isolates generally led to increases in colistin-resistant subpopulations at both the low and high inocula, with combination therapy generally resulting in a similar proportion of colistin-resistant subpopulations at 48 h as with equivalent monotherapy. While this result would appear to negate one of the major theoretical attractions of colistin combination therapy, namely a reduction in the emergence of colistin resistance, the same authors subsequently conducted a similar experiment with two of these isolates in a dynamic (PK/PD) model combining colistin with doripenem and achieved a very different result. The potential reason for this and the implications for antimicrobial combination testing are discussed in detail in the section examining PK/PD time-kill studies.

Two studies have examined colistin [46] or polymyxin B [188] combined with fosfomycin against *P. aeruginosa*. Di et al. examined this combination against 5 carbapenem-resistant but colistin-susceptible clinical isolates over 24 h [46]. Antibiotics were administered at concentrations of 0.5× and 1× MIC (range, 0.25–4 mg/L for colistin and 16–256 mg/L for fosfomycin). Neither antibiotic produced substantial bacterial killing as monotherapy with regrowth to ~10^8 CFU/mL. However, in combination at both 0.5× and 1× MIC, in all but one case no viable bacteria were detected after no later than 12 h; the only exception was against the isolate with the highest colistin MIC (4 mg/L) and only with the combination with both antibiotics at 0.5× MIC. With polymyxin B (0.5, 1 and 2 mg/L) and fosfomycin (30, 150 or 300 mg/L) combinations, Walsh et al. similarly observed enhanced bacterial killing over 24 h against 3 polymyxin B-susceptible heteroresistant isolates [189]. Though synergy was observed in only 39 (48.1%) of 81 cases (9 combinations across 3 isolates at 3 time points), this was much higher (28 [51.9%] of 54 cases) when only combinations containing polymyxin B at 1 or 2 mg/L are considered. Against 2 colistin-resistant isolates, bacterial killing was not substantially enhanced with the combination.

A. baumannii In the two studies by Pankuch et al. discussed above, colistin was also combined with either meropenem [134] or doripenem [135] against clinical isolates of *A. baumannii*; the proportion of MDR strains was not stated. Colistin (0.06–8 mg/L) and meropenem (0.03–64 mg/L) showed synergy against 49 (94.2%) of 52 isolates (13 [25%] isolates colistin-resistant; 15 [28.8%] isolates meropenem-resistant) at 24 h, whereas colistin (0.12–16 mg/L) and doripenem (0.06–32 mg/L) showed synergy against 25 (100%) of 25 isolates of *A. baumannii* (11 [44%] isolates colistin-resistant; 9 [36%] isolates doripenem-resistant). Shields et al. examined colistin plus doripenem against five XDR isolates (defined as resistant to all agents except polymyxins and tigecycline) of *A. baumannii* taken from patients who had received solid organ transplants [163]; all isolates were colistin-susceptible based on MICs. Against all five isolates doripenem monotherapy at sub-MIC concentrations resulted in virtually no antimicrobial activity, whereas colistin monotherapy (0.25× to 1× MIC) was bacteriostatic (inhibiting growth of the inocula without causing significant killing) (Fig. 16.3). However, the combination of colistin (0.125× to 0.25× MIC) plus doripenem (8 mg/L) resulted in undetectable bacterial levels at 8 h without evidence of regrowth by 24 h. Interestingly, based on this, *in vitro* data combinations of CMS (5 mg/kg/day of CBA in 2–4 divided doses) and doripenem (500 mg 8-hourly) were recommended for use in their institution for patients who have received solid organ transplants and were infected with XDR *A. baumannii*. At the time of publication four patients had been treated with this combination with a fifth patient receiving meropenem plus colistin; 4 (80%) of the 5 patients had a positive clinical response and survived.

In a follow-up study to that of Shields et al. discussed above [163], the same group compared the *in vitro* killing effects of colistin (2 mg/L), doripenem (8 mg/L) and sulbactam (4 mg/L) alone, and in combination, against isolates of XDR *A. baumannii* collected from patients with recurrent respiratory tract infections prior to (initial) and following (recurrent) treatment with intravenous CMS plus doripenem [127]; 4 (44%) of the 9 patients received additional CMS via inhalation. Patients had received the combination (doses were not stated) for a minimum of 13 days

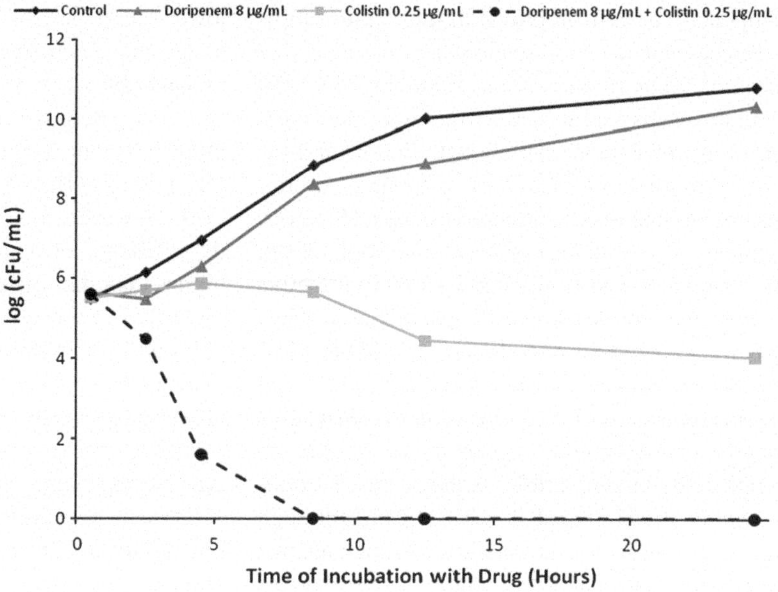

Fig. 16.3 Representative time-kill curves with colistin and doripenem alone, and in combination, against an extensively drug-resistant (XDR) isolate of *A. baumannii*. (*DOR* doripenem, *COL* colistin. Doripenem MIC alone = 64 µg/mL, Colistin MIC alone = 2 µg/mL) (Figure adapted from Shields et al. [163], with permission)

(median duration, 31 days; range, 13 to 74 days) with the median time between collection of initial and recurrent isolates being 65 days (range, 28–188 days). Nine initial and recurrent isolates (1 of each from each patient) were collected (18 isolates in total), with 8 (89%) of 9 pairs genetically indistinguishable. Time-kill studies revealed synergy at 24 h was more frequent when colistin was combined with doripenem (16 [89%] of 18 isolates) than sulbactam (9 [50%] of 18 isolates). The killing effects of the colistin/doripenem combination was attenuated against isolates previously exposed to the combination *in vivo* (mean log kill [CFU/mL] at 24 h of −5.08 $\log_{10}$ versus −2.88 $\log_{10}$ for initial and recurrent isolates, respectively), although there was no difference in the mean log kills against the initial and recurrent isolates exposed to colistin plus sulbactam. The triple combination of these agents achieved greater log kills than either colistin/doripenem or colistin/sulbactam combination among recurrent isolates (mean $\log_{10}$ kills [CFU/mL] at 24 h of −5.74 versus −2.88 and −1.51, respectively), including those that did not respond to the colistin/doripenem combination. Interestingly, although only one of nine initial isolates was colistin-resistant, five isolates were colistin-resistant following treatment. However, although colistin MICs influenced the extent of killing somewhat, colistin/doripenem combinations were equally active against colistin-susceptible and –resistant isolates. The MICs of doripenem rather than colistin were associated with the extent of killing by colistin and doripenem in combination, with each of the isolates that failed to respond to treatment having a doripenem MIC >64 mg/L. Such an association has also been demonstrated for colistin/doripenem combinations in KPC-producing *K. pneumoniae* [36] (discussed below).

Tan et al. examined colistin (at 1× MIC; range: 0.5–2 mg/L), minocycline (at 1× MIC for susceptible isolates [$n = 9$] and 4 mg/L for resistant isolates [$n = 4$]; range, 0.06–16 mg/L) and their combination against 13 imipenem-resistant isolates (MIC >8 mg/L) of *A. baumannii* across 24 h [175]. As monotherapy neither antibiotic demonstrated bactericidal activity at any time but the combination was bactericidal against 9 (69%) isolates at 24 h. Synergy was detected in 1 (8%), 2 (15%), 2 (15%) and 12 (92%) of isolates at 2, 4, 6, and 24 h, respectively. Tripodi et al. examined colistin (6 mg/L), rifampicin (5 mg/L), imipenem (20 mg/L) and ampicillin/sulbactam (50 mg/L) alone or in double (colistin plus each of the second drugs) or triple (colistin plus rifampicin plus imipenem, or colistin plus rifampicin plus ampicillin/sulbactam) combinations against nine isolates of MDR *A. baumannii* producing OXA-58 carbapenemase [180]. Colistin was the most active agent as monotherapy with double and triple combinations generally showing similar activity to that of colistin monotherapy. However, triple therapy with the combination of polymyxin B, doripenem and rifampicin was more effective against five non-MBL or KPC-producing isolates when compared to monotherapy or double combination therapy [184]. In another study, colistin at concentrations of 0.25×, 0.5× and 1× MIC plus daptomycin 10 mg/L was synergistic against ten MDR-colistin-susceptible isolates of *A. baumanni* in 16 (53.3%) of 30 cases at 24 h, although no benefit with the combination was seen against a further four MDR-colistin-resistant isolates [61]; however, it is not clear whether colistin (sulphate) or CMS was used in this investigation.

One laboratory examined colistin (1 mg/L) alone and in combination with the glycopeptide antibiotics vancomycin (20 mg/L) [66] or teicoplanin (20 mg/L) [190] against five MDR-colistin-susceptible isolates of *A. baumannii*. Colistin as monotherapy was rapidly bactericidal against all isolates with rapid regrowth to control values by 24 h. However, when combined with vancomycin regrowth was suppressed in four of the five isolates even at 48 h, with ~5–7-$\log_{10}$ CFU/mL greater killing at this time compared to colistin monotherapy. The colistin/teicoplanin combination suppressed regrowth against all isolates at 24 h, with >8-$\log_{10}$ CFU/mL greater killing compared with colistin monotherapy and a ≥ 4-fold log reduction compared with the starting inoculum at this time. Surprisingly, although experiments were conducted for 48 h only the 24 h results were reported. Despite the substantially improved bacterial killing with both glycopeptides the authors noted that, given the potential of both colistin and vancomycin to cause nephrotoxicity when either agent is used alone, there may be concern about the suitability of this combination in the clinic. Although teicoplanin has a similar mechanism of action to vancomycin, it has a more favourable effect profile including a lower incidence of renal toxicity which may make such a combination more acceptable to clinicians [29, 172]. A colistin/telavancin combination was synergistic at 24 h against a single MDR clinical isolate (representative of the epidemic UK lineage OXA-23 clone 1) of *A. baumanni* [73]. However, in contrast to teicoplanin above, the incidence of renal toxicity with telavancin is higher than that of vancomycin which may limit the utility of this combination [185]. Similarly, a polymyxin B/rifampicin combination was synergistic at 24 h against two MDR isolates of *A. baumanni* positive for OXA-23 and OXA-51 and an *Acinetobacter* sp. positive for OXA-58 and IMP-type carbapenemases [95].

Phee et al. examined colistin (1-2 mg/L) combined with fusidic acid (1 mg/L; 16 mg/L for the colistin-resistant isolate) against six isolates of *A. baumannii* across 24 h [143]. All but a single colistin-resistant isolate were colistin-heteroresistant, and all but the reference strain were either MDR, XDR or pandrug-resistant [PDR] according to the classification of Magiorakos et al. [106]. The majority of isolates contained OXA-23 clone 1 or 2, OXA-51 and OXA-23. Though bacterial killing with colistin monotherapy was virtually superimposable with that of the combination across the first 6 h for all heteroresistant isolates, by 24 h substantial regrowth had occurred with monotherapy but remained suppressed with combination therapy; bacterial killing and suppression of regrowth was also observed with the combination against the colistin-resistant isolate. Synergy was observed in all cases at 24 h (enhanced bacterial killing of ~3–8 $\log_{10}$ CFU/mL). The combination also prevented the emergence of colistin resistance, with little increase in MIC above baseline after 7 days of serial passage in the presence of both drugs compared with monotherapy. Park et al. examined the combination of colistin (2 mg/L) with doripenem (8 mg/L) or tigecycline (2 mg/L; concentration representative of achievable tissue levels) against 69 isolates of *A. baumannii* [137]. Of the isolates, 28 were MDR (100%, 0% and 25% susceptible to colistin, doripenem, and tigecycline, respectively) and 41 XDR (51.2%, 7.3%, and 29.3% susceptible to colistin, doripenem, and tigecycline, respectively). Of 35 isolates tested for the presence of the OXA carbapenemase gene, 34 (97.1%) contained OXA-23 whereas only 2 (5.7%) carried the IS*Aba*-OXA-51 gene. At 24 h, the colistin/doripenem combination showed the highest rate of synergy in both the MDR (15 [53.6%] of 28 cases) and XDR (22 [53.7%] of 41 cases) groups; the equivalent values for the colistin/tigecycline combination were 10 (35.7%) of 28 cases and 18 (43.9%) of 41 cases.

Finally, Ozbek and Mataraci [129] examined the activity of antibiotic lock therapy (ALT) with colistin plus clarithromycin against biofilm-embedded *A. baumannii* using an *in vitro* antibiotic lock model involving segments of central venous catheters; ALT involves the instillation of high concentrations of an antimicrobial agent into the lumen of an infected central venous catheter for extended periods to overcome the relative antimicrobial resistance of biofilm-embedded bacteria. Using two isolates of colistin-susceptible *A. baumannii* they found that against both strains colistin at 400× MIC completely eradicated biofilm bacteria within 3 days, whereas the combination of colistin (400× MIC) plus clarithromycin (200 mg/mL; ~100× serum concentration) sterilized the biofilm in 2 days.

K. pneumoniae and Other Enterobacteriaceae A small number of studies have examined polymyxin combinations specifically against KPC-producing bacteria, primarily *K. pneumoniae* [36, 60, 83, 148, 166, 184, 208]. Pournaras et al. examined colistin and tigecycline alone and in combination against eight KPC-producing enterobacterial clinical strains (four *K. pneumoniae*, two *Escherichia coli*, one *E. cloacae* and one *Serratia marcescens*) [148]; all produced KPC-2 carbapenemase and were colistin-susceptible. Each antibiotic was tested at 1×, 2× and 4× MIC (range, 0.5–4 mg/L for colistin and 0.25–16 mg/L for tigecycline) and experiments conducted over 24 h. The colistin/tigecycline combinations substantially improved bacterial killing across 24 h and was synergistic at 1× and 2× MIC against most organisms at 4 and 8 h; at 4× MIC, synergy was maintained at 24 h against all strains. Similar improvements in bacterial killing against four KPC-3-producing *K. pneumoniae* isolates were reported by Lee and Burgess with colistin or polymyxin B (both at 2× MIC; range, 0.125–0.5 mg/L for colistin and 0.25–0.5 mg/L for polymyxin B) combined with doripenem (6 mg/L) [83]; all isolates were polymyxin-susceptible and doripenem-resistant. In that study, none of the monotherapy regimens sustained bactericidal killing at 24 h. However, colistin or polymyxin B plus doripenem combinations maintained bactericidal activity across 24 h against all isolates, achieving synergy at this time; synergy was maintained at 48 h in 2 (50%) of 4 isolates with colistin and all isolates with polymyxin B. MIC measurements were additionally repeated at 24 h on all isolates following exposure to colistin or polymyxin B monotherapy. All isolates developed polymyxin resistance (MICs, 8–128 mg/L) and cross resistance between colistin and polymyxin B was observed. In another study triple therapy with polymyxin B, doripenem and rifampicin (all at 0.25× MIC) was most effective against five MDR isolates each of *K. pneumoniae* (two with KPC and three with ACT-1 [AMPC-type] β-lactamases) and *E. coli* (one KPC-3 and four KPC-2 β-lactamases) [184]; all isolates were polymyxin B-susceptible and doripenem-resistant. Bactericidal activity was achieved against 4 (80%) of 5 isolates of *K. pneumoniae* and 5 (100%) of 5 isolates of *E. coli* at 24 h. Monotherapy with any agent failed to produce bactericidal activity, whereas combinations utilising only two antibiotics were less effective with polymyxin B plus rifampicin bactericidal against only 1–2 (20–40%) of 5 isolates of each species; polymyxin B plus doripenem was bactericidal against only 1 (20%) of 5 *K. pneumoniae* isolates but 4 (80%) of 5 *E. coli* isolates. In another study, the combination of colistin (5 mg/L) plus fosfomycin (100 mg/L) was synergistic at 24 h against only 1 (6%) of 17 KPC-2-producing *K. pneumoniae* isolates [166].

Clancy et al. examined colistin (2 mg/L) in combination with doripenem (8 mg/L) against 23 KPC-2-producing strains of *K. pneumoniae* [36]; each strain contained a variant mutant *opmK35* porin gene. The median colistin and doripenem MICs were 4 mg/L (range, 0.125–128 mg/L) and 32 mg/L (range, 4–256 mg/L), respectively. Colistin MICs were > 2 mg/L against 14 (63%) of 23 strains. The colistin/doripenem combination was significantly more active at 12 and 24 h than either monotherapy against the four strains with doripenem MICs of ≤8 mg/L, with synergy at 24 h against all 4 strains. In contrast, there was no overall difference in median bacterial killing for strains with doripenem MICs >8 mg/L, with synergy reported at 24 h in 6 (32%) of 19 strains. There was no difference in synergy between strains with colistin MICs of ≤2 mg/L and > 2 mg/L at either 12 or 24 h. Notably, insertions encoding glycine and aspartic acid at amino acid (aa) positions 134 and 135 (ins aa134–135 GD; n = 8) and *ompK36* promoter IS5 mutations (n = 7) were associated with significantly higher doripenem MICs and diminished efficacy of colistin/doripenem combinations; in these cases, bacterial killing more closely resembled colistin monotherapy. However, other mutant/wild-type ompK36 strains demonstrated increased killing with the combination, even with elevated doripenem MICs. The authors suggested that doripenem MICs and *ompK36* genotyping of KPC-*K. pneumoniae* may be useful for identifying strains most likely to respond to colistin/doripenem combination therapy. These results suggest that despite membrane permeabilization by a polymyxin potentially increasing access of doripenem to target sites and allowing it to overcome hydrolysis by KPC, OmpK36 porins may also be necessary for synergy.

While the majority of studies (checkerboard and time-kill) examining polymyxin combination therapy against *K. pneumoniae* addressed KPC-producing strains, fewer studies address MBL-producing strains. Souli et al. examined colistin (5 mg/L) in combination with imipenem (10 mg/L) against 42 unique clinical isolates of *bla*VIM-1-type MBL-producing *K. pneumonia* [167]. After 24 h exposure to the combination, synergy was reported against 12 (50%) of 24 colistin-susceptible isolates, but antagonism was observed against 10 (55.6%) of 18 colistin-resistant isolates. Interestingly, resistance to colistin (MICs 64–256 mg/L) was observed in 7 (58.3%) of 12 isolates that were initially susceptible to colistin. In contrast, none of four isolates initially susceptible to imipenem and which showed regrowth at 24 h developed resistance to imipenem. Tangden et al. conducted more than 200 time-kill experiments with 24 antibiotic regimens including colistin (4.0 mg/L) in double and triple combinations with meropenem (6.8 mg/L), aztreonam (17 mg/L), fosfomycin (83 mg/L) and rifampicin (1.7 mg/L) against two VIM-1-type and two NDM-1-type *K. pneumoniae* strains (all colistin-susceptible; susceptibilities to the other antibiotics varied substantially) [177]. At 24 h, colistin plus fosfomycin was bactericidal and synergistic against three of the four strains (both NDM-1-types [each fosfomycin resistant] and one VIM-1-type), while the triple combination of colistin/fosfomycin/meropenem was bactericidal against three strains and synergistic against all strains. While colistin plus rifampicin was only synergistic at this time against both NDM-1-type strains, the addition of meropenem to this regimen resulted in bactericidal and synergistic activity against all strains; this triple combination was the most effective regimen overall. Double combinations of colistin with either meropenem or aztreonam produced synergy in only one strain, although the triple combination produced synergy in three of the four strains. Albur et al. reported that colistin or CMS in combination with tigecycline did not increase bacterial killing against a range of NDM-1-producing Enterobacteraceae [4]; however, the concentrations chosen in this investigation were extremely low (e.g. the maximum concentration of colistin used was 0.29 mg/L). Abdul Rahim et al. examined polymyxin B (0.5, 1 or 2 mg/L) plus chloramphenicol (8, 16 or 32 mg/L) combinations against four NDM-producing *K. pneumoniae* strains (all polymyxin B-susceptible and -heteroresistant; three susceptible to chloramphenicol) [1]. Combination therapy significantly delayed regrowth, with synergy observed in 25 (89.3%) of 28 cases at both 6 and 24 h; at 24 h, no viable bacteria were detected in 15 (53.4%) of 28 cases with various combinations across all strains. The emergence of polymyxin-resistant bacteria was also completely suppressed with combination therapy. In another study, colistin/tigecycline combinations were synergistic against a single isolate of VIM-1- and SHV-12-producing *K. pneumoniae*, although colistin/ciprofloxacin combinations were indifferent against the same isolate [37].

Corvec et al. combined colistin with tigecycline, fosfomycin or gentamicin (each at 0.5×, 1×, and 4× MIC) against a single strain of ESBL-producing *E. coli* [40]. Colistin combined with tigecycline decreased bacterial counts at 24 h by ~4.5- and 7-$\log_{10}$ CFU/mL compared with the initial inoculum and monotherapy, respectively. The colistin/fosfomycin combination was synergistic at 6 h with no viable bacteria detected at or subsequent to this time. Colistin plus gentamicin was no better than either monotherapy alone (regrowth with both monotherapies had reached control values by 24 h). A similar study by Ku et al. that employed nine ESBL-producing *K. pneumoniae* isolates (five carbapenem-resistant and four – susceptible; one colistin-resistant) examined colistin combined with either tigecycline or fosfomycin (all antibiotics at 0.25× or 0.5× MIC) [80]. With concentrations of 0.5× MIC, synergy at 24 h was reported in 8 (88.9%) and 6 (66.6%) of 9 cases for the combinations with tigecycline and fosfomycin, respectively. However, synergy was absent with both combinations when concentrations of 0.25× MIC were used.

In two further studies the combination of colistin and tigecycline had no benefit over equivalent monotherapy against a single isolate of OXA-48-producing carbapenem-resistant *K. pneumoniae* susceptible to both drugs [44], and only marginal benefit against six carbapenem-resistant isolates of Enterobacter (*E. coli* [*n* = 2], *K. pneumoniae* [*n* = 2], *E. aerogenes* [*n* = 1] and *E. cloacae* [*n* = 1]) with varying resistance determinants [18].

Other Bacteria Against one reference strain and three clinical isolates of *S. maltophilia* (all with elevated MICs to each antibiotic), colistin (2 mg/L) combined with tigecycline (1 mg/L) or rifampicin (8 mg/L) was synergistic at 24 h in all cases except against one isolate and only with the colistin/tigecycline combination (a 1.7 $\log_{10}$ CFU/mL reduction) [19].

16.2.3 PK/PD Time-Kill Studies

To date few studies have utilized PK/PD models to examine colistin in combination, while only one has employed polymyxin B. Gunderson et al. was the first to utilise a one-compartment PK/PD model to examine colistin in combination [67]. In that study colistin (steady-state peak concentration [C_{max}] of 6 or 18 mg/L every 24 h; half-life, 3 h) was combined with either ceftazidime (constant concentration of 50 mg/L) or ciprofloxacin (C_{max} 5 mg/L every 12 h; half-life, 3 h) against two colistin-susceptible MDR isolates of *P. aeruginosa*; experiments were conducted over 48 h with an inoculum of ~10^6 CFU/mL. Although the combination of colistin plus ciprofloxacin generally produced poorer bacterial killing than with either drug alone, the authors reported the combination of colistin plus ceftazidime was synergistic. However, in light of more recent understanding of colistin pharmacokinetics in both critically ill patients [63, 75, 108, 115, 146] and patients with CF [90] (Chap. 15), only one maximal concentration of colistin (6 mg/L) employed by Gunderson et al. can be considered potentially clinically achievable [67]. Additionally, although the simulated 3 h half-life of colistin is representative of that observed in patients with CF [90], colistin was administered as a single dose every 24 h. Given colistin is typically administered intermittently to patients every 8–12 h, the colistin PK profile generated across a 24-h period was not representative of that observed in CF or critically ill patients. Moreover, although synergy was defined as a ≥2-$\log_{10}$ decrease in colony count relative to the count obtained with the more active of the two antibiotics alone at 24 h, it appears that only changes in $\log_{10}$ CFU/mL between colistin monotherapy and combination therapy were considered; when data for ceftazidime monotherapy (which was performed for only one of the two isolates tested) is considered, synergy was not observed.

A small number of conference abstracts have appeared examining colistin in combination with meropenem [168], amikacin [131], and rifampicin [9] against *A. baumannii* utilising PK/PD models. While combinations with meropenem and rifampicin were reported to be synergistic, there are significant limitations with all these investigations, not least of which is that it is unclear whether 'colistin' (which was dosed every 12 h) was administered as colistin (sulphate) or CMS (sodium). Additionally, in the two studies where PK data were reported [9, 168], 'colistin' concentrations were determined using microbiological assays; as discussed in Chap. 6, microbiological assays are incapable of differentiating between colistin present in a sample at the time of collection and colistin formed *in vitro* from administered CMS during the incubation period of the microbiological assay. Finally, as for the majority of investigations examining colistin combinations using time-kill methodology, experiments were conducted for 24 h and used a single, generally lower inoculum (~5×10^5–10^6 CFU/mL). Given these limitations, while the synergy observed in these dynamic systems is interesting it is difficult to draw any firm conclusions from these studies.

More recent studies have systematically investigated polymyxin combination therapy, including the emergence of polymyxin resistance, using *in vitro* PK/PD models [5, 16, 27, 39, 45, 68, 84, 89, 100, 102, 103, 178, 201, 208]. Unfortunately, as was the case for Gunderson et al. discussed earlier [67], a number of recent studies simulated a colistin half-life more representative of that observed in patients with CF (range: 4–4.7 h), not critically ill patients (Chap. 15) [5, 27, 39, 178, 201]. Two studies were conducted over 24 h at a single, low inoculum (10^6 CFU/mL) [68, 100]. Consequently, these studies will not be considered below. Three studies utilized a 1-CM to examine colistin combinations against planktonic MDR isolates of *P. aeruginosa* [16], *K. pneumoniae* [45], and *A. baumannii* [84]. Two additional studies utilized a HFIM to examine colistin combinations against planktonic MDR isolates of *P. aeruginosa* [103] and a single KPC-producing isolate of *K. pneumoniae* [208]; one study utilized polymyxin B against a single MDR isolate of *A. baumannii* [89]. Of these six studies, three combined colistin (constant concentrations of 0.5, 2 or 5 mg/L across the studies) with doripenem (C_{max} of 2.5 or 25 mg/L every 8 h; half-life, 1.5 h) against *P. aeruginosa* (one heteroresistant reference strain and one colistin-resistant MDR clinical isolate in the 1-CM study; two heteroresistant strains and one colistin-resistant MDR clinical isolate in the HFIM study; all strains across the two studies doripenem-susceptible) [16, 103] and *K. pneumoniae* (one heteroresistant reference strain and three MDR clinical isolates [one each of colistin-susceptible, -heteroresistant, and -resistant]; three strains doripenem-susceptible) [45]. Against *A. baumannii*, one study combined colistin (constant concentrations of 0.5, 2 or 5 mg/L) with rifampicin (C_{max} of 5 mg/L every 24 h; half-life, 3 h) against one MDR-colistin-susceptible and one MDR-colistin-resistant isolate [84], whereas one combined polymyxin B (C_{max} of 3.61 mg/L at 0 h, then C_{max} of 2.41 mg/L every 12 h; half-life, 8 h) with meropenem (C_{max} of 54.8 mg/L; half-life, 1.5 h) and/or ampicillin/sulbactam (C_{max} of 132/70.2 mg/L; half-life, 1.5 h) [103]. Colistin (C_{max} of 0.46 mg/L; half-life, 7 h) and fosfomycin (C_{max} of 150 mg/L mg/L; half-life, 2 h) were combined against a single KPC-2-expressing *K. pneumoniae* isolate (colistin- and fosfomycin-susceptible) [208]. All 1-CM studies were conducted at both a low (~10^6 CFU/mL) and high (~10^8 CFU/mL) inocula to account for the attenuated activity of colistin at higher inocula [24], the latter mimicking the high bacterial densities found in some infections [107, 169]; all HFIM studies used only a single inoculum (~10^6, 10^8 or 10^9 CFU/mL). One additional study examined colistin in combination with doripenem against MDR *P. aeruginosa* growing in a biofilm [102]. In all but one case colistin was administered as a continuous infusion to simulate the 'flat' profiles of formed colistin observed in critically ill patients at steady state across a CMS dosage interval [63, 146] (see Chap. 15). The concentrations of colistin employed ranged from 0.46 mg/L to 5 mg/L. Given the bound fraction of colistin in human plasma is ~50% [115], minimal binding of colistin in the growth media [12, 102], and that total (i.e. bound and unbound) plasma colistin concentrations of ~2–3 mg/L are typically achieved at steady state (with some patients achieving concentrations of up to ~10 mg/L) [63, 108, 115, 146], these dosage regimens of colistin (and also polymyxin B) reflect clinically achievable unbound (free) plasma colistin concentration-time profiles in patients. Administration of the second, or in the case of polymyxin B, third drug (doripenem, rifampicin, meropenem, or ampicillin/sulbactam) similarly reflected unbound plasma drug concentration-time profiles achieved in patients [17, 20, 79, 101, 154]. Studies were conducted across 72–96 h (1-CM) and 10–14 days (HFIM).

Across the six above studies directed specifically against planktonic bacteria, combination therapy generally resulted in substantial improvements in bacterial killing at both inocula. In many cases improvements in bacterial killing with combination therapy were dramatic. For example, against a colistin-susceptible strain of *A. baumannii* at the 10^6 CFU/mL inoculum no viable bacteria were detected at 24 h with colistin/rifampicin combinations containing colistin 0.5 or 2 mg/L, whereas regrowth to ~8 $\log_{10}$ CFU/mL had occurred at this time with equivalent colistin monotherapy [84]. At the 10^8 CFU/mL inoculum colistin (at either 2 or 5 mg/L) plus rifampicin increased bacterial killing across 72 h by as much as ~8 $\log_{10}$ CFU/mL and, with the highest dose colistin combination regimen (5 mg/L), resulted in no viable bacteria being detected following commencement of treatment. Similar improvements were observed against the colistin-resistant isolate. In the HFIM (10^8 CFU/mL inoculum), while double polymyxin B combinations were largely ineffective against a single isolate of *A. baumannii* resistant to all investigated antibiotics, the triple combination (polymyxin B plus meropenem and ampicillin/sulbactam) resulted in no viable bacteria being detected from 96 h onwards [89]. Against a colistin-susceptible (MIC 1 mg/L) doripenem-resistant (MIC 8 mg/L) isolate of *K. pneumoniae*, the combination of colistin at 0.5 mg/L plus doripenem at C_{max} of 2.5 mg/L at the low inoculum produced ~4- to 5-$\log_{10}$-greater killing than equivalent monotherapy at 48 and 72 h, whereas colistin at 0.5 or 2 mg/L plus doripenem at C_{max} of 25 mg/L at the high inoculum produced ~5- to 7-$\log_{10}$-greater killing at 48 and 72 h (with no viable colonies detected across the 72-h period on at least one occasion) [45]. Similar improvements were observed against two colistin-heteroresistant (MIC 1 mg/L) doripenem-susceptible (MIC<0.125) isolates, although only colistin at 2 mg/L plus doripenem at C_{max} of 25 mg/L resulted in enhanced bacterial killing of the colistin-resistant isolate and only at the low inoculum. In the HFIM (inoculum 10^6 CFU/mL), a single KPC-2-producing *K. pneumoniae* isolate was completely eradicated by a colistin (C_{max} of 0.46 mg/L)/fosfomycin (C_{max} of 150 mg/L) combination [208]. Against *P. aeruginosa*, combinations containing colistin 0.5 or 2 mg/L plus doripenem at C_{max} of 25 mg/L resulted in eradication of the colistin-resistant MDR isolate at the low inoculum and substantial reductions in regrowth (including to below the limit of detection at ~50 h) at the high inoculum (Fig. 16.4) [16]. For the same combination in the HFIM (colistin 2 or 5 mg/L plus doripenem at C_{max} of 25 mg/L), markedly enhanced bacterial killing was observed with each combination against both heteroresistant (and MDR) isolates across 10 days, with only the combination containing colistin at 2 mg/L and only against one isolate failing to eradicate the bacteria [103]. Against the colistin-resistant isolate, both combinations enhanced bacterial killing by ~5–6 $\log_{10}$ cfu/mL on 3 days, with regrowth then occurring; regrowth approached control values by 10 days.

While subpopulation synergy may have contributed to enhance bacterial killing against some isolates in the above investigations, it cannot explain enhanced activity against all isolates. For example, greater bacterial killing of *P. aeruginosa* was observed with the colistin/doripenem combination against a colistin-resistant MDR isolate with near complete resistance to colistin (MIC, 128 mg/L) and which contained enzymes active against carbapenems [16, 103], and similarly with the colistin/rifampicin combination against *A. baumannii* despite rifampicin

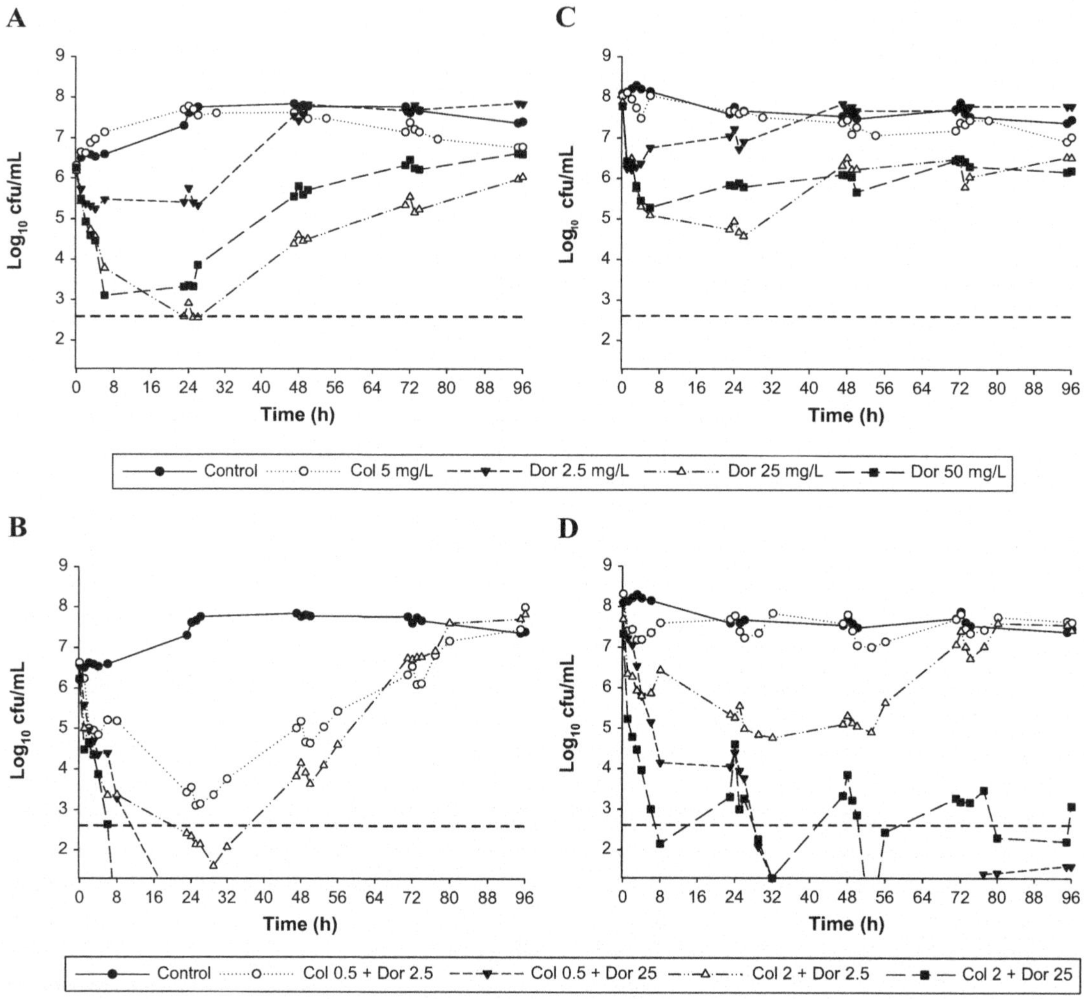

Fig. 16.4 Time-kill curves for colistin and doripenem monotherapy (Panels A and C) and the combination (Panels B and D) against a non-mucoid MDR-colistin-resistant clinical isolate (19147 n/m) of *P. aeruginosa* at an inoculum of ~10^6 CFU/mL (left-hand panels) and ~10^8 CFU/mL (right-hand panels). The *y* axis starts from the limit of detection and the limit of quantification (LOQ) is indicated by the horizontal broken line. (Figure adapted from Bergen et al. [16], with permission)

ordinarily being inactive against Gram-negative pathogens [84]. The triple combination of polymyxin B/meropenem/ampicillin/sulbactam eradicated also eradicated a clinical isolate of *A. baumannii* resistant to all antibiotics investigated [89]. In each case it may be that a form of mechanistic synergy was operative due to permeabilization of the outer membrane by colistin [207]. It is possible that increasing the permeability of the outer membrane resulted in substantially increased concentrations of β-lactam in the periplasm, facilitating access to the cytoplasmic membrane where they act on penicillin-binding proteins [125, 199]. Similarly for rifampicin, which ordinarily does not effectively penetrate the Gram-negative outer membrane [191], increased membrane permeabilization may improve access to its target site within the cytoplasm. In this latter case, the substantial changes to the outer membrane of *A. baumannii* associated with the development of colistin resistance [71, 114] may additionally facilitate access to intracellular target sites.

An important feature common to the above six studies was the substantial reduction or, in some cases, complete suppression of the emergence of colistin-resistant subpopulations with combination therapy. As observed previously against all three bacterial species monotherapy with colistin generally resulted in substantial increases in the proportion of colistin-resistant subpopulations in colistin-susceptible or -heteroresistant isolates at both high and low inocula, often by as early as 24 h. However, the addition of doripenem to colistin eliminated the emergence of colistin-resistant colonies of *K. pneumoniae* [45] except at the lowest concentration combination tested (colistin 0.5 mg/L plus doripenem 2.5 mg/L) at the high (~10^8 CFU/mL) inocula. Against *P. aeruginosa*, resistant colonies were greatly reduced in number and emerged later (following 72–96 h of treatment) with all colistin/doripenem regimens at both inocula in the 1-CM [16], with the most resistant subpopulations (i.e., those growing in the presence of colistin at 10 mg/L on the PAP plates) absent with combination therapy. In the HFIM, the same combination against *P. aeruginosa* completely eliminated colistin-resistant subpopulations [103]. All three colistin/rifampicin regimens (colistin 0.5, 2 or 5 mg/L plus rifampicin 5 mg/L) completely suppressed the emergence of colistin-resistant subpopulations in a MDR-colistin-susceptible clinical isolate of *A. baumannii* such that at 72 h no colistin-resistant colonies were detected with any colistin/rifampicin combination at either inoculum (Fig. 16.5) [84]. Two important observations arise from these investigations. First, although combination therapy with doripenem had no effect on colistin resistance of MDR-colistin-resistant isolates of *P. aeruginosa* [16, 103] and *K. pneumoniae* [45], against *A. baumannii* the colistin/rifampicin combinations containing 2- or 5-mg/L colistin reduced the pre-existing colistin-resistant subpopulations of a colistin-resistant isolate to below the limit of detection at the low inocula, indicating that this combination may suppress the emergence of *de novo* colistin resistance. Second, on the few occasions where extensive regrowth (even up to ~7-$\log_{10}$ CFU/mL) occurred with combination therapy (with both doripenem and rifampicin), no colistin-resistant colonies were detected. While the reason for the observed regrowth despite an apparent lack of colistin resistance is unknown, this important finding suggests that combining doripenem or rifampicin with colistin may reduce the emergence of colistin-resistant subpopulations.

An interesting observation to come out of the studies by Bergen et al. [16] and Ly et al. [103] and which has implication for future rational testing of antibiotic combinations generally concerns the use of dynamic antibiotic concentrations simulating human PK when assessing the efficacy of combination therapy, and the duration over which such experiments are conducted. As discussed in the static time-kill section Bergen et al. previously examined the combination of colistin and imipenem at multiple inocula (~10^6 and ~10^8 CFU/mL) against multiple strains of *P. aeruginosa* using a static time-kill model [13]. In two subsequent PK/PD (dynamic) studies investigating colistin/doripenem, both isolates investigated in the 1-CM study [16] and two of three isolates (the third isolate being an additional colistin-heteroresistant strain) in the HFIM study [103] were included in this earlier investigation. While the antibiotics and their concentrations between the three studies are not directly comparable, the activity of colistin combined with either imipenem or doripenem was broadly similar across 48 h (the duration of the earlier study) at each inoculum against heteroresistant strains. However, substantial differences were evident against a colistin-resistant MDR isolate. In the static model, combinations with concentrations as high as 32 mg/L colistin plus 16× MIC imipenem failed to reduce bacterial numbers of this isolate to below the limit of detection at any time (maximum bacterial killing of ~3.5 $\log_{10}$ CFU/mL). In stark contrast, bacterial eradication was achieved in the 1-CM (duration, 96 h) with combinations containing colistin (0.5 or 2 mg/L) and doripenem 25 mg/L no later than 24 h at the low inoculum, and bacteria reduced to below detectable levels at approximately 48 h with the same combinations at the high inoculum. With the higher initial inoculum in the HFIM (10^9 CFU/mL), progressive bacterial killing occurred over

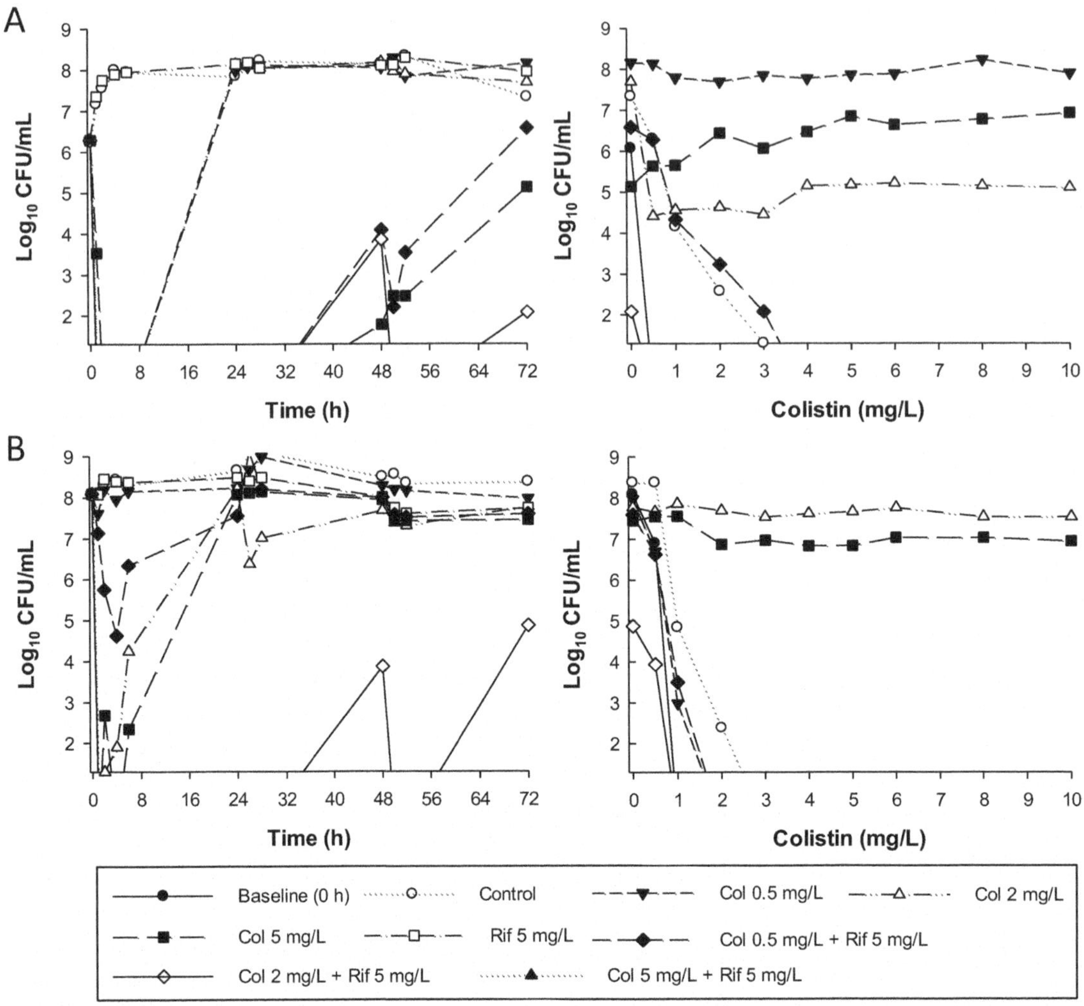

Fig. 16.5 (Left) Time-kill curves with various clinically relevant dosage regimens of colistin (Col) and rifampicin (Rif) alone and in combination at an inoculum of ~10^6 CFU/mL (Panel A) and ~10^8 CFU/mL (Panel B) against a colistin-susceptible MDR clinical isolate (FADDI-AB030) of *A. baumannii*. (Right) Population analysis profiles (PAPs) at baseline (0 h) and after 72-h exposure to colistin monotherapy, colistin-rifampicin combination therapy, or neither antibiotic (control). The *y* axis starts from the limit of detection and the limit of quantification (LOQ) is indicated by the horizontal broken line. (Figure adapted from Lee et al. [84], with permission)

72–96 h (maximum bacterial killing of ~6 $\log_{10}$ CFU/mL), but slow regrowth ultimately close to control values occurred over the subsequent 7 days. Likewise, changes in PAPs with colistin/imipenem combinations against heteroresistant isolates in the static time-kill model generally mirrored those observed with equivalent colistin monotherapy, whereas the emergence of colistin resistance was greatly reduced (1-CM) or completely suppressed (HFIM) with colistin/doripenem combinations in the PK/PD models. Loss of imipenem due to degradation in the static experiments may have contributed to this result (colistin is stable under these conditions) [16], whereas intermittent dosing of doripenem in the PK/PD models replenished concentrations and avoided the combination effectively becoming colistin monotherapy over time. These observations highlight the importance of simulating PK profiles when assessing the activity and emergence of resistance to antimicrobial therapy. Additionally, the regrowth that occurred in the HFIM following substantial initial killing across the first 72–96 h of therapy highlights the importance of longer durations of therapy to fully assess the effectiveness of combinations.

While the above studies examined bacterial killing against planktonic cells, bacteria growing in a biofilm are protected from environmental, immune system and antimicrobial threats, making them substantially more resistant to antibiotic treatment. Such resistance is evidenced by substantial increases in MICs and MBCs [41, 70, 107]. The need for very high concentrations of colistin when used as monotherapy to achieve any substantial killing of biofilm-embedded bacterial cells has been demonstrated both *in vitro* [69, 72, 132] and *in vivo* [70]. Using a mouse lung infection biofilm model, Hengzhuang et al. [70] reported a colistin serum concentration of 64× MIC (i.e. 128 mg/L) was required to achieve a 1 $\log_{10}$ decrease in CFU/lung. Such concentrations are unattainable clinically and necessitate alternative strategies such as antibiotic combinations in order to adequately treat biofilm infections.

Only one study has examined polymyxin combination therapy using dynamic antibiotic concentrations against bacteria growing in a biofilm. Using a CDC biofilm reactor Lora-Tamayo et al. examined colistin (constant concentrations of 1.25 mg/L and 3.50 mg/L) in combination with doripenem (C_{max} 25 mg/L every 8 h; half-life, 1 h) over 72 h against *P. aeruginosa* [102]. One colistin-susceptible reference strain and two MDR-colistin-susceptible-carbapenem-resistant clinical isolates were employed, with bacterial killing of both biofilm-embedded and planktonic bacteria examined; each clinical isolate had been the cause of outbreaks in the Hospital Universitario de Bellvitge in Barcelona, Spain, and contained either a VIM-2 metallo-β-lactamase or a PSE-1 β-lactamase plus a MexXY-OprM efflux-pump. Against biofilm-embedded bacteria monotherapy with colistin at 1.25 mg/L was ineffective against the reference strain and produced only modest, non-bactericidal killing of the clinical isolates; colistin at 3.5 mg/L produced greater and more rapid initial killing against all three strains, but with subsequent regrowth by 72 h such that bactericidal activity was only observed at this time against one clinical strain. The combination of colistin 1.25 mg/L plus doripenem showed some additive effects against biofilm-embedded bacteria during the first 24–32 h of treatment (Fig. 16.6, top panels), but was generally no better than colistin monotherapy against the clinical isolates. The combination of colistin 3.5 mg/L plus doripenem resulted in greater and more sustained killing than either corresponding monotherapy across 72 h. Notably, against both clinical isolates greater initial killing (of ~2–3 $\log_{10}$ CFU/cm^2 compared to equivalent monotherapy) was observed and the combination remained synergistic at 72 h (Fig. 16.6, top panels). Importantly, both colistin/doripenem combinations eliminated the emergence of colistin resistance against biofilm-embedded bacteria observed with the highest colistin monotherapy (3.50 mg/L) (Fig. 16.6, lower panels), and substantially reduced (colistin 1.25 mg/L plus doripenem) or eliminated (colistin 3.5 mg/L plus doripenem) the emergence of resistance in planktonic bacteria.

16.3 Animal Studies

Only a small number of animal studies have examined polymyxin combination therapy, providing mixed results. All these studies have utilized colistin (or CMS). Unfortunately, there are a number of shortcomings with the existing literature which makes the results difficult to interpret. Specifically, it is not always possible to ascertain whether the 'colistin' administered in these studies was colistin (sulphate) or CMS (sodium). In patients colistin is administered in the form of its inactive derivative, CMS, with the active species colistin forming *in vivo* following CMS administration (Chap. 7). However, in animal models the administration of colistin sulphate is preferable as it permits greater control over the PK profile of the active species, colistin. In a number of studies, it is unclear whether

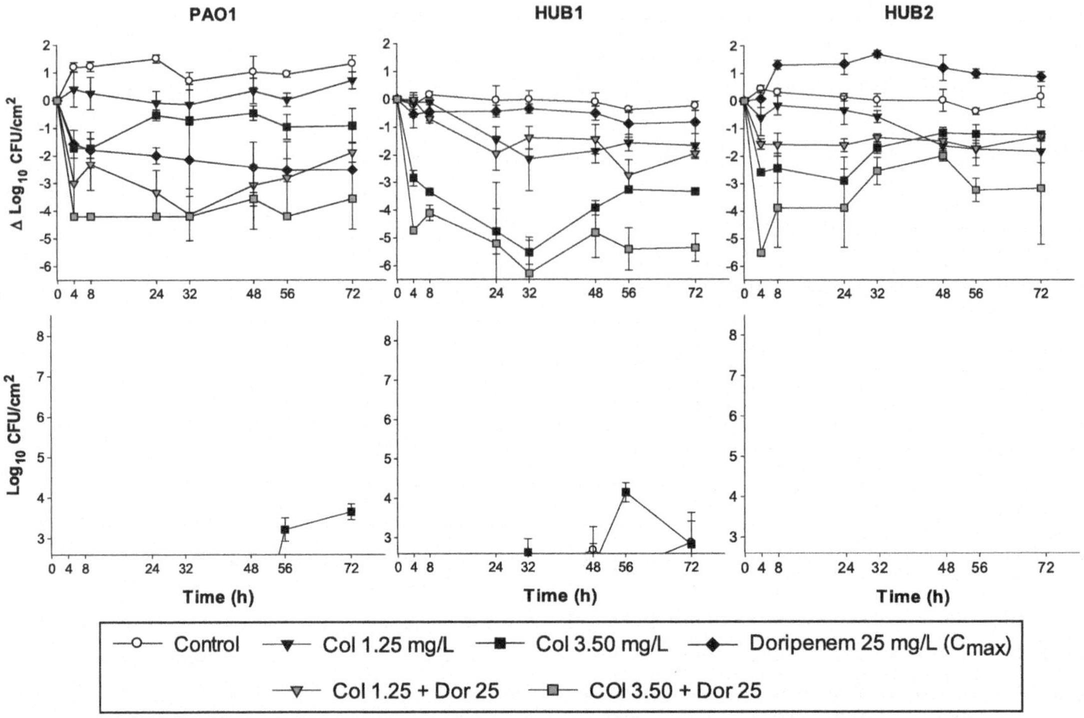

Fig. 16.6 Upper panels: Bacterial killing by colistin (Col) alone at two different clinically relevant concentrations, doripenem (Dor) alone, and in combination against biofilm-embedded cells of three different *P. aeruginosa* strains; results expressed using the log change method (log change = $\log_{10}[CFU_t] - \log_{10}[CFU_0]$). Lower panels: Emergence of colistin resistance (i.e. colonies able to grow in the presence of ≥4 mg/L colistin) among biofilm-embedded *P. aeruginosa* across the treatment period with the same treatment regimens. Results expressed as the absolute number of recovered bacteria. For the lower panels, the *y* axis starts from the limit of quantification. Data are presented as means ± standard deviation of the mean. (Figure adapted from Lora-Tamayo et al. [102], with permission)

colistin or CMS was administered [33–35, 58, 64, 136, 195, 200]. Importantly, irrespective of the form of 'colistin' utilised, few studies provide a rationale for the doses of CMS/colistin administered with the majority of administered doses apparently chosen to reflect human doses on a mg/kg basis. However, such dosing fails to recognise the importance of animal scaling that results in PK dissimilarities across species [202], resulting in substantially lower plasma concentrations in the preclinical models. Adding to this difficulty is that PK data for CMS/colistin and second antibiotic are absent from virtually all investigations, preventing comparisons with PK profiles achieved in patients; such comparisons are crucial to adequately assess the likely value of the combination in the clinical setting. Where concentrations of antibiotics are measured, antimicrobial assays are generally used for quantification of antibiotic concentrations. As previously discussed, such assays are incapable of providing accurate information on the time-course of plasma concentrations of the prodrug (CMS) and the active entity (colistin). Given these shortcomings results from animal studies will only be considered briefly here.

Yamagishi et al. used a murine thigh infection model to examine 'colistin' (16 mg/kg/12 h administered intraperitoneally [IP]) combined with aztreonam (400 mg/8 h; administered subcutaneously [SC]) against five clinical isolates (two MDR) of *P. aeruginosa* [195]. Though the authors' state the administered dosing regimens produce antimicrobial exposures similar to humans following IV administration of standard doses, the achieved concentrations of each agent were not reported. Compared to monotherapy, the combination at 24 h produced greater bacterial killing (maximum additional killing ~1 $\log_{10}$ CFU) against four of five isolates. Using mouse [34] and rat [33] sepsis models Cirioni et al. examined 'colistin' (CMS or colistin sulphate not specified; 1 mg/kg) in combination with either imipenem (mouse model; 20 mg/kg) or rifampicin (rat model; 10 mg/kg) against a reference strain and MDR clinical isolate of *P. aeruginosa*; all antibiotics were administered IV and once only. 'Colistin' plus either imipenem or rifampicin resulted in significant reductions in bacterial counts across 72 h when compared with monotherapy with either drug, although only the colistin/imipenem combination resulted in significantly lower mortality. Aoki et al. examined the effect of CMS (administered either intranasally (5 mg/kg/12 h) or subcutaneously (10 mg/kg/12 h) in combination with either imipenem (30 mg/kg/12 h SC) or rifampicin (25 mg/kg/24 h orally) against a reference strain and MDR clinical isolate of *P. aeruginosa* using a mouse pneumonia model [8]; treatment was continued for 48 h. Whereas all control mice and mice treated with CMS, imipenem or rifampicin monotherapy died within 42 h of infection with the reference strain, the CMS plus imipenem or rifampicin combinations increased survival to 62.5% and 75% at 72 h, respectively. A clear difference was observed in survival between mice treated with intranasal or SC CMS plus rifampicin (100% vs. 14%; $P < 0.01$); intranasal CMS was also superior to CMS administered SC when combined with imipenem. Similar trends were observed with the MDR clinical isolate.

Against MDR *A. baumannii*, two studies found no differences in survival or bacterial clearance from the lungs in mouse pneumonia models with rifampicin monotherapy (IP: 25 mg/kg/6 h or 25 mg/kg/24 h; rifampicin was the most active monotherapy) and rifampicin/CMS (IM; 20 mg/kg/8 h or 40 mg/kg/6 h) combination therapy [118, 130]. However, in the same model Yang et al. observed significantly fewer bacteria at 24 h in the lungs of mice treated IP with 'colistin' (10 mg/kg) and minocycline (50 mg/kg) compared to monotherapy, with the combination producing substantially greater survival at 7 days [200]. Against a single MDR isolate of *A. baumannii*, Pantopoulou et al. found little difference in survival with CMS (3 mg/kg IM) or rifampicin (5 mg/kg IV) as mono- or combination therapy in a neutropenic rat thigh infection model, although in this investigation both antibiotics were administered as single doses only at the beginning of the experiment [136]. In a much larger study in a murine thigh infection model involving 15 extensively drug-resistant (XDR) isolates of *A. baumannii*, reductions in bacterial counts of >2$\log_{10}$ CFU compared to monotherapy at 48 h were observed with the combinations of 'colistin' (20 mg/kg/8 h) and fusidic acid (500 mg/kg/8 h) or rifampicin (25 mg/kg/6 h) [58]; these combinations were superior to colistin combined with meropenem (200 mg/kg/8 h), tigecycline (50 mg/kg/24 h), fosfomycin (100 mg/kg/4 h), and sulbactam (120 mg/kg/12 h). In a mouse sepsis model, the addition of sulbactam (240 mg/kg/12 h IP) to CMS (5 mg/kg/12 h IP) had no significant effect on bacterial counts of a single carbapenem-resistant (OXA-51-, OXA-58- and PER-1-positive) isolate of *A. baumanni* [47]. However, in a mouse sepsis model involving two clinical isolates (1 MDR) of *A. baumannii*, Cirioni et al. recently showed a single a dose of 'colistin' (1 mg/kg) plus either daptomycin (7 mg/kg) or teicoplanin (7 mg/kg) administered IP substantially enhanced survival at 72 h [35]. In that study lethality rates against the susceptible isolates were 100% in the control group, 80% with daptomycin or teicoplanin alone, 50% with colistin alone, 10% with colistin/daptomycin and 15% with colistin/teicoplanin; lethality rates were similar against the MDR isolate. The combinations also significantly reduced the number of bacteria in intraabdominal fluid.

Giacometti et al. examined 'colistin' (CMS or colistin sulphate not specified; 1 mg/kg) in combination with piperacillin (60 mg/kg) against *E. coli* in a rat intraperitoneal infection model [64]. Following a single IP administration of antibiotics, mortality at 48 h was 93.3%, 33.3%, 33.3%, and 0% for controls, 'colistin' monotherapy, piperacillin monotherapy, and the 'colistin' plus piperacillin combination, respectively. More recently, Michail et al. examined several combinations of tigecycline (50 mg/kg/24 h SC) including with CMS (40 mg/kg/8 h SC) against eight clinical isolates of *K. pneumoniae* and two isolates (one clinical isolate and one reference strain) of *E. coli* in a murine thigh infection model [113]; all organisms produced KPC-2 carbapenemase and were susceptible to colistin. As monotherapy, CMS exhibited substantially less bacterial killing than tigecycline. In combination, bacterial killing at 48 h was either essentially the same as tigecycline monotherapy or, in 4 (40%) of 10 cases, antagonistic. However, as antagonism was broadly defined as simply a lower $\log_{10}$ CFU reduction with combination therapy compared to monotherapy, the magnitude of this antagonism is unclear. Demiraslan et al. similarly examined the combination of CMS (5 mg/kg/12 h IP) and tigecycline (20 mg/kg/12 h IP) against a single OXA-48-producing carbapenem-resistant isolate of *K. pneumoniae* using a sepsis mouse model [44]; this strain was also positive for *bla*$_{\text{TEM-1}}$ and *bla*$_{\text{CTX-M-15}}$ genes and was susceptible to both colistin and tigecycline. The combination was tested against both immunocompetent and immunosuppressed mice. In both sets of mice, bacterial counts at 24 and 48 h in liver and lung samples were decreased by both CMS and tigecycline monotherapy compared to controls, however there was no significant difference between the most active monotherapy (CMS) and combination therapy at this time. Mutlu Yilmaz et al. likewise found no differences in efficacy between CMS (1.25 mg/kg/6 h IP) and tigecycline (10 mg/kg/12 h IP) monotherapy and combination therapy across 48 h against a single MDR strain of *A. baumannii* using a rat pneumonia model [122].

Only one study has specifically examined polymyxin combinations against biofilms *in vivo* [102], most likely due to a lack of suitable models. Bacterial cells growing within a biofilm are often substantially more resistant than planktonic cells to antibiotic treatment due to the self-produced polymeric matrix that protects the cells from environmental, immune system and antimicrobial threats [41, 48, 107, 121]. With increased MICs and minimum bactericidal concentrations (MBCs) of polymyxins associated with biofilm infections [69, 70], and increasing multidrug-resistance generally, alternative strategies such as polymyxin combination therapy have been suggested for treatment of biofilm infections [102]. Corvec et al. employed a foreign-body infection model involving the implantation of Teflon cages into guinea pigs (four cages/guinea pig) to investigate the activity of antibiotic combinations including colistin (15 mg/kg) in combination with fosfomycin (150 mg/kg), gentamicin (10 mg/kg) and tigecycline (10 mg/kg) [40]; all antibiotics were administered 12-hourly IP for 4 days. A single extended-spectrum-β-lactamase (ESBL)-producing clinical strain of *E. coli* susceptible to all antibiotics tested was employed. Although the authors reported significantly lower (>3 $\log_{10}$ CFU/mL) bacterial counts (and therefore greater bacterial killing) with each combination immediately and 5 days after the treatment period against planktonic bacteria aspirated from cage fluid, it appears that comparisons of combination therapy were only made against gentamicin or tigecycline monotherapy without including colistin or fosfomycin monotherapy. When the latter are included the differences in bacterial killing appear not to be as great immediately following therapy, although in all cases combinations did result in substantially improved bacterial killing relative to monotherapy 5 days following cessation of treatment. Against biofilm-embedded bacteria 5 days following discontinuation of antibiotic therapy, only monotherapy with fosfomycin was able to eradicate some biofilms (cure rate of 17%; cure rate defined as the percentage of total cages without *E. coli* growth). However, the combinations of colistin with fosfomycin, tigecycline and gentamicin significantly increased the cure rate to 67%, 50% and 33%, respectively.

Two research group have employed an invertebrate model of the wax moth caterpillar *Galleria mellonella* which has been proposed as an inexpensive an easy alternative to mammalian models to generate reliable and reproducible data on microbial virulence similar to that obtained using higher animals [31, 76, 159]. One group examined the activities of colistin (2.5 mg/kg)/glycopeptide (vancomycin and teicoplanin, 10 mg/kg) [74] or colistin (2.5 mg/kg)/telavancin

(10 mg/kg) [73] combinations against *A. baumannii*-infected caterpillars (1 ATCC reference strain and 1 MDR clinical isolate) over 96 h. In 5 (83%) of 6 cases (3 combinations across 2 isolates) combinations significantly enhanced the survival of larvae compared with monotherapy. Other similar experiments by the same group examined a colistin (0.25 mg/kg)/tigecycline (1 mg/kg) combination against carbapenem-resistant Enterobacteriaceae (six strains comprising *E. coli* (*n* = 2), *Enterobacter aerogenes* (*n* = 1), *Enterobacter cloacae* (*n* = 1) and *K. pneumoniae* (*n* = 2)) [18], and the same combination plus a colistin (0.25 mg/kg)/rifampicin (10 mg/kg) combination against two strains of *Stenotrophomonas maltophilia* [19]. The colistin/tigecycline combination significantly improved survival against all *Enterobacteriaceae* isolates and 1 (50%) of 2 *S. maltophilia* isolates, while the colistin/rifampicin combination significantly improved survival in both *S. maltophilia* isolates. More recently, another group has undertaken similar experiments over 96 with colistin (2.5 mg/kg) combined with vancomycin (15 mg/kg; *n* = 4) [198], levofloxacin (6.7 mg/kg; *n* = 4) [192] and daptomycin (4 mg/kg; *n* = 2) [196] against *A. baumannii* and colistin (2.5 mg/kg) combined with imipenem (15 mg/kg; *n* = 2) against *E. cloacae* [197]; all studies included at least 1 MDR isolate. With the exception of the colistin/vancomycin combination that was less effective than vancomycin monotherapy against a colistin-resistant isolate, in all cases combination therapy significantly improved survival compared to monotherapy.

Clearly, future animal studies investigating polymyxin combination therapy which administer colistin (sulphate) or polymyxin B and which provide the crucial PK data currently lacking in existing studies are urgently required. Such investigations will be crucial to build on the knowledge gained from *in vitro* studies (discussed above) and are essential to optimise polymyxin therapy.

16.4 Clinical Studies of CMS or Polymyxin B Combination Therapy

Very few studies have formally assessed the benefit of CMS (the sulphomethylated derivative of colistin and the form administered intravenously [IV]) or polymyxin B combinations, and those that have are commonly retrospective in nature. Although a small number of investigations have been undertaken prospectively, these tend to contain small patient numbers and are thus low powered. Additionally, the doses of antibiotics administered, including polymyxins, are often not stated and PK data is absent. The majority of the data reviewed here is taken from studies seeking to ascertain their general benefit in patients. Studies that have assessed CMS or polymyxin B for a variety of MDR Gram-negative pathogens and infection sites combined into single studies have been inconclusive in differentiating between the value of monotherapy and combination therapy [54–57, 145, 181]. This section will focus on those studies that provide the greatest insight into specific situations where polymyxin combination therapy appears to be of promise or significant value.

Klebsiella pneumoniae A retrospective cohort analysis by Qureshi et al. examined the utility of combination therapy in treating KPC-producing *K. pneumoniae* bacteraemia [150]. In total, 41 patients with genetically confirmed infections were included with a majority (32 [78%] of 41) being hospital acquired and the remainder (9 [22%] of 41) health care associated. The primary outcome was 28-day mortality which, among all patients that received definitive antibiotic therapy for >48 h, was 38.2% (12/34; 7 patients did not receive definitive antibiotic therapy). Treatments varied extensively. Nineteen patients received monotherapy with most receiving CMS or polymyxin B (*n* = 7), tigecycline (*n* = 5), or a carbapenem (imipenem or meropenem; *n* = 4); 15 patients received combination antibiotics. For combination therapy, CMS or polymyxin B were combined with unspecified carbapenems (*n* = 5), tigecycline (*n* = 1) or a fluoroquinolone (*n* = 1) while the most common polymyxin-free combination was tigecycline with either a carbapenem ($n = 3$) or aminoglycoside ($n = 2$). The doses of each antibiotic administered were not reported. Combination treatment was the only significant predictor of survival ($p = 0.02$) with a 28-day mortality of 13.3% (2/15) compared to 57.8% (11/19) for monotherapy. Of specific interest, 1 patient receiving CMS or polymyxin B (which polymyxin was not stated) in combination died compared with 4 (57.1%) of 7 patients that received polymyxin monotherapy. The incidence of mortality in patients receiving polymyxin monotherapy was higher than that reported by Dubrovskaya et al. with polymyxin B monotherapy against KPC producing *K. pneumoniae* (57.1% [4/7] vs. 18% [7/40]) [49]. This difference is likely due to the greater severity of infection in the patients in the former study who were mostly critically ill. All of the deaths in this studied occurred despite *K. pneumoniae* having MICs within the susceptible range for each of the respective antibiotics administered, highlighting the suboptimal use of CMS and polymyxin B especially as monotherapy.

A case control study conducted in Greece produced similar results for KPC producing *K. pneumoniae* bloodstream infections. Zarkotou et al. identified 35 patients that received appropriate antimicrobial therapy (considered susceptible to the respective antibiotic using EUCAST Clinical Breakpoints), a subset of which received CMS [203]. None of 20 patients administered multiple antibiotics died compared to 7 (46.7%) of 15 patients receiving monotherapy. Of the patients that received combination treatment, 14 were administered CMS whereas 7 received CMS as monotherapy; in this latter group mortality was 66.7% (4/7). The most common combination was CMS plus tigecycline ($n = 9$), while unspecified carbapenems were combined with CMS in an additional 2 patients. Unfortunately, the doses of each antibiotic administered were not specified. Nevertheless, this data provides qualified support for the use of combination regimens including colistin (administered as CMS) against KPC-producing *K. pneumoniae* bacteraemia. Another study conducted in Italy similarly compared monotherapy to combination treatment in a larger population of 125 patients with KPC-producing *K. pneumoniae* bacteraemia [182]. CMS was administered as monotherapy in 22 patients and in combination in 51 patients (combined with [No. of patients]: tigecycline [23], gentamicin [7], meropenem [4], tigecycline plus meropenem [16], gentamicin plus meropenem [1]). The dose of CMS administered in both groups was six to nine million international units (IU; equivalent to 180–270 mg of colistin base activity [CBA]) IV every 8–12 h following an unspecified loading dose. Thirty-day mortality was significantly reduced with combination therapy (34.1%; 27/79) compared to monotherapy (54.3%; 25/46). Of the 22 patients that received CMS monotherapy, 11 (50%) died; unfortunately, individual mortality rates for each combination regimen were not stated. The triple combination of colistin, tigecycline, and meropenem was the only drug regimen reported as significantly more common in the survivor group. However, it must not be overlooked that this finding may be the result of the triple combination also being the most common carbapenem-containing combination regimen. This study again confirms the importance of combination therapy in KPC-producing *K. pneumoniae* bacteraemia and emphasizes the benefit of including a carbapenem with CMS. Further studies are warranted to optimize specific combination regimens.

Overall, the available clinical data supports the use of combination antibiotic regimens over monotherapy for KPC-producing *K. pneumoniae* bacteraemia, especially those containing either CMS or polymyxin B in combination with a carbapenem or tigecycline [150, 182, 203]. Since KPC strains hydrolyze carbapenems, the evidence that mortality is reduced by the combination of a carbapenem and a polymyxin is of interest. Results from an investigation by Daikos et al. further support the use of a carbapenem in addition to another agent to treat KPC-producing *K. pneumoniae*, suggesting that if the infecting pathogen has a carbapenem MIC of ≤4 mg/L, combination therapy may reduce mortality compared to other non-carbapenem combinations

[42]. The type and severity of infection caused by KPC-producing *K. pneumoniae* may be an important factor in dictating the utility of polymyxin combination therapy. More severe infections (i.e. bacteraemia) have benefited from combinations with these drugs [150, 182, 203], whereas the cumulative assessment of all types of infection including patients who were considerably less ill, suggests monotherapy with a polymyxin may be sufficient [49]. Further, KPC-producing *K. pneumoniae* pneumonia and bacteraemia with pneumonia as its source of infection have both been associated with higher mortality and underline clinical scenarios where monotherapy appears insufficient for most patients. This lack of success in treating pneumonia based infections with monotherapy may be the result of low polymyxin concentrations at the site of infection in the lungs where supplemental antibiotics would in theory be useful [75, 209]. Further studies in this regard are warranted.

Pseudomonas aeruginosa Conway et al. prospectively treated patients with cystic fibrosis (CF) chronically colonized with *P. aeruginosa* and experiencing an acute respiratory tract exacerbation with CMS monotherapy or combination therapy in an effort to define the benefit of multiple *P. aeruginosa* coverage in these patients [38]. Patients treated with monotherapy ($n = 36$) received 160 mg CMS (two million IU [equivalent to 60 mg CBA]) IV every 8 h while those receiving combination therapy ($n = 35$) received the same CMS dose with additional aztreonam, azlocillin, piperacillin, ceftazidime, imipenem or ciprofloxacin. By Day 12 all patients showed clinical improvement based on clinical measurement, patient weight, Shwachman-Kulczycki score, Chrispin-Norman and Northern chest radiograph scores. However, combination treatment resulted in significantly more patients returning to a normal C-reactive protein level at this time suggesting less inflammatory activity in the lungs. The authors concluded that IV CMS was effective in treating acute respiratory exacerbations of *P. aeruginosa* as monotherapy or combination therapy.

Linden et al. conducted a prospective study that compared treatment efficacy of CMS monotherapy ($n = 10$) and combination therapy ($n = 13$) in 23 patients infected with MDR *P. aeruginosa* [98]; 21 patients were critically ill, defined as having at least 2 major organ system failures during the study. The types of infection varied with the most common being pneumonia ($n = 18$), bacteraemia ($n = 8$) and intra-abdominal infections ($n = 6$). For patients in both monotherapy and combination treatment groups, CMS was administered IV based on ideal body weight and estimated creatinine clearance (range: ~2.7–13.3 mg/kg/day; equivalent to ~33,000–167,000 IU/kg/day or 1–5 mg CBA/kg/day). Amikacin or an antipseudomonal β-lactam was added to CMS for patients in the combination group. An unfavourable response, defined as persistence or worsening of presenting signs and symptoms or death, was reported for 4 (40%) of 10 patients receiving only CMS and 5 (38.5%) of 13 patients on combination therapy. However, 11 patients had other co-infecting pathogens which may have confounded the results. Based on this data it is evident that colistin provides an important 'salvage' option for patients who have failed or are resistant to other antipseudomonal therapies, but it cannot support the use of combination treatment. In a similar study by Furtado et al., polymyxin B combinations (most commonly combined with imipenem) did not provide additional benefit over polymyxin B monotherapy for pneumonia caused by MDR *P. aeruginosa* [59]. Polymyxin B was dosed based on creatinine clearance (1.5–2.5 mg/kg/day when CrCl ≥80 mL/min; 2.5 mg/kg on Day 1, then 1.0–1.5 mg/kg/day thereafter when CrCl 30–80 mL/min; 2.5 mg/kg on Day 1, then 1.0–1.5 mg/kg every 2–3 days thereafter when CrCl <30 mL/min; 2.5 mg/kg on Day 1, then 1.0 mg/kg every 5–7 days thereafter) and, unusually, was administered by continuous infusion over 24 h rather than in divided intervals (usually every 12 h). There was no difference in favourable outcomes (defined as partial resolution of signs and symptoms by the end of treatment; unfavourable was the persisting or worsening of signs and symptoms or death during treatment) between the groups (14 [50.0%] of 28 vs. 21 [45.7%] of 46 in patients receiving combination therapy and monotherapy, respectively). Based on these data, the authors suggested polymyxin B monotherapy would be an appropriate 'salvage' option for MDR *P. aeruginosa* pneumonia, although the overall low favourable outcome rate (47.3%) relative to other studies may suggest against administering it as a continuous infusion.

To our knowledge, no clinical studies to date support the use of CMS or polymyxin B based combinations in favour of polymyxin monotherapy for treatment of infections caused by MDR *P. aeruginosa*. Existing data regarding CMS or polymyxin B combinations in humans is limited with studies frequently pooling patients with many types and sites of infection and varying degrees of severity, limiting the usefulness of the results obtained [38, 59, 98]. Further, more focussed studies are warranted which may assist to identify subsets of patients that benefit from combination therapy.

Acinetobacter baumannii In a recent prospective study, Aydemir et al. compared CMS monotherapy (n = 22) to a combination of CMS and rifampicin (n = 21) for ventilator-associated pneumonia (VAP) caused by carbapenem resistant *A. baumannii* [10]. CMS was administered at 300 mg CBA/day IV in three divided doses adjusted for renal impairment based on the manufacturer's recommendations; rifampicin was administered nasogastrically at a dose of 600 mg/day. No difference in the primary endpoint of clinical response was observed between the two groups (40.9% for monotherapy, 52.4% for combination; p = 0.654), however microbiological clearance (a secondary endpoint) was obtained significantly more quickly with combination therapy (4.5 ± 1.7 days for monotherapy, 3.1 ± 0.5 days for combination; P = 0.029).

It is important to note that in the studies discussed above, CMS was dosed according to the product information which likely cannot achieve high enough plasma concentrations to optimally treat severe infections for all patients. In order to more rapidly attain higher plasma concentrations recent studies have suggested the use of a loading dose of nine million IU of CMS (equivalent to ~270 mg of CBA) followed by nine million IU per day in divided doses instead of the six million IU (equivalent to ~180 mg of CBA) received by many of the patients reviewed above [43, 55, 108]; administration of higher doses of polymyxin B have also been suggested [52]. Non-traditional 'front loaded' or 'burst' polymyxin regimens (e.g. high dose, short duration polymyxin at the start of therapy, with lower overall exposure), especially in combination, require further analysis in patients in order to fully define their therapeutic role in the management of MDR Gram-negative infections. Since the mortality rate remains high for infections with KPC-producing *K. pneumoniae*, *P. aeruginosa* and *A. baumannii*, it is critical to continue to investigate optimal dosing strategies for polymyxins, including the role of combination therapy. Given the limitations associated with existing clinical data future randomized controlled trials with robust study designs are urgently required to more fully understand the utility of CMS or polymyxin B based combinations [138].

16.5 Randomized Controlled Trials Evaluating Polymyxin Combinations

Although there are no adequately powered published randomized controlled trials (RCTs) to examine whether therapy with polymyxins (polymyxin B or colistin) administered in combination with another active agent is superior to polymyxin B or colistin monotherapy against carbapenem-resistant Enterobacteriaceae or carbapenem-resistant *P. aeruginosa* infections, there are recent RCTs in evaluating polymyxin combinations against MDR *A. baumannii*. The first open label RCT comparing synergistic combinations with monotherapy was a prospective study by Durante-Mangoni et al. who conducted a larger (n = 209) multi-centre prospective study examining CMS/rifampicin combinations against extensively drug resistant *A. baumannii* [51]; extensively drug resistant was defined as an MIC

≥16 mg/L for carbapenems and resistant to all other antibiotics except colistin. Patients were allocated to receive either CMS (160 mg or two million units; equivalent to ~60 mg CBA; n = 105) every 8 h IV as monotherapy or CMS (same dose) plus rifampicin 600 mg every 12 h IV (n = 105). Most patients had VAP (69.8%) while the remainder had bloodstream infections (20.1%), hospital acquired pneumonia (8.6%), or intra-abdominal infections (2.4%). Although there was no difference between monotherapy and combination therapy for the primary endpoint of 30-day mortality, eradication of *A. baumannii* was significantly higher with the addition of rifampicin (60.6% vs 44.8%, P = 0.034). Additionally, the risk of death within 30 days was similar between combination therapy and monotherapy (OR = 0.88, 95% CI 0.46–1.69; P = 0.71) despite a significantly improved microbiological cure rate in patients receiving colistin + rifampin (P = 0.034) with no resistance developing in either arm. No colistin loading dose was administered and the maximum daily maintenance dose was low by current standards.

In another, open label, prospective, randomized trial of 94 patients with carbapenem-resistant *A. baumannii* (CRAB) infections, subjects were randomised to receive colistin alone or colistin + fosfomycin for 7–14 days [164]. Some patients in both groups received other antibiotics; for example, 17.0% and 8.5% of patients in the monotherapy and combination groups, respectively, received a carbapenem. There was no difference between monotherapy and combination therapy arms in infection-related (23.1% vs. 16.3%; P = 0.507) or all-cause mortality (57.4% vs. 46.8%; P = 0.41). However, the patients who received combination therapy had a significantly more favourable microbiological response than those who received colistin alone. Interestingly, microbiological cure in the first 72 h (65.7% vs. 78.8%; P = 0.028) and at the end of treatment (84.5% vs. 100%; P = 0.023) was greater in the combination arm.

Recently, Paul et al. conducted a randomized controlled superiority trial in 406 patients comparing colistin monotherapy with colistin (nine MIU or 300 mg CBA/day) + high dose extended infusion meropenem combination therapy for the treatment of carbapenem-resistant Gram-negative bacilli [139]. Patients with bacteraemia, ventilator-associated pneumonia, hospital-acquired pneumonia, or urosepsis caused by carbapenem-non-susceptible Gram-negative bacteria were included. Patients received either intravenous colistin (9-million unit loading dose, followed by 4.5 million units twice per day) or colistin with meropenem (2-g prolonged infusion three times per day). The primary outcome was clinical failure, defined as not meeting all success criteria by intention-to-treat analysis, at 14 days after randomisation. Most infections were caused by *A. baumannii* (312/406, 77%), although some infections were due to CRE and carbapenem-resistant *P. aeruginosa*. No significant difference between colistin monotherapy (156/198, 79%) and combination therapy (152/208, 73%) was observed for clinical failure at 14 days (risk difference – 5.7%, 95% CI -13.9 to 2.4; risk ratio [RR] 0.93, 95% CI 0.83–1.03). Results were similar among patients with *A. baumannii* infections (RR 0.97, 95% CI 0.87–1.09). No differences were noted in clinical failure (76% vs. 71%; P = 0.22) or 28-day mortality (41% vs. 41%; P = 0.84) in comparing monotherapy and combination therapy arms. All-cause 28-day mortality was 86 (43%) of 198 patients treated with colistin monotherapy and 94 (45%) of 208 patients treated with combination therapy. Combination therapy increased the incidence of diarrhea (56 [27%] vs 32 [16%] patients) and decreased the incidence of mild renal failure (37 [30%] of 124 vs 25 [20%] of 125 patients at risk of or with kidney injury). There were no significant differences (6% for monotherapy versus 5% for combination therapy; P = 0.77) noted as it relates to colistin-resistance during or after therapy or isolation of new carbapenem-resistant bacteria. As it relates to infection type, most patients had hospital-acquired or ventilator associated pneumonia or bacteraemia (355/406, 87%).

Finally, there is an ongoing RCT comparing colistin monotherapy to colistin plus meropenem combination therapy for the management of invasive infections due to carbapenem-resistant Gram-negative organisms (https://clinicaltrials.

gov/ct2/show/NCT01597973). Data from this study, should further elucidate the role of polymyxin combinations. Furthermore, given the potential advantages of polymyxin B over colistin, clinical data assessing the impact of polymyxin B-based combination regimens are needed. Future studies should also address the impact of infection site and resistance mechanisms on the effectiveness of combination therapy.

16.6 Conclusions and Future Directions

In general, the *in vitro* data for polymyxin combination therapy suggests a potential benefit with many drug combinations, particularly so when only the more sophisticated PK/PD models are considered. A common finding is that low, sub-MIC (yet clinically achievable) concentrations of polymyxins (e.g. 0.5 mg/L) in combination with another agent may significantly enhance bacterial killing even when resistance to one or more of the drugs in combination is present. This may be true not only when the second drug would normally be active against the particular bacterial species but also with agents such as the glycopeptides that should ordinarily have no effect on Gram-negative organisms due to the relative impermeability of the outer membrane. Such an observation is important as total (i.e. bound and unbound) plasma concentrations of colistin (following IV administration of CMS) and polymyxin B are typically in the range of ~2–3 mg at steady state, although a proportion of patients will achieve lower plasma concentrations (Chap. 15) [63, 81, 85, 90, 115, 146, 156, 157, 204]. Given this situation it may nevertheless be possible to enhance bacterial killing with polymyxin combination therapy even in patients who achieve low plasma concentrations with standard dosage regimens. Alternatively, it may be possible to take advantage of increased bacterial killing at low plasma concentrations by using lower-than-normal doses of polymyxins, especially given the toxicity concerns associated with their use (discussed in Chap. 17).

A close look at the existing *in vitro* data on combination therapy also reveals that even when improvements in bacterial killing were not observed at later time points (e.g. 24 or 48 h), in many cases there were improvement in initial killing (e.g. up to 6 h). While regrowth obviously occurred in these situations it must be remembered that the *in vitro* models used lack the immune components present *in vivo*. Thus, in an immunocompetent host combination therapy at the commencement of treatment may help to quickly reduce bacterial levels to facilitate clearance by the immune system. Importantly, the few studies undertaken in PK/PD models have shown a substantial reduction in the emergence of polymyxin-resistant subpopulations. Given the increasing emergence of polymyxin resistance since their reintroduction into clinical practice [3, 7, 77, 86, 110, 112, 170] and their role as a last-line therapeutic option, combination therapy could potentially play an important role in minimising further resistance development.

Finally, the data would suggest that a 'one-size-fits-all' approach to identifying optimal combination regimens is not appropriate. This can be illustrated by the study conducted by Clancy et al. where isolates of *K. pneumoniae* responded differently to a colistin/doripenem combination depending on the presence or absence of particular resistance mechanisms [36]. Thus, the specific resistance mechanisms manifested by different isolates of a bacterial species may dictate the efficacy of particular combination regimens. Ultimately the true value of combination therapy must be evaluated in well-designed, well powered, randomized clinical trials in critically ill patients which are urgently required in order to define the clinical benefit of polymyxin combination therapy. Future advances in rapid diagnostics and next-generation omics technologies will guide optimal use of polymyxin combinations which are precise to each patient, infection site, pathogen and resistance mechanism.

References

1. Abdul Rahim N, Cheah SE, Johnson MD, Yu H, Sidjabat HE, Boyce J, Butler MS, Cooper MA, Fu J, Paterson DL, Nation RL, Bergen PJ, Velkov T, Li J (2015) Synergistic killing of NDM-producing MDR Klebsiella pneumoniae by two 'old' antibiotics-polymyxin B and chloramphenicol. J Antimicrob Chemother 70:2589–2597
2. Adams EK, Ashcraft DS, Pankey GA (2016) In vitro synergistic activity of caspofungin plus polymyxin B against fluconazole-resistant Candida glabrata. Am J Med Sci 351:265–270
3. Al-Sweih NA, Al-Hubail MA, Rotimi VO (2011) Emergence of tigecycline and colistin resistance in acinetobacter species isolated from patients in Kuwait hospitals. J Chemother 23:13–16
4. Albur M, Noel A, Bowker K, Macgowan A (2012) Bactericidal activity of multiple combinations of tigecycline and colistin against NDM-1-producing Enterobacteriaceae. Antimicrob Agents Chemother 56:3441–3443
5. Albur MS, Noel A, Bowker K, Macgowan A (2015) The combination of colistin and fosfomycin is synergistic against NDM-1-producing Enterobacteriaceae in in vitro pharmacokinetic/pharmacodynamic model experiments. Int J Antimicrob Agents 46:560–567
6. Anonymous (2011) Instructions to authors. Antimicrob Agents Chemother 55:1–23
7. Antoniadou A, Kontopidou F, Poulakou G, Koratzanis E, Galani I, Papadomichelakis E, Kopterides P, Souli M, Armaganidis A, Giamarellou H (2007) Colistin-resistant isolates of Klebsiella pneumoniae emerging in intensive care unit patients: first report of a multiclonal cluster. J Antimicrob Chemother 59:786–790
8. Aoki N, Tateda K, Kikuchi Y, Kimura S, Miyazaki C, Ishii Y, Tanabe Y, Gejyo F, Yamaguchi K (2009) Efficacy of colistin combination therapy in a mouse model of pneumonia caused by multidrug-resistant Pseudomonas aeruginosa. J Antimicrob Chemother 63:534–542
9. Attridge RT, Carden MF, Padilla S, Nathisuwan S, Burgess DS (2008) Colistin and rifampicin alone and in combination against multidrug-resistant *Acinetobacter baumannii* using an in vitro PK-PD model (abstract C1-1053, p69). In: Abstracts of the 48th annual interscience conference on antimicrobial agents and chemotherapy (ICAAC), Washington, DC, October 25–28. American Society for Microbiology
10. Aydemir H, Akduman D, Piskin N, Comert F, Horuz E, Terzi A, Kokturk F, Ornek T, Celebi G (2013) Colistin vs. the combination of colistin and rifampicin for the treatment of carbapenem-resistant Acinetobacter baumannii ventilator-associated pneumonia. Epidemiol Infect 141:1214–1222
11. Batirel A, Balkan I, Karabay O, Agalar C, Akalin S, Alici O, Alp E, Altay FA, Altin N, Arslan F, Aslan T, Bekiroglu N, Cesur S, Celik AD, Dogan M, Durdu B, Duygu F, Engin A, Engin DO, Gonen I, Guclu E, Guven T, Hatipoglu CA, Hosoglu S, Karahocagil MK, Kilic AU, Ormen B, Ozdemir D, Ozer S, Oztoprak N, Sezak N, Turhan V, Turker N, Yilmaz H (2014) Comparison of colistin-carbapenem, colistin-sulbactam, and colistin plus other antibacterial agents for the treatment of extremely drug-resistant Acinetobacter baumannii bloodstream infections. Eur J Clin Microbiol Infect Dis 33(8):1311–1322
12. Bergen PJ, Bulitta JB, Forrest A, Tsuji BT, Li J, Nation RL (2010) Pharmacokinetic/pharmacodynamic investigation of colistin against Pseudomonas aeruginosa using an in vitro model. Antimicrob Agents Chemother 54:3783–3789
13. Bergen PJ, Forrest A, Bulitta JB, Tsuji BT, Sidjabat HE, Paterson DL, Li J, Nation RL (2011) Clinically relevant plasma concentrations of colistin in combination with imipenem enhance pharmacodynamic activity against multidrug-resistant Pseudomonas aeruginosa at multiple inocula. Antimicrob Agents Chemother 55:5134–5142
14. Bergen PJ, Li J, Nation RL, Turnidge JD, Coulthard K, Milne RW (2008) Comparison of once-, twice- and thrice-daily dosing of colistin on antibacterial effect and emergence of resistance: studies with Pseudomonas aeruginosa in an in vitro pharmacodynamic model. J Antimicrob Chemother 61:636–642
15. Bergen PJ, Li J, Rayner CR, Nation RL (2006) Colistin methanesulfonate is an inactive prodrug of colistin against Pseudomonas aeruginosa. Antimicrob Agents Chemother 50:1953–1958
16. Bergen PJ, Tsuji BT, Bulitta JB, Forrest A, Jacob J, Sidjabat HE, Paterson DL, Nation RL, Li J (2011) Synergistic killing of multidrug-resistant Pseudomonas aeruginosa at multiple inocula by colistin combined with doripenem in an in vitro pharmacokinetic/pharmacodynamic model. Antimicrob Agents Chemother 55:5685–5695
17. Betrosian AP, Frantzeskaki F, Xanthaki A, Georgiadis G (2007) High-dose ampicillin-sulbactam as an alternative treatment of late-onset VAP from multidrug-resistant Acinetobacter baumannii. Scand J Infect Dis 39:38–43
18. Betts JW, Phee LM, Hornsey M, Woodford N, Wareham DW (2014) In vitro and in vivo activity of tigecycline/colistin combination therapies against carbapenem resistant Enterobacteriaceae. Antimicrob Agents Chemother 58:3541–3546
19. Betts JW, Phee LM, Woodford N, Wareham DW (2014) Activity of colistin in combination with tigecycline or rifampicin against multidrug-resistant Stenotrophomonas maltophilia. Eur J Clin Microbiol Infect Dis 33(9):1565–1572
20. Bhavnani SM, Hammel JP, Cirincione BB, Wikler MA, Ambrose PG (2005) Use of pharmacokinetic-pharmacodynamic target attainment analyses to

support phase 2 and 3 dosing strategies for doripenem. Antimicrob Agents Chemother 49:3944–3947
21. Blaser J (1985) In-vitro model for simultaneous simulation of the serum kinetics of two drugs with different half-lives. J Antimicrob Chemother 15(Suppl A):125–130
22. Bulitta JB, Landersdorfer CB, Forrest A, Brown SV, Neely MN, Tsuji BT, Louie A (2011) Relevance of pharmacokinetic and pharmacodynamic modeling to clinical care of critically ill patients. Curr Pharm Biotechnol 12:2044–2061
23. Bulitta JB, Li J, Poudyal A, Yu HH, Owen RJ, Tsuji BT, Nation RL, Forrest A (2009) Quantifying synergy of colistin combinations against MDR Gram-negatives by mechanism-based models (abstract A1-573, p41). In: Abstracts of the 49th annual interscience conference of antimicrobial agents and chemotherapy (ICAAC), San Francisco, CA, September 12–15. American Society for Microbiology
24. Bulitta JB, Yang JC, Yohonn L, Ly NS, Brown SV, D'Hondt RE, Jusko WJ, Forrest A, Tsuji BT (2010) Attenuation of colistin bactericidal activity by high inoculum of Pseudomonas aeruginosa characterized by a new mechanism-based population pharmacodynamic model. Antimicrob Agents Chemother 54:2051–2062
25. Bulman ZP, Ly NS, Lenhard JR, Holden PN, Bulitta JB, Tsuji BT (2017) Influence of rhlR and lasR on polymyxin pharmacodynamics in Pseudomonas aeruginosa and implications for quorum sensing inhibition with azithromycin. Antimicrob Agents Chemother 61(4):pii: e00096-16
26. Bulman ZP, Sutton MD, Ly NS, Bulitta JB, Holden PN, Nation RL, Li J, Tsuji BT (2015) Emergence of polymyxin B resistance influences pathogenicity in Pseudomonas aeruginosa mutators. Antimicrob Agents Chemother 59:4343–4346
27. Cai X, Yang Z, Dai J, Chen K, Zhang L, Ni W, Wei C, Cui J (2017) Pharmacodynamics of tigecycline alone and in combination with colistin against clinical isolates of multidrug-resistant Acinetobacter baumannii in an in vitro pharmacodynamic model. Int J Antimicrob Agents 49:609–616
28. Cappelletty DM, Rybak MJ (1996) Comparison of methodologies for synergism testing of drug combinations against resistant strains of Pseudomonas aeruginosa. Antimicrob Agents Chemother 40:677–683
29. Cavalcanti AB, Goncalves AR, Almeida CS, Bugano DD, Silva E (2010) Teicoplanin versus vancomycin for proven or suspected infection. Cochrane Database Syst Rev. https://doi.org/10.1002/14651858.CD007022.pub2
30. Ceccarelli G, Falcone M, Giordano A, Mezzatesta ML, Caio C, Stefani S, Venditti M (2013) Successful ertapenem-doripenem combination treatment of bacteremic ventilator-associated pneumonia due to colistin-resistant KPC-producing Klebsiella pneumoniae. Antimicrob Agents Chemother 57:2900–2901
31. Champion OL, Cooper IA, James SL, Ford D, Karlyshev A, Wren BW, Duffield M, Oyston PC, Titball RW (2009) Galleria mellonella as an alternative infection model for Yersinia pseudotuberculosis. Microbiology 155:1516–1522
32. Choi MJ, Ko KS (2013) Mutant prevention concentrations of colistin for Acinetobacter baumannii, Pseudomonas aeruginosa and Klebsiella pneumoniae clinical isolates. J Antimicrob Chemother 69(1):275–277
33. Cirioni O, Ghiselli R, Orlando F, Silvestri C, Mocchegiani F, Rocchi M, Chiodi L, Abbruzzetti A, Saba V, Scalise G, Giacometti A (2007) Efficacy of colistin/rifampin combination in experimental rat models of sepsis due to a multiresistant Pseudomonas aeruginosa strain. Crit Care Med 35:1717–1723
34. Cirioni O, Ghiselli R, Silvestri C, Kamysz W, Orlando F, Mocchegiani F, Di Matteo F, Riva A, Lukasiak J, Scalise G, Saba V, Giacometti A (2007) Efficacy of tachyplesin III, colistin, and imipenem against a multiresistant Pseudomonas aeruginosa strain. Antimicrob Agents Chemother 51:2005–2010
35. Cirioni O, Simonetti O, Pierpaoli E, Barucca A, Ghiselli R, Orlando F, Pelloni M, Trombettoni MM, Guerrieri M, Offidani A, Giacometti A, Provinciali M (2016) Colistin enhances therapeutic efficacy of daptomycin or teicoplanin in a murine model of multiresistant Acinetobacter baumannii sepsis. Diagn Microbiol Infect Dis 86:392–398
36. Clancy CJ, Chen L, Hong JH, Cheng S, Hao B, Shields RK, Farrell AN, Doi Y, Zhao Y, Perlin DS, Kreiswirth BN, Nguyen MH (2013) Mutations of the ompK36 porin gene and promoter impact responses of sequence type 258, KPC-2-producing Klebsiella pneumoniae strains to doripenem and doripenem-colistin. Antimicrob Agents Chemother 57:5258–5265
37. Cobo J, Morosini MI, Pintado V, Tato M, Samaranch N, Baquero F, Canton R (2008) Use of tigecycline for the treatment of prolonged bacteremia due to a multiresistant VIM-1 and SHV-12 beta – lactamase-producing Klebsiella pneumoniae epidemic clone. Diagn Microbiol Infect Dis 60:319–322
38. Conway SP, Pond MN, Watson A, Etherington C, Robey HL, Goldman MH (1997) Intravenous colistin sulphomethate in acute respiratory exacerbations in adult patients with cystic fibrosis. Thorax 52:987–993
39. Cordoba J, Coronado-Alvarez NM, Parra D, Parra-Ruiz J (2015) In vitro activities of novel antimicrobial combinations against extensively drug-resistant Acinetobacter baumannii. Antimicrob Agents Chemother 59:7316–7319
40. Corvec S, Furustrand Tafin U, Betrisey B, Borens O, Trampuz A (2013) Activities of fosfomycin, tigecycline, colistin, and gentamicin against extended-spectrum-beta-lactamase-producing Escherichia coli in a foreign-body infection model. Antimicrob Agents Chemother 57:1421–1427

41. Costerton JW, Stewart PS, Greenberg EP (1999) Bacterial biofilms: a common cause of persistent infections. Science 284:1318–1322
42. Daikos G, Markogiannakis A (2011) Carbapenemase-producing Klebsiella pneumoniae: (when) might we still consider treating with carbapenems? Clin Microbiol Infect 17:1135–1141
43. Dalfino L, Puntillo F, Mosca A, Monno R, Spada ML, Coppolecchia S, Miragliotta G, Bruno F, Brienza N (2012) High-dose, extended-interval colistin administration in critically ill patients: is this the right dosing strategy? A preliminary study. Clin Infect Dis 54:1720–1726
44. Demiraslan H, Dinc G, Ahmed SS, Elmali F, Metan G, Alp E, Doganay M (2013) Carbapenem-resistant Klebsiella pneumoniae sepsis in corticosteroid receipt mice: tigecycline or colistin monotherapy versus tigecycline/colistin combination. J Chemother 26(5):276–281. https://doi.org/10.1179/1973947813Y.0000000143
45. Deris ZZ, Yu HH, Davis K, Soon RL, Jacob J, Ku CK, Poudyal A, Bergen PJ, Tsuji BT, Bulitta JB, Forrest A, Paterson DL, Velkov T, Li J, Nation RL (2012) The combination of colistin and doripenem is synergistic against Klebsiella pneumoniae at multiple inocula and suppresses colistin resistance in an in vitro pharmacokinetic/pharmacodynamic model. Antimicrob Agents Chemother 56:5103–5112
46. Di X, Wang R, Liu B, Zhang X, Ni W, Wang J, Liang B, Cai Y, Liu Y (2015) In vitro activity of fosfomycin in combination with colistin against clinical isolates of carbapenem-resistant Pseudomas aeruginosa. J Antibiot (Tokyo) 68:551–555
47. Dinc G, Demiraslan H, Elmali F, Ahmed SS, Metan G, Alp E, Doganay M (2014) Efficacy of sulbactam and its combination with imipenem, colistin and tigecycline in an experimental model of carbapenem-resistant Acinetobacter baumannii sepsis. Chemotherapy 59:325–329
48. Donlan RM (2002) Biofilms: microbial life on surfaces. Emerg Infect Dis 8:881–890
49. Dubrovskaya Y, Chen TY, Scipione MR, Esaian D, Phillips MS, Papadopoulos J, Mehta SA (2013) Risk factors for treatment failure of polymyxin B monotherapy for carbapenem-resistant Klebsiella pneumoniae infections. Antimicrob Agents Chemother 57:5394–5397
50. Dudhani RV, Turnidge JD, Nation RL, Li J (2010) fAUC/MIC is the most predictive pharmacokinetic/pharmacodynamic index of colistin against Acinetobacter baumannii in murine thigh and lung infection models. J Antimicrob Chemother 65:1984–1990
51. Durante-Mangoni E, Signoriello G, Andini R, Mattei A, De Cristoforo M, Murino P, Bassetti M, Malacarne P, Petrosillo N, Galdieri N, Mocavero P, Corcione A, Viscoli C, Zarrilli R, Gallo C, Utili R (2013) Colistin and rifampicin compared with colistin alone for the treatment of serious infections due to extensively drug-resistant Acinetobacter baumannii: a multicenter, randomized clinical trial. Clin Infect Dis 57:349–358
52. Elias LS, Konzen D, Krebs JM, Zavascki AP (2010) The impact of polymyxin B dosage on in-hospital mortality of patients treated with this antibiotic. J Antimicrob Chemother 65:2231–2237
53. Entenza JM, Moreillon P (2009) Tigecycline in combination with other antimicrobials: a review of in vitro, animal and case report studies. Int J Antimicrob Agents 34(8):e1–e9
54. Falagas ME, Bliziotis IA, Kasiakou SK, Samonis G, Athanassopoulou P, Michalopoulos A (2005) Outcome of infections due to pandrug-resistant (PDR) Gram-negative bacteria. BMC Infect Dis 5:24
55. Falagas ME, Rafailidis PI, Ioannidou E, Alexiou VG, Matthaiou DK, Karageorgopoulos DE, Kapaskelis A, Nikita D, Michalopoulos A (2010) Colistin therapy for microbiologically documented multidrug-resistant Gram-negative bacterial infections: a retrospective cohort study of 258 patients. Int J Antimicrob Agents 35:194–199
56. Falagas ME, Rafailidis PI, Kasiakou SK, Hatzopoulou P, Michalopoulos A (2006) Effectiveness and nephrotoxicity of colistin monotherapy vs. colistin-meropenem combination therapy for multidrug-resistant Gram-negative bacterial infections. Clin Microbiol Infect 12:1227–1230
57. Falagas ME, Rafailidis PI, Matthaiou DK, Virtzili S, Nikita D, Michalopoulos A (2008) Pandrug-resistant Klebsiella pneumoniae, Pseudomonas aeruginosa and Acinetobacter baumannii infections: characteristics and outcome in a series of 28 patients. Int J Antimicrob Agents 32:450–454
58. Fan B, Guan J, Wang X, Cong Y (2016) Activity of colistin in combination with meropenem, tigecycline, fosfomycin, fusidic acid, rifampin or sulbactam against extensively drug-resistant Acinetobacter baumannii in a murine thigh-infection model. PLoS One 11:e0157757
59. Furtado GHC, D'Azevedo PA, Santos AF, Gales AC, Pignatari ACC, Medeiros EAS (2007) Intravenous polymyxin B for the treatment of nosocomial pneumonia caused by multidrug-resistant *Pseudomonas aeruginosa*. Int J Antimicrob Agents 30:315–319
60. Gaibani P, Lombardo D, Lewis RE, Mercuri M, Bonora S, Landini MP, Ambretti S (2014) In vitro activity and post-antibiotic effects of colistin in combination with other antimicrobials against colistin-resistant KPC-producing Klebsiella pneumoniae bloodstream isolates. J Antimicrob Chemother 69(7):1856–1865
61. Galani I, Orlandou K, Moraitou H, Petrikkos G, Souli M (2014) Colistin/daptomycin: an unconventional antimicrobial combination synergistic in vitro against multidrug-resistant Acinetobacter baumannii. Int J Antimicrob Agents 43:370–374
62. Garnacho-Montero J, Amaya-Villar R, Gutierrez-Pizarraya A, Espejo-Gutierrez De Tena E, Artero-

Gonzalez ML, Corcia-Palomo Y, Bautista-Paloma J (2013) Clinical efficacy and safety of the combination of colistin plus vancomycin for the treatment of severe infections caused by carbapenem-resistant Acinetobacter baumannii. Chemotherapy 59:225–231
63. Garonzik SM, Li J, Thamlikitkul V, Paterson DL, Shoham S, Jacob J, Silveira FP, Forrest A, Nation RL (2011) Population pharmacokinetics of colistin methanesulfonate and formed colistin in critically ill patients from a multicenter study provide dosing suggestions for various categories of patients. Antimicrob Agents Chemother 55:3284–3294
64. Giacometti A, Cirioni O, Ghiselli R, Orlando F, Mocchegiani F, D'Amato G, Silvestri C, Riva A, Del Prete MS, Saba V, Scalise G (2003) Antiendotoxin activity of antimicrobial peptides and glycopeptides. J Chemother 15:129–133
65. Gloede J, Scheerans C, Derendorf H, Kloft C (2010) In vitro pharmacodynamic models to determine the effect of antibacterial drugs. J Antimicrob Chemother 65:186–201
66. Gordon NC, Png K, Wareham DW (2010) Potent synergy and sustained bactericidal activity of a vancomycin-colistin combination versus multidrug-resistant strains of Acinetobacter baumannii. Antimicrob Agents Chemother 54:5316–5322
67. Gunderson BW, Ibrahim KH, Hovde LB, Fromm TL, Reed MD, Rotschafer JC (2003) Synergistic activity of colistin and ceftazidime against multiantibiotic-resistant Pseudomonas aeruginosa in an in vitro pharmacodynamic model. Antimicrob Agents Chemother 47:905–909
68. Hagihara M, Housman ST, Nicolau DP, Kuti JL (2014) In vitro pharmacodynamics of polymyxin B and tigecycline alone and in combination against carbapenem-resistant Acinetobacter baumannii. Antimicrob Agents Chemother 58:874–879
69. Hengzhuang W, Wu H, Ciofu O, Song Z, Hoiby N (2011) Pharmacokinetics/pharmacodynamics of colistin and imipenem on mucoid and nonmucoid Pseudomonas aeruginosa biofilms. Antimicrob Agents Chemother 55:4469–4474
70. Hengzhuang W, Wu H, Ciofu O, Song Z, Hoiby N (2012) In vivo pharmacokinetics/pharmacodynamics of colistin and imipenem in Pseudomonas aeruginosa biofilm infection. Antimicrob Agents Chemother 56:2683–2690
71. Henry R, Vithanage N, Harrison P, Seemann T, Coutts S, Moffatt JH, Nation RL, Li J, Harper M, Adler B, Boyce JD (2012) Colistin-resistant, lipopolysaccharide-deficient Acinetobacter baumannii responds to lipopolysaccharide loss through increased expression of genes involved in the synthesis and transport of lipoproteins, phospholipids, and poly-beta-1,6-N-acetylglucosamine. Antimicrob Agents Chemother 56:59–69
72. Herrmann G, Yang L, Wu H, Song Z, Wang H, Hoiby N, Ulrich M, Molin S, Riethmuller J, Doring G (2010) Colistin-tobramycin combinations are superior to monotherapy concerning the killing of biofilm Pseudomonas aeruginosa. J Infect Dis 202:1585–1592
73. Hornsey M, Phee L, Longshaw C, Wareham DW (2013) In vivo efficacy of telavancin/colistin combination therapy in a Galleria mellonella model of Acinetobacter baumannii infection. Int J Antimicrob Agents 41:285–287
74. Hornsey M, Wareham DW (2011) In vivo efficacy of glycopeptide-colistin combination therapies in a Galleria mellonella model of Acinetobacter baumannii infection. Antimicrob Agents Chemother 55:3534–3537
75. Imberti R, Cusato M, Villani P, Carnevale L, Iotti GA, Langer M, Regazzi M (2010) Steady-state pharmacokinetics and BAL concentration of colistin in critically Ill patients after IV colistin methanesulfonate administration. Chest 138:1333–1339
76. Jander G, Rahme LG, Ausubel FM (2000) Positive correlation between virulence of Pseudomonas aeruginosa mutants in mice and insects. J Bacteriol 182:3843–3845
77. Johansen HK, Moskowitz SM, Ciofu O, Pressler T, Hoiby N (2008) Spread of colistin resistant non-mucoid Pseudomonas aeruginosa among chronically infected Danish cystic fibrosis patients. J Cyst Fibros 7:391–397
78. Kasiakou SK, Michalopoulos A, Soteriades ES, Samonis G, Sermaides GJ, Falagas ME (2005) Combination therapy with intravenous colistin for management of infections due to multidrug-resistant Gram-negative bacteria in patients without cystic fibrosis. Antimicrob Agents Chemother 49:3136–3146
79. Kaye KS, Pogue JM, Tran TB, Nation RL, Li J (2016) Agents of last resort: polymyxin resistance. Infect Dis Clin N Am 30:391–414
80. Ku YH, Chen CC, Lee MF, Chuang YC, Tang HJ, Yu WL (2017) Comparison of synergism between colistin, fosfomycin and tigecycline against extended-spectrum beta-lactamase-producing Klebsiella pneumoniae isolates or with carbapenem resistance. J Microbiol Immunol Infect 50(6):931–939
81. Kwa AL, Lim TP, Low JG, Hou J, Kurup A, Prince RA, Tam VH (2008) Pharmacokinetics of polymyxin B1 in patients with multidrug-resistant Gram-negative bacterial infections. Diagn Microbiol Infect Dis 60:163–167
82. Lagerback P, Khine WW, Giske CG, Tangden T (2016) Evaluation of antibacterial activities of colistin, rifampicin and meropenem combinations against NDM-1-producing Klebsiella pneumoniae in 24 h in vitro time-kill experiments. J Antimicrob Chemother 71:2321–2325
83. Lee GC, Burgess DS (2013) Polymyxins and doripenem combination against KPC-producing Klebsiella pneumoniae. J Clin Med Res 5:97–100
84. Lee HJ, Bergen PJ, Bulitta JB, Tsuji B, Forrest A, Nation RL, Li J (2013) Synergistic activity of colistin and rifampin combination against

multidrug-resistant Acinetobacter baumannii in an in vitro pharmacokinetic/pharmacodynamic model. Antimicrob Agents Chemother 57:3738–3745

85. Lee J, Han S, Jeon S, Hong T, Song W, Woo H, Yim DS (2013) Population pharmacokinetic analysis of colistin in burn patients. Antimicrob Agents Chemother 57:2141–2146
86. Lee JY, Song JH, Ko KS (2011) Identification of nonclonal Pseudomonas aeruginosa isolates with reduced colistin susceptibility in Korea. Microb Drug Resist 17:299–304
87. Leite GC, Oliveira MS, Perdigao-Neto LV, Rocha CK, Guimaraes T, Rizek C, Levin AS, Costa SF (2016) Antimicrobial combinations against pan-resistant Acinetobacter baumannii isolates with different resistance mechanisms. PLoS One 11:e0151270
88. Lenhard JR, Smith NM, Bulman ZP, Tao X, Thamlikitkul V, Shin BS, Nation RL, Li J, Bulitta JB, Tsuji BT (2017) High dose ampicillin/sulbactam combinations combat polymyxin-resistant Acinetobacter baumannii in a hollow-fiber infection model. Antimicrob Agents Chemother 61(3):pii: e01268-16
89. Lenhard JR, Thamlikitkul V, Silveira FP, Garonzik SM, Tao X, Forrest A, Soo Shin B, Kaye KS, Bulitta JB, Nation RL, Li J, Tsuji BT (2017) Polymyxin-resistant, carbapenem-resistant Acinetobacter baumannii is eradicated by a triple combination of agents that lack individual activity. J Antimicrob Chemother 72:1415–1420
90. Li J, Coulthard K, Milne R, Nation RL, Conway S, Peckham D, Etherington C, Turnidge J (2003) Steady-state pharmacokinetics of intravenous colistin methanesulphonate in patients with cystic fibrosis. J Antimicrob Chemother 52:987–992
91. Li J, Milne RW, Nation RL, Turnidge JD, Coulthard K (2003) Stability of colistin and colistin methanesulfonate in aqueous media and plasma as determined by high-performance liquid chromatography. Antimicrob Agents Chemother 47:1364–1370
92. Li J, Nation RL, Owen RJ, Wong S, Spelman D, Franklin C (2007) Antibiograms of multidrug-resistant clinical Acinetobacter baumannii: promising therapeutic options for treatment of infection with colistin-resistant strains. Clin Infect Dis 45:594–598
93. Li J, Rayner CR, Nation RL, Owen RJ, Spelman D, Tan KE, Liolios L (2006) Heteroresistance to colistin in multidrug-resistant Acinetobacter baumannii. Antimicrob Agents Chemother 50:2946–2950
94. Liang W, Liu XF, Huang J, Zhu DM, Li J, Zhang J (2011) Activities of colistin- and minocycline-based combinations against extensive drug resistant Acinetobacter baumannii isolates from intensive care unit patients. BMC Infect Dis 11:109
95. Lim TP, Tan TY, Lee W, Sasikala S, Tan TT, Hsu LY, Kwa AL (2009) In vitro activity of various combinations of antimicrobials against carbapenem-resistant Acinetobacter species in Singapore. J Antibiot (Tokyo) 62:675–679
96. Lin KH, Chuang YC, Lee SH, Yu WL (2010) In vitro synergistic antimicrobial effect of imipenem and colistin against an isolate of multidrug-resistant Enterobacter cloacae. J Microbiol Immunol Infect 43:317–322
97. Lin L, Nonejuie P, Munguia J, Hollands A, Olson J, Dam Q, Kumaraswamy M, Rivera H Jr, Corriden R, Rohde M, Hensler ME, Burkart MD, Pogliano J, Sakoulas G, Nizet V (2015) Azithromycin synergizes with cationic antimicrobial peptides to exert bactericidal and therapeutic activity against highly multidrug-resistant Gram-negative bacterial pathogens. EBioMedicine 2:690–698
98. Linden PK, Kusne S, Coley K, Fontes P, Kramer DJ, Paterson D (2003) Use of parenteral colistin for the treatment of serious infection due to antimicrobial-resistant Pseudomonas aeruginosa. Clin Infect Dis 37:e154–e160
99. Lister PD, Wolter DJ, Hanson ND (2009) Antibacterial-resistant Pseudomonas aeruginosa: clinical impact and complex regulation of chromosomally encoded resistance mechanisms. Clin Microbiol Rev 22:582–610
100. Liu X, Zhao M, Chen Y, Bian X, Li Y, Shi J, Zhang J (2016) Synergistic killing by meropenem and colistin combination of carbapenem-resistant Acinetobacter baumannii isolates from Chinese patients in an in vitro pharmacokinetic/pharmacodynamic model. Int J Antimicrob Agents 48:559–563
101. Loos U, Musch E, Jensen JC, Mikus G, Schwabe HK, Eichelbaum M (1985) Pharmacokinetics of oral and intravenous rifampicin during chronic administration. Klin Wochenschr 63:1205–1211
102. Lora-Tamayo J, Murillo O, Bergen PJ, Nation RL, Poudyal A, Luo X, Yu HY, Ariza J, Li J (2014) Activity of colistin combined with doripenem at clinically relevant concentrations against multidrug-resistant Pseudomonas aeruginosa in an in vitro dynamic biofilm model. J Antimicrob Chemother 69(9):2434–2442
103. Ly NS, Bulitta JB, Rao GG, Landersdorfer CB, Holden PN, Forrest A, Bergen PJ, Nation RL, Li J, Tsuji BT (2015) Colistin and doripenem combinations against Pseudomonas aeruginosa: profiling the time course of synergistic killing and prevention of resistance. J Antimicrob Chemother 70:1434–1442
104. Ly NS, Bulman ZP, Bulitta JB, Baron C, Rao GG, Holden PN, Li J, Sutton MD, Tsuji BT (2016) Optimization of polymyxin B in combination with doripenem to combat mutator Pseudomonas aeruginosa. Antimicrob Agents Chemother 60:2870–2880
105. Maccallum DM, Desbois AP, Coote PJ (2013) Enhanced efficacy of synergistic combinations of antimicrobial peptides with caspofungin versus Candida albicans in insect and murine models of systemic infection. Eur J Clin Microbiol Infect Dis 32:1055–1062

106. Magiorakos AP, Srinivasan A, Carey RB, Carmeli Y, Falagas ME, Giske CG, Harbarth S, Hindler JF, Kahlmeter G, Olsson-Liljequist B, Paterson DL, Rice LB, Stelling J, Struelens MJ, Vatopoulos A, Weber JT, Monnet DL (2012) Multidrug-resistant, extensively drug-resistant and pandrug-resistant bacteria: an international expert proposal for interim standard definitions for acquired resistance. Clin Microbiol Infect 18:268–281
107. Mah TF, O'Toole GA (2001) Mechanisms of biofilm resistance to antimicrobial agents. Trends Microbiol 9:34–39
108. Markou N, Markantonis SL, Dimitrakis E, Panidis D, Boutzouka E, Karatzas S, Rafailidis P, Apostolakos H, Baltopoulos G (2008) Colistin serum concentrations after intravenous administration in critically ill patients with serious multidrug-resistant, Gram-negative bacilli infections: a prospective, open-label, uncontrolled study. Clin Ther 30:143–151
109. Mendes RE, Fritsche TR, Sader HS, Jones RN (2008) Increased antimicrobial susceptibility profiles among polymyxin-resistant Acinetobacter baumannii clinical isolates. Clin Infect Dis 46:1324–1326
110. Menuet M, Bittar F, Stremler N, Dubus JC, Sarles J, Raoult D, Rolain JM (2008) First isolation of two colistin-resistant emerging pathogens, Brevundimonas diminuta and Ochrobactrum anthropi, in a woman with cystic fibrosis: a case report. J Med Case Rep 2:373
111. Mezzatesta ML, Caio C, Gona F, Zingali T, Salerno I, Stefani S (2016) Colistin Increases the cidal activity of antibiotic combinations against multidrug-resistant Klebsiella pneumoniae: an in vitro model comparing multiple combination bactericidal testing at one peak serum concentration and time-kill method. Microb Drug Resist 22:360–363
112. Mezzatesta ML, Gona F, Caio C, Petrolito V, Sciortino D, Sciacca A, Santangelo C, Stefani S (2011) Outbreak of KPC-3-producing, and colistin-resistant, Klebsiella pneumoniae infections in two Sicilian hospitals. Clin Microbiol Infect 17:1444–1447
113. Michail G, Labrou M, Pitiriga V, Manousaka S, Sakellaridis N, Tsakris A, Pournaras S (2013) Activity of tigecycline in combination with colistin, meropenem, rifampin, or gentamicin against KPC-producing Enterobacteriaceae in a murine thigh infection model. Antimicrob Agents Chemother 57:6028–6033
114. Moffatt JH, Harper M, Harrison P, Hale JD, Vinogradov E, Seemann T, Henry R, Crane B, St Michael F, Cox AD, Adler B, Nation RL, Li J, Boyce JD (2010) Colistin resistance in Acinetobacter baumannii is mediated by complete loss of lipopolysaccharide production. Antimicrob Agents Chemother 54:4971–4977
115. Mohamed AF, Karaiskos I, Plachouras D, Karvanen M, Pontikis K, Jansson B, Papadomichelakis E, Antoniadou A, Giamarellou H, Armaganidis A, Cars O, Friberg LE (2012) Application of a loading dose of colistin methanesulfonate in critically ill patients: population pharmacokinetics, protein binding, and prediction of bacterial kill. Antimicrob Agents Chemother 56:4241–4249
116. Moland ES, Craft DW, Hong SG, Kim SY, Hachmeister L, Sayed SD, Thomson KS (2008) In vitro activity of tigecycline against multidrug-resistant Acinetobacter baumannii and selection of tigecycline-amikacin synergy. Antimicrob Agents Chemother 52:2940–2942
117. Moneib NA (1995) In-vitro activity of commonly used antifungal agents in the presence of rifampin, polymyxin B and norfloxacin against Candida albicans. J Chemother 7:525–529
118. Montero A, Ariza J, Corbella X, Domenech A, Cabellos C, Ayats J, Tubau F, Borraz C, Gudiol F (2004) Antibiotic combinations for serious infections caused by carbapenem-resistant Acinetobacter baumannii in a mouse pneumonia model. J Antimicrob Chemother 54:1085–1091
119. Moore NM, Flaws ML (2011) Treatment strategies and recommendations for Pseudomonas aeruginosa infections. Clin Lab Sci 24:52–56
120. Morelli P, Ferrario A, Tordato F, Piazza A, Casari E (2014) Successful treatment of post-neurosurgical multidrug-resistant Pseudomonas aeruginosa meningo-encephalitis with combination therapy of colistin, rifampicin and doripenem. J Antimicrob Chemother 69:857–859
121. Moskowitz SM, Foster JM, Emerson J, Burns JL (2004) Clinically feasible biofilm susceptibility assay for isolates of Pseudomonas aeruginosa from patients with cystic fibrosis. J Clin Microbiol 42:1915–1922
122. Mutlu Yilmaz E, Sunbul M, Aksoy A, Yilmaz H, Guney AK, Guvenc T (2012) Efficacy of tigecycline/colistin combination in a pneumonia model caused by extensively drug-resistant Acinetobacter baumannii. Int J Antimicrob Agents 40:332–336
123. Napier BA, Band V, Burd EM, Weiss DS (2014) Colistin heteroresistance in Enterobacter cloacae is associated with cross-resistance to the host antimicrobial lysozyme. Antimicrob Agents Chemother 58:5594–5597
124. Nastro M, Rodriguez CH, Monge R, Zintgraff J, Neira L, Rebollo M, Vay C, Famiglietti A (2013) Activity of the colistin-rifampicin combination against colistin-resistant, carbapenemaseproducing Gram-negative bacteria. J Chemother 26(4):211–216
125. Nicolau DP (2008) Carbapenems: a potent class of antibiotics. Expert Opin Pharmacother 9:23–37
126. Norden CW, Wentzel H, Keleti E (1979) Comparison of techniques for measurement of in vitro antibiotic synergism. J Infect Dis 140:629–633
127. Oleksiuk LM, Nguyen MH, Press EG, Updike CL, O'Hara JA, Doi Y, Clancy CJ, Shields RK (2014) In vitro responses of Acinetobacter baumannii to two- and three-drug combinations following exposure to colistin and doripenem. Antimicrob Agents Chemother 58:1195–1199

128. Owen RJ, Li J, Nation RL, Spelman D (2007) In vitro pharmacodynamics of colistin against Acinetobacter baumannii clinical isolates. J Antimicrob Chemother 59:473–477
129. Ozbek B, Mataraci E (2013) In vitro effectiveness of colistin, tigecycline and levofloxacin alone and combined with clarithromycin and/or heparin as lock solutions against embedded Acinetobacter baumannii strains. J Antimicrob Chemother 68:827–830
130. Pachon-Ibanez ME, Docobo-Perez F, Lopez-Rojas R, Dominguez-Herrera J, Jimenez-Mejias ME, Garcia-Curiel A, Pichardo C, Jimenez L, Pachon J (2010) Efficacy of rifampin and its combinations with imipenem, sulbactam, and colistin in experimental models of infection caused by imipenem-resistant Acinetobacter baumannii. Antimicrob Agents Chemother 54:1165–1172
131. Padilla S, Carden MF, Attridge RT, Burgess DS (2008) In-vitro activity of colistin, amikacin and in combination against multi-drug resistant Acinetobacter baumannii in a pharmacodynamic-pharmacokinetic model. 4th annual Louis C. Littlefield celebrating pharmacy research excellence day in professional, graduate and postgraduate programs, University of Texas, Austin, TX, April 17
132. Pamp SJ, Gjermansen M, Johansen HK, Tolker-Nielsen T (2008) Tolerance to the antimicrobial peptide colistin in Pseudomonas aeruginosa biofilms is linked to metabolically active cells, and depends on the pmr and mexAB-oprM genes. Mol Microbiol 68:223–240
133. Pankey G, Ashcraft D, Kahn H, Ismail A (2014) Time-kill assay and Etest evaluation for synergy with polymyxin B and fluconazole against Candida glabrata. Antimicrob Agents Chemother 58:5795–5800
134. Pankuch GA, Lin G, Seifert H, Appelbaum PC (2008) Activity of meropenem with and without ciprofloxacin and colistin against Pseudomonas aeruginosa and Acinetobacter baumannii. Antimicrob Agents Chemother 52:333–336
135. Pankuch GA, Seifert H, Appelbaum PC (2010) Activity of doripenem with and without levofloxacin, amikacin, and colistin against Pseudomonas aeruginosa and Acinetobacter baumannii. Diagn Microbiol Infect Dis 67:191–197
136. Pantopoulou A, Giamarellos-Bourboulis EJ, Raftogannis M, Tsaganos T, Dontas I, Koutoukas P, Baziaka F, Giamarellou H, Perrea D (2007) Colistin offers prolonged survival in experimental infection by multidrug-resistant Acinetobacter baumannii: the significance of co-administration of rifampicin. Int J Antimicrob Agents 29:51–55
137. Park GC, Choi JA, Jang SJ, Jeong SH, Kim CM, Choi IS, Kang SH, Park G, Moon DS (2016) In vitro interactions of antibiotic combinations of colistin, tigecycline, and doripenem against extensively drug-resistant and multidrug-resistant Acinetobacter baumannii. Ann Lab Med 36:124–130
138. Paul M, Carmeli Y, Durante-Mangoni E, Mouton JW, Tacconelli E, Theuretzbacher U, Mussini C, Leibovici L (2014) Combination therapy for carbapenem-resistant Gram-negative bacteria. J Antimicrob Chemother 69(9):2305–2309. https://doi.org/10.1093/jac/dku168
139. Paul M, Daikos GL, Durante-Mangoni E, Yahav D, Carmeli Y, Benattar YD, Skiada A, Andini R, Eliakim-Raz N, Nutman A, Zusman O, Antoniadou A, Pafundi PC, Adler A, Dickstein Y, Pavleas I, Zampino R, Daitch V, Bitterman R, Zayyad H, Koppel F, Levi I, Babich T, Friberg LE, Mouton JW, Theuretzbacher U, Leibovici L (2018) Colistin alone versus colistin plus meropenem for treatment of severe infections caused by carbapenem-resistant Gram-negative bacteria: an open-label, randomised controlled trial. Lancet Infect Dis 18:391–400
140. Percin D, Akyol S, Kalin G (2014) In vitro synergism of combinations of colistin with selected antibiotics against colistin-resistant Acinetobacter baumannii. GMS Hyg Infect Control 9(2):Doc14
141. Petrosillo N, Giannella M, Antonelli M, Antonini M, Barsic B, Belancic L, Inkaya AC, De Pascale G, Grilli E, Tumbarello M, Akova M (2014) Clinical experience of colistin-glycopeptide combination in critically ill patients infected with Gram-negative bacteria. Antimicrob Agents Chemother 58:851–858
142. Petrosillo N, Ioannidou E, Falagas ME (2008) Colistin monotherapy vs. combination therapy: evidence from microbiological, animal and clinical studies. Clin Microbiol Infect 14:816–827
143. Phee LM, Betts JW, Bharathan B, Wareham DW (2015) Colistin and fusidic acid: a novel potent synergistic combination for the treatment of multi-drug resistant Acinetobacter baumannii infections. Antimicrob Agents Chemother 59(8):4544–4550
144. Pillai SK, Moellering RC, Eliopoulos GM (2005) Antimicrobial combinations. In: Lorian V (ed) Antibiotics in laboratory medicine, 5th edn. Lippincott Williams & Wilkins, Philadelphia
145. Pintado V, San Miguel LG, Grill F, Mejia B, Cobo J, Fortun J, Martin-Davila P, Moreno S (2008) Intravenous colistin sulphomethate sodium for therapy of infections due to multidrug-resistant gram-negative bacteria. J Infect 56:185–190
146. Plachouras D, Karvanen M, Friberg LE, Papadomichelakis E, Antoniadou A, Tsangaris I, Karaiskos I, Poulakou G, Kontopidou F, Armaganidis A, Cars O, Giamarellou H (2009) Population pharmacokinetic analysis of colistin methanesulphonate and colistin after intravenous administration in critically ill patients with Gram-negative bacterial infections. Antimicrob Agents Chemother 53:3430–3436
147. Poudyal A, Howden BP, Bell JM, Gao W, Owen RJ, Turnidge JD, Nation RL, Li J (2008) In vitro pharmacodynamics of colistin against multidrug-resistant Klebsiella pneumoniae. J Antimicrob Chemother 62:1311–1318
148. Pournaras S, Vrioni G, Neou E, Dendrinos J, Dimitroulia E, Poulou A, Tsakris A (2011) Activity of tigecycline alone and in combination with colistin and meropenem against Klebsiella pneumoniae car-

bapenemase (KPC)-producing Enterobacteriaceae strains by time-kill assay. Int J Antimicrob Agents 37:244–247
149. Principe L, Capone A, Mazzarelli A, D'Arezzo S, Bordi E, Di Caro A, Petrosillo N (2013) In vitro activity of doripenem in combination with various antimicrobials against multidrug-resistant Acinetobacter baumannii: possible options for the treatment of complicated infection. Microb Drug Resist 19:407–414
150. Qureshi ZA, Paterson DL, Potoski BA, Kilayko MC, Sandovsky G, Sordillo E, Polsky B, Adams-Haduch JM, Doi Y (2012) Treatment outcome of bacteremia due to KPC-producing Klebsiella pneumoniae: superiority of combination antimicrobial regimens. Antimicrob Agents Chemother 56:2108–2113
151. Rahal JJ (2006) Novel antibiotic combinations against infections with almost completely resistant Pseudomonas aeruginosa and Acinetobacter species. Clin Infect Dis 43(Suppl 2):S95–S99
152. Rodriguez-Avial I, Pena I, Picazo JJ, Rodriguez-Avial C, Culebras E (2015) In vitro activity of the next-generation aminoglycoside plazomicin alone and in combination with colistin, meropenem, fosfomycin or tigecycline against carbapenemase-producing Enterobacteriaceae strains. Int J Antimicrob Agents 46:616–621
153. Rodriguez CH, Nastro M, Vay C, Famiglietti A (2015) In vitro activity of minocycline alone or in combination in multidrug-resistant Acinetobacter baumannii isolates. J Med Microbiol 64:1196–1200
154. Ruslami R, Nijland HM, Alisjahbana B, Parwati I, Van Crevel R, Aarnoutse RE (2007) Pharmacokinetics and tolerability of a higher rifampin dose versus the standard dose in pulmonary tuberculosis patients. Antimicrob Agents Chemother 51:2546–2551
155. Safarika A, Galani I, Pistiki A, Giamarellos-Bourboulis EJ (2015) Time-kill effect of levofloxacin on multidrug-resistant Pseudomonas aeruginosa and Acinetobacter baumannii: synergism with imipenem and colistin. Eur J Clin Microbiol Infect Dis 34:317–323
156. Sandri AM, Landersdorfer CB, Jacob J, Boniatti MM, Dalarosa MG, Falci DR, Behle TF, Bordinhao RC, Wang J, Forrest A, Nation RL, Li J, Zavascki AP (2013) Population pharmacokinetics of intravenous polymyxin B in critically ill patients: implications for selection of dosage regimens. Clin Infect Dis 57:524–531
157. Sandri AM, Landersdorfer CB, Jacob J, Boniatti MM, Dalarosa MG, Falci DR, Behle TF, Saitovitch D, Wang J, Forrest A, Nation RL, Zavascki AP, Li J (2013) Pharmacokinetics of polymyxin B in patients on continuous venovenous haemodialysis. J Antimicrob Chemother 68:674–677
158. Schemuth H, Dittmer S, Lackner M, Sedlacek L, Hamprecht A, Steinmann E, Buer J, Rath PM, Steinmann J (2013) In vitro activity of colistin as single agent and in combination with antifungals against filamentous fungi occurring in patients with cystic fibrosis. Mycoses 56:297–303
159. Seed KD, Dennis JJ (2008) Development of Galleria mellonella as an alternative infection model for the Burkholderia cepacia complex. Infect Immun 76:1267–1275
160. Selwyn S (1983) The history of antimicrobial agents. In: Clinical chemotherapy. fundamentals, vol 1. Thieme-Stratton, New York
161. Sheng WH, Wang JT, Li SY, Lin YC, Cheng A, Chen YC, Chang SC (2011) Comparative in vitro antimicrobial susceptibilities and synergistic activities of antimicrobial combinations against carbapenem-resistant Acinetobacter species: Acinetobacter baumannii versus Acinetobacter genospecies 3 and 13TU. Diagn Microbiol Infect Dis 70:380–386
162. Shields RK, Clancy CJ, Gillis LM, Kwak EJ, Silveira FP, Massih RC, Eschenauer GA, Potoski BA, Nguyen MH (2012) Epidemiology, clinical characteristics and outcomes of extensively drug-resistant Acinetobacter baumannii infections among solid organ transplant recipients. PLoS One 7:e52349
163. Shields RK, Kwak EJ, Potoski BA, Doi Y, Adams-Haduch JM, Silviera FP, Toyoda Y, Pilewski JM, Crespo M, Pasculle AW, Clancy CJ, Nguyen MH (2011) High mortality rates among solid organ transplant recipients infected with extensively drug-resistant Acinetobacter baumannii: using in vitro antibiotic combination testing to identify the combination of a carbapenem and colistin as an effective treatment regimen. Diagn Microbiol Infect Dis 70:246–252
164. Sirijatuphat R, Thamlikitkul V (2014) Preliminary study of colistin versus colistin plus fosfomycin for treatment of carbapenem-resistant Acinetobacter baumannii infections. Antimicrob Agents Chemother 58:5598–5601
165. Smith NM, Bulman ZP, Sieron AO, Bulitta JB, Holden PN, Nation RL, Li J, Wright GD, Tsuji BT (2017) Pharmacodynamics of dose-escalated 'front-loading' polymyxin B regimens against polymyxin-resistant mcr-1-harbouring Escherichia coli. J Antimicrob Chemother 72:2297–2303
166. Souli M, Galani I, Boukovalas S, Gourgoulis MG, Chryssouli Z, Kanellakopoulou K, Panagea T, Giamarellou H (2011) In vitro interactions of antimicrobial combinations with fosfomycin against KPC-2-producing Klebsiella pneumoniae and protection of resistance development. Antimicrob Agents Chemother 55:2395–2397
167. Souli M, Rekatsina PD, Chryssouli Z, Galani I, Giamarellou H, Kanellakopoulou K (2009) Does the activity of the combination of imipenem and colistin in vitro exceed the problem of resistance in metallo-beta-lactamase-producing Klebsiella pneumoniae isolates? Antimicrob Agents Chemother 53:2133–2135
168. Srisupha-Olarn W, Burgess DS (2010) Activity of meropenem and colistin alone and in combina-

tion against multidrug-resistant *Acinetobacter baumannii* in a pharmacokinetic-pharmacodynamic model (abstract E-1591, p125). In: Abstracts of the 50th annual interscience conference on antimicrobial agents and chemotherapy (ICAAC), Boston, MA, September 12–15. American Society for Microbiology

169. Stressmann FA, Rogers GB, Marsh P, Lilley AK, Daniels TW, Carroll MP, Hoffman LR, Jones G, Allen CE, Patel N, Forbes B, Tuck A, Bruce KD (2011) Does bacterial density in cystic fibrosis sputum increase prior to pulmonary exacerbation? J Cyst Fibros 10:357–365
170. Suh JY, Son JS, Chung DR, Peck KR, Ko KS, Song JH (2010) Nonclonal emergence of colistin-resistant Klebsiella pneumoniae isolates from blood samples in South Korea. Antimicrob Agents Chemother 54:560–562
171. Sun HY, Shields RK, Cacciarelli TV, Muder RR, Singh N (2010) A novel combination regimen for the treatment of refractory bacteremia due to multidrug-resistant Pseudomonas aeruginosa in a liver transplant recipient. Transpl Infect Dis 12:555–560
172. Svetitsky S, Leibovici L, Paul M (2009) Comparative efficacy and safety of vancomycin versus teicoplanin: systematic review and meta-analysis. Antimicrob Agents Chemother 53:4069–4079
173. Tam VH, Schilling AN, Vo G, Kabbara S, Kwa AL, Wiederhold NP, Lewis RE (2005) Pharmacodynamics of polymyxin B against Pseudomonas aeruginosa. Antimicrob Agents Chemother 49:3624–3630
174. Tan CH, Li J, Nation RL (2007) Activity of colistin against heteroresistant Acinetobacter baumannii and emergence of resistance in an in vitro pharmacokinetic/pharmacodynamic model. Antimicrob Agents Chemother 51:3413–3415
175. Tan TY, Ng LS, Tan E, Huang G (2007) In vitro effect of minocycline and colistin combinations on imipenem-resistant Acinetobacter baumannii clinical isolates. J Antimicrob Chemother 60:421–423
176. Tang HJ, Ku YH, Lee MF, Chuang YC, Yu WL (2015) In vitro activity of imipenem and colistin against a carbapenem-resistant Klebsiella pneumoniae isolate coproducing SHV-31, CMY-2, and DHA-1. Biomed Res Int 2015:568079
177. Tangden T, Hickman RA, Forsberg P, Lagerback P, Giske CG, Cars O (2014) Evaluation of double- and triple-antibiotic combinations for VIM- and NDM-producing Klebsiella pneumoniae by in vitro time-kill experiments. Antimicrob Agents Chemother 58:1757–1762
178. Tangden T, Karvanen M, Friberg LE, Odenholt I, Cars O (2017) Assessment of early combination effects of colistin and meropenem against Pseudomonas aeruginosa and Acinetobacter baumannii in dynamic time-kill experiments. Infect Dis (Lond) 49:521–527
179. Tascini C, Gemignani G, Ferranti S, Tagliaferri E, Leonildi A, Lucarini A, Menichetti F (2004) Microbiological activity and clinical efficacy of a colistin and rifampin combination in multidrug-resistant Pseudomonas aeruginosa infections. J Chemother 16:282–287
180. Tripodi MF, Durante-Mangoni E, Fortunato R, Utili R, Zarrilli R (2007) Comparative activities of colistin, rifampicin, imipenem and sulbactam/ampicillin alone or in combination against epidemic multidrug-resistant Acinetobacter baumannii isolates producing OXA-58 carbapenemases. Int J Antimicrob Agents 30:537–540
181. Tsioutis C, Kritsotakis EI, Maraki S, Gikas A (2010) Infections by pandrug-resistant Gram-negative bacteria: clinical profile, therapeutic management, and outcome in a series of 21 patients. Eur J Clin Microbiol Infect Dis 29:301–305
182. Tumbarello M, Viale P, Viscoli C, Trecarichi EM, Tumietto F, Marchese A, Spanu T, Ambretti S, Ginocchio F, Cristini F (2012) Predictors of mortality in bloodstream infections caused by Klebsiella pneumoniae carbapenemase-producing K. pneumoniae: importance of combination therapy. Clin Infect Dis 55:943–950
183. Turnidge J (2014) Drug-drug combinations. In: Vinks AA, Derendorf H, Mouton J (eds) Fundamentals of antimicrobial pharmacokinetics and pharmacodynamics. Springer, New York
184. Urban C, Mariano N, Rahal JJ (2010) In vitro double and triple bactericidal activities of doripenem, polymyxin B, and rifampin against Multidrug-resistant Acinetobacter baumannii, Pseudomonas aeruginosa, Klebsiella pneumoniae, and Escherichia coli. Antimicrob Agents Chemother 54:2732–2734
185. Vardakas KZ, Mavros MN, Roussos N, Falagas ME (2012) Meta-analysis of randomized controlled trials of vancomycin for the treatment of patients with gram-positive infections: focus on the study design. Mayo Clin Proc 87:349–363
186. Velkov T, Bergen PJ, Lora-Tamayo J, Landersdorfer CB, Li J (2013) PK/PD models in antibacterial development. Curr Opin Microbiol 16(5):573–579
187. Vidaillac C, Benichou L, Duval RE (2012) In vitro synergy of colistin combinations against colistin-resistant Acinetobacter baumannii, Pseudomonas aeruginosa, and Klebsiella pneumoniae isolates. Antimicrob Agents Chemother 56:4856–4861
188. Walsh CC, Landersdorfer CB, McIntosh MP, Peleg AY, Hirsch EB, Kirkpatrick CM, Bergen PJ (2016) Clinically relevant concentrations of fosfomycin combined with polymyxin B, tobramycin or ciprofloxacin enhance bacterial killing of Pseudomonas aeruginosa, but do not suppress the emergence of fosfomycin resistance. J Antimicrob Chemother 71:2218–2229
189. Walsh CC, McIntosh MP, Peleg AY, Kirkpatrick CM, Bergen PJ (2015) In vitro pharmacodynamics of fosfomycin against clinical isolates of Pseudomonas aeruginosa. J Antimicrob Chemother 70(11):3042–3050

190. Wareham DW, Gordon NC, Hornsey M (2011) In vitro activity of teicoplanin combined with colistin versus multidrug-resistant strains of Acinetobacter baumannii. J Antimicrob Chemother 66:1047–1051
191. Wehrli W (1983) Rifampin: mechanisms of action and resistance. Rev Infect Dis 5(Suppl 3):S407–S411
192. Wei W, Yang H, Hu L, Ye Y, Li J (2015) Activity of levofloxacin in combination with colistin against Acinetobacter baumannii: in vitro and in a Galleria mellonella model. J Microbiol Immunol Infect 50(6):821–830
193. Wei WJ, Yang HF (2017) Synergy against extensively drug-resistant Acinetobacter baumannii in vitro by two old antibiotics: colistin and chloramphenicol. Int J Antimicrob Agents 49:321–326
194. Wootton M, Hedges AJ, Bowker KE, Holt HA, Reeves DS, Macgowan AP (1995) A critical assessment of the agar dilution chequerboard technique for studying in-vitro antimicrobial interactions using a representative beta-lactam, aminoglycoside and fluoroquinolone. J Antimicrob Chemother 35:569–576
195. Yamagishi Y, Hagihara M, Kato H, Hirai J, Nishiyama N, Koizumi Y, Sakanashi D, Suematsu H, Nakai H, Mikamo H (2017) In vitro and in vivo pharmacodynamics of colistin and aztreonam alone and in combination against multidrug-resistant Pseudomonas aeruginosa. Chemotherapy 62:105–110
196. Yang H, Chen G, Hu L, Liu Y, Cheng J, Li H, Ye Y, Li J (2015) In vivo activity of daptomycin/colistin combination therapy in a Galleria mellonella model of Acinetobacter baumannii infection. Int J Antimicrob Agents 45:188–191
197. Yang H, Chen G, Hu L, Liu Y, Cheng J, Ye Y, Li J (2016) Enhanced efficacy of imipenem-colistin combination therapy against multiple-drug-resistant Enterobacter cloacae: in vitro activity and a Galleria mellonella model. J Microbiol Immunol Infect 51(1):70–75
198. Yang H, Lv N, Hu L, Liu Y, Cheng J, Ye Y, Li J (2016) In vivo activity of vancomycin combined with colistin against multidrug-resistant strains of Acinetobacter baumannii in a Galleria mellonella model. Infect Dis (Lond) 48:189–194
199. Yang Y, Bhachech N, Bush K (1995) Biochemical comparison of imipenem, meropenem and biapenem: permeability, binding to penicillin-binding proteins, and stability to hydrolysis by beta-lactamases. J Antimicrob Chemother 35:75–84
200. Yang YS, Lee Y, Tseng KC, Huang WC, Chuang MF, Kuo SC, Lauderdale TL, Chen TL (2016) In vivo and in vitro efficacy of minocycline-based combination therapy for minocycline-resistant Acinetobacter baumannii. Antimicrob Agents Chemother 60:4047–4054
201. Yu W, Zhou K, Guo L, Ji J, Niu T, Xiao T, Shen P, Xiao Y (2017) In vitro pharmacokinetics/pharmacodynamics evaluation of fosfomycin combined with amikacin or colistin against KPC2-producing Klebsiella pneumoniae. Front Cell Infect Microbiol 7:246
202. Zak O, O'Reilly T (1991) Animal models in the evaluation of antimicrobial agents. Antimicrob Agents Chemother 35:1527–1531
203. Zarkotou O, Pournaras S, Tselioti P, Dragoumanos V, Pitiriga V, Ranellou K, Prekates A, Themeli-Digalaki K, Tsakris A (2011) Predictors of mortality in patients with bloodstream infections caused by KPC-producing Klebsiella pneumoniae and impact of appropriate antimicrobial treatment. Clin Microbiol Infect 17:1798–1803
204. Zavascki AP, Goldani LZ, Cao G, Superti SV, Lutz L, Barth AL, Ramos F, Boniatti MM, Nation RL, Li J (2008) Pharmacokinetics of intravenous polymyxin B in critically ill patients. Clin Infect Dis 47:1298–1304
205. Zeidler U, Bougnoux ME, Lupan A, Helynck O, Doyen A, Garcia Z, Sertour N, Clavaud C, Munier-Lehmann H, Saveanu C, D'Enfert C (2013) Synergy of the antibiotic colistin with echinocandin antifungals in Candida species. J Antimicrob Chemother 68:1285–1296
206. Zhai B, Zhou H, Yang L, Zhang J, Jung K, Giam CZ, Xiang X, Lin X (2010) Polymyxin B, in combination with fluconazole, exerts a potent fungicidal effect. J Antimicrob Chemother 65:931–938
207. Zhang L, Dhillon P, Yan H, Farmer S, Hancock RE (2000) Interactions of bacterial cationic peptide antibiotics with outer and cytoplasmic membranes of Pseudomonas aeruginosa. Antimicrob Agents Chemother 44:3317–3321
208. Zhao M, Bulman ZP, Lenhard JR, Satlin MJ, Kreiswirth BN, Walsh TJ, Marrocco A, Bergen PJ, Nation RL, Li J, Zhang J, Tsuji BT (2017) Pharmacodynamics of colistin and fosfomycin: a 'treasure trove' combination combats KPC-producing Klebsiella pneumoniae. J Antimicrob Chemother 72(7):1985–1990
209. Ziv G, Nouws JF, Van Ginneken CA (1982) The pharmacokinetics and tissue levels of polymyxin B, colistin and gentamicin in calves. J Vet Pharmacol Ther 5:45–58

Toxicity in Patients

17

Jason M. Pogue and Vincent H. Tam

Abstract

Polymyxin toxicity remains a significant concern that limits the clinical utility of this class of antibacterials for patient care. The most notable adverse event is the dose- and treatment-limiting nephrotoxicity that occurs in roughly 30–60% of patients receiving a systemic polymyxin. This chapter focuses on this adverse event with a detailed assessment of the incidence of, and risk factors for, polymyxin-associated nephrotoxicity. In particular, the text focuses on the impact of dose, serum concentrations, and polymyxin selection on nephrotoxicity. Additionally, less common, but clinically important adverse events are discussed.

Keywords

Colistin · Colistimethate · Polymyxin B · Nephrotoxicity · Non-renal toxicities · Gram-negative bacteria

J. M. Pogue (✉)
Department of Pharmacy Services, Sinai-Grace Hospital, Detroit Medical Center, Detroit, MI, USA
e-mail: jpogue@dmc.org

V. H. Tam
Department of Pharmacological and Pharmaceutical Sciences, Department of Pharmacy Practice and Translational Research, University of Houston College of Pharmacy, Houston, TX, USA

17.1 Introduction

Toxicity is an important consideration in evaluating the clinical utility of the polymyxins, and more remains to be learned on how to optimally use these agents. Originally introduced for use in the 1950s, polymyxin data that were published throughout the 1960s and into the early 1970s showed high rates of adverse events, notably nephrotoxicity and neurotoxicity. Although definitions were rarely given, nephrotoxicity rates of 10–50% were described and these findings were compounded by neurotoxicity rates, largely manifested as parasthesias, that, in some cases, exceeded 25% [1]. These seemingly unacceptable rates of toxicity, when combined with the new availability of less toxic antibiotics such as the aminoglycosides, and eventually the second and third generation cephalosporins, led to the polymyxins being rarely used clinically from the 1970s until the early 1990s.

In the early 1990s, starting in the cystic fibrosis population, the polymyxins (primarily colistin, formulated as its inactive prodrug colistimethate or CMS), started to have a resurgence of use because of the rise of resistant Gram-negative organisms. With the turn of the century, the spread of carbapenem-resistant *Acinetobacter baumannii* (CRAB) and multi-drug resistant *Pseudomonas aeruginosa* throughout intensive care units (ICUs) in both Europe and the United States, necessitated polymyxin

J. Li et al. (eds.), *Polymyxin Antibiotics: From Laboratory Bench to Bedside*, Advances in Experimental Medicine and Biology 1145, https://doi.org/10.1007/978-3-030-16373-0_17

use in non-cystic fibrosis patients. This spread ultimately went worldwide, and was joined by outbreaks of carbapenem-resistant enterobacteriaceae (CRE). Therefore, the last 15 years, as previously described in this book, has led to a renaissance of the polymyxins. As these data from the "modern era" have more clear definitions of toxicity, dosing regimens utilized, and descriptions of adverse events, they will be the focus of this chapter. For the purposes of this chapter the "modern era" will consist of polymyxin literature from approximately the year 2000 except where specifically noted.

Undoubtedly, the main toxicity of concern with the polymyxins is nephrotoxicity, and it will be the major emphasis of the chapter. This dose-limiting toxicity is well studied and clinically relevant. The development of acute kidney injury, particularly in critically ill patients, can lead to increased mortality. Other toxicities that will be discussed are neurotoxicity, hypersensitivity, and potential respiratory toxicities seen with inhaled CMS.

17.2 Nephrotoxicity

Since their re-emergence, the true incidence of nephrotoxicity with the polymyxins remains highly controversial. The modern era consists of over 60 publications assessing nephrotoxicity rates with the polymyxins (over 80% of these papers relate to colistin). Initial reports in the modern era began with multiple analyses looking at the safety (and efficacy) of colistin in cystic fibrosis patients, with the reports published between 1990 and 2000. These four publications in cystic fibrosis patients suggested low incidences of acute kidney injury (0–25%) [2–5]. Furthermore, when patients did develop toxicity it was mild (although no clear definitions were given) and reversible upon discontinuation. These findings were strengthened by initial data from Europe, largely from Greece, showing colistin nephrotoxicity rates less than 20%, with many publications showing toxicity to be in the 5–19% range [6–15]. These data led many to perceive the agent to be less toxic than previously believed. However, soon thereafter, multiple studies, largely from the United States, were published showing higher incidences of colistin-associated nephrotoxicity with rates often in the 30–60% range [16–20]. Although data are more limited, similarly wide ranges of 4–60% [21–35] have been reported as the incidence of polymyxin B associated nephrotoxicity. When combining all these studies, nephrotoxicity is seen in 795/3036 (26%) of patients receiving colistin, and 364/1075 (34%) of patients receiving polymyxin B. Undoubtedly, a complication in interpreting the nephrotoxicity literature is that in the modern era polymyxin use is largely in critically ill patients often suffering from life-threatening infections. These patients have multiple risk factors (e.g. severe sepsis/septic shock, concomitant nephrotoxins) for acute kidney injury and the contribution of the polymyxin to that injury is difficult to ascertain. The primary drivers of discordant results in these data are dose of colistin/polymyxin B given, and definition of nephrotoxicity. The following sections will look at: clinical features of polymyxin nephrotoxicity; the importance of definition and dose on incidence of nephrotoxicity; the comparative nephrotoxicity of the polymyxins relative to other agents as well as each other; the impact of colistin serum levels on nephrotoxicity; the impact of a loading dose on toxicity; and, finally other risk factors identified for toxicity.

17.2.1 Clinical Features of Nephrotoxicity

Although detailed descriptions of the clinical features of acute kidney injury are lacking in the currently available literature, there are some analyses that give us insight into the onset and reversibility in patients who develop nephrotoxicity while on polymyxin therapy. In the 12 studies reporting on onset of colistin-associated nephrotoxicity, 54–100% of cases occur in the first week, with median times until onset ranging from 4 to 12 days [17–20, 24, 26, 28, 36–40]. Although not as well described, the median onset of nephrotoxicity in the nine polymyxin B studies ranged from

6 to 11 days [24, 26, 28–33]. Rates of reversibility vary widely in the literature and are complicated by whether or not authors consider deaths in their reversibility analysis. In general, if a patient survives the acute event, reversibility rates range from 20% to 100% in 18 colistin studies [12, 16–18, 26, 36, 39, 41–50], and 69–100% in five polymyxin B studies [26, 30, 31, 34, 35] that report on this feature. Further data are needed addressing the long term outcomes of patients who develop acute kidney injury.

17.2.2 Impact of Definition

One of the primary features of polymyxin literature in the "old era" that made toxicity difficult to interpret was the lack of definitions for toxicity endpoints. While nephrotoxicity definitions in the modern era are well described in most analyses, the actual definition varies greatly, which significantly impacts both the incidence of and risk factors for nephrotoxicity. In the ~50 colistin-associated nephrotoxicity papers in the modern era, four definitions predominate. Two of these four definitions are commonly utilized in the 11 unique polymyxin B toxicity analyses.

The most common definition seen in 14 (29%) of the colistin analyses are the RIFLE criteria [16–20, 26, 28, 40–43, 50–52]. The RIFLE criteria present a grading system for toxicity, and represent a relatively sensitive measure for detecting modest decreases in renal function. The minimum criteria for acute kidney injury with the RIFLE criteria are a serum creatinine rise to 1.5× the baseline creatinine or a decrease in creatinine clearance of 25% in order to meet the "Risk" stage. In the 14 colistin studies using this definition, nephrotoxicity was seen in 465/1232 (38%), which is a similar rate to the (12/40) 30% rate seen with three studies using similar criteria for defining toxicity of an increase of serum creatinine of 0.5 mg/dL from baseline [36, 53, 54].

Interestingly, when those same criteria (increase in serum creatinine of 0.5 mg/dL) are applied with additional conditions (e.g., serum creatinine has to be above the upper limit of normal), the rates significantly decrease. In the nine analyses (18% of total toxicity papers) using this definition, nephrotoxicity rates are much lower, and seen in 60/541 (11%) of patients [10–14, 55–58].

The fourth commonly utilized toxicity definition seen in 8 (16%) of the colistin nephrotoxicity papers requires a much more significant rise in serum creatinine for toxicity to be met if a patient has normal baseline renal function (usually defined as a serum creatinine of ≤1.2 mg/dL), than if a patient has some degree of baseline renal insufficiency [15, 38, 44, 59–63]. The most common version of this definition requires the serum creatinine to rise to ≥2.0 mg/dL in normal renal function, while in patients with abnormal renal function an increase in serum creatinine of 1.5 times the baseline value is needed. Therefore, both a patient with a baseline creatinine of 0.6 mg/dL and one with a baseline of 1.3 mg/dL would need a rise to ≥2.0 mg/dL to reach the toxicity endpoint, despite the fact that would be a greater than tripling of creatinine in one instance. Using this definition, nephrotoxicity rates were lower and seen in 76/507 (15%) of patients. Importantly, as the toxicity endpoint is easier to meet in those with baseline renal insufficiency with this definition, it often leads to conclusions that chronic kidney disease (baseline creatinine greater than the upper limit of normal) is a risk factor for colistin-associated nephrotoxicity. Using this definition Montero showed nephrotoxicity in 5/107 (5%) of patients with normal renal function compared to 5/14 (36%) of patients with baseline renal insufficiency [61]. Similar results by Bassetti [62] and Betrosian [63] show the importance of this definition on both lowering the overall incidence of nephrotoxicity (poor detection of mild-moderate toxicity in patients with low baseline creatinine), and identification of baseline renal insufficiency as a risk factor.

While the number of nephrotoxicity analyses with polymyxin B (n = 15) limit the ability to robustly perform a similar analysis, the data, as scant as they are, support similar conclusions. When limiting only to definitions used in three or more different studies, analyses that used the more sensitive RIFLE criteria showed toxicity in 122/310 (39%) of patients [26–28, 35], whereas analyses using a much less sensitive measure of requiring a doubling of serum creatinine +/− additional conditions (e.g. to a creatinine ≥2.0 mg/dL) showed a lower cumulative toxicity incidence of 16/96 (17%) [23, 30, 34].

17.2.3 Impact of Dose

While dosing in patients with normal renal function is relatively consistent in the polymyxin B literature (1.5–2.5 mg/kg/day), this is not the case with colistin. Because of substantially different daily dose recommendations in the package inserts of the different colistimethate products used around the world, until recently daily doses utilized in Europe have commonly been ~50 to 75% of the daily dose used in the United States, Korea, Thailand, Brazil and Australia. The last few years has seen doses in Europe more similar to those used in other countries. Additionally, as the dose outside of Europe is a weight-based recommendation without clear instruction of what dosing weight to use (ideal body weight, total body weight, or adjusted body weight), the actual doses that patients receive can vary significantly and are often poorly described. This is of particular importance as multiple studies have shown a dose-dependent toxicity with both polymyxin B and colistin.

17.2.3.1 Different Scheduled Doses of Colistin and Rates of Nephrotoxicity

In the colistin literature five different dosing schedules predominate. They are 3–6 MU (100–200 mg CBA)/day, 9 MU (300 mg CBA/day), 3–9 MU (100–300 mg/day), 5 mg/kg/day CBA (for 70 kg patient, 350 mg CBA/day or 10.5 MU/day), and 2.5–5 mg/kg/day CBA (175–350 mg CBA/day or 5.3–10.5 MU/day). The fact that inconsistent or poorly described renal dosing strategies were employed further complicates these data, however, the impact of these different scheduled dosing strategies on nephrotoxicity rates is very apparent. In 19 studies including 1358 patients receiving 2.5–5 mg/kg of CBA a day, nephrotoxicity was seen in 500 (37%) of patients [8, 16–20, 24, 26, 28, 36, 37, 44, 47–50, 52, 55, 61]. This is in contrast to lower toxicity rates seen in studies where patients received daily doses of 3–6 MU (7%, n = 259) [7, 9, 10, 12, 13, 53, 56, 62, 64], 9 MU (16%. n = 315) [6, 15, 39, 45, 60, 63, 65], and 3–9 MU (21%, n = 415) [38, 42, 52], respectively. This stepwise increase in toxicity rates as daily doses rise from 3–6 MU/day to 9 MU/day to doses greater than 9 MU/day seen in patients being dosed on mg/kg of CBA/day shows the importance of dose used on the incidence of nephrotoxicity reported.

17.2.3.2 Individual Studies Assessing the Association Between Dose and Nephrotoxicity for Colistin

In the 14 studies assessing risk factors for colistin-associated nephrotoxicity, 6 showed an association between either daily dose (n = 5) or cumulative exposure/duration of therapy (n = 3). Hartzell and colleagues found that patients with toxicity had a cumulative colistin exposure of 6454 ± 3421 mg of CBA as compared to an exposure of 4727 ± 3263 mg in those who did not (p = 0.005) [16]. In a multivariate analysis, Rattanaumpawan showed that duration of colistin (OR. 1.1 (95% confidence interval 1.03–1.19)), a CBA dose of 3–5 mg/kg/day (OR 3.1 95% CI 1.3–7.5), and a dose of >5 mg/kg/day (OR 15.3 (3.9–60.6) were risk factors for nephrotoxicity [37]. Pogue and colleagues showed a similar stepwise increase in risk of nephrotoxicity from 3.3 (0.8–13.0) to 23.4 (5.3–103.6) when the dose went from 3 to 4.9 mg/kg/day to ≥5 mg/kg/day [18]. Similar findings were seen in three other analyses [17, 24, 52], and highlight the dose-dependent nature of colistin-associated nephrotoxicity.

17.2.3.3 The Impact of Dose on Toxicity with Polymyxin B

In the nine studies analyzing risk factors for polymyxin B associated nephrotoxicity in the "modern-era", two showed an association between dose (n = 1) and duration (n = 1) of polymyxin B and toxicity. Elias and colleagues analyzed predictors of nephrotoxicity in 235 patients eligible for the toxicity endpoint. Patients receiving ≥200 mg/day of polymyxin B had an adjusted odds ratio of 4.5 (1.6–12.9) for the development of severe renal impairment [21]. Mostardiero and colleagues analyzed polymyxin use in 92 (90 received polymyxin B, 2 received colistin) solid-organ transplant patients [29]. In multivariate analysis, duration of polymyxin therapy (OR 1.06 (1.00–1.12) was independently associated with renal dysfunction.

Two more recent analyses have also assessed the impact of polymyxin B dose and incidence of nephrotoxicity. Nelson and colleagues assessed safety and efficacy endpoints related to polymyxin B dose with 109 patients able to be assessed for the safety endpoint [66]. In this analysis receipt of daily doses ≥250 mg were associated with higher rates of acute kidney injury (8/12 (67%) receiving this dose developed AKI, versus 31/97 (32%) of those who received lower doses; $p = 0.03$) and in multivariate analysis daily doses ≥250 mg were an independent predictor of AKI (OR 4.32, 95% CI 1.15–16.25.) Similarly, Rigatto and colleagues assessed risk factors, including dose, for AKI in patients receiving polymyxin B therapy [67]. In bivariate analysis, AKI developed in 33/103 (32%), 109/202 (54%), and 47/105 (44%) of patients receiving <150 mg, 150–199 mg, and ≥200 mg of polymyxin B daily, respectively ($p = 0.001$). In accordance with these results, polymyxin B doses ≥150 mg/day were highly associated with AKI in multivariate modeling (HR 9.81, 95% CI 2.37–40.62), but no additional risk was seen with doses ≥200 mg/day.

Much like difference in toxicity definitions, dosing variability driven by differences in package insert recommendations, contribute considerably to the discordant results seen for nephrotoxicity in the polymyxin literature. This is much more apparent in the colistin literature, as the doses vary more greatly than in the polymyxin B literature. However, with recent clinical practices moving towards the upper end (or even slightly beyond the upper end) of the package insert dosing recommendation for polymyxin B, a similar association is becoming apparent.

17.2.4 Polymyxin Nephrotoxicity Rates in Comparison to Other Antimicrobials

Another difficulty in interpreting the polymyxin nephrotoxicity literature is that most analyses are descriptive in nature, thus making it difficult, if not impossible, to assess the independent impact of the polymyxin exposure on toxicity. There are 12 studies (11 colistin, 1 polymyxin B) comparing the safety of these agents with other antimicrobial classes, and the findings of these studies are summarized in Table 17.1.

In general, the comparative data suffer from similar limitations to the ones previously discussed; namely inconsistent dosing, definitions of nephrotoxicity, and small sample sizes. There are three studies that show a statistically significant difference between a polymyxin and a comparator. Paul and colleagues compared nephrotoxicity in patients on colistin to those receiving other active agents for infections due to *A. baumannii*, *P. aeruginosa*, or enterobacteriaceae [60]. Using a definition for nephrotoxicity that differed in patients with normal baseline creatinine (serum creatinine >2 mg/dL, a decrease in creatinine clearance of 50% or the need for renal replacement therapy) from those with baseline renal insufficiency (increase in serum creatinine of 50%, decrease in creatinine clearance of 50%, or the need for renal replacement therapy), the authors showed an increase of toxicity with colistin (26/168 (16%) vs. 17/244 (7%) for comparators, $p = 0.006$.) The second analysis by Kvitko and colleagues was a comparison of toxicity in patients receiving polymyxin B compared to those receiving other anti-pseudomonal agents for the treatment of *P. aeruginosa* bacteremia [22]. In this analysis, nephrotoxicity, defined as an increase in serum creatinine ≥50% for the baseline value, occurred in 16/45 (36%) of patients on polymyxin B compared with 10/88 (11%) of patients on other anti-pseudomonals ($p = 0.002$). Interestingly, the third analysis showing a significant difference between a polymyxin and comparator showed rates of nephrotoxicity, using a definition of doubling of serum creatinine or a decrease in creatinine clearance of 30%, to

Table 17.1 Polymyxin nephrotoxicity rates in comparison to other antimicrobials

Author	Polymyxin, n	Comparator, n	Scheduled Polymyxin dose	Nephrotoxicity definition	Toxicity
Rocco [27]	147 colistin or colistin + vancomycin/aminoglycoside	132 vancomycin or aminoglycoside	3.9 mg/kg/day CBA	RIFLE	57 (41) colistin vs. 54 (41) other; p = NS
Chan [21]	7 "polymyxin"	30 aminoglycoside	2.5–5 mg/kg/day CBA	Increase Scr 0.5 or decrease Clcr 50%	4/7 (58) polymyxin vs. 6/30 (20) aminoglycoside; p = 0.07
Durakovic [38]	26 colistin	26 "other anti-pseudomonals"	3 MU q8h	Scr ≥ 1.7 or increase ≥50% if pre-existing renal insufficiency	3 (10) colistin vs. 0 for "other" p = 0.07
Lim [50]	20 colistin	35 inactive antimicrobials	2.5–5 mg/kg/day CBA	Inc Scr 50% to a value ≥1.3 or RRT	10/20 (50) colistin vs. 10/35 (29) inactive; p = 0.10
Paul [55]	168 colistin	244 other agents for A. baumannii, P. aeruginosa, or enterobacteriaceae	6–9 MU/day	Baseline Scr ≤ 1.2: Scr >2 or dec Clcr 50% or RRT baseline >1.2: 50% increase in Scr, 50% decrease in Clcr or RRT	26/168 (16) colistin vs. 17/244 (7) comparators
Gounden [51]	21 colistin	23 tobramycin	2 MU q8h	Increase Scr to >50% the upper limit of normal	4/21 (19) colistin vs. 2/23(9) tobramycin; p = 0.07
Koomanachai [40]	78 colistin	15 inactive antimicrobials	5 mg/kg/day CBA	Doubling of Scr or decrease of 30% in Clcr	24/78 (31) colistin vs. 10/15 (67) inactive; p = 0.02
Betrosian [58]	15 colistin	13 ampicillin/sulbactam	3 MU q8	Baseline Scr ≤ 1.2: Scr >2 or decrease Clcr 50% or RRT baseline >1.2: 50% increase in Scr, 50% decrease in Clcr or RRT	5/15 (33) colistin vs. 2/13 (15) ampicillin/sulbactam; p = 0.4
Kallel [7]	60 colistin	60 imipenem	2 MU q8	Scr > 1.7, BUN >28	0% in each group
Montero [56]	21 colistin	14 imipenem	2.5–5 mg/kg/day CBA	Baseline Scr ≤ 1.2: Scr >2 or decrease Clcr 50% or RRT baseline >1.2: 50% increase in Scr, 50% decrease in Clcr or RRT	5/21 (24) colistin, 6/14 (42) imipenem; p = NS
Reina [8]	55 colistin	130 others for A. baumannii or P. aeruginosa	5 mg/kg/day	Scr ≥ 2, 50% decrease in Clcr or RRT	0% in each group
Kvitko [62]	45 polymyxin B	88 other anti-pseudomonals	Not listed	Increase Scr ≥ 50%	16/45 (36) polymyxin B vs. 10/88 (11) others; p = 0.002

Scr serum creatinine, *Clcr* creatinine clearance, *BUN* blood urea nitrogen, *RRT* renal replacement therapy be higher with inactive therapy (i.e. agents lacking in vitro activity against the causative pathogen) than colistin (24/78 (31) colistin vs. 10/15 (67) inactive; p = 0.02.) [47] It should be noted however, that mortality was 80% in the inactive therapy group, and thus worsening sepsis due to inactive agents likely influenced the development of acute kidney injury. While the other analyses do not show statistically significant increases in toxicity with polymyxins, they are often numerically higher, and the failure to see statistical significance is often due to small sample sizes. Based on these data, it is reasonable to conclude that in general the polymyxins are more nephrotoxic than other antimicrobials.

17.2.5 Comparative Toxicity of Colistin and Polymyxin B

As previously discussed, one of the primary drivers between preferential use of colistin over polymyxin B in both the "old" and "modern" era of the polymyxins was the belief that colistin was less nephrotoxic than polymyxin B. This theory was largely debunked when data showed that larger doses of colistin (in the form of CMS) were needed for efficacy, and when the two were "on equal terms" that toxicity would be equal. To date there are six analyses and one meta analysis available in the literature attempting to assess the comparative nephrotoxicity of the polymyxins.

The first analysis, published in 2009 by Oliveira and colleagues, compared rates of nephrotoxicity, defined as a twofold increase in serum creatinine at any time during the treatment or an increase by 1 mg/dL if the patient had a baseline creatinine >1.4 mg/dL, between 39 patients receiving colistin and 30 receiving polymyxin B [23]. Median daily dose in the study was 6 MU (range 1–9 MU) for CMS (200 mg CBA (range 33–300 mg) and 100 mg (range 40–150 mg) for polymyxin B. The onset of renal impairment occurred in 10/39 (26%) and 8/30 (27%) (p = 0.92) of patients receiving colistin and polymyxin B, respectively, and the authors concluded there was no difference in toxicity between the two.

The second study, by Tuon and colleagues in 2013, analyzed risk factors for acute kidney injury, defined by the AKIN criteria, in patients receiving colistin and polymyxin B [24]. In bivariate analysis, the incidence of acute kidney injury was numerically higher with colistin (14/36, 39%) than polymyxin B (20/96, 21%), p = 0.06. However, when controlling for polymyxin dose and concomitant vancomycin in the multivariate model, colistin (compared to polymyxin B) use was not significantly associated with an increased risk for toxicity (adjusted odds ratio 1.74 [95% confidence interval 0.82–3.69]).

The third analysis, also published in 2013 by Akajagbor and colleagues [26], compared nephrotoxicity rates, defined by the RIFLE criteria, between 173 patients receiving one of the two polymyxins. Nephrotoxicity was seen in 64/106 (60%) of patients receiving colistin, and 28/67 (41.8%) of patients receiving polymyxin B, p = 0.03. When controlling for age, hypertension, vasopressors, and concomitant nephrotoxins, colistin use was independently associated with an increased risk for nephrotoxicity (Hazard Ratio 2.27 (1.35–3.82); p = 0.002).

While the previous two analyses suggested that colistin might in fact be associated with higher rates of nephrotoxicity they also suffered from the same major limitation. Since it was only recently appreciated that polymyxin B is not renally eliminated, and therefore should not have renal dose adjustments, patients with baseline renal insufficiency (likely those with creatinine clearances ≤80 mL/min) underwent unnecessary dose adjustments, and therefore likely had lower polymyxin B exposure. With both polymyxins showing a dose-dependent toxicity, this makes it extremely difficult to interpret these findings.

Phe and colleagues were the first to attempt to address this limitation. They compared nephrotoxicity rates, defined by the RIFLE criteria, of the polymyxins, in a multicenter cohort study limiting inclusion to those with stable, normal (baseline creatinine ≤1.5 mg/dL) renal function. In the overall cohort of 225 patients, nephrotoxicity was seen in 41/121 (34%) of patients receiving colistin, compared to 24/104 (23%) of patients receiving polymyxin B, p = 0.08) [28].

The authors then provided a matched cohort analysis controlling for the factors associated with colistin and polymyxin B nephrotoxicity. In this well-matched analysis (n = 38 in each group) median daily doses were 291 mg CBA (5.0 mg/kg/day of ideal body weight) for colistin (median dose 8.8 MU, dosed at 0.152 MU/kg/day) and 126 mg (2.1 mg/kg/day of ideal body weight) for polymyxin B. Nephrotoxicity was seen in 21 (55%) of patients receiving colistin compared to 8 (21%) on polymyxin B, p = 0.003.

Rigatto and colleagues published data from a large cohort (n = 491, including 81 receiving colistin and 410 receiving polymyxin B) of patients that also overcame the previous limitations, as the institutions involved in this analysis did not recommend renal dose adjustments for polymyxin B [68]. This more optimal dosing strategy was reflected in the median doses used with both colistin (median dose 300 mg CBA interquartile range (IQR) 253–300) and polymyxin B (150 mg IQR 140–187) in this analysis. Using these dosing strategies, which better reflect currently recommended doses of both polymyxins, the authors found that the rate of renal failure (the "F" category of the RIFLE criteria, or a rise in creatinine three times the baseline or a decrease in creatinine clearance ≥75%) to be significantly higher with colistin than polymyxin B (38.3% vs. 12.7%, $p < 0.001$), with colistin being an independent predictor of renal failure in the multivariate model (HR 3.35 95% CI 2.05–5.48).

The five aforementioned studies were included in a meta-analysis by Vardakas and Falagas [69]. The authors concluded that when combining these data colistin was associated with risk ratio of 1.55 (95% CI 1.36–1.78) for nephrotoxicity when compared to polymyxin B. While the findings of this meta-analysis are interesting and support the emerging conclusion that colistin is associated with increased toxicity when compared to polymyxin B, it is worth mentioning that the same limitations from the first three studies described above (inappropriate renal dosing of polymyxin B) do play a role in the findings of this meta analysis.

The sixth and final analysis assessing comparative nephrotoxicity rates in both a non-cystic fibrosis (n = 194; 45 polymyxin B and 149 colistin) and a cystic fibrosis (n = 220; 29 polymyxin B and 191 colistin) population found no association between polymyxin choice and AKI [70]. Acute Kidney Injury occurred in 21/49 (43%) and 73/145 (50%) of polymyxin B and colistin patients in the non-cystic fibrosis population (p = 0.46). Similarly there was no difference in AKI rates in the cystic fibrosis patient population (10/29 (35%) and 57/191 (30%); p = 0.77). It is worth mentioning that due to the temporal nature of this study (polymyxin B recently became the formulary preferred agent with the advent of recent pharmacokinetic and safety data) that while polymyxin B was dosed in what would be considered an optimal manner (loading dose used in 74% of non-cystic fibrosis patients followed by a median daily maintenance dose of 200 mg ± 83 mg), colistin loading doses were used less frequently (14% of non CF patients) and maintenance doses were lower than generally recommended (226 ± 106.1 mg/day CBA). Because of these dosing differences between the two polymyxins, these data are not as strong as the two aforementioned studies which showed an association between polymyxin selection and toxicity.

Although it is not a universal finding, and the data are limited by study design there is a strong suggestion in the literature that polymyxin B might be less toxic to the kidneys than colistin. Prospective studies, looking at both pharmacodynamic and toxicodynamic effects of achievable concentrations of both polymyxins are urgently needed to assure that the polymyxin with the superior benefit-to-cost ratio is being utilized.

17.2.6 Serum Levels and Nephrotoxicity

There have been four analyses with colistin reporting on serum concentrations obtained and rates of nephrotoxicity, only two of which looked directly at the association between concentrations and incidence of acute kidney injury. Markou [65] and Karnik [64] reported on pharmacokinetics of colistin after administration of intravenous CMS in 14 and 15 critically ill patients, respectively. Patients in these analyses had maximum serum concentrations 2.93 ± 1.24 and 4.6 (2.5–23.2) mcg/mL, respectively. Although nephrotoxicity was not clearly defined or incidence stated in either of these analyses, zero patients had any "clinically significant changes in laboratory values" related to renal parameters. Conversely, in the largest currently available pharmacokinetic study in critically ill patients where the median steady state colistin level was 2.36 (0.48–9.38) mcg/mL, Garonzik and colleagues reported that 43/89 (48%) of patients who did not have pre-existing need for renal replacement therapy had a rise in serum creatinine of ≥50% [71].

Sorli and colleagues published the first analysis assessing the association between serum levels and nephrotoxicity defined by the RIFLE criteria [42]. The investigators performed colistin trough sampling after 3 days of treatment and looked at the influence of those levels on nephrotoxicity at day 7 and the end of therapy. A concentration-dependent toxicity was seen at both endpoints with a 2% toxicity rate at day 7 if the day 3 serum colistin concentration was ≤1.04 mcg/mL, compared to a 32% rate if the concentration was between 1.05 and 2.2 mcg/mL, and a 65% rate if the concentration was >2.2 mcg/mL. A similar concentration dependent effect was seen at the end of therapy, with day 3 concentrations >2.2 mcg/mL being associated with an 85% chance of toxicity at the end of therapy.

More recently, Forrest and colleagues published a toxicodynamic analysis from a pharmacokinetic study which included 153 critically ill patients who could be assessed for a nephrotoxicity endpoint [72]. In this analysis the authors demonstrated a clear association between average colistin steady state concentrations, baseline renal function, and both the incidence and severity of colistin-associated nephrotoxicity. For patients with a baseline creatinine clearance <80 mL/min, average colistin steady state concentrations of 1.88 mcg/mL or higher increased the incidence and severity of acute kidney injury, whereas in patients with creatinine clearances ≥80 mL/min concentrations ≥2.25 mcg/mL increased this risk. These toxicity thresholds identified are consistent with those demonstrated in the aforementioned study by Sorli and colleagues. Furthermore, much like in the Sorli analysis, the rates of acute kidney injury were very high when these thresholds were met with roughly 50% and 65% of patients in the different baseline renal function groups demonstrating a ≥ 50% reduction in creatinine clearance at or above these concentrations.

17.2.7 Polymyxin Loading Dose and Nephrotoxicity

Recent pharmacokinetic data have stressed the importance of a loading dose, usually in the 270–360 mg CBA range, in order to rapidly obtain target serum concentrations with colistin, and more limited data suggest that, although not as crucial, a loading dose of 2–2.5 mg/kg can help more rapidly achieve steady state concentrations with polymyxin B. The chief concern, in light of the dose-dependent toxicity described in this chapter is the impact that a one-time large dose might have on nephrotoxicity. To date, limited evidence exists exploring the safety (and efficacy) of a polymyxin loading dose. The four studies to date assessing the impact of a polymyxin loading dose are limited by differing definitions of loading dose, different polymyxins being used, small numbers, and the fact that analyzing the impact of the loading dose on AKI rates was not the primary objective of the study.

Nelson and colleagues found that rates of nephrotoxicity in patients receiving a polymyxin B loading dose (defined as initial dose ≥2.5 mg/kg) was not associated with an increased risk of AKI [66] (AKI occurred in 9/19 (47%) of patients who received a loading dose versus 30/90 (33%) of those who did not; p = 0.30). Conversely, in an analysis of 81 colistin patients Rigatto and colleagues found that renal failure occurred in 17/22 (77%) of patients who received loading doses compared to 14/59 (24%) who did not ($p < 0.001$) [68]. It is worth mentioning that the primary intent of this analysis, as described above, was to compare AKI rates in patients receiving colistin and polymyxin B, and a post hoc analysis of these 81 colistin patients showed significant differences between patients who received loading doses and those who did not, including differences in baseline renal function and chronic comorbidities. Nonetheless, when controlling for these differences in the post hoc analysis, receipt of a colistin loading dose was associated with an increased risk of AKI (HR 5.2; 95% CI, 2.3–12.0).

Crass and colleagues assessed the incidence of AKI in the 56 patients who received a loading dose with either colistin or polymyxin B and 138 who did not, and found no association in either bivariate (HR 0.67 95% CI 0.39–1.17) or multivariate (HR 0.78 95% CI 0.42–1.46) analyses between receipt of a polymyxin loading dose and AKI [70]. Finally, while Shields and colleagues found an association with colistin loading dose and AKI on day 7 in bivariate analysis [73] (42/118 (36%) vs. 31/131 (24%); p = 0.05), this did not persist on multivariate analysis when concomitant vancomycin and higher maintenance dose strategies were controlled for (p = 0.28).

The safety of a polymyxin loading dose remains unclear. The data presented here are limited by small numbers as well as the lack of uniform definition of a loading dose (three of the four studies did not clearly define what constituted a loading dose). Further data, in larger populations with clearly defined (and pharmacokinetically optimized) loading doses are clearly warranted to further assess this strategy.

17.2.8 Other Risk Factors for Polymyxin Nephrotoxicity

While dose and subsequent concentration are important predictors of nephrotoxicity, several other risk factors have been identified in the literature. In 18 publications looking at risk factors for nephrotoxicity with either colistin or polymyxin B, 15 identified at least one additional factor in either bivariate or multivariate analyses. In addition to many variables related to polymyxin therapy (daily dose, cumulative dose, duration of therapy), concomitant nephrotoxins [17, 18, 25, 35, 37, 50, 61] (including vancomycin), chronic kidney disease [21, 25, 50, 61], age [17, 30, 37], and body mass index [19, 35] are seen repeatedly as predictors of toxicity. Other risk factors that have been identified are malignancy [48], length of stay [48], concomitant rifampin [18], hypoalbuminemia [46], and site of infection [25].

17.2.9 Conclusion

While debate continues to exist about just how nephrotoxic the polymyxins are, there is little doubt that (a) they are nephrotoxic and (b) they are more nephrotoxic than other agents used for Gram negative infections. Importantly, the relationship between dose, serum/plasma levels, and rates of toxicity suggest that there is potential for minimizing this toxicity with therapeutic drug monitoring. Unfortunately, the feasibility of doing that with colistin is difficult, due to issues with continued conversion from the prodrug (CMS) to active colistin unless samples are very carefully collected, processed, stored and analysed. This represents a potential advantage for polymyxin B, for which therapeutic drug monitoring would be more straightforward.

In addition to identifying the appropriate or optimal dose, other strategies to minimize polymyxin nephrotoxicity would include minimizing the use of concomitant nephrotoxins. Another potential strategy for limiting toxicity would be the co-administration of anti-oxidants; however, clarity on this preventative strategy is urgently needed. As oxidative stress is considered to have a key role in tubular cell apoptosis, interest surrounding the possible protective role of anti-oxidants has emerged. Animal data have suggested that co-administration of ascorbic acid can mitigate colistin-associated nephrotoxicity [74]; however, clinical data to date have been mixed [75, 76]. Adequately powered and well controlled studies are needed to clearly address the question.

Furthermore, it will be interesting to see if dose frequency strategies can mitigate nephrotoxicity. A study in rats suggested dividing daily CMS doses thrice daily could decrease the incidence and severity of renal lesions when compared to twice daily administration of the same daily dose [77] These data, in addition to the in vitro data suggesting the potential for resistance suppression with more frequent dosing [78], are the basis for expert recommendations of dividing the daily dose of CMS (e.g. divided daily dose administered 8 hourly). While this is a reasonable approach, the relevance of these findings is questionable given that these analyses had significantly different Cmax and Cmin concentrations (high C_{max}, low C_{min}), particularly with less frequent dosing, and clinical pharmacokinetics from critically ill patients suggest a relatively constant, flat, concentration-time profile in human patients given the slow conversion from CMS to colistin [71]. Conversely, data with polymyxin B suggest a saturable toxicity similar to the aminoglycosides which would more lend itself towards a once-daily dosing strategy [79]. Future analyses should analyze dose frequency strategies on the incidence of toxicity in patients. Finally, further research is needed to identify if one polymyxin, when dosed optimally, truly is less nephrotoxic than the other.

17.3 Neurotoxicity

Neurotoxicity is much less commonly reported in the modern literature, and even when investigated, rates tend to be much lower than rates in the "old" polymyxin era. This seemingly safer use of the agents is undoubtedly related to the patient population now treated with polymyxins. In the old era polymyxins were used as first-line agents for treating a wide variety of infections, including infections in relatively healthy individuals. Conversely, in the modern era, use is often limited (due to concerns relating to toxicity and emergence of resistance) to critically ill patients with no other treatment options; therefore, many of the toxicities mentioned in the old literature are often not evaluable or are undetected due to heavily sedated or otherwise unresponsive patients. Therefore, all incidences of neurotoxicity in the modern literature must be analyzed with this understanding.

In total 19 studies assess possible colistin-associated neurological toxicity [6, 7, 10, 13–16, 23, 43, 45, 47–49, 54, 59, 61–64], with 8 of them reporting at least one case of an adverse event. Four studies have investigated neurological toxicity possibly related to polymyxin B with three of them reporting two instances each of an adverse event. No analysis showed greater than a 7% neurotoxicity rate, and not all neurological adverse events were considered associated with the polymyxin.

Manifestations of neurotoxicity possibly associated with colistin have varied greatly. Averbuch and colleagues reported on two patients who had convulsions, although neither was considered drug related as one patient had uncontrolled epilepsy, and the other had multifocal encephalopathy [43]. Hartzell and colleagues reported on two patients (out of a cohort of 66) who had parasthesias [16], while Sabuda [54] and colleagues described four cases of neurological adverse events in patients receiving colistin. In this analysis patient 1 had somnolence, but was on gabapentin, baclofen, tizanidine and these agents were considered more likely as a root cause. Patient 2 suffered from dizziness, but MRI showed progression of cancer. The authors only stated that patient 3 had "neurotoxicity". Patient 4 was probably the most convincing, as this patient was on 500 mg CBA/day (15 MU/day) and had encephalopathy, respiratory muscle weakness with no other obvious cause. Kasaikou reported on a patient who developed polyneuropathy on day 25 of colistin therapy [10], however the patient continued treatment for 11 more days and it gradually subsided. Durakovic described a patient who developed Jackson's partial epilepsy with a secondary generalization in which the investigators reduced the dose and the seizure activity ceased [45]. Cheng and colleagues described neurotoxicity in 4 (3.5%) of 115 patients manifesting as focal seizures in the extremity in 3 patients, and as altered mental status in the fourth [48]. Encouragingly, none of the patients had permanent sequelae. Kallel reported on one patient receiving colistin who had muscular weakness during hospitalization that recovered at 1 month follow up [7]. Finally, Linden and colleagues described a case that manifested as diffuse weakness on day 10 of colistin therapy that resolved 1 week after cessation of colistin therapy [80].

Polymyxin B neurotoxicity, while much less commonly described, showed similar variations in its clinical presentation. Sobieszczyk reported on two cases of neurotoxicity, which manifested as seizures in one patient and neuromuscular weakness in the other [34]. Holloway described one patient with altered mental status, and one with distal parasthesias [31]. Finally, Weinstein and colleagues reported on two patients who experienced parasthesias (both oral, and one lower extremity as well) which resolved after discontinuation [81].

17.4 Other Toxicities

Although infrequent, incidences of other adverse events potentially related to colistin therapy have been reported in the modern literature. Pintado and colleagues described two cases of potential hypersensitivity reactions to colistin where one patient had a mild self-limiting rash, and the second had angioedema that led to discontinuation of colistin [59]. Additionally, these authors mentioned one patient who had vomiting that was temporally related to colistin administration [59]. Similarly, Durakovic reported on occurrence of an "allergic reaction" to colistin [45]. Unfortunately, no more details of this case were given. Karnik described three patients who had elevated liver enzymes, one of which was thought to be possibly related to colistin administration [64]. Additionally, the authors described one patient who had hypokalemia and hyponatremia [64].

Interestingly, an emerging body of evidence has suggested an association between polymyxin B usage and skin hyperpigmentation. Two initial reports suggested a possible association between long-term (>21 days) polymyxin B exposure and hyperpigmentation [82, 83] which was described as a darkening of the skin [82] or "gray-skin discoloration" [83]. More recently Zavascki and colleagues described a case where a patient developed a "head and neck skin darkness" that was evident by day 14 of polymyxin B which had not resolved 3 months later [84]. According to the Naranjo Adverse Drug Reaction probability scale, there was a probable association between polymyxin B and the adverse event. The same investigator published an additional case report where hyperpigmentation occurred 5 days into polymyxin B therapy and consisted of skin darkening and the emergence of round hyperchromic spots. This report included longer term follow up and the hyperpigmentation had somewhat resolved at 3 months and nearly completely resolved at the 6 month follow up visit [85]. Further investigation into this adverse event is clearly warranted.

Two analyses have looked closely at respiratory toxicities following nebulized colistin. Dominguez-Ortega described a 63-year-old man who developed severe bronchospasm after administration of nebulized colistin (in the form of CMS). Interestingly, the authors were able to successfully administer inhaled colistin to this patient in the future by inducing tolerance with a graded challenge [86]. Rattanaumpawan and colleagues reported a rate of bronchospasm 7.8% in 49 patients who received nebulized colistimethate as adjunctive therapy for ventilator-associated pneumonia, but this was not statistically higher than the 2.0% rate seen in the control arm ($p = 0.36$) [87]. Importantly, in 2007, the United States Food and Drug Administration issued a warning regarding a cystic fibrosis patient who had a potential fatal adverse reaction to nebulized colistin [88]. Within hours of receipt of nebulized colistin (in the form of CMS) the patient developed respiratory distress, which progressed to respiratory failure, and ultimately death. The analysis concluded that a component of active colistin (polymyxin E1) is toxic to lung tissue, and since the inhaled colistimethate had been premixed well before administration, significant conversion to colistin had already occurred, and this might have been the cause of the toxicity. The FDA cautioned that in the future doses of nebulized colistin should be reconstituted immediately before administration, and clinicians be aware of this potentially fatal adverse event. Pereira and colleagues published the only analysis in the modern era looking at respiratory toxicities with inhaled polymyxin B (50 mg twice daily administration) [89]. All patients were given pretreatment with beta-agonists in order to minimize respiratory adverse events; however, the authors reported that 4 of 19 patients (21%) suffered adverse events with three described as bronchospasm and one as "cough." All four patients were able to tolerate inhalation of polymyxin B with dose reduction, although the dose reduction used was not described. The role of beta-agonists for prevention of bronchospasm warrants further evaluation.

17.5 Summary and Conclusions

By far, the most commonly manifested toxicity seen with both polymyxins is nephrotoxicity. While the reported rates might vary widely, significant attention should be given by clinicians to strategies to minimize renal toxicity. These strategies include dose optimization, minimizing risk factors (namely receipt of concomitant nephrotoxins), and in the future, selection of the optimal polymyxin. Future studies on the topic should also describe management of patients with mild-moderate nephrotoxicity to help guide clinicians (for example: in the setting of acute kidney injury, should polymyxin be continued with a dose reduction?). Clinicians should also be aware of the various manifestations of neurotoxicity described in this chapter, and alternative antibiotics (if available) should be administered if possible. Encouragingly, multiple analyses reported tolerance and resolution of mild neurotoxicity over time without discontinuation of therapy. As hypersensitivity reactions have been described, monitoring is warranted upon initiation of polymyxin therapy in patients. Finally, if patients receive inhaled colistin (in the form of CMS), clinicians should be aware of the potential for bronchospasm, and doses should be reconstituted immediately prior to administration.

References

1. Falagas ME, Kasiakou SK (2006) Toxicity of polymyxins: a systematic review of the evidence from old and recent studies. Crit Care 10(1):R27
2. Bosso JA, Liptak CA, Seilheimer DK, Harrison GM (1991) Toxicity of colistin in cystic fibrosis patients. DICP 25(11):1168–1170
3. Conway SP, Etherington C, Munday J et al (2000) Safety and efficacy of bolus intravenous colistin in acute respiratory exacerbations in adults with cystic fibrosis. Ann Pharmacother 34:1238–1242
4. Ledson MJ, Gallagher MJ, Cowperthwaite C et al (1998) Four years experience of intravenous colomycin in an adult cystic fibrosis unit. Eur Respir J 12:592–594
5. Conway SP, Pond MN, Watson A et al (1997) Intravenous colistin sulphomethate in acute respiratory exacerbations in adult patients with cystic fibrosis. Thorax 52:987–993
6. Markou N, Apostolakos H, Koumoudiou C et al (2003) Intravenous colistin in the treatment of sepsis from multiresistant Gram-negative bacilli in critically ill patients. Crit Care 7(5):R78–R83
7. Kallel H, Hergafi L, Bahloul M et al (2007) Safety and efficacy of colistin compared with imipenem in the treatment of ventilator-associated pneumonia: a matched case-control study. Intensive Care Med 33(7):1162–1167
8. Reina R, Estenssoro E, Sáenz G et al (2005) Safety and efficacy of colistin in Acinetobacter and Pseudomonas infections: a prospective cohort study. Intensive Care Med 31(8):1058–1065
9. Petrosillo N, Chinello P, Proietti MF et al (2005) Combined colistin and rifampicin therapy for carbapenem-resistant Acinetobacter baumannii infections: clinical outcome and adverse events. Clin Microbiol Infect 11(8):682–683
10. Kasiakou SK, Michalopoulos A, Soteriades ES et al (2005) Combination therapy with intravenous colistin for management of infections due to multidrug-resistant Gram-negative bacteria in patients without cystic fibrosis. Antimicrob Agents Chemother 49(8):3136–3146
11. Falagas ME, Rafailidis PI, Kasiakou SK et al (2006) Effectiveness and nephrotoxicity of colistin monotherapy vs. colistin-meropenem combination therapy for multidrug-resistant Gram-negative bacterial infections. Clin Microbiol Infect 12(12):1227–1230
12. Falagas ME, Kasiakou SK, Kofteridis DP et al (2006) Effectiveness and nephrotoxicity of intravenous colistin for treatment of patients with infections due to polymyxin-only-susceptible (POS) Gram-negative bacteria. Eur J Clin Microbiol Infect Dis 25(9):596–599
13. Falagas ME, Rizos M, Bliziotis IA et al (2005) Toxicity after prolonged (more than four weeks) administration of intravenous colistin. BMC Infect Dis 5:1

14. Falagas ME, Fragoulis KN, Kasiakou SK et al (2005) Nephrotoxicity of intravenous colistin: a prospective evaluation. Int J Antimicrob Agents 26(6):504–507
15. Michalopoulos AS, Tsiodras S, Rellos K et al (2005) Colistin treatment in patients with ICU-acquired infections caused by multiresistant Gram-negative bacteria: the renaissance of an old antibiotic. Clin Microbiol Infect 11(2):115–121
16. Hartzell JD, Neff R, Ake J et al (2009) Nephrotoxicity associated with intravenous colistin (colistimethate sodium) treatment at a tertiary care medical center. Clin Infect Dis 48(12):1724–1728
17. Deryke CA, Crawford AJ, Uddin N, Wallace MR (2010) Colistin dosing and nephrotoxicity in a large community teaching hospital. Antimicrob Agents Chemother 54(10):4503–4505
18. Pogue JM, Lee J, Marchaim D et al (2011) Incidence of and risk factors for colistin-associated nephrotoxicity in a large academic health system. Clin Infect Dis 53(9):879–884
19. Gauthier TP, Wolowich WR, Reddy A et al (2012) Incidence and predictors of nephrotoxicity associated with intravenous colistin in overweight and obese patients. Antimicrob Agents Chemother 56(5):2392–2396
20. Ko HJ, Jeon MH, Choo EJ et al (2011) Early acute kidney injury is a risk factor that predicts mortality in patients treated with colistin. Nephron Clin Pract 117(3):c284–c288
21. Elias LS, Konzen D, Krebs JM, Zavascki AP (2010) The impact of polymyxin B dosage on in-hospital mortality of patients treated with this antibiotic. J Antimicrob Chemother 65(10):2231–2237
22. Kvitko CH, Rigatto MH, Moro AL, Zavascki AP (2011) Polymyxin B versus other antimicrobials for the treatment of *Pseudomonas aeruginosa* bacteraemia. J Antimicrob Chemother 66(1):175–179
23. Oliveira MS, Prado GV, Costa SF et al (2009) Polymyxin B and colistimethate are comparable as to efficacy and renal toxicity. Diagn Microbiol Infect Dis 65(4):431–434
24. Tuon FF, Rigatto MH, Lopes CK et al (2014) Risk factors for acute kidney injury in patients treated with polymyxin B or colistin methanesulfonate sodium. Int J Antimicrob Agents 43(4):349–352
25. Mendes CA, Cordeiro JA, Burdmann EA (2009) Prevalence and risk factors for acute kidney injury associated with parenteral polymyxin B use. Ann Pharmacother 43(12):1948–1955
26. Akajagbor DS, Wilson SL, Shere-Wolfe KD et al (2013) Higher incidence of acute kidney injury with intravenous colistimethate sodium compared with polymyxin B in critically ill patients at a tertiary care medical center. Clin Infect Dis 57(9):1300–1303
27. Esaian D, Dubrovskaya Y, Phillips M, Papadopoulos J (2012) Effectiveness and tolerability of a polymyxin B dosing protocol. Ann Pharmacother 46(3):455–456
28. Phe K, Lee Y, McDaneld PM et al (2014) In vitro assessment and multicenter cohort study of comparative nephrotoxicity rates associated with colistimethate versus polymyxin B therapy. Antimicrob Agents Chemother 58(5):2740–2746
29. Mostardeiro MM, Pereira CA, Marra AR et al (2013) Nephrotoxicity and efficacy assessment of polymyxin use in 92 transplant patients. Antimicrob Agents Chemother 57(3):1442–1446
30. Ouderkirk JP, Nord JA, Turett GS, Kislak JW (2003) Polymyxin B nephrotoxicity and efficacy against nosocomial infections caused by multiresistant Gram-negative bacteria. Antimicrob Agents Chemother 47(8):2659–2662
31. Holloway KP, Rouphael NG, Wells JB et al (2006) Polymyxin B and doxycycline use in patients with multidrug-resistant Acinetobacter baumannii infections in the intensive care unit. Ann Pharmacother 40(11):1939–1945
32. Ramasubban S, Majumdar A, Das PS (2008) Safety and efficacy of polymyxin B in multidrug resistant Gram-negative severe sepsis and septic shock. Indian J Crit Care Med 12(4):153–157
33. Pastewski AA, Caruso P, Parris AR et al (2008) Parenteral polymyxin B use in patients with multidrug-resistant Gram-negative bacteremia and urinary tract infections: a retrospective case series. Ann Pharmacother 42(9):1177–1187
34. Sobieszczyk ME, Furuya EY, Hay CM et al (2004) Combination therapy with polymyxin B for the treatment of multidrug-resistant Gram-negative respiratory tract infections. J Antimicrob Chemother 54(2):566–569
35. Kubin CJ, Ellman TM, Phadke V et al (2012) Incidence and predictors of acute kidney injury associated with intravenous polymyxin B therapy. J Infect 65(1):80–87
36. Chan JD, Graves JA, Dellit TH (2010) Antimicrobial treatment and clinical outcomes of carbapenem-resistant Acinetobacter baumannii ventilator-associated pneumonia. J Intensive Care Med 25(6):343–348
37. Rattanaumpawan P, Ungprasert P, Thamlikitkul V (2011) Risk factors for colistin-associated nephrotoxicity. J Infect 62(2):187–190
38. Turkoglu M, Dizbay M, Ciftçi A et al (2012) Colistin therapy in critically ill patients with chronic renal failure and its effect on development of renal dysfunction. Int J Antimicrob Agents 39(2):142–145
39. Dalfino L, Puntillo F, Mosca A et al (2012) High-dose, extended-interval colistin administration in critically ill patients: is this the right dosing strategy? A preliminary study. Clin Infect Dis 54(12):1720–1726
40. Rocco M, Montini L, Alessandri E et al (2013) Risk factors for acute kidney injury in critically ill patients receiving high intravenous doses of colistin methanesulfonate and/or other nephrotoxic antibiotics: a retrospective cohort study. Crit Care 17(4):R174.
41. Kalin G, Alp E, Coskun R et al (2012) Use of high-dose IV and aerosolized colistin for the treatment of multidrug-resistant Acinetobacter baumannii ventilator-associated pneumonia: do we really need this treatment? J Infect Chemother 18(6):872–877

42. Sorlí L, Luque S, Grau S et al (2013) Trough colistin plasma level is an independent risk factor for nephrotoxicity: a prospective observational cohort study. BMC Infect Dis 13:380
43. Averbuch D, Horwitz E, Strahilevitz J et al (2013) Colistin is relatively safe in hematological malignancies and hematopoietic stem cell transplantation patients. Infection 41(5):991–997
44. Garnacho-Montero J, Ortiz-Leyba C, Jiménez-Jiménez FJ et al (2003) Treatment of multidrug-resistant Acinetobacter baumannii ventilator-associated pneumonia (VAP) with intravenous colistin: a comparison with imipenem-susceptible VAP. Clin Infect Dis 36(9):1111–1118
45. Durakovic N, Radojcic V, Boban A et al (2011) Efficacy and safety of colistin in the treatment of infections caused by multidrug-resistant Pseudomonas aeruginosa in patients with hematologic malignancy: a matched pair analysis. Intern Med 50(9):1009–1013
46. Kim J, Lee KH, Yoo S, Pai H (2009) Clinical characteristics and risk factors of colistin-induced nephrotoxicity. Int J Antimicrob Agents 34(5):434–438
47. Koomanachai P, Tiengrim S, Kiratisin P, Thamlikitkul V (2007) Efficacy and safety of colistin (colistimethate sodium) for therapy of infections caused by multidrug-resistant Pseudomonas aeruginosa and Acinetobacter baumannii in Siriraj Hospital, Bangkok, Thailand. Int J Infect Dis 11(5):402–406
48. Cheng CY, Sheng WH, Wang JT et al (2010) Safety and efficacy of intravenous colistin (colistin methanesulphonate) for severe multidrug-resistant Gram-negative bacterial infections. Int J Antimicrob Agents 35(3):297–300
49. Levin AS, Barone AA, Penço J et al (1999) Intravenous colistin as therapy for nosocomial infections caused by multidrug-resistant Pseudomonas aeruginosa and Acinetobacter baumannii. Clin Infect Dis 28(5):1008–1011
50. Doshi NM, Mount KL, Murphy CV (2011) Nephrotoxicity associated with intravenous colistin in critically ill patients. Pharmacotherapy 31(12):1257–1264
51. Petrosillo N, Giannella M, Antonelli M et al (2014) Clinical experience of colistin-glycopeptide combination in critically ill patients infected with Gram-negative bacteria. Antimicrob Agents Chemother 58(2):851–858
52. Vicari G, Bauer SR, Neuner EA, Lam SW (2013) Association between colistin dose and microbiologic outcomes in patients with multidrug-resistant Gram-negative bacteremia. Clin Infect Dis 56(3):398–404
53. Yılmaz GR, Baştuğ AT, But A et al (2013) Clinical and microbiological efficacy and toxicity of colistin in patients infected with multidrug-resistant Gram-negative pathogens. J Infect Chemother 19(1):57–62
54. Sabuda DM, Laupland K, Pitout J et al (2008) Utilization of colistin for treatment of multidrug-resistant Pseudomonas aeruginosa. Can J Infect Dis Med Microbiol 19(6):413–418
55. Lim SK, Lee SO, Choi SH et al (2011) The outcomes of using colistin for treating multidrug resistant Acinetobacter species bloodstream infections. J Korean Med Sci 26(3):325–331
56. Gounden R, Bamford C, van Zyl-Smit R et al (2009) Safety and effectiveness of colistin compared with tobramycin for multi-drug resistant Acinetobacter baumannii infections. BMC Infect Dis 9:26
57. Falagas ME, Rafailidis PI, Ioannidou E et al (2010) Colistin therapy for microbiologically documented multidrug-resistant Gram-negative bacterial infections: a retrospective cohort study of 258 patients. Int J Antimicrob Agents 35(2):194–199
58. Santamaría C, Mykietiuk A, Temporiti E et al (2009) Nephrotoxicity associated with the use of intravenous colistin. Scand J Infect Dis 41(10):767–769
59. Pintado V, San Miguel LG, Grill F et al (2008) Intravenous colistin sulphomethate sodium for therapy of infections due to multidrug-resistant Gram-negative bacteria. J Infect 56(3):185–190
60. Paul M, Bishara J, Levcovich A et al (2010) Effectiveness and safety of colistin: prospective comparative cohort study. J Antimicrob Chemother 65(5):1019–1027
61. Montero M, Horcajada JP, Sorlí L et al (2009) Effectiveness and safety of colistin for the treatment of multidrug-resistant Pseudomonas aeruginosa infections. Infection 37(5):461–465
62. Bassetti M, Repetto E, Righi E et al (2008) Colistin and rifampicin in the treatment of multidrug-resistant Acinetobacter baumannii infections. J Antimicrob Chemother 61(2):417–420
63. Betrosian AP, Frantzeskaki F, Xanthaki A, Douzinas EE (2008) Efficacy and safety of high-dose ampicillin/sulbactam vs. colistin as monotherapy for the treatment of multidrug resistant Acinetobacter baumannii ventilator-associated pneumonia. J Infect 56(6):432–436
64. Karnik ND, Sridharan K, Jadhav SP et al (2013) Pharmacokinetics of colistin in critically ill patients with multidrug-resistant Gram-negative bacilli infection. Eur J Clin Pharmacol 69(7):1429–1436
65. Markou N, Markantonis SL, Dimitrakis E et al (2008) Colistin serum concentrations after intravenous administration in critically ill patients with serious multidrug-resistant, Gram-negative bacilli infections: a prospective, open-label, uncontrolled study. Clin Ther 30(1):143–151
66. Nelson BC, Eiras DP, Gomez-Simmonds A et al (2015) Clinical outcomes associated with polymyxin B dose in patients with bloodstream infections due to carbapenem-resistant Gram-negative rods. Antimicrob Agents Chemother 59(11):7000–7006
67. Rigatto MH, Behle TF, Falci DR et al (2015) Risk factors for acute kidney injury (AKI) in patients treated with polymyxin B and influence of AKI on mortality: a multicentre prospective cohort study. J Antimicrob Chemother 70(5):1552–1557
68. Rigatto MH, Oliveira MS, Perdigão-Neto LV et al (2016) Multicenter prospective cohort study of

renal failure in patients treated with colistin versus polymyxin B. Antimicrob Agents Chemother 60(4):2443–2449

69. Vardakas KZ, Falagas ME (2017) Colistin versus polymyxin B for the treatment of patients with multidrug-resistant Gram-negative infections: a systematic review and meta-analysis. Int J Antimicrob Agents 49(2):233–238
70. Crass RL, Rutter WC, Burgess DR et al (2017) Nephrotoxicity in patients with or without cystic fibrosis treated with polymyxin B compared to colistin. Antimicrob Agents Chemother 61(4). pii: e02329-16
71. Garonzik SM, Li J, Thamlikitkul V et al (2011) Population pharmacokinetics of colistin methanesulfonate and formed colistin in critically ill patients from a multicenter study provide dosing suggestions for various categories of patients. Antimicrob Agents Chemother 55(7):3284–3294
72. Forrest A, Garonzik SM, Thamlikitkul V et al (2017) Pharmacokinetic/toxicodynamic analysis of colistin-associated acute kidney injury in critically ill patients. Antimicrob Agents Chemother 61(11). pii: e01367-17
73. Shields RK, Anand R, Clarke LG et al (2017) Defining the incidence and risk factors of colistin-induced acute kidney injury by KDIGO criteria. PLoS One 12(3):e0173286
74. Yousef JM, Chen G, Hill PA, Nation RL, Li J (2012) Ascorbic acid protects against the nephrotoxicity and apoptosis caused by colistin and affects its pharmacokinetics. J Antimicrob Chemother 67(2):452–459
75. Sirijatuphat R, Limmahakhun S, Sirivatanauksorn V et al (2015) Preliminary clinical study of the effect of ascorbic acid on colistin-associated nephrotoxicity. Antimicrob Agents Chemother 59(6):3224–3232
76. Dalfino L, Puntillo F, Ondok MJ et al (2015) Colistin-associated acute kidney injury in severely ill patients: a step toward a better renal care? A prospective cohort study. Clin Infect Dis 61(12):1771–1777
77. Wallace SJ, Li J, Nation RL et al (2008) Subacute toxicity of colistin methanesulfonate in rats: comparisons of various intravenous dosage regimens. Antimicrob Agents Chemother 52(3):1159–1161
78. Bergen PJ, Li J, Nation RL et al (2008) Comparison of once-, twice, and thrice-daily dosing of colistin on antibacterial effect and emergence of resistance: studies with Pseudomonas aeruginosa in an in vitro pharmacodynamic model. J Antimicrob Chemother 61:636–642
79. Abdelraouf K, Chang KT, Yin T et al (2014) Uptake of polymyxin B into renal cells. Antimicrob Agents Chemother 58(7):4200–4202
80. Linden PK, Kusne S, Coley K et al (2003) Use of parenteral colistin for the treatment of serious infection due to antimicrobial-resistant Pseudomonas aeruginosa. Clin Infect Dis 37(11):e154–e160
81. Weinstein L, Doan TL, Smith MA (2009) Neurotoxicity in patients treated with intravenous polymyxin B: two case reports. Am J Health Syst Pharm 66(4):345–347
82. Knueppel RC, Rahimian J (2007) Diffuse cutaneous hyperpigmentation due to tigecycline or polymyxin B. Clin Infect Dis 45(1):136–138
83. Bergamasco MD, Barroso Barbosa M, de Oliveira Garcia D et al (2012) Infection with Klebsiella pneumoniae carbapenemase (KPC)-producing K. pneumoniae in solid organ transplantation. Transpl Infect Dis 14(2):198–205
84. Zavascki AP, Manfro RC, Maciel RA, Falci DR (2015) Head and neck hyperpigmentation probably associated with polymyxin B therapy. Ann Pharmacother 49(10):1171–1172
85. Zavascki AP, Schuster LF, Duquia RP (2016) Histopathological findings of pigmented lesion and recovery of natural skin colour in a patient with polymyxin B-associated diffuse hyperpigmentation. Int J Antimicrob Agents 48(5):579–580
86. Domínguez-Ortega J, Manteiga E, Abad-Schilling C et al (2007) Induced tolerance to nebulized colistin after severe reaction to the drug. J Investig Allergol Clin Immunol 17(1):59–61
87. Rattanaumpawan P, Lorsutthitham J, Ungprasert P et al (2010) Randomized controlled trial of nebulized colistimethate sodium as adjunctive therapy of ventilator-associated pneumonia caused by Gram-negative bacteria. J Antimicrob Chemother 65(12):2645–2649
88. FDA alert 06.28.2007. Accessed from https://wayback.archiveit.org/7993/20170406044905/https://www.fda.gov/downloads/Drugs/DrugSafety/PostmarketDrugSafetyInformationforPatientsandProviders/UCM124894.pdf. Last access 19 Jun 2019
89. Pereira GH, Muller PR, Levin AS (2007) Salvage treatment of pneumonia and initial treatment of tracheobronchitis caused by multidrug-resistant Gram-negative bacilli with inhaled polymyxin B. Diagn Microbiol Infect Dis 58:235–240

18 Mechanisms of Polymyxin-Induced Nephrotoxicity

Mohammad A. K. Azad, Roger L. Nation,
Tony Velkov, and Jian Li

Abstract

Polymyxin-induced nephrotoxicity is the major dose-limiting factor and can occur in up to 60% of patients after intravenous administration. This chapter reviews the latest literature on the mechanisms of polymyxin-induced nephrotoxicity and its amelioration. After filtration by glomeruli, polymyxins substantially accumulate in renal proximal tubules via receptor-mediated endocytosis mainly by megalin and PEPT2. It is believed that subsequently, a cascade of interconnected events occur, including the activation of death receptor and mitochondrial apoptotic pathways, mitochondrial damage, endoplasmic reticulum stress, oxidative stress and cell cycle arrest. The current literature shows that oxidative stress plays a key role in polymyxin-induced kidney damage. Use of antioxidants have a potential in the attenuation of polymyxin-induced nephrotoxicity, thereby widening the therapeutic window. Mechanistic findings on polymyxin-induced nephrotoxicity are critical for the optimization of their use in the clinic and the discovery of safer polymyxin-like antibiotics.

Keywords

Polymyxin · Nephrotoxicity · Apoptosis · Oxidative stress · Cell cycle

As reviewed in Chap. 17, the incidence of polymyxin-associated nephrotoxicity is up to 60% in patients with the currently recommended dosage regimens [1–7]. Recent pharmacological studies have indicated that polymyxin-associated nephrotoxicity is the major dose-limiting adverse effect after parenteral administration [8–16] (also Chap. 15). The key features of polymyxin-associated nephrotoxicity include acute tubular damage, decreased creatinine clearance (CrCL),

M. A. K. Azad
Biomedicine Discovery Institute, Infection & Immunity Program and Department of Microbiology, Monash University, Clayton Campus, Melbourne, VIC, Australia

R. L. Nation
Drug Delivery, Disposition and Dynamics, Monash Institute of Pharmaceutical Sciences, Monash University, Parkville Campus, Melbourne, VIC, Australia

T. Velkov
Department of Pharmacology & Therapeutics, School of Biomedical Sciences, Faculty of Medicine, Dentistry and Health Sciences, The University of Melbourne, Melbourne, VIC, Australia

J. Li (✉)
Biomedicine Discovery Institute, Infection & Immunity Program and Department of Microbiology, Monash University, Clayton Campus, Melbourne, VIC, Australia
e-mail: jian.li@monash.edu

J. Li et al. (eds.), *Polymyxin Antibiotics: From Laboratory Bench to Bedside*, Advances in Experimental Medicine and Biology 1145, https://doi.org/10.1007/978-3-030-16373-0_18

and increased serum urea and creatinine concentrations [9] (also Chap. 17). This chapter focuses on the latest progress in understanding the mechanisms of polymyxin-associated nephrotoxicity.

18.1 Renal Disposition of Polymyxins

18.1.1 Differential Renal Handling of Colistin, Polymyxin B and CMS

Although both colistin and polymyxin B are available for clinical use, they differ in their forms for parenteral administration. Polymyxin B is available as the sulfate salt, whereas colistin is available as the prodrug colistimethate sodium (CMS). After intravenous administration of colistin (sulfate) in rats, the urinary recovery of colistin was less than 1% of the administered dose [17, 18]. In comparison to its anticipated clearance by glomerular filtration (2.3 mL/min/kg), the much lower renal clearance of colistin (0.010 ± 0.008 mL/min/kg) indicates extensive tubular reabsorption in rats [17]. In contrast, the urinary recovery of CMS (as CMS and formed colistin in the kidney and urinary tract) after intravenous administration was approximately 60–70% in rats [19–21] and humans [22–24]. The greater renal clearance of CMS compared to its anticipated clearance by glomerular filtration indicates net tubular secretion into the urine [19]. As the major structural difference between colistin and CMS is due to the modification of the primary amines of colistin with negatively charged methanesulfonate groups in CMS, the significantly different renal handling and urinary recovery of colistin and CMS are due to the different charges of the two chemical entities. Similar to colistin, very low urinary recovery of polymyxin B following intravenous administration also suggests that non-renal elimination predominates in both rodents [22–24] and humans [18, 25–27]. Indeed, it has been suggested that polymyxin B undergoes very extensive tubular reabsorption in patients [27, 28].

The very different renal disposition of colistin/polymyxin B and CMS is illustrated in Fig. 18.1. The extensive reabsorption of colistin and polymyxin B from glomerular filtrate to peritubular capillaries would expose tubular cells to high concentrations of these molecules. The net tubular secretion [19] of CMS from peritubular capillaries into the tubular lumen through the epithelial tubular cells may result in intracellular conversion of CMS to colistin [19]. This may enhance the exposure of tubular cells to colistin [29]. In summary, the difference in renal excretion mechanisms of CMS and formed colistin versus polymyxin B is an important factor to modulate the exposure of renal tubular cells to polymyxins and the degree of nephrotoxicity following intravenous polymyxin treatments.

18.1.2 Significant Accumulation of Polymyxins in Renal Tubular Cells

Several recent studies have revealed significant renal accumulation of polymyxins using immunostaining, mass spectrometry imaging, fluorescence microscopy and X-ray fluorescence microscopy (XFM) [30–36]. As CMS is not stable and is a very complex mixture of numerous methanesulfonated derivatives [37–40], its disposition in renal tubular cells has not been examined and the studies in the literature employed colistin, polymyxin B or novel polymyxin analogues. In a mouse study, the distribution of polymyxin B in the kidney tissue was examined after intravenous administration using immunostaining with a polymyxin-specific monoclonal antibody [32]. Predominant accumulation of polymyxin B was evident in the renal cortex, in particular the renal proximal tubular cells, but much less in the distal tubular cells (Fig. 18.2) [30, 32, 41]. Furthermore, matrix-assisted laser desorption/ionizing mass spectroscopy (MALDI-MS) imaging revealed that following subcutaneous administration polymyxins largely accumulated in the renal cortex (Fig. 18.2), but not in the lungs, liver or heart [41].

Abdelraouf et al. employed a commercial product, boron-dipyrromethene (BODIPY)-polymyxin B to examine the uptake of polymyxin B by mammalian renal tubular cells

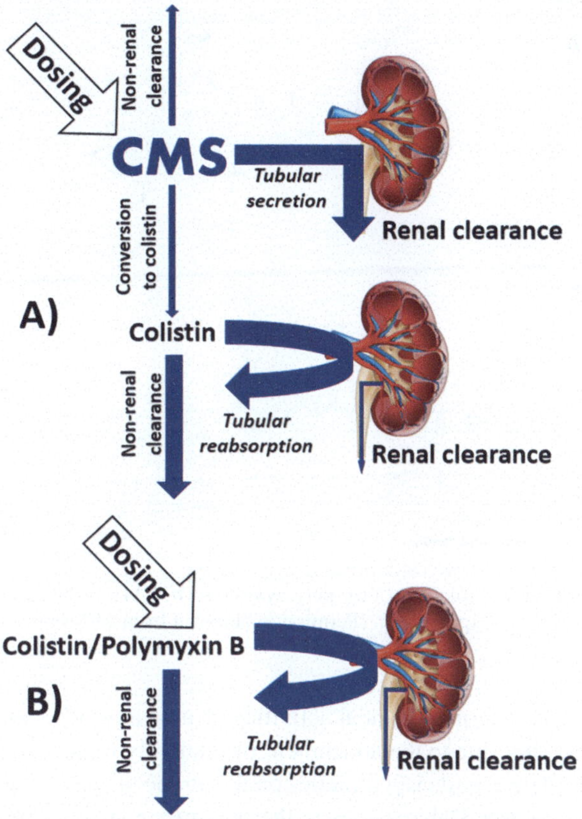

Fig. 18.1 Schematic representation of the renal disposition of (**a**) CMS and formed colistin and (**b**) colistin/polymyxin B. Thickness of the arrows corresponds to the magnitude of the processes involved in the renal deposition. (Figure adapted from Zavascki et al. [29]. Permission obtained from the American Society of Microbiology [ASM])

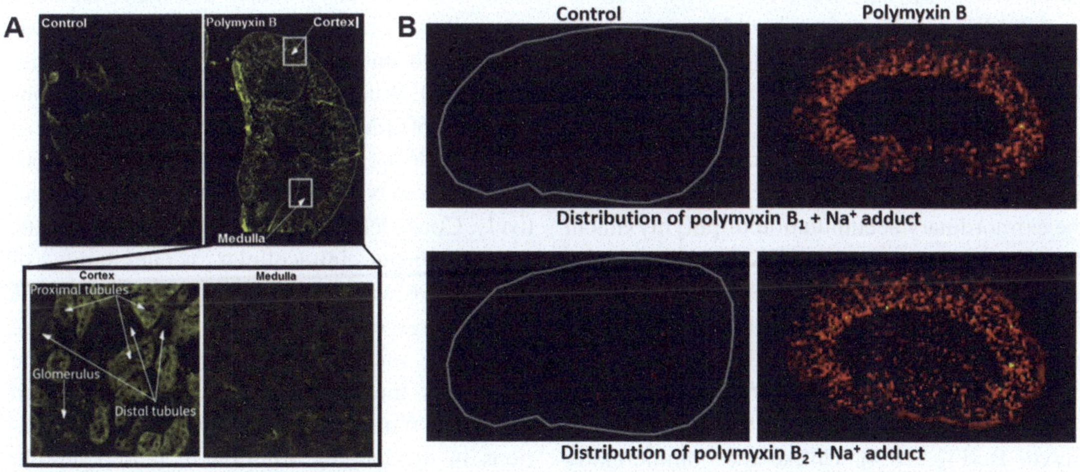

Fig. 18.2 (**a**) Immunostaining demonstrates the distribution of polymyxin B within mouse kidneys following subcutaneous administration (Insert: 10× magnification of the cortex and medulla) [32]. (**b**) Targeted MALDI-MS imaging for detecting polymyxin B_1 and B_2 as Na^+ adduct in the kidneys of mice treated with polymyxin B [41]. (Permission obtained from Oxford University Press)

(LLC-PK1) [42]. Saturable uptake of polymyxin B into LLC-PK1 cells suggested transporter-mediated uptake of polymyxin B. However, it is important to note that commercially available fluorescent polymyxin probes (e.g. dansyl-polymyxin B and BODIPY-polymyxin B) are devoid

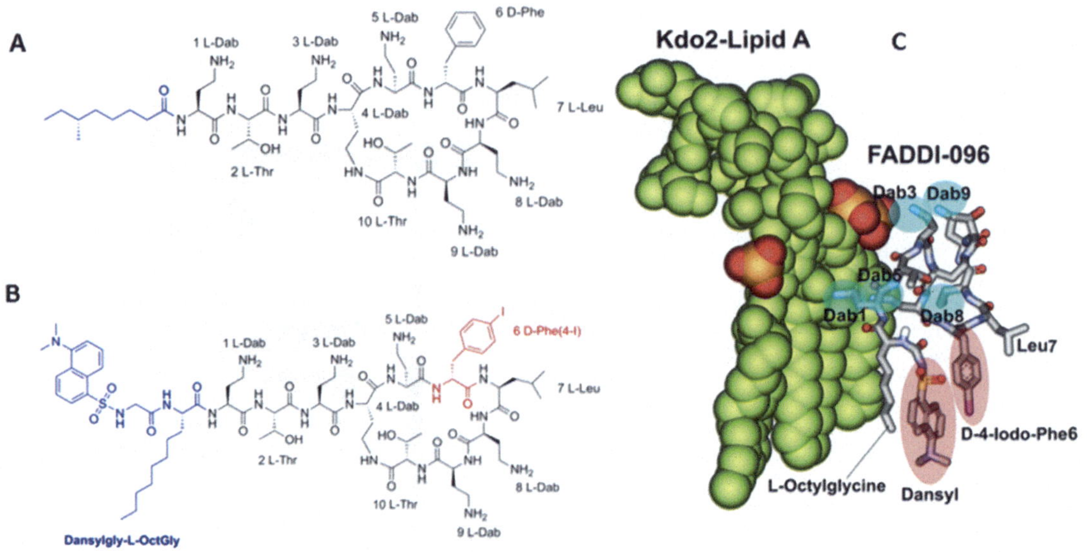

Fig. 18.3 Structures of (**a**) polymyxin B1, (**b**) FADDI-096, and the molecular model of FADDI-096 with *Escherichia coli* Kdo2-Lipid A [30]. (Permission obtained from ACS Publications)

of the pharmacological activities of native polymyxins, due to the attachment of relatively large BODIPY or dansyl moieties on the amine groups of the five Dab residues in the polymyxin structure [30, 43–45]. The structure-activity relationship (SAR) of polymyxins should be considered when using polymyxin probes for pharmacological research.

Using synchrotron X-ray fluorescence (XFM), fluorescence, and scanning electron microscopy, a recent correlative microscopic study discovered the extraordinary accumulation of polymyxins in rat (NRK-52E) and human (HK-2) kidney proximal tubular cells [30]. Based upon the polymyxin SAR model [45], a novel dual-module fluorescent probe, FADDI-096, was designed, consisting of a dansyl group in the *N*-terminus and an iodine fluorophore at position 6 D-Phe of polymyxin B (Fig. 18.3). Unlike the commercially available fluorescent polymyxin probes BODIPY-polymyxin B and dansyl-polymyxin B, FADDI-096 has the structural features required for the biological activity of natural polymyxins. For example, similar to polymyxin B, FADDI-096 displayed antibacterial activity (MIC 8 mg/L against *P. aeruginosa* ATCC 27853) and the ability to induce oxidative stress in both NRK-52E and HK-2 cells [30]. Therefore, it is a valid probe to investigate the nephrotoxicity of polymyxins.

Quantitative mapping of polymyxin distributions in single rat (NRK-52E) and human (HK-2) kidney tubular cells revealed that the remarkable intracellular accumulation of FADDI-096 was both concentration- and time-dependent (Fig. 18.4). With the extracellular concentrations of 5 and 50 μM, intracellular concentrations of FADDI-096 were approximately 1,930- to 4,760-fold higher in NRK-52E cells at 1 and 4 h, respectively. Consistent with the XFM imaging results, the significant intracellular accumulation of FADDI-096 was also observed in the same cells using fluorescence microscopy (Fig. 18.4). These correlative microscopy results demonstrate the overlap of the dansyl and iodine signals from FADDI-096 itself. While FADDI-096 concentrations in the bathing solution increased tenfold (i.e. 5 vs 50 μM), its intracellular concentrations (23.8 ± 6.63 mM *vs* 110 ± 28.2 mM, respectively) only increased approximately 4.62-fold in NRK-52E cells. This finding indicated that the significant accumulation of polymyxins by NRK-52E cells was saturable and likely carrier-mediated [46]. In HK-2 cells, the intracellular concentration of FADDI-096 (31.0 ± 5.69 mM) was

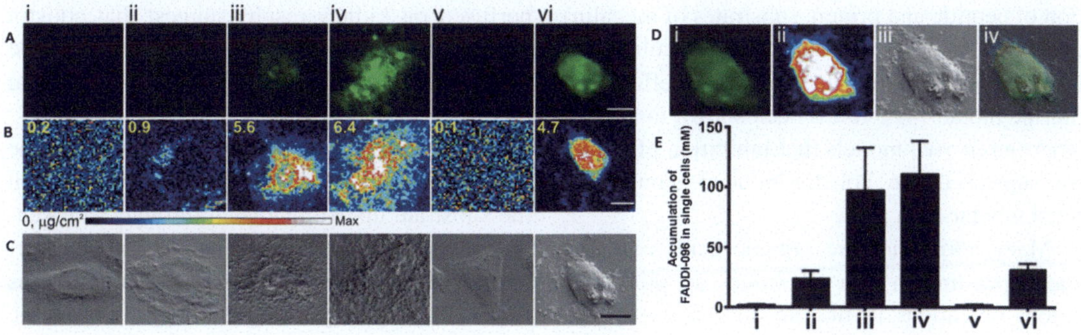

Fig. 18.4 Single-cell correlative microscopy results demonstrate the accumulation of FADDI-096 in NRK-52E and HK-2 cells [30]. (**a**) Fluorescence images of NRK-52E cells (*i*) without treatment, (*ii*) treated with 5 μM FADDI-096 for 4 h, (*iii*) treated with 50 μM FADDI-096 for 1 h, (*iv*) treated with 50 μM FADDI-096 for 4 h; and HK-2 cells (*v*) without treatment, (*vi*) treated with 10 μM FADDI-096 for 4 h. (**b**) Iodine distribution within the same NRK-52E and HK-2 cells as shown in panel A; iodine concentrations (μg/cm^2) are shown using a linear scale from zero to the maximum value; the yellow numbers note the maximum iodine concentration in each sample. (**c**) SEM images of the same NRK-52E and HK-2 cells identified in panel A. (**d**) Correlation of signals from fluorescence microscopy (*i*: Green), XFM (*ii*: Blue to red), and surface morphology from SEM (*iii*: Grey); and their superposition. (**e**) Accumulation of FADDI-096 in single NRK-52E and HK-2 cells measured via iodine content using XFM as shown in panel A (mean ± SD; n = 10). Scale bar: 10 μm. (Permission obtained from ACS Publications)

approximately 3,100-fold higher than the concentration in the bathing solution (10 μM for 4 h). Interestingly, the XFM results also revealed a significant increase in the intracellular calcium concentration, which is a potential stimulus to trigger apoptosis [47]. No correlation was observed between the localization of FADDI-096 and other elements including phosphorus and sodium [30].

Collectively, the immunostaining, mass spectrometry imaging and XFM results all demonstrate the very substantial uptake of polymyxins by renal tubular cells and the potential involvement of transporters; these results are consistent with the pharmacokinetic findings from rats and humans [17, 28, 29]. Further investigations are required to elucidate the detail mechanisms of polymyxin accumulation in renal tubular cells.

18.1.3 Roles of Transporters in the Uptake of Polymyxins by Kidney Tubular Cells

The significant accumulation of polymyxins in renal tubular cells indicates that transporters play an important role in the uptake of colistin and polymyxin B in kidneys [30, 48, 49]. Different transport mechanisms exist in the elimination of drugs, toxins, and endogenous compounds by kidney tubular cells [50–57]. Megalin is a key endocytic receptor for reabsorption of the proteins and small bioactive molecules present in the glomerular filtrate [58], and has been demonstrated to mediate the significant reabsorption of polymyxins by renal tubular cells [46, 59, 60]. Moreover, colistin displays competitive inhibition for binding to megalin with cytochrome *c* (a known substrate for megalin) [46]. In megalin-shed rats, decreased accumulation of colistin in the kidneys and increased excretion in urine suggest that megalin is important for the reabsorption of colistin by tubular cells [46]. Co-administration of colistin with cytochrome *c* or fragment of albumin (FRALB) caused a decreased urinary excretion of *N*-acetyl-β-D-glucosaminidase (NAG), a marker of tubular damage; this suggested the prevention of colistin-induced tubular damage by blocking megalin-mediated uptake [46]. The key role of megalin in the reabsorption of polymyxins is also supported by the finding that co-administration of colistin with succinylated bovine gelatin polypeptides (known competitive inhibitors of the reabsorption of peptide and protein substrates of megalin) decreased both the accumulation of colistin in kidney tissue and also its nephrotoxic effect in a murine model [49]. It has been shown in both *in vitro* and *in vivo* models that inhibition of megalin supressed the colistin-induced damage to renal tubular cells [60].

Many antibiotics are organic acids or bases and, depending on their pK_a values, are present as anions or cations in the physiological environment. Recently, carrier-mediated renal tubular reabsorption of colistin has been suggested from studies conducted *ex vivo* [61]. Using isolated perfused rat kidney, Ma et al. examined the renal disposition and the potential role of kidney transporters in the disposition of colistin [61]. A considerable amount of colistin (administered as colistin sulfate) was removed from the perfusate, but only a relatively low proportion (<10%) was ultimately excreted into the urine, indicating the accumulation of colistin in the kidney tissue [61]. The extensive reabsorption of colistin was inhibited by tetraethylammonium (TEA, a typical substrate of rat OCTN1 [62]), glycine-glycine (Gly-Gly), and hydrochloric acid, suggesting that the renal reabsorption of colistin was mediated by organic transporters and peptide transporters (e.g. OCTN1 and OCTN2) and might be sensitive to the pH of urine [61]. Since colistin is a peptide and the di-peptide Gly-Gly is a typical substrate/inhibitor for PEPT [63], the isolated perfused rat kidney results suggest that colistin might undergo reabsorption via polypeptide transporters (PEPT1 and PEPT2) in the renal tubular cells [61].

A recent study systematically investigated the inhibitory effects of colistin and polymyxin B on the substrate uptake mediated through 15 essential solute carrier transporters (SLCs) in overexpressing HEK293 cells [64]. Both polymyxins had no or only very mild inhibitory effect on the transport activity of the SLCs examined, except human peptide transporter 2 (PEPT2). The concentrations of colistin and polymyxin B required to inhibit 50% uptake (IC_{50}) of the specific human PEPT2 substrate [^{3}H]glycyl-sarcosine were 11.4 ± 3.1 and 18.3 ± 4.2 μM, respectively (Fig. 18.5). PEPT2 is a key SLC expressed particularly in the kidneys and brain [64]. It is a low-capacity high-affinity proton-coupled cotransporter, mainly involved in the renal reabsorption of peptides and peptide-like substrates (including drugs) to maintain systemic nitrogen homeostasis [65]. [^{3}H]Polymyxin B_1 and a fluorescent polymyxin probe MIPS-9541 were also employed as a complementary approach to examine the cellular uptake by PEPT2. The results revealed a significant inhibition of PEPT2-mediated uptake by glycyl-sarcosine, colistin or polymyxin B [64]. Collectively, it is very likely that PEPT2 also plays a critical role in the renal tubular accumulation of polymyxins.

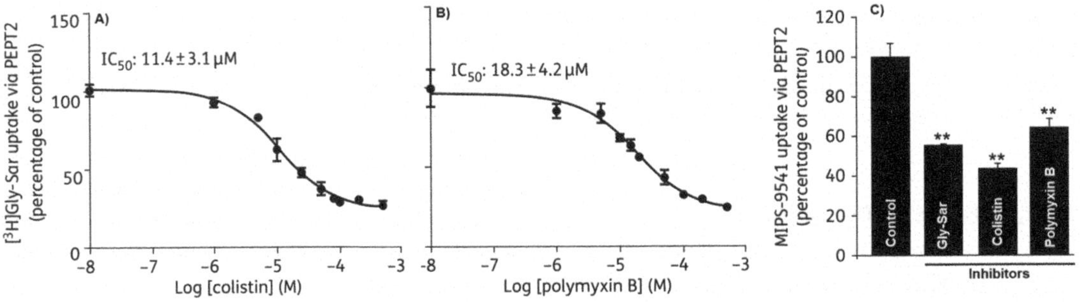

Fig. 18.5 Inhibitory effect of polymyxins on PEPT2-mediated uptake of [^{3}H]Gly-Sar [64]. Cellular uptake of [^{3}H]Gly-Sar was measured in the absence or presence of (**a**) colistin and (**b**) polymyxin B. (**c**) Inhibition of MIPS-9541 uptake by Gly-Sar, colistin or polymyxin B in PEPT2 transfected HEK293 cells. (Permission obtained from the Oxford University Press)

18.1.4 Localisation of Polymyxins in Renal Tubular Cells

There is limited information on the co-localisation of polymyxins with different organelles in renal tubular cells. By incorporating a single dansyl fluorophore in the hydrophobic regions of the polymyxin core structure, we designed, synthesised, and evaluated four novel regioselectively labeled monodansylated polymyxin B probes (MIPS-9541, MIPS-9542, MIPS-9543, and MIPS-9544) for intracellular localisation studies [31]. We examined their antimicrobial activities, cellular uptake, and apoptotic effects on NRK-52E cells. It became evident that incorporation of a dansyl group at position 6 or 7 (e.g. MIPS-9543 and MIPS-9544) of polymyxins is appropriate for generating fluorescent polymyxin probes for intracellular imaging and mechanistic studies. Confocal fluorescence imaging experiments conducted with MIPS-9543 and MIPS-9544 reveal partial co-localisation of polymyxins with both endoplasmic reticulum and mitochondria in NRK-52E cells. Super-resolution imaging is required to elucidate the intracellular localisation of polymyxins in renal tubular cells and the toxic effect on subcellular organelles [31].

In summary, the accumulation, intra-cellular trafficking and localisation of polymyxins in renal tubular cells have not been fully elucidated, and the mechanistic findings may lead to novel approaches to attenuate polymyxin-induced nephrotoxicity.

18.2 Effects of Polymyxins on Renal Tubular Cells

The execution of renal tubular cell death is usually highly orchestrated and interconnected between cell cycle, apoptosis, necrosis and autophagy [66–68]. Depending on the insult and stimulus, tubular cell death can simultaneously trigger multiple pathways and lead to the activation of common downstream cascades [69, 70]. The current literature shows that polymyxin treatment can cause cell cycle arrest, apoptosis and autophagy in renal tubular cells *in vitro* and *in vivo*.

18.2.1 Polymyxins Induce Cell Cycle Arrest

Eadon et al. reported that cell cycle arrest is associated with colistin-induced nephrotoxicity in a murine model using microarray [71]. C57/BL6 mice were intraperitoneally administered with saline or 16 mg/kg/day colistin (in two divided doses), and kidneys were collected after 3 and 15 days. Gene expression microarray analysis of kidney tissues identified 21 differentially expressed genes during the colistin treatment. Up-regulation of the differentially expressed genes from both microarray and RT-PCR results suggested that the cellular injury induced by colistin was mediated through p53 pathway to inhibit cell cycle progression. Up-regulation of CCNB1, CDC2 and the indifferent expression of CDK2, CCND, CCNE genes following colistin treatment indicated G2/M as the point of arrest in the cell cycle. Moreover, translocation of cyclin B1 to the nucleus is another indicator of cell cycle arrest at the G2/M phase induced by colistin [71]. It was also demonstrated that the expression of galectin-3 was up-regulated, supporting the cell cycle arrest through G1/S and G2/M [71]. The up-regulation of galectin-3 is potentially an early marker of the colistin-induced kidney injury. The detection of the proliferating cell nuclear antigen following exposure to colistin for 3 days indicates the emergence of subclinical kidney injury through the blockade of DNA replication at S phase, and subtle pathogenic injury was also observed. Cell cycle arrest may represent a protective mechanism for recovering from colistin-induced nephrotoxicity. However, activation of p53 and galectin-3 can also lead to the apoptotic cell death if the cellular damage is non-recoverable [72, 73]. We examined polymyxin-induced cell death in HK-2 cells and a mouse model using biochemical and molecular approaches. Interestingly, our results indicate the association of DNA damage with polymyxin B induced nephrotoxicity, leading to chromosome mis-segregation and genome instability [74, 75]. There is still much to be learned on polymyxin-induced nephrotoxicity and systems investigations are required to elucidate the complex interplay of major biochemical pathways in polymyxin-induced toxicity in renal tubular cells.

18.2.2 Polymyxins Induce Apoptosis and Oxidative Stress In Vitro and In Vivo

Recent studies revealed that colistin-induced renal tubular apoptosis *in vitro* and in animals [48, 76, 77]. After colistin treatment (cumulative dose of 20.5 mg/kg over 5 days) in rats, Yousef et al., discovered in the kidneys increased TUNEL positive nuclei (%) and fragmentation of DNA, a biochemical hallmark of apoptosis (Fig. 18.6) [48]. Similar results were observed in rat proximal tubular cells (NRK-52E) treated with colistin (0.1 mM for 24 h). Dai et al. revealed the involvement of the death receptor, mitochondrial and endoplasmic reticular pathways in colistin-induced apoptosis in mouse kidney tissues [78]. Colistin was intravenously administered to mice (7.5 or 15 mg of colistin/kg/day in two doses) for 7 days. After 7 days, a significant decrease of Bcl-2 and a concomitant increase of Cytc, AIF, cleaved caspase-9 and cleaved caspase-3 were observed. These findings confirmed that both mitochondria-dependent and -independent pathways are involved in colistin-induced apoptosis in mouse kidneys [78]. Furthermore, significantly increased expression of Fas, FasL, and FADD, and cleavage of caspase-8 were also revealed in the colistin-treated mouse kidneys, demonstrating the involvement of death receptor mediated pathway in colistin-induced apoptosis [78]. Interestingly, the increased expression of tBid indicated the cross-talk between the death receptor and mitochondria apoptotic pathways. In addition, significantly increased concentrations of Grp78/Bip, cleaved ATF6, GADD153/CHOP and caspase-12 were observed in mice following colistin treatment, suggesting that the endoplasmic reticulum pathway is also involved in colistin-induced apoptosis. To date, it is still unknown how each apoptosis pathway is triggered and the interplay among them.

Using cell culture, the activation of caspase-3/8/9, DNA damage and translocation of membrane phosphatidylserine following polymyxin B treatment has been demonstrated in rat (NRK-52E) kidney tubular cells (Fig. 18.7) [79]. In NRK-52E cells treated with polymyxin B (1.0 mM for 24 h), positive labelling with the caspase substrate Red-VAD-FMK showed the presence of activated caspase-3, 8 and 9. Polymyxin-induced apoptosis in NRK-52E cells was also confirmed by positive labelling TUNEL assay and annexin V-PI double staining. Polymyxin-induced apoptosis was both concentration- and time-dependent in NRK-52E and HK-2 cells. Interestingly, HK-2 cells displayed

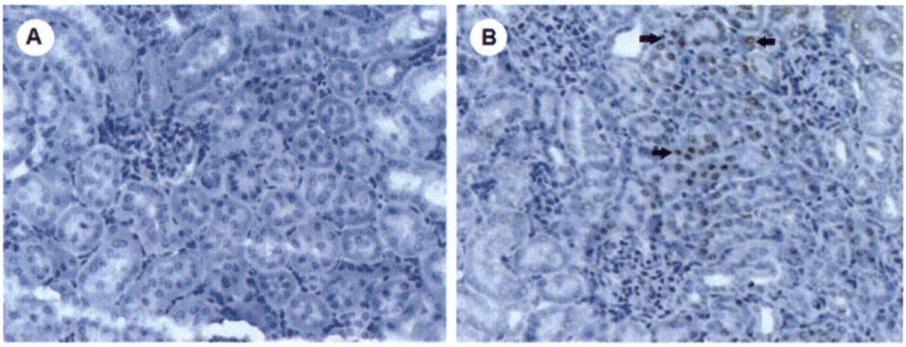

Fig. 18.6 TUNEL positive nulcei (black arrows) after immunohistochemical staining in kidney sections of rats treated for 5 days with (**a**) saline and (**b**) colistin (cumulative dose of 20.5 mg/kg). (Figure modified from Yousef et al. [48] and permission obtained from Oxford University Press)

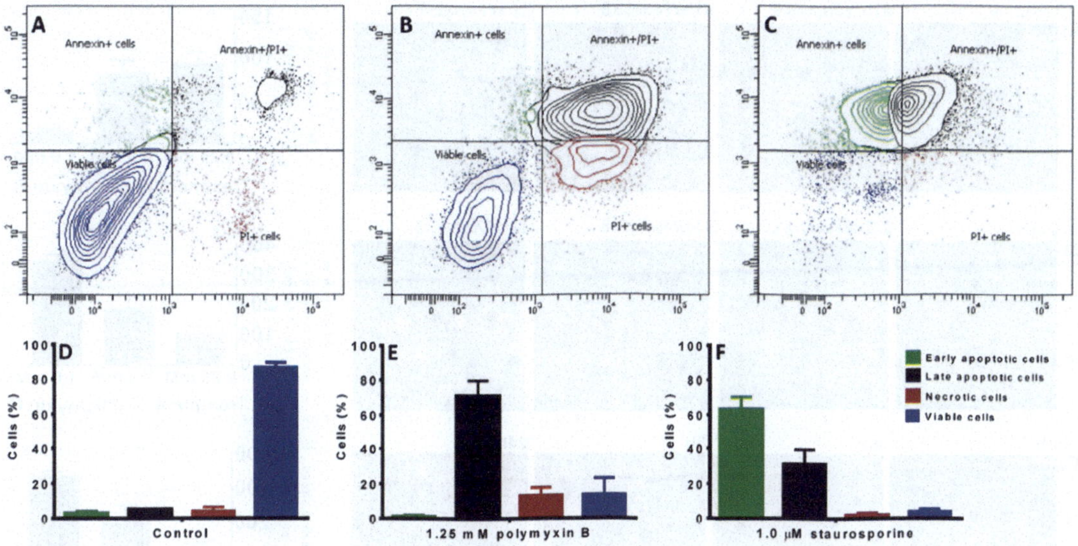

Fig. 18.7 Double staining with annexin V and PI in NRK-52E cells [79]. (**a**) Control cells. (**b**) Cells treated with 1.25 mM polymyxin B for 24 h. (**c**) Cells treated with 1.0 μM staurosporine. In each panel, the upper left quadrant represents cells stained by annexin V (early-apoptotic cells), the bottom right quadrant represents cells stained by PI (necrotic cells), the upper right quadrant represents cells stained by both annexin V and PI (late-apoptotic cells), and the bottom left quadrant represents cells not stained by annexin V or PI (viable cells). (**d–f**) Viability data for panels A to C. (**d**) Control cells. (**e**) Cells treated with 1.25 mM polymyxin B. (**f**) Cells treated with 1.0 μM staurosporine. The error bars represent SD. (Permission obtained from the American Society of Microbiology [ASM])

higher sensitivity to polymyxin B induced toxicity than NRK-52E cells [79]. Mingeot-Leclercq et al. and Vaara et al. also demonstrated dose-dependent cytotoxic activity of polymyxins in porcine renal proximal tubular cells (LLC-PK1) and HK-2 cells, respectively [7, 80].

The relative toxic effect of polymyxin B_1, polymyxin B_2, colistin A and colistin B were examined in HK-2 cells and mice [81]. Comparable nephrotoxicity was observed in mice with mild to moderate histological damage; however, polymyxin B_1 and colistin A showed >3-fold higher *in vitro* apoptotic effect on HK-2 cells than polymyxin B_2 and colistin B, respectively. As there is only one carbon difference in the *N*-terminal fatty acyl group between the two major components of polymyxin B and colistin (Fig. 1.6), these results indicate that the hydrophobicity of the *N*-terminal fatty acyl group of polymyxins plays an important role in polymyxin-induced apoptosis. As shown in Fig. 1.6, the only difference between polymyxin B_1 and colistin A (also polymyxin B_2 and colistin B) is position 6 (i.e. D-Phe *versus* D-Leu); therefore, the hydrophobicity at position 6 is also important to the toxicity on renal tubular cells [81]. The lack of differences in their *in vivo* nephrotoxicity may be due to the sensitivity of the mouse model or the slightly different PK of the two major components of both polymyxins [17, 18].

Mitochondrial stress occurred during polymyxin-induced apoptosis in NRK-52E cells (Fig. 18.8) [82]. In healthy rat kidney tubular cells NRK-52E, mitochondria predominantly were filamentous, whereas in cells undergoing apoptotic cell death mitochondria became fragmented. Concentration- and time-dependent transitions of the mitochondrial morphology from the filamentous (regular) to fragment (stressed) were observed in NRK-52E cells following polymyxin B treatment (1.0 and 2.0 mM up to 24 h) [82]. A concentration-dependent perturbation of mitochondrial morphology was associated with the loss of mitochondrial membrane potential ($\Delta\psi$). Furthermore, it was also evident that polymyxin B induced toxicity was associated with the gen-

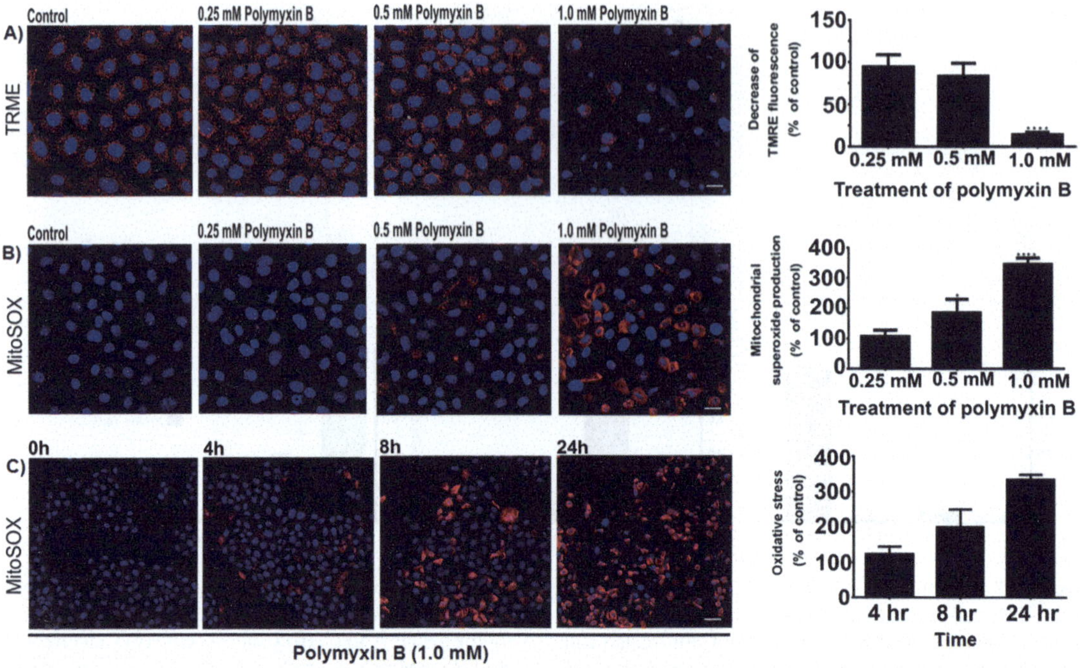

Fig. 18.8 (**a**) Loss of mitochondrial membrane potential measured by fluorescence microscopy using tetramethylrhodamine ethyl ester in NRK-52E cells treated with polymyxin B (0.25, 0.5 and 1.0 mM for 24 h). (**b–c**) Polymyxin B treatments caused concentration- and time-dependent production of mitochondrial superoxide in NRK-52E cells [82]. (Permission obtained from the American Society of Microbiology [ASM])

eration of reactive oxygen species (ROS). Our recent metabolomic study discovered the perturbation of taurine-hypotaurine pathway in polymyxin-treated kidney HK-2 and NRK-52E cells, indicating a loss of cellular capacity to scavenge ROS [41].

Collectively, a working model (Fig. 18.9) was proposed based on the recent literature to understand the complex mechanism of polymyxin-induced apoptosis in renal tubular cells [78]. The precise mechanisms of polymyxin-induced nephrotoxicity remain unknown and require further studies.

18.3 Amelioration of Polymyxin-Induced Nephrotoxicity

Current efforts to minimise the incidence and impact of polymyxin-associated nephrotoxicity in patients rely on monitoring of renal function and electrolyte balance, avoidance of concurrent nephrotoxic agents (if feasible) and optimization of the polymyxin dose [24]. These have been discussed in Chap. 17 and in the literature [29]. Significant efforts have been made over the last decade to attenuate polymyxin-induced nephrotoxicity using different approaches, including decreasing the uptake by renal tubular cells, attenuating polymyxin-induced oxidative stress, and modifying the polymyxin structure (Chap. 20) [45, 48, 49, 61, 76, 77, 83, 84].

A number of animal studies investigated the potential role of co-administered agents to ameliorate polymyxin-induced nephrotoxicity; the majority of these studies involved antioxidants. Ozyilmaz et al., demonstrated that *N*-acetylcysteine (NAC) ameliorated polymyxin-induced oxidative stress and nephrotoxicity in rats [76]. Yousef et al., reported decreased excretion of urinary NAG and less histopathological damage in rat kidneys following co-administration of ascorbic acid (50 or 200 mg/kg) with colistin (cumulative dose, 36.5 mg/kg), compared to rats

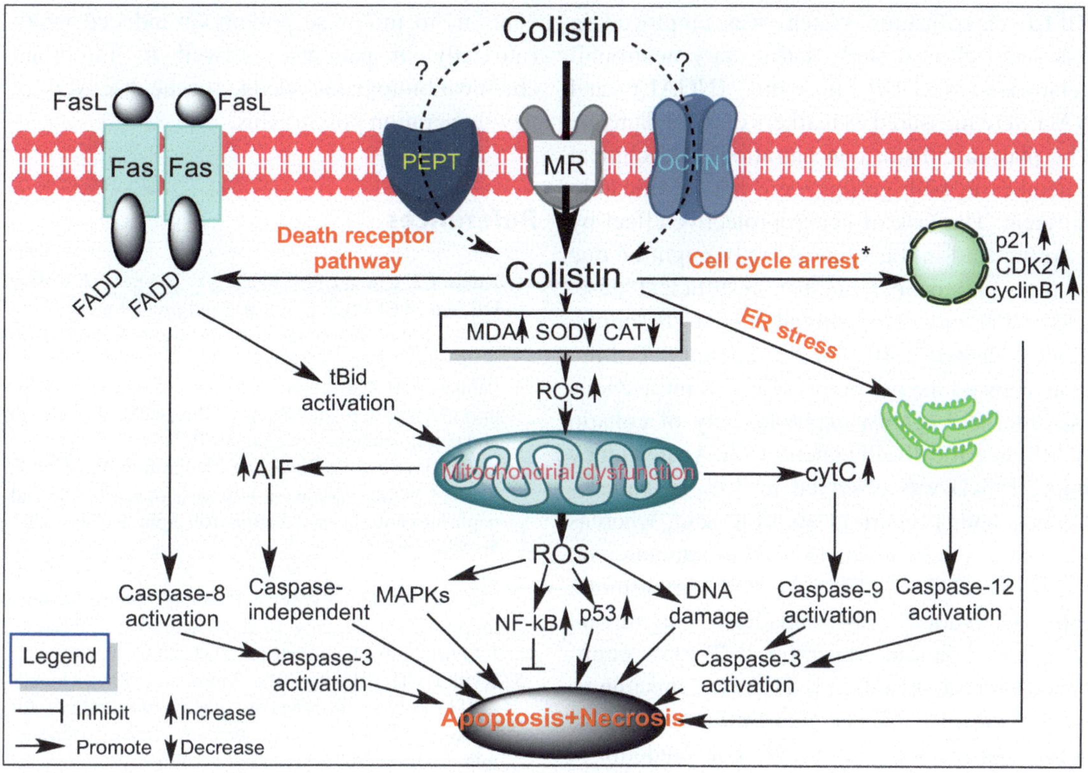

Fig. 18.9 Schematic diagram of the proposed mechanisms of polymyxin-induced apoptosis [78]. (Permission obtained from the American Society of Microbiology [ASM])

treated with colistin or ascorbic acid alone [48]. Similar results have been reported with the co-administration of melatonin, polyaspartic acid, grape seed extract, and methionine [77, 83–85]. Methionine (100 or 400 mg/kg co-administered) protected against polymyxin-induced kidney damage in mice (polymyxin B 35 mg/kg, twice daily over 3.5 days) and significantly attenuated mitochondrial oxidative stress in NRK-52E cells [84]. Interestingly, the pharmacokinetics of polymyxin B in rats were not affected by co-administration of methionine [84]. Ozkan et al., also reported that colistin-induced oxidative stress and apoptosis in rat kidney tissues were attenuated by co-administration of grape seed proanthocyanidin extract, using kidney function estimates from blood urea nitrogen (BUN), creatinine plasma levels and renal histopathological scores [83]. Similarly, protection against colistin-induced apoptosis by proanthocyanidin extract was observed by measuring apoptotic index, caspase-1, caspase-3, and calpain-1 in the kidney tissues [83]. It should be noted that considering animal scaling, a relatively low dose of CMS (300,000 IU/kg/day by intraperitoneal administration, equal to 9 mg colistin base activity/kg/day) was used in the study [83]. Whereas the above co-administered agents probably rely on their antioxidant effects for nephroprotection, the ameliorating effect of co-administered succinylated bovine gelatin polypeptides (Gelofusine) appears to rely on the ability of these peptides to decrease accumulation of polymyxins in renal tissue [49].

Thus far, there is little information on the protection from polymyxin-associated nephrotoxicity in patients. A preliminary randomized controlled study was conducted in 28 patients to investigate the potential nephroprotective effect of intravenous ascorbic acid (2 g every 12 h) against colistin-associated nephrotoxicity in patients requiring intravenous colistin [86]. The

RIFLE classification system was employed in this small clinical study and urinary neutrophil gelatinase-associated lipocalin (NGAL) and NAG were measured as markers of renal damage. The plasma colistin concentrations and clinical outcomes in both groups were not significantly different. The lack of nephroprotective effect by ascorbic acid in this clinical study might be due to the small patient number, insufficient dose, and/or the failure of animal models to mimic clinical disease [86]. On the contrary, Dalfino et al. showed the protective effect of intravenous ascorbic acid against nephrotoxicity of colistin (CMS) in critically-ill patients [87]. Acute kidney injury (AKI) was observed in 30% of patients treating with concurrent ascorbic acid, whereas the rate of AKI was about 67% in patients who did not receive ascorbic acid [87]. Furthermore, this observation was statistically significant ($P < 0.05$, adjusted odds ratio, 0.27 [95% confidence interval, 0.13–0.57]). However, it is important to consider the potential limitations of this study, particularly the small, non-randomized nature and the lack of characterization of patients between the groups. It is also critical to consider the possible effect of ascorbic acid on polymyxin pharmacokinetics/pharmacodynamics in patients [48]. Nevertheless, well-designed clinical studies are warranted to develop novel approaches to attenuate polymyxin-induced nephrotoxicity.

18.4 Conclusions

Significant progress has been made over the last two decades in understanding the mechanism of polymyxin-induced nephrotoxicity. It is clear that polymyxins are substantially accumulated in renal tubular cells, causes oxidative stress and apoptosis via the activation of the death receptor, mitochondria and endoplasmic reticulum mediated pathways. However, the complex interplay of multiple pathways remains undefined in polymyxin-induced nephrotoxicity, and systems investigations on the mechanisms of polymyxin-induced nephrotoxicity are required. The mechanistic findings will provide key pharmacological information for the development of novel interventions to minimise polymyxin-induced nephrotoxicity in patients, as well as important chemical biology knowledge for the discovery of new-generation polymyxins.

References

1. Kubin CJ, Ellman TM, Phadke V, Haynes LJ, Calfee DP, Yin MT (2012) Incidence and predictors of acute kidney injury associated with intravenous polymyxin B therapy. J Infect 65:80–87
2. Falagas ME, Kasiakou SK (2006) Toxicity of polymyxins: a systematic review of the evidence from old and recent studies. Crit Care 10:R27
3. Dezoti Fonseca C, Watanabe M, Vattimo MF (2012) Role of heme oxygenase-1 in polymyxin B-induced nephrotoxicity in rats. Antimicrob Agents Chemother 56:5082–5087
4. Elias LS, Konzen D, Krebs JM, Zavascki AP (2010) The impact of polymyxin B dosage on in-hospital mortality of patients treated with this antibiotic. J Antimicrob Chemother 65:2231–2237
5. Kvitko CH, Rigatto MH, Moro AL, Zavascki AP (2011) Polymyxin B versus other antimicrobials for the treatment of *Pseudomonas aeruginosa* bacteraemia. J Antimicrob Chemother 66:175–179
6. Pastewski AA, Caruso P, Parris AR, Dizon R, Kopec R, Sharma S, Mayer S, Ghitan M, Chapnick EK (2008) Parenteral polymyxin B use in patients with multidrug-resistant gram-negative bacteremia and urinary tract infections: a retrospective case series. Ann Pharmacother 42:1177–1187
7. Mingeot-Leclercq MP, Tulkens PM, Denamur S, Vaara T, Vaara M (2012) Novel polymyxin derivatives are less cytotoxic than polymyxin B to renal proximal tubular cells. Peptides 35:248–252
8. Evans ME, Feola DJ, Rapp RP (1999) Polymyxin B sulfate and colistin: old antibiotics for emerging multiresistant gram-negative bacteria. Ann Pharmacother 33:960–967
9. Falagas ME, Kasiakou SK (2005) Colistin: the revival of polymyxins for the management of multidrug-resistant gram-negative bacterial infections. Clin Infect Dis 40:1333–1341
10. Garonzik SM, Li J, Thamlikitkul V, Paterson DL, Shoham S, Jacob J, Silveira FP, Forrest A, Nation RL (2011) Population pharmacokinetics of colistin methanesulfonate and formed colistin in critically ill patients from a multicenter study provide dosing suggestions for various categories of patients. Antimicrob Agents Chemother 55:3284–3294
11. Crass RL, Rutter WC, Burgess DR, Martin CA, Burgess DS (2017) Nephrotoxicity in patients with or without cystic fibrosis treated with polymyxin B compared to colistin. Antimicrob Agents Chemother 61:e02329-02316

12. Li J, Nation RL, Milne RW, Turnidge JD, Coulthard K (2005) Evaluation of colistin as an agent against multi-resistant in Gram-negative bacteria. Int J Antimicrob Agents 25:11–25
13. Phe K, Shields RK, Tverdek FP, Aitken SL, Guervil DJ, Lam WYM, Musgrove RJ, Luce AM, Tam VH (2016) Predicting the risk of nephrotoxicity in patients receiving colistimethate sodium: a multicentre, retrospective, cohort study. J Antimicrob Chemother 71:3585–3587
14. John JF, Falci DR, Rigatto MH, Oliveira RD, Kremer TG, Zavascki AP (2018) Severe infusion-related adverse events and renal failure in patients receiving high-dose intravenous polymyxin B. Antimicrob Agents Chemother 62:e01617-01617
15. Benattar YD, Omar M, Zusman O, Yahav D, Zak-Doron Y, Altunin S, Elbaz M, Daitch V, Granot M, Leibovici L, Paul M (2016) The effectiveness and safety of high-dose colistin: prospective cohort study. Clin Infect Dis 63:1605–1612
16. Babic JT, Manchandani P, Ledesma KR, Tam VH (2017) Evaluation of urinary KIM-1 for prediction of polymyxin B-induced nephrotoxicity. Antimicrob Agents Chemother 61:e01735-01717
17. Li J, Milne RW, Nation RL, Turnidge JD, Smeaton TC, Coulthard K (2003) Use of high-performance liquid chromatography to study the pharmacokinetics of colistin sulfate in rats following intravenous administration. Antimicrob Agents Chemother 47:1766–1770
18. Sivanesan S, Roberts K, Wang JP, Chea SE, Thompson PE, Li J, Nation RL, Velkov T (2017) Pharmacokinetics of the individual major components of polymyxin B and colistin in rats. J Nat Prod 80:225–229
19. Li J, Milne RW, Nation RL, Turnidge JD, Smeaton TC, Coulthard K (2004) Pharmacokinetics of colistin methanesulphonate and colistin in rats following an intravenous dose of colistin methanesulphonate. J Antimicrob Chemother 53:837–840
20. Marchand S, Lamarche I, Gobin P, Couet W (2010) Dose-ranging pharmacokinetics of colistin methanesulphonate (CMS) and colistin in rats following single intravenous CMS doses. J Antimicrob Chemother 65:1753–1758
21. He H, Li JC, Nation RL, Jacob J, Chen G, Lee HJ, Tsuji BT, Thompson PE, Roberts K, Velkov T, Li J (2013) Pharmacokinetics of four different brands of colistimethate and formed colistin in rats. J Antimicrob Chemother 68:2311–2317
22. Zhao M, Wu XJ, Fan YX, Zhang YY, Guo BN, Yu JC, Cao GY, Chen YC, Wu JF, Shi YG, Li J, Zhang J (2018) Pharmacokinetics of colistin methanesulfonate (CMS) in healthy Chinese subjects after single and multiple intravenous doses. Int J Antimicrob Agents 51:714–720
23. Couet W, Gregoire N, Gobin P, Saulnier PJ, Frasca D, Marchand S, Mimoz O (2011) Pharmacokinetics of colistin and colistimethate sodium after a single 80-mg intravenous dose of CMS in young healthy volunteers. Clin Pharmacol Ther 89:875–879
24. Hartzell JD, Neff R, Ake J, Howard R, Olson S, Paolino K, Vishnepolsky M, Weintrob A, Wortmann G (2009) Nephrotoxicity associated with intravenous colistin (colistimethate sodium) treatment at a tertiary care medical center. Clin Infect Dis 48:1724–1728
25. Abdelraouf K, He J, Ledesma KR, Hu M, Tam VH (2012) Pharmacokinetics and renal disposition of polymyxin B in an animal model. Antimicrob Agents Chemother 56:5724–5727
26. Manchandani P, Dubrovskaya Y, Gao S, Tam VH (2016) Comparative pharmacokinetic profiling of different polymyxin B components. Antimicrob Agents Chemother 60:6980–6982
27. Zavascki AP, Goldani LZ, Cao GY, Superti SV, Lutz L, Barth AL, Ramos F, Boniatti MM, Nation RL, Li J (2008) Pharmacokinetics of intravenous polymyxin B in critically ill patients. Clin Infect Dis 47:1298–1304
28. Sandri AM, Landersdorfer CB, Jacob J, Boniatti MM, Dalarosa MG, Falci DR, Behle TF, Bordinhao RC, Wang J, Forrest A, Nation RL, Li J, Zavascki AP (2013) Population pharmacokinetics of intravenous polymyxin B in critically ill patients: implications for selection of dosage regimens. Clin Infect Dis 57:524–531
29. Zavascki AP, Nation RL (2017) Nephrotoxicity of Polymyxins: is there any difference between Colistimethate and Polymyxin B? Antimicrob Agents Chemother 61:e02319-02316
30. Azad MA, Roberts KD, Yu HH, Liu B, Schofield AV, James SA, Howard DL, Nation RL, Rogers K, de Jonge MD, Thompson PE, Fu J, Velkov T, Li J (2015) Significant accumulation of polymyxin in single renal tubular cells: a medicinal chemistry and triple correlative microscopy approach. Anal Chem 87:1590–1595
31. Yun B, Azad MAK, Nowell CJ, Nation RL, Thompson PE, Roberts KD, Velkov T, Li J (2015) Cellular uptake and localization of polymyxins in renal tubular cells using rationally designed fluorescent probes. Antimicrob Agents Chemother 59:7489–7496
32. Yun B, Azad MA, Wang J, Nation RL, Thompson PE, Roberts KD, Velkov T, Li J (2014) Imaging the distribution of polymyxins in the kidney. J Antimicrob Chemother 70:827–829
33. Nilsson A, Goodwin RJA, Swales JG, Gallagher R, Shankaran H, Sathe A, Pradeepan S, Xue AX, Keirstead N, Sasaki JC, Andren PE, Gupta A (2015) Investigating nephrotoxicity of polymyxin derivatives by mapping renal distribution using mass spectrometry imaging. Chem Res Toxicol 28:1823–1830
34. Manchandani P, Zhou J, Ledesma KR, Truong LD, Chow DS, Eriksen JL, Tam VH (2016) Characterization of polymyxin B biodistribution and disposition in an animal model. Antimicrob Agents Chemother 60:1029–1034
35. Velkov T, Yun B, Schneider EK, Azad MA, Dolezal O, Morris FC, Nation RL, Wang J, Chen K, Yu HH, Wang L, Thompson PE, Roberts KD, Li J (2016) A novel chemical biology approach for mapping of

polymyxin lipopeptide antibody binding epitopes. ACS Infect Dis 2:341–351
36. Vattimo Mde F, Watanabe M, da Fonseca CD, Neiva LB, Pessoa EA, Borges FT (2016) Polymyxin B nephrotoxicity: from organ to cell damage. PLoS One 11:e0161057
37. Li J, Milne RW, Nation RL, Turnidge JD, Coulthard K (2003) Stability of colistin and colistin methanesulfonate in aqueous media and plasma as determined by high-performance liquid chromatography. Antimicrob Agents Chemother 47:1364–1370
38. Nation RL, Li J, Cars O, Couet W, Dudley MN, Kaye KS, Mouton JW, Paterson DL, Tam VH, Theuretzbacher U, Tsuji BT, Turnidge JD (2015) Framework for optimisation of the clinical use of colistin and polymyxin B: the Prato polymyxin consensus. Lancet Infect Dis 15:225–234
39. Metcalf AP, Hardaker LEA, Hatley RHM (2017) A simple method for assaying colistimethate sodium in pharmaceutical aerosol samples using high performance liquid chromatography. J Pharm Biomed Anal 142:15–18
40. British Pharmacopoeia (2018) Monographs for colistimethate sodium. In: British Pharmacopoeia 2018. Stationery Office, London, pp I-675–I-678
41. Azad MAK, Cheah SE, Yun B, Maifiah MHM, Johnson MD, Wang J, Boughton BA, Chapman R, Gould J, Hertzog P, Velkov T, Creek DJ, Li J (2015) Understanding polymyxin-induced nephrotoxicity: combination of transcriptomics, metabolomics and mass spectrometry imaging. Interscience Conference of Antimicrobial Agents and Chemotherapy/ International Congress of Chemotherapy, San Diego, CA, USA, A-942: Poster
42. Abdelraouf K, Chang KT, Yin T, Hu M, Tam VH (2014) Uptake of polymyxin B into renal cells. Antimicrob Agents Chemother 58:4200–4202
43. Azad MA, Yun B, Roberts KD, Nation RL, Thompson PE, Velkov T, Li J (2014) Measuring polymyxin uptake by renal tubular cells: is BODIPY-polymyxin B an appropriate probe? Antimicrob Agents Chemother 58:6337–6338
44. Velkov T, Roberts KD, Nation RL, Thompson PE, Li J (2013) Pharmacology of polymyxins: new insights into an 'old' class of antibiotics. Future Microbiol 8:711–724
45. Velkov T, Thompson PE, Nation RL, Li J (2010) Structure-activity relationships of polymyxin antibiotics. J Med Chem 53:1898–1916
46. Suzuki T, Yamaguchi H, Ogura J, Kobayashi M, Yamada T, Iseki K (2013) Megalin contributes to kidney accumulation and nephrotoxicity of colistin. Antimicrob Agents Chemother 57:6319–6324
47. Mattson MP, Chan SL (2003) Calcium orchestrates apoptosis. Nat Cell Biol 5:1041–1043
48. Yousef JM, Chen G, Hill PA, Nation RL, Li J (2012) Ascorbic acid protects against the nephrotoxicity and apoptosis caused by colistin and affects its pharmacokinetics. J Antimicrob Chemother 67:452–459
49. Sivanesan S, Azad MAK, Schneider EK, Ahmed MU, Huang J, Wang J, Li J, Nation RL, Velkov T (2017) Gelofusine ameliorates colistin-induced nephrotoxicity. Antimicrob Agents Chemother 61:e00985-17
50. Jacquemin E, Hagenbuch B, Stieger B, Wolkoff AW, Meier PJ (1994) Expression cloning of a rat liver Na(+)-independent organic anion transporter. Proc Natl Acad Sci U S A 91:133–137
51. Kekuda R, Prasad PD, Wu X, Wang H, Fei YJ, Leibach FH, Ganapathy V (1998) Cloning and functional characterization of a potential-sensitive, polyspecific organic cation transporter (OCT3) most abundantly expressed in placenta. J Biol Chem 273:15971–15979
52. Busch AE, Quester S, Ulzheimer JC, Waldegger S, Gorboulev V, Arndt P, Lang F, Koepsell H (1996) Electrogenic properties and substrate specificity of the polyspecific rat cation transporter rOCT1. J Biol Chem 271:32599–32604
53. Okuda M, Saito H, Urakami Y, Takano M, Inui K (1996) cDNA cloning and functional expression of a novel rat kidney organic cation transporter, OCT2. Biochem Biophys Res Commun 224:500–507
54. Nigam SK (2015) What do drug transporters really do? Nat Rev Drug Discov 14:29–44
55. Petzinger E, Geyer J (2006) Drug transporters in pharmacokinetics. Naunyn Schmiedeberg's Arch Pharmacol 372:465–475
56. Hua WJ, Hua WX, Jian Z, Wei PH, Ni LY, Hua LY, Wen CD, Ying Z, Li C (2016) The role of drug transporters in the pharmacokinetics of antibiotics. Curr Drug Metab 17:799–805
57. International Transporter Consortium, Giacomini KM, Huang SM, Tweedie DJ, Benet LZ, Brouwer KL, Chu X, Dahlin A, Evers R, Fischer V, Hillgren KM, Hoffmaster KA, Ishikawa T, Keppler D, Kim RB, Lee CA, Niemi M, Polli JW, Sugiyama Y, Swaan PW, Ware JA, Wright SH, Yee SW, Zamek-Gliszczynski MJ, Zhang L (2010) Membrane transporters in drug development. Nat Rev Drug Discov 9:215–236
58. Eshbach ML, Weisz OA (2017) Receptor-mediated endocytosis in the proximal tubule. Annu Rev Physiol 79:425–448
59. Manchandani P, Zhou J, Babic JT, Ledesma KR, Truong LD, Tam VH (2017) Role of renal drug exposure in polymyxin B-induced nephrotoxicity. Antimicrob Agents Chemother 61:e02391-16
60. Hori Y, Aoki N, Kuwahara S, Hosojima M, Kaseda R, Goto S, Iida T, De S, Kabasawa H, Kaneko R, Aoki H, Tanabe Y, Kagamu H, Narita I, Kikuchi T, Saito A (2017) Megalin blockade with cilastatin suppresses drug-induced nephrotoxicity. J Am Soc Nephrol 28:1783–1791
61. Ma Z, Wang J, Nation RL, Li J, Turnidge JD, Coulthard K, Milne RW (2009) Renal disposition of colistin in the isolated perfused rat kidney. Antimicrob Agents Chemother 53:2857–2864
62. Wu X, George RL, Huang W, Wang H, Conway SJ, Leibach FH, Ganapathy V (2000) Structural and functional characteristics and tissue distribution pattern

of rat OCTN1, an organic cation transporter, cloned from placenta. Biochim Biophys Acta 1466:315–327
63. Shen H, Smith DE, Yang T, Huang YG, Schnermann JB, Brosius FC 3rd (1999) Localization of PEPT1 and PEPT2 proton-coupled oligopeptide transporter mRNA and protein in rat kidney. Am J Phys 276:F658–F665
64. Lu XX, Chan T, Xu CH, Zhu L, Zhou QT, Roberts KD, Chan HK, Li J, Zhou FF (2016) Human oligopeptide transporter 2 (PEPT2) mediates cellular uptake of polymyxins. J Antimicrob Chemother 71:403–412
65. Daniel H, Rubio-Aliaga I (2003) An update on renal peptide transporters. Am J Physiol Renal 284:F885–F892
66. Chaabane W, User SD, El-Gazzah M, Jaksik R, Sajjadi E, Rzeszowska-Wolny J, Los MJ (2013) Autophagy, apoptosis, mitoptosis and necrosis: interdependence between those pathways and effects on cancer. Arch Immunol Ther Exp 61:43–58
67. Wang S, Zhang C, Hu L, Yang C (2016) Necroptosis in acute kidney injury: a shedding light. Cell Death Dis 7:e2125
68. Linkermann A (2016) Nonapoptotic cell death in acute kidney injury and transplantation. Kidney Int 89:46–57
69. Pabla N, Dong Z (2008) Cisplatin nephrotoxicity: mechanisms and renoprotective strategies. Kidney Int 73:994–1007
70. Hanigan MH, Devarajan P (2003) Cisplatin nephrotoxicity: molecular mechanisms. Cancer Ther 1:47–61
71. Eadon MT, Hack BK, Alexander JJ, Xu C, Dolan ME, Cunningham PN (2013) Cell cycle arrest in a model of colistin nephrotoxicity. Physiol Genomics 45:877–888
72. Sutton TA, Hato T, Mai E, Yoshimoto M, Kuehl S, Anderson M, Mang H, Plotkin Z, Chan RJ, Dagher PC (2013) p53 is renoprotective after ischemic kidney injury by reducing inflammation. J Am Soc Nephrol 24:113–124
73. Chen SC, Kuo PL (2016) The role of galectin-3 in the kidneys. Int J Mol Sci 17:565
74. Yun B, Zhang T, Azad MAK, Wang J, Nowell CJ, Kalitsis P, Velkov T, Hudson DF, Li J (2016) Targeting the 'Achilles heel': investigating the mechanisms of polymyxin-induced nephrotoxicity. ComBio 2016, Brisbane, Australia: Poster-101
75. Yun B, Zhang T, Azad MAK, Wang J, Nowell CJ, Kalitsis P, Velkov T, Hudson DF, Li J (2018) Polymyxin B causes DNA damage in HK-2 cells and mice. Arch Toxicol 92:2259–2271
76. Ozyilmaz E, Ebinc FA, Derici U, Gulbahar O, Goktas G, Elmas C, Oguzulgen IK, Sindel S (2011) Could nephrotoxicity due to colistin be ameliorated with the use of *N*-acetylcysteine? Intens Care Med 37:141–146
77. Yousef JM, Chen G, Hill PA, Nation RL, Li J (2011) Melatonin attenuates colistin-induced nephrotoxicity in rats. Antimicrob Agents Chemother 55:4044–4049
78. Dai C, Li J, Tang S, Li J, Xiao X (2014) Colistin-induced nephrotoxicity in mice involves the mitochondrial, death receptor, and endoplasmic reticulum pathways. Antimicrob Agents Chemother 58:4075–4085
79. Azad MA, Finnin BA, Poudyal A, Davis K, Li J, Hill PA, Nation RL, Velkov T, Li J (2013) Polymyxin B induces apoptosis in kidney proximal tubular cells. Antimicrob Agents Chemother 57:4329–4335
80. Vaara M, Vaara T (2013) The novel polymyxin derivative NAB739 is remarkably less cytotoxic than polymyxin B and colistin to human kidney proximal tubular cells. Int J Antimicrob Agents 41:292–293
81. Roberts KD, Azad MAK, Wang JP, Horne AS, Thompson PE, Nation RL, Velkov T, Li J (2015) Antimicrobial activity and toxicity of the major lipopeptide components of polymyxin B and colistin: last-line antibiotics against multidrug-resistant Gram-negative bacteria. ACS Infect Dis 1:568–575
82. Azad MA, Akter J, Rogers K, Nation RL, Velkov T, Li J (2015) Major pathways of polymyxin-induced apoptosis in rat kidney proximal tubular cells. Antimicrob Agents Chemother 59:2136–2143
83. Ozkan G, Ulusoy S, Orem A, Alkanat M, Mungan S, Yulug E, Yucesan FB (2013) How does colistin-induced nephropathy develop and can it be treated? Antimicrob Agents Chemother 57:3463–3469
84. Azad MAK, Sivanesan S, Wang J, Chen K, Nation RL, Thompson PE, Roberts KD, Velkov T, Li J (2017) Methionine ameliorates polymyxin-induced nephrotoxicity by attenuating cellular oxidative stress. Antimicrob Agents Chemother 62:e01254-01217
85. Azad MAK, Zhu Y, Han ML, Wang J, Creek DJ, Velkov T, Li J (2016) Effect of poly-L-aspartic acid on polymyxin-induced nephrotoxicity: a systems pharmacology approach. ASM Microbe 2017:Poster-217
86. Sirijatuphat R, Limmahakhun S, Sirivatanauksorn V, Nation RL, Li J, Thamlikitkul V (2015) Preliminary clinical study of the effect of ascorbic acid on colistin-associated nephrotoxicity. Antimicrob Agents Chemother 59:3224–3232
87. Dalfino L, Puntillo F, Ondok MJM, Mosca A, Monno R, Coppolecchia S, Spada ML, Bruno F, Brienza N (2015) Colistin-associated acute kidney injury in severely ill patients: a step toward a better renal care? A prospective cohort study. Clin Infect Dis 61:1771–1777

19 Anti-endotoxin Properties of Polymyxin B-immobilized Fibers

Tohru Tani, Tomoharu Shimizu, Masaji Tani, Hisataka Shoji, and Yoshihiro Endo

Abstract

Polymyxin B is an antibiotic that shows strong bactericidal activity against Gram-negative bacteria, by binding to and inactivating endotoxin. Systemic administration of polymyxin B in humans is restricted because of its nephrotoxicity and neurotoxicity, and this compound was therefore considered a strong candidate ligand for the extracorporeal selective adsorption of circulating endotoxin in the blood. Toraymyxin® is a direct hemoperfusion column that uses polymyxin B attached to an insoluble carrier to bind endotoxin in the blood. In 1994, the Japanese National Health Insurance system approved the use of Toraymyxin for the treatment of endotoxemia and septic shock.

In this chapter, we will review the development, clinical use, and efficacy of Toraymyxin, examine the structure of the Toraymyxin column, and comment on the current position of Toraymyxin in the treatment of severe sepsis and septic shock. We will also highlight some potential new applications of Toraymyxin for pulmonary diseases.

Keywords

Polymyxin · Lipopolysaccharide · Endotoxemia · Septic shock · Toraymyxin

19.1 Introduction

Endotoxin is one of major constituents of the cell wall in Gram-negative bacteria. Endotoxin is recognized through toll-like receptor-4 as a pathogen-associated molecular component that activates macrophages and other leukocytes to produce various inflammatory mediators. Endotoxin plays a major role in the development of toxic symptoms in Gram-negative bacterial infections. A recent review showed that co-detection of Gram-negative bacteria and endotoxemia is strongly predictive of increased risk of mortality compared to detection of neither [1]. It is therefore a reasonable therapeutic approach to evaluate and remove circulating endotoxin in

Authors Tohru Tani and Tomoharu Shimizu have equally contributed to this chapter.

T. Tani (✉)
Biomedical Innovation Center, Shiga University of Medical Science, Otsu, Shiga, Japan

Department of Surgery, Shiga University of Medical Science, Otsu, Shiga, Japan
e-mail: tan@belle.shiga-med.ac.jp

T. Shimizu · M. Tani · H. Shoji
Department of Surgery, Shiga University of Medical Science, Otsu, Shiga, Japan

Y. Endo
Department of Clinical Nursing, Shiga University of Medical Science, Otsu, Shiga, Japan

J. Li et al. (eds.), *Polymyxin Antibiotics: From Laboratory Bench to Bedside*, Advances in Experimental Medicine and Biology 1145, https://doi.org/10.1007/978-3-030-16373-0_19

the blood of patients with Gram-negative bacterial infections.

Polymyxin B is a polycationic antibiotic that binds to and inactivates endotoxin, thereby neutralizing endotoxin-associated toxicity in humans [2]. Systemic administration of polymyxin B in humans is restricted because of its nephrotoxicity and neurotoxicity, and this compound was considered a strong candidate ligand for the extracorporeal selective adsorption of circulating endotoxin in the blood.

The selective removal of endotoxin from blood has been discussed since the 1970s, when Nolan and Ali first demonstrated the adsorption of endotoxin via the ion-exchange resin cholestyramine [3]. Various non-selective adsorbents such as charcoal or ion-exchange resins have since been considered, but a new selective adsorbent was needed for the specific and efficient removal of circulating endotoxin from blood. Sepharose 4B beads to which polymyxin B was covalently immobilized were shown to be effective adsorbents for the removal of endotoxin in a rat model and in human hemodialysis systems [4, 5]. Polyethylenimine [6] and albumin [7], rather than polymyxin B, have also been used as ligands for the selective adsorption of endotoxin.

Initial research into polymyxin B-immobilized fibers began in 1981 at the Department of Surgery, Shiga University of Medical Science, Japan, as a collaboration between our research group and Toray Medical Co., Ltd. (Chiba, Japan) (Fig. 19.1). In 1982, the preliminary results of this research were published, demonstrating that polymyxin B-immobilized fibers reduce endotoxin activity in saline solution, as measured using a semi-quantitative limulus amebocyte lysate (LAL) assay.

In 1983, we began *in vitro* and *in vivo* experimental studies. A pre-clinical study showed that selective removal of endotoxin from the blood by extracorporeal direct hemoperfusion with

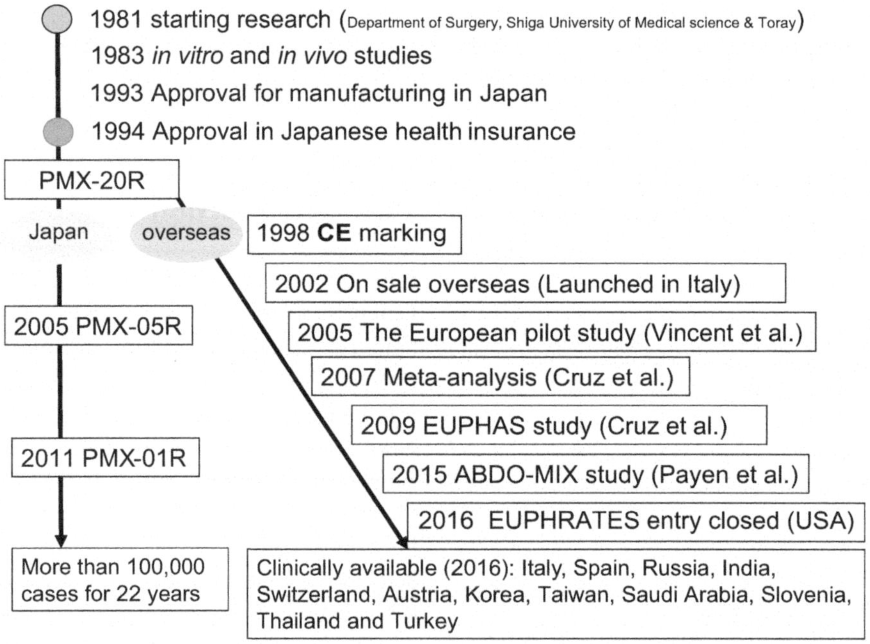

Fig. 19.1 History of the development of Toraymyxin
EUPHAS: Early use of polymyxin B hemoperfusion in abdominal sepsis
ABDO-MIX trial: Effects of hemoperfusion with a polymyxin B membrane in peritonitis with septic shock
EUPHRATES trial: Evaluating the use of polymyxin B hemoperfusion in a randomized controlled trial of adults treated for endotoxemia and septic shock

polymyxin B-immobilized fibers resulted in improved survival in an endotoxemia canine model [8]. Survival also improved in a Gram-negative bacterium (*Escherichia coli*) intravenous administration canine model after extracorporeal hemoperfusion with polymyxin B-immobilized fibers, indicating the potential for use thereof in humans [9].

The polymyxin B-immobilized fibers blood purification column, Toraymyxin®, was developed as a medical device to be used in conjunction with direct hemoperfusion to remove circulating endotoxin from the blood in humans. A phase I clinical study was initiated in 1989, and in 1994 the first clinical report of 16 patients with septic multiple organ failure treated with direct hemoperfusion with Toraymyxin (PMX) was published [10]. PMX for 2 h significantly decreased the level of circulating endotoxin from 76 to 21 pg/mL. Furthermore, the patients' hyperdynamic state (in terms of cardiac index), which is characteristic of endotoxic shock, returned to normal after treatment [10]. Regarding the technical term for Toraymyxin, we have suggested defining "PMX" as "direct hemoperfusion with the polymyxin B-immobilized fiber column (Toraymyxin)".

In 1994, Toraymyxin was adopted by the Japanese National Health Insurance system for the treatment of endotoxemia and septic shock [11]. Since then, Toraymyxin has been used safely in more than 100,000 cases in emergency and intensive care units in Japan. Three different columns are currently available for clinical use in Japan: PMX-20R for use in adults, PMX-05R for use in children, and PMX-01R for use in babies (Fig. 19.1).

In 1998, Toraymyxin received CE mark approval in Europe. In 2005, the results of the first randomized controlled trial (RCT) in Europe were published, showing that treatment with Toraymyxin is safe, and improves cardiac and renal dysfunction due to sepsis or septic shock [12]. In 2007, the findings of a meta-analysis demonstrated the favorable effects of PMX [13]. In 2009, the results of the second RCT in Europe, the EUPHAS study (Early Use of Polymyxin B Hemoperfusion in Abdominal Sepsis), which was conducted in Italy, were published, showing that Toraymyxin treatment results in a significant reduction in sepsis-associated mortality [14]. A large RCT, the ABDO-MIX trial in France, has failed to show survival benefit and improvement in organ failure with Toraymyxin treatment compared to conventional treatment of peritonitis-induced septic shock [15]. Recruitment into another large RCT, the EUPHRATES trial in the USA, is now closed and the results showed that polymyxin B hemoperfusion plus conventional medical therapy did not reduce mortality at 28 days in patients with septic shock and high endotoxin activity (Fig. 19.1, ClinicalTrials.gov Identifier: NCT01046669). Toraymyxin is currently available for use in clinical settings in Italy, Spain, Russia, India, Switzerland, Austria, Korea, Taiwan, Saudi Arabia, Slovenia, Thailand, and Turkey.

In this chapter, we will review the development, clinical use, and efficacy of Toraymyxin, examine the structure of the Toraymyxin column, and comment on the current position of Toraymyxin in the treatment of severe sepsis and septic shock. We will also highlight some potential new applications for Toraymyxin.

19.2 Development of Polymyxin B-immobilized Fibers

19.2.1 Polymyxin B and Endotoxin

Polymyxin B has strong bactericidal activity against Gram-negative bacteria. The bactericidal properties of polymyxin B are discussed in more detail in other chapters. Polymyxin B also neutralizes the lethal toxicity, limulus gelation activity, and hemodynamic effects of endotoxin, a major component of the outer membrane of Gram-negative bacteria, by binding to the lipid A domain, which is the active center of the endotoxin molecule [16]. The protective effects of polymyxin B against endotoxemia and septicemia have been demonstrated in various animal models [17–19].

Electron microscopy imaging has revealed that when endotoxin is exposed to polymyxin B, its usual ribbon-like structure partially or fully disaggregates [20]. The detergent-like activity of polymyxin B arises from its ability to prompt the dissociation of endotoxin's micellar structure [21]. Polymyxin B has been shown to bind to the lipopolysaccharide of *Salmonella typhimurium* at the negatively charged 2-keto-3-deoxyoctulosonate lipid A region of the lipopolysaccharide molecule. This is achieved via electrostatic, and possibly hydrophobic, interactions at a stoichiometric ratio of one polymyxin B molecule to one lipopolysaccharide monomer unit [22]. Furthermore, Vesentini et al. demonstrated that short-range interactions between polymyxin B and endotoxin are mediated mainly via hydrophobic forces, whereas long-term complex formation is mediated via ionic forces. The interaction energy occurring in each molecular complex was calculated at different intermolecular distances, and the binding forces were estimated by fitting interaction energy data. Maximum binding forces calculated via molecular mechanics for the polymyxin B–endotoxin complex range from 1.39 to 3.79 nN [23].

Together, these studies have clarified that polymyxin B binds to and detoxifies endotoxin, and that it is through this binding that polymyxin B exerts its anti-microbial activity.

19.2.2 Polymyxin B-immobilized Fibers

To enable selective adsorption of circulating endotoxin in the blood, polymyxin B was covalently immobilized on the surface of polystyrene-derived, polypropylene-reinforced conjugated carrier fibers (Fig. 19.2). α-Chloroacetoamide methyl groups were chemically introduced into the polystyrene molecule to provide a moiety to which the polymyxin B could be fixed [24]. The endotoxin adsorption capacity of polymyxin B-immobilized fibers has been shown to be almost the same as that of the ion-exchange resin IRA-938 (Fig. 19.3a, b) [24].

The endotoxin detoxification capacity of polymyxin B-immobilized fibers *in vitro* changes depending on the number of residual primary amino groups in the immobilized polymyxin B molecule

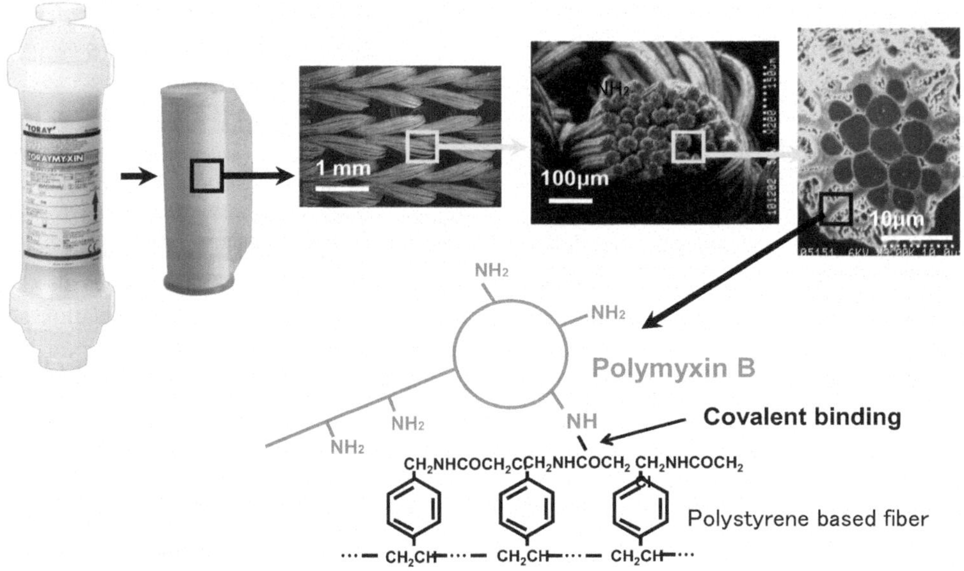

Fig. 19.2 Physical structure of a Toraymyxin cartridge and of the knitted fabric roll of polymyxin B-immobilized fibers. (Provided by Toray Medical Co., Ltd.)
The Toraymyxin cartridge contains a roll of knitted fibers. Each fiber consists of a bundle of ultra-fine fibers with a diameter of approximately 20 μm. The polymyxin B molecules are covalently bound onto the fiber surface and therefore do not leak into the patient
Molecular conformation is shown. Polymyxin B was covalently bound to polystyrene-based fibers

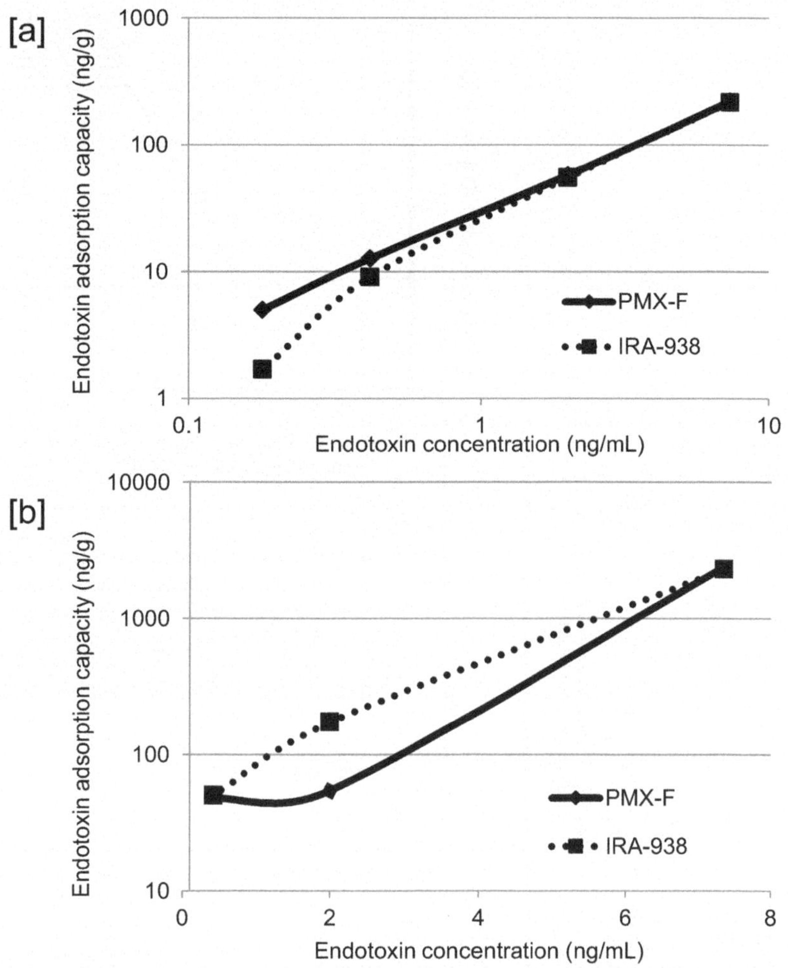

Fig. 19.3 Endotoxin adsorption capacity of polymyxin B-immobilized fiber or IRA-938 data from Shoji et al. [24]. Two grams of polymyxin B-immobilized fibers (polymyxin B-immobilized fibers, solid line) or ion-exchange resin (IRA-938, dotted line) were incubated with 30 mL of various concentrations of endotoxin in (**a**) aqueous solution or (**b**) bovine serum solution

(Fig. 19.4a) [24]. In an *in vivo* canine experiment, the survival rate of endotoxin-challenged dogs was higher when the immobilized polymyxin B had a large number of residual primary amino groups (Fig. 19.4b) [24]. These results suggested that polymyxin B-immobilized fibers had a maximal endotoxin detoxification capacity when three primary amino groups were left unbound from the carrier fiber. Furthermore, concentrations of immobilized polymyxin B above 3.5 μg/mg of fiber were suggested to be optimal. Because the primary amino groups are positively charged, it is likely that they play a major role in the ionic binding of polymyxin B to endotoxin [24].

In addition to detoxifying the limulus activity of endotoxin, polymyxin B-immobilized fibers have also been shown to neutralize the pyrogenic activity of endotoxin solution in a rabbit pyrogen test. Therefore, polymyxin B-immobilized fibers are also effective in removing pyrogenic agents from serum (Fig. 19.5) [24].

19.2.3 Endotoxin Adsorption Capacity and Bactericidal Activity of Polymyxin B-immobilized Fibers *In Vitro*

When polymyxin B-immobilized fibers, or carrier fibers alone, were incubated with synthetic lipid A *in vitro*, polymyxin B-immobilized fibers effectively adsorbed synthetic lipid A, whereas the carrier fibers alone did not, as assessed using an LAL assay, showing that the capability of polymyxin B to bind to lipid A is retained even when polymyxin B is immobilized (Fig. 19.6a)

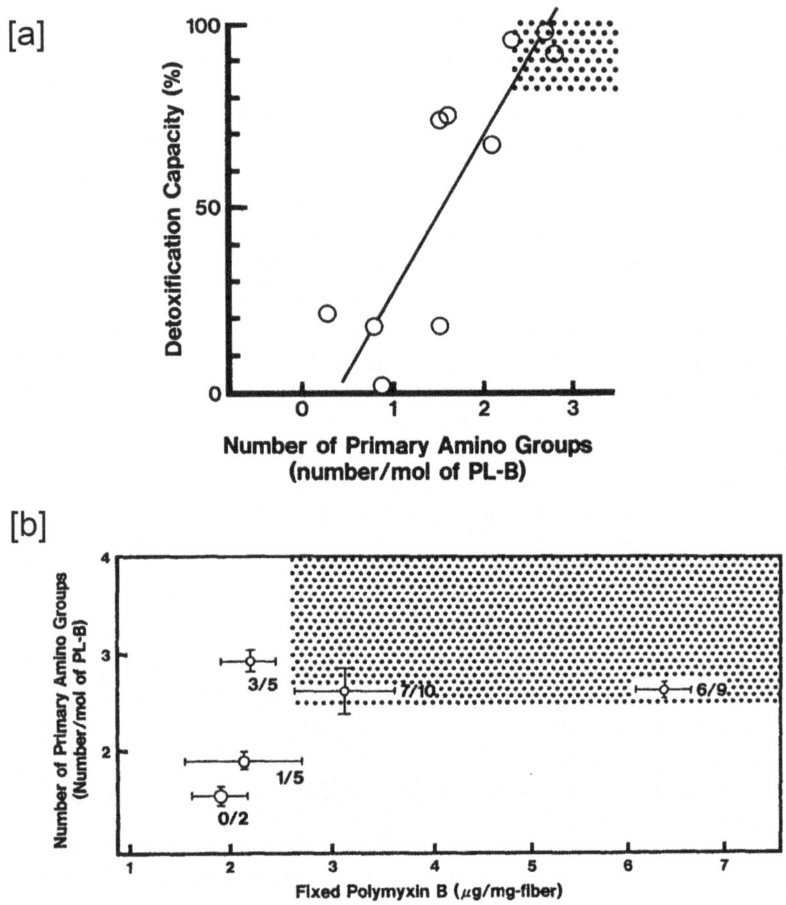

Fig. 19.4 Relationship between the residual number of primary amino groups in the immobilized polymyxin B molecule
(**a**) Relationship between the residual number of primary amino groups in the immobilized polymyxin B molecule and its detoxification capacity
Polymyxin B-immobilized fibers were added to a bovine serum solution containing lipopolysaccharides, and incubated with continuous stirring for 2 h. Endotoxin concentration was measured via limulus amebocyte lysate assay. Polymyxin B-immobilized fibers had the best endotoxin detoxification capacity when three amino groups were left in the immobilized polymyxin B molecule bound to the carrier fiber. (Reproduced from Shoji et al. [24])
(**b**) Relationship between the number of residual amino groups in fixed immobilized polymyxin B and the rate of survival in endotoxin-challenged dogs
The number of residual primary amino groups in the immobilized polymyxin B had a greater overall effect on the survival of dogs than did the quantity of immobilized polymyxin B used. (Reproduced from Shoji et al. [24])

[24]. The lipid A portion of endotoxin is located in the least variable region of the lipopolysaccharide molecule, meaning that it does not change from species to species, or from strain to strain. Polymyxin B-immobilized fibers can adsorb various types of endotoxins with different O-side chains or chemotypes, indicating that immobilized polymyxin B binds specifically to the lipid A portion of the LPS molecule (Fig. 19.6b) [24].

Polymyxin B-immobilized fibers also show bactericidal activity. Polymyxin B-immobilized fibers were added to a phosphate buffer solution containing *Pseudomonas aeruginosa*, and changes in bacterial cell count were assessed under continual stirring for 9 h. Bacterial cell counts decreased sharply from approximately 10^7 to 10^3 CFU/mL (Fig. 19.6c) [24].

Together, these results clearly demonstrate that immobilized polymyxin B retains the endo-

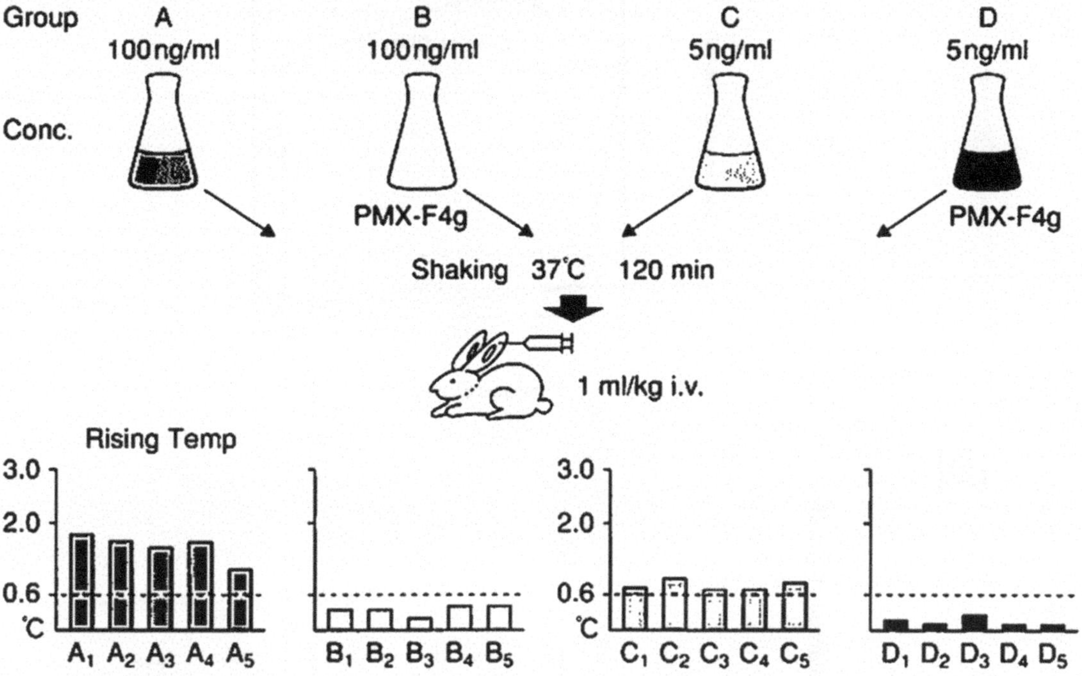

Fig. 19.5 Endotoxin removal by polymyxin B-immobilized fibers examined in a rabbit pyrogen test Polymyxin B-immobilized fibers (PMX-F) were incubated with various concentrations of endotoxin in an aqueous solution at 37 °C for 120 min, and then intravenously administered to rabbits. Treatment with polymyxin B-immobilized fibers suppressed the elevation of body temperature

toxin binding and bactericidal properties of free polymyxin B [24].

19.2.4 Endotoxin Adsorption Capacity of Polymyxin B-immobilized Fibers *In Vivo*

Direct hemoperfusion under heparin infusion for 2 h with an animal-use polymyxin B-immobilized fiber column improved mortality and hemodynamic parameters in endotoxemic dogs receiving intravenous endotoxin administration, compared with carrier fiber alone, charcoal, and ion-exchange resin (Fig. 19.7) [25]. Direct hemoperfusion with an animal-use polymyxin B-immobilized fiber column has also been shown to improve hemodynamic parameters in dogs with sepsis induced by an intravenous infusion of *E. coli*. Blood lactate levels were improved in dogs treated with polymyxin B-immobilized fibers compared with non-treated dogs; however, platelet counts were significantly lower, which is an adverse effect of direct hemoperfusion with a polymyxin B-immobilized fiber column for animal use (Fig. 19.8) [9].

19.3 Toraymyxin, a Polymyxin B-immobilized Fiber Column for Human Use

19.3.1 Physical Structure of Toraymyxin Cartridge

The Toraymyxin cartridge comprises a plastic case containing a knitted roll of polymyxin B-immobilized fiber fabric for human use. The Toraymyxin cartridge is sterilized via autoclave and filled with physiological saline. The use of a thin, fibrous carrier produces a hemoadsorption cartridge with a large surface area that does not cause a large pressure drop in the blood flow compartment (Fig. 19.2). Flow through the

[a]

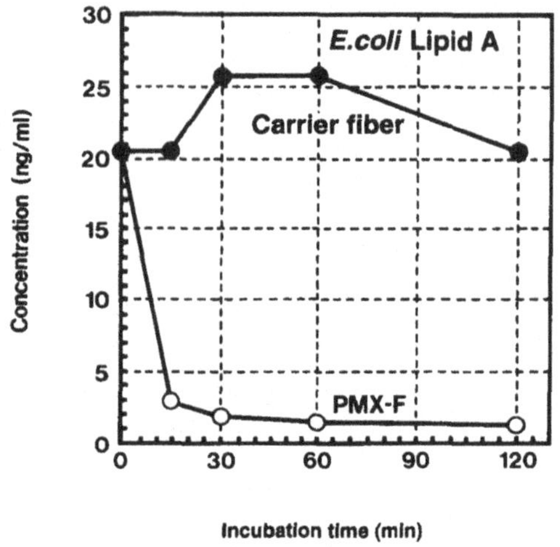

[b]

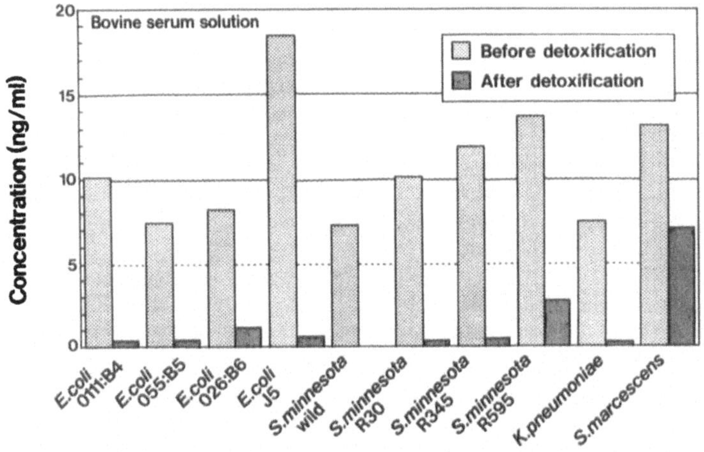

[c]

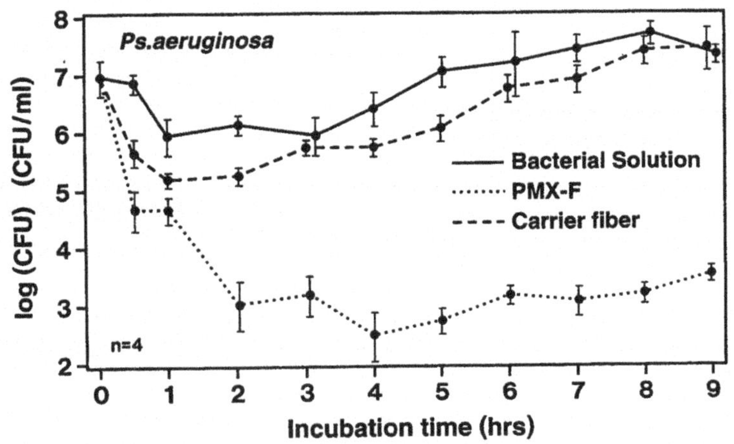

Fig. 19.6 The endotoxin adsorption ability and bactericidal activity of polymyxin B-immobilized fibers. (Reproduced from Shoji et al. [24])
(**a**) Alteration in lipid A concentration by polymyxin B-immobilized fibers
Polymyxin B-immobilized fibers (PMX-F) were incubated with a bovine serum solution containing 20 ng/mL of synthetic *Escherichia coli* lipid A with continuous stirring. Lipid A concentration was measured via limulus amebocyte lysate assay
(**b**) Detoxification of endotoxin extracted from different species or strains of Gram-negative bacteria by use of polymyxin B-immobilized fibers
(**c**) Bactericidal activity of polymyxin B-immobilized fibers against *Pseudomonas aeruginosa*
Polymyxin B-immobilized fibers (PMX-F) or carrier fibers alone were incubated with a solution of *Pseudomonas aeruginosa* for 9 h and residual bacterial numbers were assessed

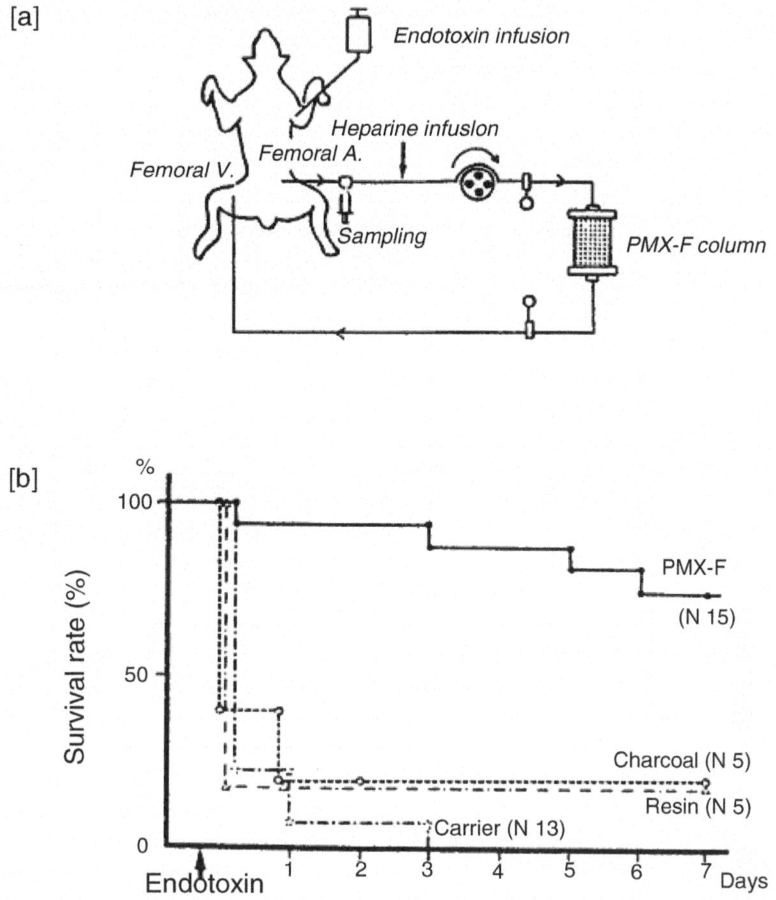

Fig. 19.7 Endotoxin adsorption by polymyxin B-immobilized fibers in an endotoxemia dog model. (Reproduced from Kodama et al. [25]) (**a**) Schematic of the experimental model: endotoxin was infused intravenously to dogs. Extracorporeal direct-hemoperfusion with a polymyxin B-immobilized fiber (PMX-F) column or control column was initiated (**b**) Survival rates after PMX-F treatment were compared with those after treatment with a PMX-F column, or a charcoal, resin, or carrier fiber column

Toraymyxin column is unidirectional and moves radially from the center to the outside of the roll of polymyxin B-immobilized fiber fabric, which improves adsorption capacity by ensuring a homogeneous distribution of blood within the column (Fig. 19.9a). Therefore, direct hemoperfusion can be performed in the clinical setting without the need for a complicated extracorporeal blood perfusion circuit.

The endotoxin adsorption capacity of the Toraymyxin column was evaluated in a bovine serum perfusion model. An endotoxin-containing bovine serum solution (10 ng/mL of endotoxin, total 1.5 L) was perfused through the Toraymyxin cartridge at a flow rate of 100 mL/min, and the change in endotoxin level was monitored using an LAL assay. Endotoxin concentration reached equilibrium after 2 h of perfusion (Fig. 19.9b) [11].

19.3.2 Toraymyxin as a Medical Device

There are three types of Toraymyxin columns currently available for clinical use in Japan: PMX-20R® (volume: 135 mL), PMX-05R® (40 mL), and PMX-01R® (8 mL) (Fig. 19.10a). The first Toraymyxin column released was PMX-20R, which was developed for the treatment of adults with septic shock [11]. The second column released was PMX-05R, which was developed for use in pediatric or elderly patients with smaller circulating blood volumes or a body weight less than 40 kg [26]. Several reports in Japanese literature have shown acceptable results by using PMX-05R in pediatric or elderly patients with sepsis. In 2011, PMX-01R was released for use in newborn or premature babies, and there are case reports describing the clinical use of PMX-

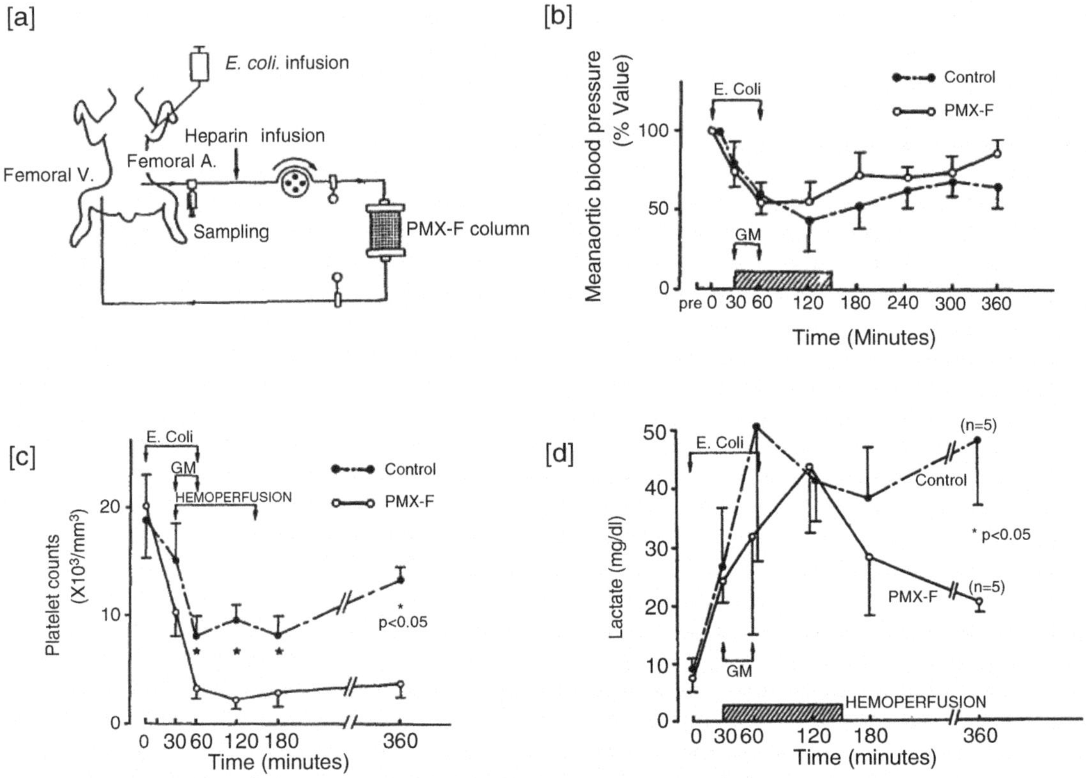

Fig. 19.8 Efficacy of polymyxin B-immobilized fibers in a dog bacterial infusion model. (Reproduced from Hanasawa et al. [9])
(**a**) Schematic of the experimental model: the solution containing *Escherichia coli* was infused intravenously to dogs. Extracorporeal direct-hemoperfusion with a polymyxin B-immobilized fiber (PMX-F) column or control column was initiated
(**b**) Alteration in mean aortic blood pressure: mean aortic blood pressure after PMX-F treatment increased compared to that after treatment with the control column
(**c**) Alteration in platelet counts: platelet count significantly decreased following PMX-F treatment compared that following treatment with the control column
(**d**) Alteration in blood lactate levels: blood lactate levels significantly decreased following PMX-F treatment compared to those following treatment with the control column

01R in patients with a body weight less than 1000 g with severe sepsis [27, 28].

19.3.3 Use of the Toraymyxin Column in Hemoperfusion

Prior to use, Toraymyxin columns must be rinsed with physiological saline and then primed. For example, a PMX-20R column must be washed out with 4 L of physiological saline and then primed with 1 L of physiological saline containing 20 mg of nafamostat mesilate or 1000 U of unfractionated heparin. For patients in an extremely critical condition or with an extremely low blood pressure, an albumin solution containing anticoagulation agents may be used as the final priming solution.

Vascular access for hemoperfusion is usually via a central vein (internal cervical vein, subclavian vein, or femoral vein). The standard blood flow rate through the Toraymyxin column depends on the size of the column: 80–120 mL/min for PMX-20R, 20–40 mL/min for PMX-05R, and 8–12 mL/min for PMX-01R. An anticoagulant (20–30 mg/h nafamostat mesilate or bolus 40–60 U/kg and 40–60 U/kg/h unfractionated heparin, depends on patients' coagulation condition) should also be administered to prevent coagulation within the blood circuit (Fig. 19.10b). Since coagulation dysfunction is common in patients with septic shock, the short half-lives of

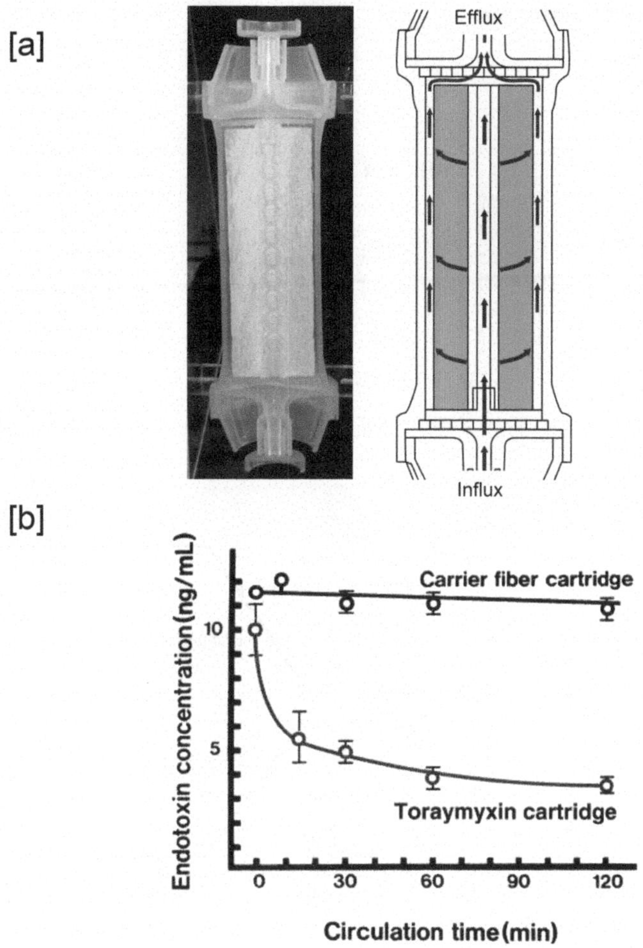

Fig. 19.9 The inside structure and endotoxin adsorption capacity of the Toraymyxin cartridge
(**a**) Structure of a Toraymyxin cartridge and schematic of the blood flow inside the adsorption cartridge. (Provided by Toray Medical Co., Ltd.)
(**b**) Endotoxin adsorption capacity of the Toraymyxin cartridge compared with the carrier fiber cartridge alone. Bovine serum solution containing 10 ng/mL of endotoxin was circulated through the Toraymyxin and carrier fiber cartridges. (Reproduced from Shoji et al. [11])

these regional anticoagulants in the blood mean that they are safe for use during hemoperfusion. Nafamostat mesilate is the most commonly used anticoagulant for this purpose in Japan.

Since 1994, Toraymyxin has been used in more than 100,000 cases in Japan. Despite adverse effects such as thrombocytopenia or hypotension being reported, their incidence is rare, and there have been no reports of any serious adverse effects [11].

19.4 Mechanism of Action of PMX Therapy

19.4.1 Removal of Endotoxin

The Toraymyxin column was originally designed to specifically adsorb endotoxin. Plasma endotoxin levels significantly decrease immediately after PMX treatment compared with pre-treatment levels [10]. By using data from a multicenter study, Tani reported that PMX treatment significantly reduced plasma endotoxin concentrations from 83.7 ± 26.7 pg/mL to 56.4 ± 27.9 pg/mL after 2 h of direct hemoperfusion with a Toraymyxin column [29]. Although almost all of the studies conducted so far have demonstrated a significant decrease in plasma endotoxin levels following PMX treatment, a recent RCT failed to show such an effect [12]. However, this could be due to the use of the LAL assay, which is prone to contamination and lacks precision and accuracy resulting in both false positive and false negative results [30]. Indeed, our previous study showed that sensitivity for the detection of endotoxin in the turbidimetric LAL assay was very low (26.9%; 14 of 52 patients) in patients with severe sepsis and sep-

[a]

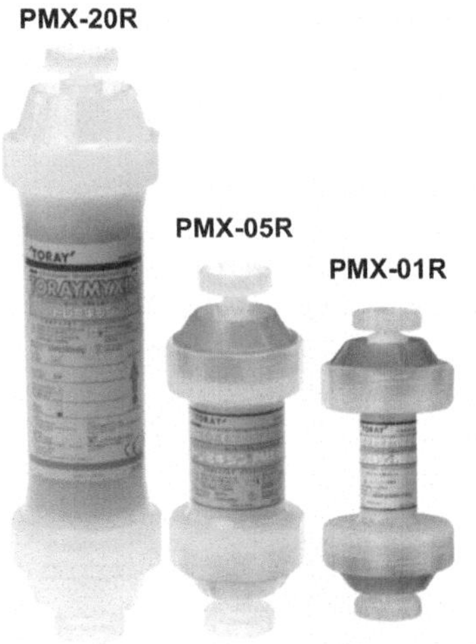

	PMX-20R	PMX-05R	PMX-01R
length (mm)	224	133	133
Max. diameter (mm)	63	55	55
Shell diameter(mm	49	40	25
Fiber weight (g)	56 ± 3	15 ± 2	4 ± 1
Blood volume (mL)	135 ± 3	40 ± 3	8 ± 2.5
rinsing volume (L)	4.0	2.0	0.5
Blood flow (mL/min)	80-120	20-40	8-12
Operation time (hr)	2.0	2.0	2.0

[b]

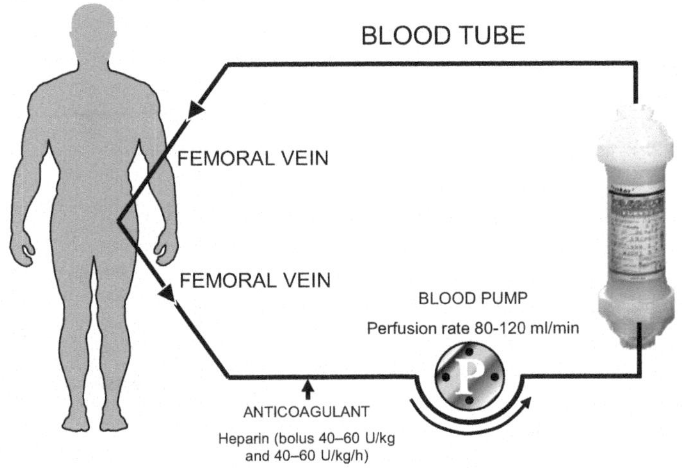

Fig. 19.10 The specifications of the Toraymyxin cartridge
(**a**) Overview of the three Toraymyxin cartridges currently available for clinical use. (Provided by Toray Medical Co., Ltd.)
(**b**) Schematic of direct hemoperfusion with a Toraymyxin cartridge. (Provided by Toray Medical Co., Ltd.)

tic shock who required PMX treatment [31]. A new method for the detection and quantitation of endotoxin is therefore immediately needed.

Recently, Romaschin developed a rapid assay, called the endotoxin activity assay (EAA), that detects endotoxin in whole blood by using autologous neutrophil-dependent chemiluminescence [32]. The EEA has been approved by the United States Food and Drug Administration as a clinical method for the diagnosis of endotoxemia, and was used in the EUPHRATES trial (see Sect. 19.5.1) [33].

A rapid LAL assay for detecting endotoxin by using a laser light-scattering particle-counting method called endotoxin scattering photometry (ESP) has also been developed. Endotoxin binds to and activates factor C, a clotting enzyme, early in the LAL cascade, which ultimately results in the production of the insoluble protein coagulin. Coagulin spontaneously forms a polymer, and it is this polymer formation that is detected in the turbidimetric LAL assay. Obata et al. reported that by using ESP they were able to detect coagulin particles formed under agitation of the LAL cascade, and that the transferring period of the

LAL cascade after endotoxin stimulation depends on the concentration of endotoxin before gelation [34]. The ESP method allows for more rapid and sensitive detection of endotoxin compared with the turbidimetric LAL method.

Furthermore, ESP is able to discriminate between patients with sepsis and those with septic shock undergoing gastrointestinal emergency surgery compared with the standard turbidimetric LAL assay [35]. After undergoing PMX therapy for a longer-duration (> 2 h), the hemodynamic condition of patients with septic shock due to intra-abdominal infection improved, and the reduction in plasma endotoxin concentration could be detected via ESP but not through the standard turbidimetric method. We have demonstrated the ability of endotoxin adsorption in longer-duration PMX treatment [36]. Using the ESP method, we observed a reduction in endotoxin after passing through the Toraymyxin column even when PMX duration was greater than 2 h. Therefore, ESP appears to be sensitive enough to detect circulating endotoxin in the plasma of patients with septic shock who require PMX therapy.

19.4.2 Other Potential Mechanism of Action of PMX Therapy

Although Toraymyxin was originally developed for the adsorption of circulating endotoxin from the blood, recent studies have shown other potential applications. Jaber reported that it may be possible to use Toraymyxin to remove lipoteichoic acid (LTA) from peripheral blood [47]. In an *in vitro* study, polymyxin B significantly suppressed LTA-induced tumor necrosis factor (TNF)-α production by peripheral blood mononuclear cells, suggesting that LTA may also be removed during direct hemoperfusion with PMX and that this removal may be due to binding of LTA to polymyxin B.

Other recent studies have reported a possible relationship between immune cells and the reduction in inflammatory mediators. Ono demonstrated that the expression levels of HLA-DR, a major histocompatibility complex class II cell surface receptor on monocytes, and CD16 on granulocytes, markedly decrease in patients with septic shock. However, PMX treatment had beneficial effects by increasing leukocyte expression of these surface antigens [37]. The number of $CD16^{+}CD14^{+}$ monocytes and the expression level of monocytic toll-like receptor 4 dramatically increase in patients with severe infection, and recent studies have shown that PMX treatment is effective in reducing both of these in patients with septic shock [38]. Moreover, Ono demonstrated that the removal of surplus circulating $CD4^{+}CD25^{+}Foxp3^{+}$ regulatory T cells in patients with septic shock via hemoperfusion with Toraymyxin might represent a novel strategy for inducing recovery from sepsis-associated immunosuppression [39]. Tsuzuki have demonstrated immediate inhibition of NF-κB binding activity and suppression of TNF-α secretion after endotoxin neutralization with polymyxin B regardless of whether the peripheral blood mononuclear cells were already producing TNF-α or not [40]. Reducing the number of circulating monocytes or modulating their function may contribute to improving pro-inflammatory responses following PMX treatment.

Anandamide and 2-arachidonyl glyceride are endogenous cannabinoids that are released by activated macrophages and platelets during endotoxic shock, and play a crucial role in the induction of shock-related hypotension [41]. Wang showed that anandamide was efficiently adsorbed by a polymyxin B-immobilized bead column *in vitro* [42]. Therefore, the adsorption of cannabinoids by polymyxin B may be an important mechanism for improving hemodynamic dysfunction following PMX treatment.

Macrophage migration inhibitory factor (MIF) is constitutively expressed by monocytes, macrophages, T cells, B cells, endocrine cells, and epithelial cells. Microbial toxins and cytokines are powerful inducers of MIF release by immune cells, and up-regulation of MIF expression during the course of inflammatory and infectious diseases plays an important role in the pathogenesis of sepsis and septic shock [43]. In a recent study, we demonstrated that Toraymyxin directly adsorbed MIF *in vitro* (unpublished data).

High-mobility group box 1 protein (HMGB1), which is a protein previously known only as a nuclear transcription factor, is now implicated as a mediator of delayed endotoxin lethality and systemic inflammation [44]. An experimental study demonstrated that serum levels of HMGB1 were lower in PMX-treated patients than in controls [45]. The reduction in circulating HMGB1 level may also contribute to the beneficial effects of PMX treatment in patients with sepsis [46–48].

Further studies are required to investigate the clinical efficacy of PMX treatment and elucidate the precise mechanism. Nevertheless, in Japan, PMX treatment is a widely accepted method to improve the condition of patients with septic shock, and this useful clinical device is gaining acceptance as its use becomes more widespread.

19.5 Clinical Use of PMX Therapy

19.5.1 PMX Therapy for the Treatment of Sepsis

The first clinical report from a phase I study of PMX therapy for the treatment of sepsis was published in 1994 [10]. Sixteen patients with septic multiple organ failure were treated with direct hemoperfusion with a Toraymyxin column over 2 h. PMX treatment significantly decreased the endotoxin level from 76 to 21 pg/mL (Fig. 19.11a). Furthermore, the patients' hyperdynamic state (in terms of cardiac index), which is characteristic of endotoxic shock, returned to normal after treatment (Fig. 19.11b) [10].

The results of the first prospective, multicenter, observational clinical trial of Toraymyxin for the treatment of patients with sepsis in Japan were published in 1998 [29]. The survival rate was significantly higher in the treatment (PMX) group (54.0%) compared with the control group who received standard therapies (36.4%). In the treatment group, the mean plasma endotoxin concentration was significantly lowered from 83.7 pg/mL before treatment to 56.4 pg/mL immediately after treatment. The day after treatment, the mean plasma endotoxin concentration was found to be 28.5 pg/mL. Post treatment, the mean plasma endotoxin concentration was lower in those who survived (18.8 pg/mL) compared with those who died (88 pg/mL). Other cardiac function parameters also improved after treatment.

Nemoto demonstrated that the rate of overall survival in patients with sepsis significantly improved after PMX treatment compared with standard therapies without PMX (41% vs. 11%) in a RCT of 98 patients with sepsis in Japan [31]. PMX improved the rate of survival in patients with an APACHE II (Acute Physiology and Chronic Health Evaluation II) score <30, but not in those with scores >30, demonstrating that Toraymyxin treatment is most effective at improving patient outcome when applied during the early stages of sepsis [49].

In 1998, Toraymyxin received CE mark approval in Europe. In 2005, Vincent et al. published the results of a multicenter, open-label, pilot, randomized controlled study conducted in the intensive care units of six academic medical centers in Europe, which was the first RCT of Toraymyxin conducted outside of Japan [11]. Thirty-six postoperative patients with severe sepsis or septic shock secondary to intra-abdominal infection were randomized to either PMX treatment for 2 h ($n = 17$) or standard therapy ($n = 19$). PMX treatment was well tolerated and no significant adverse effects were observed. Patients treated with PMX showed significant increases in cardiac index ($P = 0.012$ and 0.032 at days 1 and 2, respectively), left ventricular stroke work index ($P = 0.015$ at day 2), and oxygen delivery index ($P = 0.007$ at day 2) compared with controls. Furthermore, the need for continuous renal replacement therapy after study entry was significantly reduced in the PMX group ($P = 0.043$). There were no statistically significant differences in endotoxin and interleukin six levels, and organ dysfunction as assessed by the Sequential Organ Failure Assessment (SOFA) score, or 28-day mortality. Note that sepsis in the patients enrolled in this study was likely too severe to find statistical significance in the mortality outcome. Together, these results show that PMX treatment

[a]

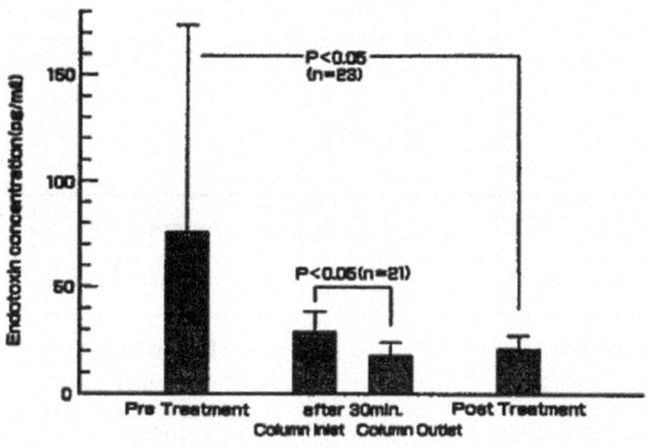

[b]

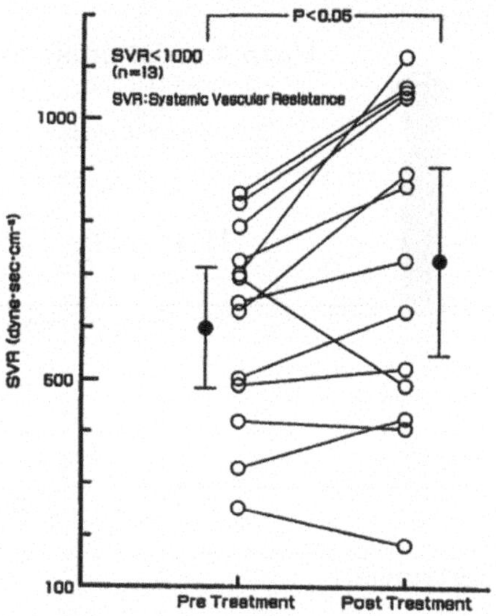

Fig. 19.11 Results from the first clinical trial of Toraymyxin in Japan
(**a**) Endotoxin concentrations before and after Toraymyxin treatment
(**b**) Systemic vascular resistance (SVR) before and after Toraymyxin treatment. Toraymyxin treatment decreased the concentration of endotoxin in the blood and improved hemodynamic status in patients with severe sepsis and septic shock. (Reproduced from Aoki et al. [10])

is safe and that it improves cardiac and renal dysfunction due to sepsis or septic shock.

In 2007, Cruz et al. published a systematic review of the effectiveness of Toraymyxin for the treatment of sepsis [12]. They included 28 publications that reported at least one of the specified outcome measures for PMX therapy. In this meta-analysis of 1425 patients (PMX therapy, 978 patients; conventional therapy, 447 patients), PMX therapy was associated with a significantly lower risk of mortality compared with conventional therapy (PMX, 33.5% vs. conventional treatment, 66.5%; risk ratio, 0.53; 95% CI, 0.43–0.65). A 33–80% reduction in plasma endotoxin levels compared to pre-treatment levels was also observed. The large decrease in mortality was associated with an improvement in hemodynamic condition: after PMX therapy, mean arterial pressure significantly increased by 19 mmHg (mean increase, 26%; range, 14–42%), and dopamine/dobutamine dose was decreased by 1.8 μg/kg/min. Improvement in pulmonary function was also demonstrated: the mean ratio of partial pressure arterial oxygen to fraction of inspired oxygen (PaO_2/FiO_2) increased by 32 units (95% CI, 23–41 units; $P < 0.001$) (Fig. 19.12).

Cruz et al. have also published the results of the EUPHAS (Early Use of Polymyxin B Hemoperfusion in Abdominal Sepsis) trial, which was a prospective, multicenter, RCT conducted at the intensive care units of ten Italian tertiary care hospitals in 2009 [14]. Sixty-four patients with severe sepsis or septic shock who had undergone emergency surgery for intra-abdominal infection were enrolled in this study. Patients were randomized within 6 h after open abdominal surgery to either conventional therapy ($n = 30$) or conventional therapy plus two sessions of 2 h PMX with an interval of 24 h between sessions ($n = 34$). Twenty-eight-day mortality was significantly improved in the PMX group (32%; 11/ 34 patients) compared with that in the conventional therapy group (53%; 16/30 patients). In the PMX group, mean arterial pressure significantly increased from 76 mmHg (before treatment) to 84 mmHg (72 h after treatment; $P = 0.001$) and vasopressor requirement (measured as inotropic score) significantly decreased from 29.9 (before treatment) to 6.8 (72 h after treatment; $P < 0.001$). In contrast, in the conventional therapy group, mean arterial pressure (before, 74 mmHg; 72 h after, 77 mmHg; $P = 0.37$) and inotropic score

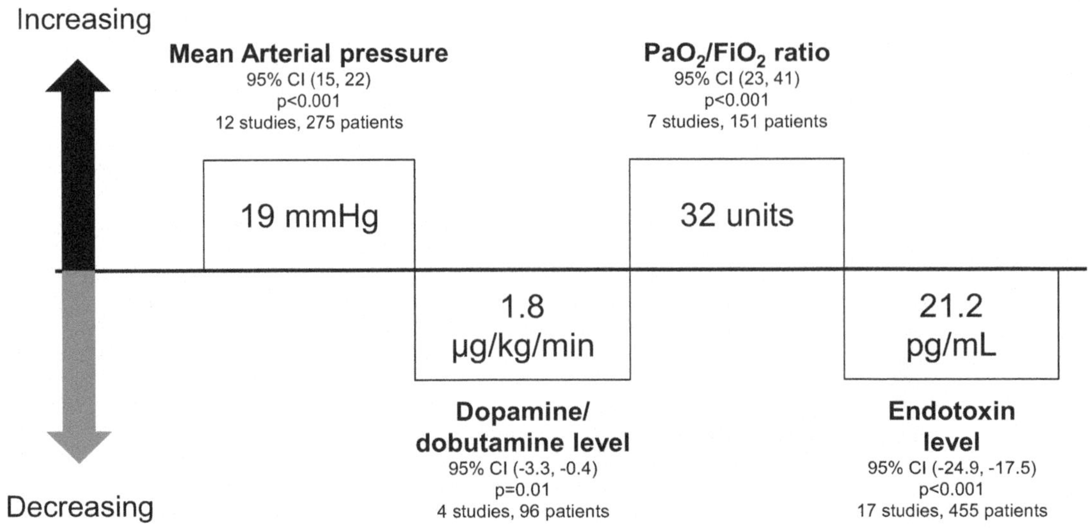

Fig. 19.12 Main results of a meta-analysis of Toraymyxin treatment. (Provided by Toray Medical Co., Ltd.) *MAP* mean arterial pressure, *PaO_2/FIO_2* ratio of partial pressure arterial oxygen and fraction of inspired oxygen, *CI* confidence interval

(before, 28.6; after, 22.4; $P = 0.37$) did not change significantly with treatment. Ratio of partial pressure arterial oxygen and fraction of inspired oxygen (PaO_2/FIO_2 ratio) significantly increased (before, 235; after, 264; $P = 0.049$) in the PMX group but not in the conventional therapy group (before, 217; after, 228; $P = 0.79$). Moreover, SOFA score, which is an indicator of the severity of organ dysfunction, improved in the PMX group compared with the conventional therapy group (change in SOFA score, −3.4 vs. −0.1; $P < 0.001$).

Although the goal was planned to enroll 120 patients in this study, the interim analysis revealed that the risk of mortality in the conventional treatment group was significantly higher than that in the PMX group; thus, the study was terminated midway. Despite the high mortality rate in the conventional treatment group and the lack of evaluation of circulating endotoxin levels, this was the first report showing an improved rate of survival in a RCT conducted outside of Japan.

A multicenter randomized controlled trial, The ABDOMIX trial (Effects of Hemoperfusion with a Polymyxin B Membrane in Peritonitis with Septic Shock), was conducted in France [15]. A total of 243 patients with peritonitis-induced septic shock from abdominal infections were enrolled and the primary end point of the study was 28-day mortality. This multicenter randomized controlled study demonstrated a non-significant increase in mortality and no improvement in organ failure with PMX treatment compared to conventional treatment of peritonitis-induced septic shock. However, Antonelli et al. pointed out that major differences in mortality and completion rates of two scheduled sessions of PMX compared to the EUPHAS study may jeopardize their comparability, suggesting that any definitive conclusion be put on hold [50]. The 28-day mortality rate recorded in both groups was significantly lower than that reported in larger studies (between 32.7% and 53% for similar patient cohorts) [51–53]. Furthermore, only 81 of the 119 treated patients completed the two scheduled sessions of PMX. In the previously published EUPHAS study, all patients enrolled had completed the two planned sessions of PMX, and a significantly higher mortality rate was recorded in the control group [50].

The EUPHRATES trial (Evaluating the Use of Polymyxin B Hemoperfusion in a Randomized Controlled Trial of Adults Treated for Endotoxemia and Septic Shock), which is ongoing in the USA and Canada [33], is a trial designed to address the criticisms of previous studies. Circulating endotoxin levels in patients with septic shock were evaluated by using a novel endotoxin detection method called the endotoxin activity assay (EAA). A total of 432 patients with septic shock and high EAA activity (> 0.6) were enrolled and randomized to either PMX or conventional treatment. Dellinger et al. reported that the primary endpoint for mortality rate (44.3% in placebo group and 43.75% in PMX group) was not met in the full intention-to-treat population. Their interim results showed that the mortality rate in the per protocol population (36.9% in placebo group and 31.9% in PMX group) was 5% in favor of PMX treatment ($P = 0.407$). Interestingly, the final results of the EUPHRATES trial did not show reduced mortality at 28 days in patients with PMX treatment (ClinicalTrials.gov Identifier: NCT01046669).

Direct hemoperfusion sessions with Toraymyxin column usually last 2 h; however, the obvious clinical efficacy of longer-duration PMX (> 6 h) in such cases has been reported [36, 54–56]. No adverse effects, such as thrombocytopenia, have been reported with longer duration Toraymyxin treatment [36, 54]. Further studies to clarify the suitability of longer-duration Toraymyxin treatment are expected.

The combination of PMX and continuous hemodiafiltration (CHDF) has been reported to be more beneficial for patients with septic renal dysfunction than CHDF alone. Combination treatment significantly decreases the concentration of circulating interleukin-6 and improves patient survival (Fig. 19.13a) [57]. Moreover, PMX followed by CHDF with a polymethyl methacrylate (PMMA) membrane significantly decreased the concentrations of plasminogen activator inhibitor-1, protein C, interleukin-6, and endogenous anandamide compared with CHDF with a polyacrylonitrile membrane in patients with septic shock [58]. Therefore, the combination of PMX with PMMA–CHDF is beneficial for patients with septic shock and septic renal dysfunction. In combination therapy, the Toraymyxin column is generally placed before or after CHDF on a single circuit. However, when CHDF cannot be interrupted, for example in patients with renal failure, the Toraymyxin circuit can be connected in parallel or in series with the CHDF circuit (Fig. 19.13b) [59].

19.5.2 PMX Therapy for the Treatment of Acute Exacerbation of Idiopathic Pulmonary Fibrosis and Acute Respiratory Distress Syndrome

Idiopathic pulmonary fibrosis (IPF) is one of a heterogeneous group of diffuse parenchymal lung disorders of unknown etiology. Acute exacerbation of IPF is characterized by severe worsening dyspnea and high mortality. IPF is one of the most common presentations of idiopathic interstitial pneumonia. Previous reports have suggested that PMX treatment improves oxygenation in patients with acute lung injury or acute respiratory distress syndrome (ARDS) due to severe sepsis [60, 61]. The potential benefit of PMX therapy for the treatment of acute exacerbation of IPF has also been reported [62]. A multicenter retrospective analysis of 160 patients showed that PMX therapy improves oxygenation and survival in patients experiencing acute exacerbation of IPF [63]. A large-scale, prospective, RCT in patients experiencing acute exacerbation of IPF will be conducted in the near future.

The successful treatment of severe ARDS due to influenza virus infection with PMX has been reported recently. Yokoyama et al. reported a case of severe ARDS caused by novel swine-origin influenza virus (A/H1N1pdm) [64]. The patient underwent PMX, after which her hypoxemia improved and she survived. This is the first report of severe, life-threatening ARDS due to a novel influenza virus in which PMX was beneficial. Yatera et al. reported a case of ARDS due to influenza A infection that was successfully treated with PMX [65]. Kudo et al. has reported cases of severe pneumonia due to highly pathogenic avian influenza A (H5N1) in Vietnam successfully treated with CHDF coupled with Toraymyxin hemoperfusion, suggesting that it is

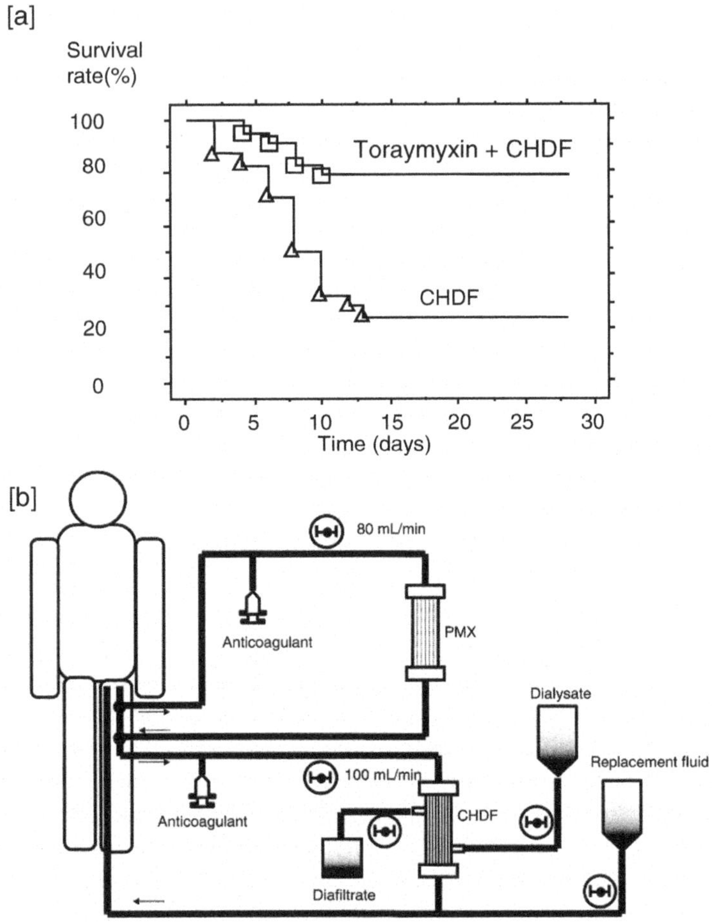

Fig. 19.13 Use of Toraymyxin in combination with continuous hemodiafiltration (**a**) Survival of patients receiving Toraymyxin therapy in combination with continuous hemodiafiltration. (Reproduced from Suzuki et al. [57]) The use of this combination significantly improved survival rate in patients with sepsis and acute renal failure (**b**) Schematic of the combination of direct hemoperfusion with Toraymyxin and continuous hemodiafiltration in a series-parallel circuit

an effective candidate treatment for ARDS due to the H5N1 virus if applied early in the disease [66].

These data show that new indications for PMX treatment in the field of pulmonary medicine will likely be available in the near future.

References

1. Hurley JC, Guidet B, Offenstadt G, Maury E (2012) Endotoxemia and mortality prediction in ICU and other settings: underlying risk and co-detection of Gram-negative bacteremia are confounders. J Crit Care 16(4):R148. https://doi.org/10.1186/cc11462
2. Anspach FB (2001) Endotoxin removal by affinity sorbents. J Biochem Biophys Methods 49(1–3):665–681
3. Nolan JP, Ali MV (1972) Effect of cholestyramine on endotoxin toxicity and absorption. Am J Dig Dis 17(2):161–166
4. Cohen J, Aslam M, Pusey CD, Ryan CJ (1987) Protection from endotoxemia: a rat model of plasmapheresis and specific adsorption with polymyxin B. J Infect Dis 155(4):690–695
5. Umeda M, Niwa M, Yamagami S, Kishimot T, Maekawa M, Sawada Y (1990) Novel endotoxin adsorbing materials, polymyxin-sepharose and polyporous polyethylene membrane for removal of endotoxin from dialysis systems. Biomater Artif Cells Artif Organs 18(4):491–497
6. Mitzner S, Schneidewind J, Falkenhagen D, Loth F, Klinkmann H (1993) Extracorporeal endotoxin removal by immobilized polyethylenimine. Artif Organs 17(9):775–781
7. Zimmermann M, Busch K, Kuhn S, Zeppezauer M (1999) Endotoxin adsorbent based on immobilized human serum albumin. Clin Chem Lab Med 37(3):373–379. https://doi.org/10.1515/CCLM.1999.062
8. Hanasawa K, Tani T, Oka T, Yoshioka T, Aoki H, Endo Y, Kodama M (1988) Selective removal of endotoxin from the blood by extracorporeal hemoperfusion with

polymyxin B-immobilized fiber. Prog Clin Biol Res 264:337–341
9. Hanasawa K, Tani T, Kodama M (1989) New approach to endotoxic and septic shock by means of polymyxin B-immobilized fiber. Surg Gynecol Obstet 168(4):323–331
10. Aoki H, Kodama M, Tani T, Hanasawa K (1994) Treatment of sepsis by extracorporeal elimination of endotoxin using polymyxin B-immobilized fiber. Am J Surg 167(4):412–417
11. Shoji H (2003) Extracorporeal endotoxin removal for the treatment of sepsis: endotoxin adsorption cartridge (Toraymyxin). Ther Apher Dial 7(1):108–114
12. Vincent JL, Laterre PF, Cohen J, Burchardi H, Bruining H, Lerma FA, Wittebole X, De Backer D, Brett S, Marzo D, Nakamura H, John S (2005) A pilot-controlled study of a polymyxin B-immobilized hemoperfusion cartridge in patients with severe sepsis secondary to intra-abdominal infection. Shock 23(5):400–405
13. Cruz DN, Perazella MA, Bellomo R, de Cal M, Polanco N, Corradi V, Lentini P, Nalesso F, Ueno T, Ranieri VM, Ronco C (2007) Effectiveness of polymyxin B-immobilized fiber column in sepsis: a systematic review. Crit Care 11(2):R47. https://doi.org/10.1186/cc5780
14. Cruz DN, Antonelli M, Fumagalli R, Foltran F, Brienza N, Donati A, Malcangi V, Petrini F, Volta G, Bobbio Pallavicini FM, Rottoli F, Giunta F, Ronco C (2009) Early use of polymyxin B hemoperfusion in abdominal septic shock: the EUPHAS randomized controlled trial. JAMA 301(23):2445–2452. https://doi.org/10.1001/jama.2009.856
15. Payen DM, Guilhot J, Launey Y, Lukaszewicz AC, Kaaki M, Veber B, Pottecher J, Joannes-Boyau O, Martin-Lefevre L, Jabaudon M, Mimoz O, Coudroy R, Ferrandiere M, Kipnis E, Vela C, Chevallier S, Mallat J, Robert R, ABDOMIX Group (2015) Early use of polymyxin B hemoperfusion in patients with septic shock due to peritonitis: a multicenter randomized control trial. Intensive Care Med 41(6):975–984. https://doi.org/10.1007/s00134-015-3751-z
16. Morrison DC, Jacobs DM (1976) Binding of polymyxin B to the lipid A portion of bacterial lipopolysaccharides. Immunochemistry 13(10):813–818
17. Corrigan JJ Jr, Kiernat JF (1979) Effect of polymyxin B sulfate on endotoxin activity in a Gram-negative septicemia model. Pediatr Res 13(1):48–51. https://doi.org/10.1203/00006450-197901000-00011
18. Flynn PM, Shenep JL, Stokes DC, Fairclough D, Hildner WK (1987) Polymyxin B moderates acidosis and hypotension in established, experimental Gram-negative septicemia. J Infect Dis 156(5):706–712
19. Ingoldby CJ (1980) The value of polymixin B in endotoxaemia due to experimental obstructive jaundice and mesenteric ischaemia. Br J Surg 67(8):565–567
20. Lopes J, Inniss WE (1969) Electron microscopy of effect of polymyxin on *Escherichia coli* lipopolysaccharide. J Bacteriol 100(2):1128–1129
21. Ribi E, Anacker RL, Brown R, Haskins WT, Malmgren B, Milner KC, Rudbach JA (1966) Reaction of endotoxin and surfactants. I. Physical and biological properties of endotoxin treated with sodium deoxycholate. J Bacteriol 92(5):1493–1509
22. Bader J, Teuber M (1973) Action of polymyxin B on bacterial membranes. 1. Binding to the O-antigenic lipopolysaccharide of *Salmonella typhimurium*. Z Naturforsch 28(7):422–430
23. Vesentini S, Soncini M, Fiore GB, Redaelli A (2010) Mechanisms of polymyxin B endotoxin removal from extracorporeal blood flow: molecular interactions. Contrib Nephrol 167:45–54. https://doi.org/10.1159/000315918
24. Shoji H, Tani T, Hanasawa K, Kodama M (1998) Extracorporeal endotoxin removal by polymyxin B immobilized fiber cartridge: designing and antiendotoxin efficacy in the clinical application. Ther Apher 2(1):3–12
25. Kodama M, Hanasawa K, Tani T (1990) New therapeutic method against septic shock—removal of endotoxin using extracorporeal circulation. Adv Exp Med Biol 256:653–664
26. Nakamura T, Kawagoe Y, Suzuki T, Shoji H, Ueda Y, Koide H (2007) Polymyxin B-immobilized fiber hemoperfusion with the PMX-05R column in elderly patients suffering from septic shock. Am J Med Sci 334(4):244–247. https://doi.org/10.1097/MAJ.0b013e3180a5e8d8
27. Nishizaki N, Nakagawa M, Hara S, Oda H, Kantake M, Obinata K, Uehara Y, Hiramatsu K, Shimizu T (2016) Effect of PMX-DHP for sepsis due to ESBL-producing *E. coli* in an extremely low-birthweight infant. Pediatr Int 58(5):411–414. https://doi.org/10.1111/ped.12825
28. Tokumasu H, Watabe S, Tokumasu S (2016) Effect of hemodiafiltration therapy in a low-birthweight infant with congenital sepsis. Pediatr Int 58(3):237–240. https://doi.org/10.1111/ped.12776
29. Tani T, Hanasawa K, Endo Y, Yoshioka T, Kodama M, Kaneko M, Uchiyama Y, Akizawa T, Takahasi K, Sugai K (1998) Therapeutic apheresis for septic patients with organ dysfunction: hemoperfusion using a polymyxin B-immobilized column. Artif Organs 22(12):1038–1044
30. Cohen J (2000) The detection and interpretation of endotoxaemia. Intensive Care Med 26(Suppl 1):S51–S56
31. Shimizu T, Hanasawa K, Sato K, Umeki M, Koga N, Naganuma T, Sato S, Shimonishi T, Ikeda T, Matsuno N, Ono S, Saitoh H, Satoh K, Otani Y, Endo Y, Eguchi Y, Tani T, Group PMXTS (2009) Direct hemoperfusion with polymyxin B-immobilized fiber columns improves septic hypotension and reduces inflammatory mediators in septic patients with colorectal perforation. Langenbeck's Arch Surg 394(2):303–311. https://doi.org/10.1007/s00423-008-0395-2
32. Romaschin AD, Harris DM, Ribeiro MB, Paice J, Foster DM, Walker PM, Marshall JC (1998) A rapid

assay of endotoxin in whole blood using autologous neutrophil dependent chemiluminescence. J Immunol Methods 212(2):169–185
33. Foster D, Klein DJ, Guadagni G, Walker PM (2014) Endotoxin removal: bringing the mission to North America. Blood Purif 37(Suppl 1):14–17. https://doi.org/10.1159/000356832
34. Obata T, Nomura M, Kase Y, Sasaki H, Shirasawa Y (2008) Early detection of the Limulus amebocyte lysate reaction evoked by endotoxins. Anal Biochem 373(2):281–286. https://doi.org/10.1016/j.ab.2007.09.018
35. Shimizu T, Obata T, Sonoda H, Akabori H, Miyake T, Yamamoto H, Tabata T, Eguchi Y, Tani T (2013) Diagnostic potential of endotoxin scattering photometry for sepsis and septic shock. Shock 40(6):504–511. https://doi.org/10.1097/SHK.0000000000000056
36. Shimizu T, Obata T, Sonoda H, Akabori H, Tabata T, Eguchi Y, Endo Y, Tani T (2013) The ability of endotoxin adsorption during a longer duration of direct hemoperfusion with a polymyxin B-immobilized fiber column in patients with septic shock. Transfus Apher Sci 49(3):499–503. https://doi.org/10.1016/j.transci.2013.04.042
37. Ono S, Tsujimoto H, Matsumoto A, Ikuta S, Kinoshita M, Mochizuki H (2004) Modulation of human leukocyte antigen-DR on monocytes and CD16 on granulocytes in patients with septic shock using hemoperfusion with polymyxin B-immobilized fiber. Am J Surg 188(2):150–156. https://doi.org/10.1016/j.amjsurg.2003.12.067
38. Tsujimoto H, Ono S, Hiraki S, Majima T, Kawarabayashi N, Sugasawa H, Kinoshita M, Hiraide H, Mochizuki H (2004) Hemoperfusion with polymyxin B-immobilized fibers reduced the number of CD16+ CD14+ monocytes in patients with septic shock. J Endotoxin Res 10(4):229–237. https://doi.org/10.1179/096805104225005814
39. Ono S, Kimura A, Hiraki S, Takahata R, Tsujimoto H, Kinoshita M, Miyazaki H, Yamamoto J, Hase K, Saitoh D (2013) Removal of increased circulating CD4+CD25+Foxp3+ regulatory T cells in patients with septic shock using hemoperfusion with polymyxin B-immobilized fibers. Surgery 153(2):262–271. https://doi.org/10.1016/j.surg.2012.06.023
40. Tsuzuki H, Tani T, Ueyama H, Kodama M (2001) Lipopolysaccharide: neutralization by polymyxin B shuts down the signaling pathway of nuclear factor kappa B in peripheral blood mononuclear cells, even during activation. J Surg Res 100(1):127–134. https://doi.org/10.1006/jsre.2001.6227
41. Varga K, Wagner JA, Bridgen DT, Kunos G (1998) Platelet- and macrophage-derived endogenous cannabinoids are involved in endotoxin-induced hypotension. FASEB J 12(11):1035–1044
42. Wang Y, Liu Y, Sarker KP, Nakashima M, Serizawa T, Kishida A, Akashi M, Nakata M, Kitajima I, Maruyama I (2000) Polymyxin B binds to anandamide and inhibits its cytotoxic effect. FEBS Lett 470(2):151–155
43. Froidevaux C, Roger T, Martin C, Glauser MP, Calandra T (2001) Macrophage migration inhibitory factor and innate immune responses to bacterial infections. Crit Care Med 29(7 Suppl):S13–S15
44. Wang H, Yang H, Czura CJ, Sama AE, Tracey KJ (2001) HMGB1 as a late mediator of lethal systemic inflammation. Am J Respir Crit Care Med 164(10 Pt 1):1768–1773. https://doi.org/10.1164/ajrccm.164.10.2106117
45. Hussein MH, Kato T, Sugiura T, Daoud GA, Suzuki S, Fukuda S, Sobajima H, Kato I, Togari H (2005) Effect of hemoperfusion using polymyxin B-immobilized fiber on IL-6, HMGB-1, and IFN gamma in a neonatal sepsis model. Pediatr Res 58(2):309–314. https://doi.org/10.1203/01.PDR.0000169995.25333.F7
46. Sakamoto Y, Mashiko K, Matsumoto H, Hara Y, Kutsukata N, Yamamoto Y (2007) Relationship between effect of polymyxin B-immobilized fiber and high-mobility group box-1 protein in septic shock patients. ASAIO J 53(3):324–328. https://doi.org/10.1097/MAT.0b013e3180340301
47. Ueno T, Ikeda T, Ikeda K, Taniuchi H, Suda S, Yeung MY, Matsuno N (2011) HMGB-1 as a useful prognostic biomarker in sepsis-induced organ failure in patients undergoing PMX-DHP. J Surg Res 171(1):183–190. https://doi.org/10.1016/j.jss.2009.11.708
48. Abe S, Hayashi H, Seo Y, Matsuda K, Kamio K, Saito Y, Usuki J, Azuma A, Kudo S, Gemma A (2011) Reduction in serum high mobility group box-1 level by polymyxin B-immobilized fiber column in patients with idiopathic pulmonary fibrosis with acute exacerbation. Blood Purif 32(4):310–316. https://doi.org/10.1159/000330325
49. Nemoto H, Nakamoto H, Okada H, Sugahara S, Moriwaki K, Arai M, Kanno Y, Suzuki H (2001) Newly developed immobilized polymyxin B fibers improve the survival of patients with sepsis. Blood Purif 19(4):361–368. discussion 368–369. https://doi.org/10.1159/000046966
50. Antonelli M, Ronco C (2015) Polymyxin B hemoperfusion in sepsis: growing body of evidence and occasional conflicting results. Blood Purif 39(1–3):I–II. https://doi.org/10.1159/000431018
51. Antonelli M, Fumagalli R, Cruz DN, Brienza N, Giunta F, EUPHAS Study Group (2010) PMX endotoxin removal in the clinical practice: results from the EUPHAS trial. Contrib Nephrol 167:83–90. https://doi.org/10.1159/000315922
52. Cruz DN, de Cal M, Piccinni P, Ronco C (2010) Polymyxin-B hemoperfusion and endotoxin removal: lessons from a review of the literature. Contrib Nephrol 167:77–82. https://doi.org/10.1159/000315921
53. Early Use of Polymyxin BHitASCG (2014) Polymyxin B hemoperfusion in clinical practice: the picture from an unbound collaborative registry. Blood Purif 37(Suppl 1):22–25. https://doi.org/10.1159/000356835
54. Mitaka C, Tsuchida N, Kawada K, Nakajima Y, Imai T, Sasaki S (2009) A longer duration of polymyxin

B-immobilized fiber column hemoperfusion improves pulmonary oxygenation in patients with septic shock. Shock 32(5):478–483. https://doi.org/10.1097/SHK.0b013e3181a2a978

55. Ueno T, Sugino M, Nemoto H, Shoji H, Kakita A, Watanabe M (2005) Effect over time of endotoxin adsorption therapy in sepsis. Ther Apher Dial 9(2):128–136. https://doi.org/10.1111/j.1774-9987.2005.00230.x
56. Yamashita C, Hara Y, Kuriyama N, Nakamura T, Nishida O (2015) Clinical effects of a longer duration of polymyxin B-immobilized fiber column direct hemoperfusion therapy for severe sepsis and septic shock. Ther Apher Dial 19(4):316–323. https://doi.org/10.1111/1744-9987.12339
57. Suzuki H, Nemoto H, Nakamoto H, Okada H, Sugahara S, Kanno Y, Moriwaki K (2002) Continuous hemodiafiltration with polymyxin-B immobilized fiber is effective in patients with sepsis syndrome and acute renal failure. Ther Apher 6(3):234–240
58. Sakamoto Y, Mashiko K, Obata T, Matsumoto H, Hara Y, Kutsukata N, Yamamoto Y (2008) Effectiveness of continuous hemodiafiltration using a polymethylmethacrylate membrane hemofilter after polymyxin B-immobilized fiber column therapy of septic shock. ASAIO J 54(1):129–132. https://doi.org/10.1097/MAT.0b013e31815d2f01
59. Naka T, Shinozaki M, Akizawa T, Shima Y, Takaesu H, Nasu H (2006) The effect of continuous veno-venous hemofiltration or direct hemoperfusion with polymyxin B-immobilized fiber on neutrophil respiratory oxidative burst in patients with sepsis and septic shock. Ther Apher Dial 10(1):7–11. https://doi.org/10.1111/j.1744-9987.2006.00339.x
60. Ebihara I, Hirayama K, Nagai M, Kakita T, Sakai K, Tajima R, Sato C, Kurosawa H, Togashi A, Okada A, Usui J, Yamagata K, Kobayashi M (2011) Angiopoietin balance in septic shock patients with acute lung injury: effect of direct hemoperfusion with polymyxin B-immobilized fiber. Ther Apher Dial 15(4):349–354. https://doi.org/10.1111/j.1744-9987.2011.00963.x
61. Tsushima K, Kubo K, Koizumi T, Yamamoto H, Fujimoto K, Hora K, Kan-Nou Y (2002) Direct hemoperfusion using a polymyxin B immobilized column improves acute respiratory distress syndrome. J Clin Apher 17(2):97–102. https://doi.org/10.1002/jca.10019
62. Seo Y, Abe S, Kurahara M, Okada D, Saito Y, Usuki J, Azuma A, Koizumi K, Kudoh S (2006) Beneficial effect of polymyxin B-immobilized fiber column (PMX) hemoperfusion treatment on acute exacerbation of idiopathic pulmonary fibrosis. Intern Med 45(18):1033–1038
63. Abe S, Azuma A, Mukae H, Ogura T, Taniguchi H, Bando M, Sugiyama Y (2012) Polymyxin B-immobilized fiber column (PMX) treatment for idiopathic pulmonary fibrosis with acute exacerbation: a multicenter retrospective analysis. Intern Med 51(12):1487–1491
64. Yokoyama T, Tateishi K, Tsushima K, Agatsuma T, Yamamoto H, Koizumi T, Kubo K (2010) A case of severe ARDS caused by novel swine-origin influenza (A/H1N1pdm) virus: a successful treatment with direct hemoperfusion with polymyxin B-immobilized fiber. J Clin Apher 25(6):350–353. https://doi.org/10.1002/jca.20252
65. Yatera K, Yamasaki K, Kawanami T, Tokuyama S, Ogoshi T, Kouzaki M, Nagata S, Nishida C, Yoshii C, Mukae H (2011) A case of successful treatment with polymyxin B-immobilized fiber column direct hemoperfusion in acute respiratory distress syndrome after influenza A infection. Intern Med 50(6):601–605
66. Kudo K, Binh NG, Manabe T, Co DX, Tuan ND, Izumi S, Takasaki J, Minh DH, Thuy PT, Van VT, Hanh TT, Chau NQ (2012) Clinical preparedness for severe pneumonia with highly pathogenic avian influenza A (H5N1): experiences with cases in Vietnam. Respir Investig 50(4):140–150. https://doi.org/10.1016/j.resinv.2012.08.005

Discovery of Novel Polymyxin-Like Antibiotics

20

Tony Velkov and Kade D. Roberts

Abstract

The antimicrobial lipopeptides polymyxin B and colistin (polymyxin E) are used as a 'last-line' therapy for infections caused by multidrug-resistant (MDR) Gram-negative pathogens. However, their effective use as antibiotic drugs in the clinical setting is still plagued by significant toxicity issues, in particular their potential for nephrotoxicity. Furthermore, resistance to the polymyxins has begun to emerge in the clinic, which implies a total lack of antibiotics for the treatment of life-threatening infections caused by the Gram-negative 'superbugs'. This chapter details our current understanding of polymyxin structure-activity relationships as well as recent pre-clinical and clinical drug development efforts aimed at generating new polymyxin antibiotics with improved safety and efficacy.

T. Velkov (✉)
Department of Pharmacology and Therapeutics, University of Melbourne, Parkville, VIC, Australia
e-mail: tony.velkov@unimelb.edu.au

K. D. Roberts (✉)
Biomedicine Discovery Institute and Department of Microbiology, Monash University, Clayton, VIC, Australia
e-mail: kade.roberts@monash.edu

Keywords

Polymyxin · Llipid A · Structure-activity relationship · Nephrotoxicity · Drug discovery

20.1 The Structure-Activity-Relationships (SAR) Underlying the Antibacterial Activity of the Polymyxins

The polymyxins are a family of structurally related non-ribosomal polybasic cyclic lipopeptides produced by the soil bacterium *Paenibacillus polymyxa*. They were first discovered in the late 1940s, and in the late 1950s the antibiotic drugs polymyxin B and colistin (Fig. 20.1) were introduced into clinical practice for treating infections caused by Gram-negative bacteria [1, 2]. In Chap. 3 we discussed in detail the chemistry of the polymyxins; their nomenclature, chemical structures, unique structural features as well as the chemical compositions of the clinically used drugs polymyxin B and colistin. In this chapter, we focus on our current understanding of the fundamental structure-activity relationships (SAR) of the polymyxins and the use of this information to develop new polymyxin antibiotics with improved safety and efficacy. Although the polymyxin class of lipopeptide antibiotics was discovered over 70 years ago, no new polymyxin drugs have been approved for clinical use since

J. Li et al. (eds.), *Polymyxin Antibiotics: From Laboratory Bench to Bedside*, Advances in Experimental Medicine and Biology 1145, https://doi.org/10.1007/978-3-030-16373-0_20

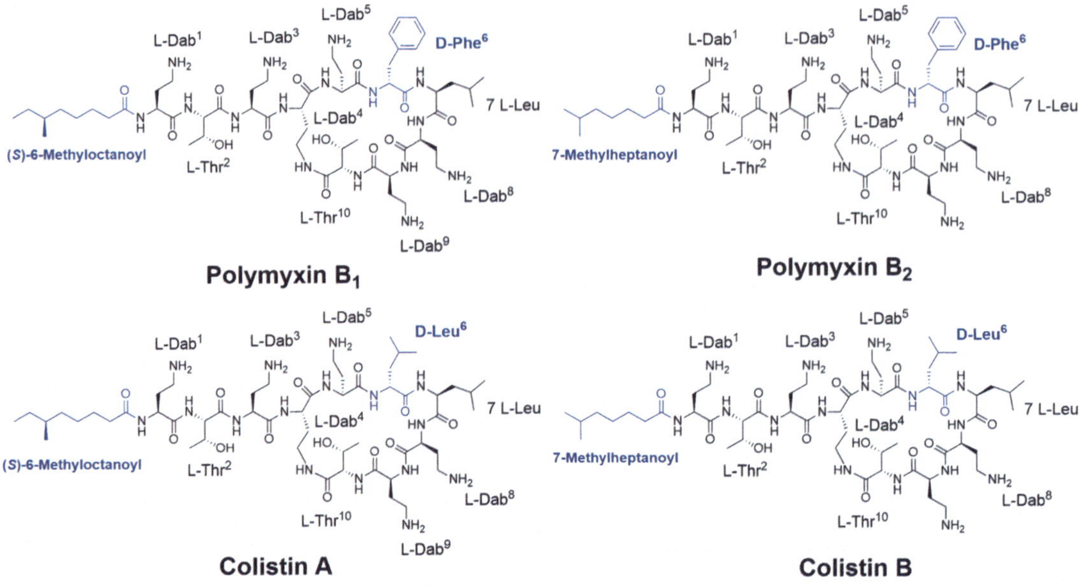

Fig. 20.1 The chemical structures of the major components in the clinically used polymyxin B and colistin. The structural differences are highlighted in blue

the 1960s. Attempts to explore polymyxin SAR and develop new polymyxin-like lipopeptides with improved pharmacological properties had been limited for most of this time up until the late 1990s. This was due in part to limitations in the chemical technology available (e.g. appropriate peptide synthesis and purification-analysis techniques) that allowed for full synthetic preparation of modified forms of these complex lipopeptides. Since then an increasing number of papers have been published exploring polymyxin SAR, which lead us to publish the first comprehensive review of polymyxin SAR studies [2]. Based on our extensive analysis of all reported polymyxin analogues in the literature and pharmacophore development studies, we have proposed that polymyxin SAR data are best interpreted based on a mechanistic model of the interaction of the polymyxin molecule with lipopolysaccharide (LPS), its primary target in the Gram-negative outer membrane (Fig. 20.2) [1, 2]. Modeling of the polymyxin-LPS interaction utilizing NMR data shows that a single polymyxin molecule specifically binds with the lipid A component of LPS and that this binding is stabilized by a combina-

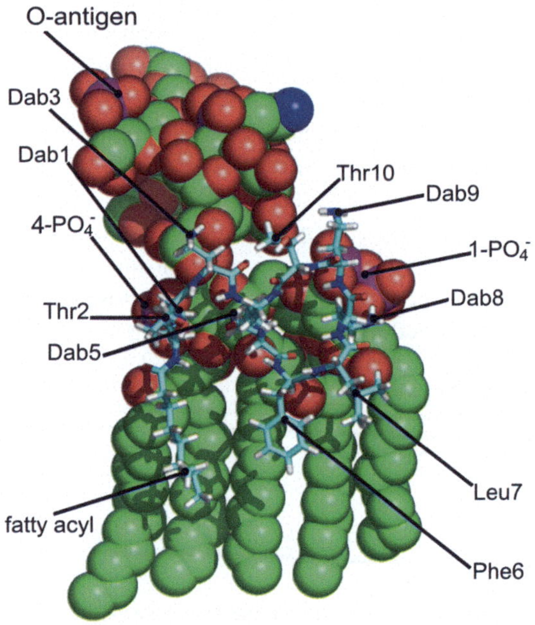

Fig. 20.2 Molecular model of the complex between polymyxin B_1 and lipopolysaccharide (LPS) from *E. coli* showing the binding of the polymyxin molecule with the lipid A component of LPS. The LPS is shown in space filled representation, while Polymyxins B_1 is shown in stick representation

tion of electrostatic and hydrophobic interactions (Fig. 20.2) [2]. Specifically, the positively charged side chains of Dab[1] and Dab[5] interact with the negatively charged 4′-phosphate group of lipid A, while those of Dab[8] and Dab[9] interact with the 1′-phosphate group of lipid A. The hydrophobic *N*-terminal fatty-acyl group and the hydrophobic residues at positions 6/7 (D-Phe[6]-Leu[7] in polymyxin B) form important hydrophobic contacts with the fatty-acyl chains of lipid A. This binding of the polymyxin molecule to LPS ultimately leads to destabilization of the outer membrane of the bacteria [2]. In its LPS-bound state the polymyxin backbone adopts an envelope-like fold separating the polar/charged residues from the hydrophobic residues, such that the polymyxin molecule is divided into a set of polar and hydrophobic domains. The exo-cyclic linear tripeptide sequence and cyclic heptapeptide ring serves to maintain the optimal distance between each domain, giving the structure its amphipathicity, a property that is essential for antimicrobial activity [3, 4].

Understanding of how the polymyxin molecule specifically interacts with LPS along with the findings provided from SAR studies has lead us to identify five key structural features of the polymyxin molecular scaffold that contribute to its antibacterial activity (Fig. 20.3). These five key structural features are: (i) the hydrophobic *N*-terminal fatty-acyl chain; (ii) five L-2,4-diaminobutyric acid (Dab) residues (positively charged at physiological pH); (iii) an exo-cyclic linear tripeptide sequence; (iv) the hydrophobic motif at positions 6 and 7 in the polymyxin scaffold; and (v) the heptapeptide cyclic ring [2]. The specific SAR of each of these key structural features, are summarised in the following paragraphs.

20.1.1 The Hydrophobic *N*-Terminal Fatty-Acyl Chain

The availability of large quantities of polymyxin B and colistin as a cheap source of starting material and the ease of enzymatically removing the *N*-terminal fatty-acyl groups has meant that most studies on the SAR of the polymyxins have focused on generating new *N*-terminal analogues of polymyxin B or colistin [5–10]. A comparison of these *N*-terminal analogues reveals that antimicrobial activity appears to correlate with the hydrophobicity, length and steric bulk of the

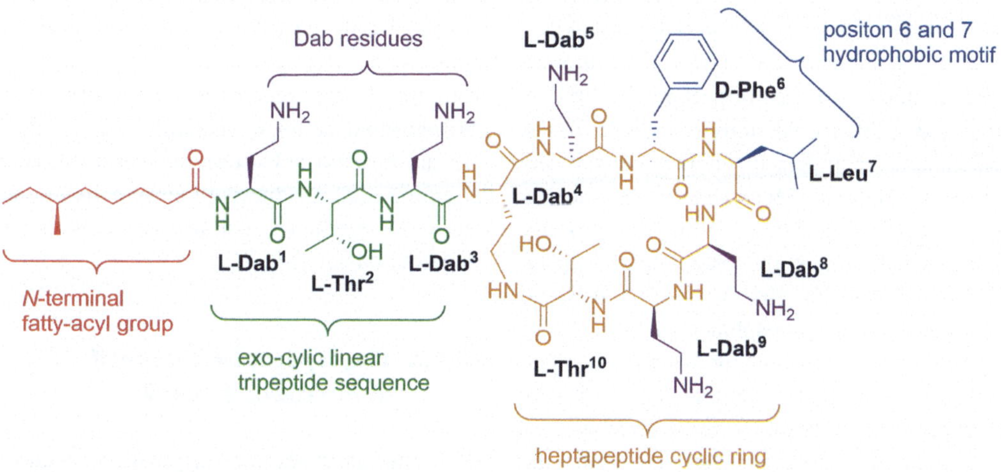

Fig. 20.3 The five key structural features of the polymyxins that contribute to polymyxin SAR as highlighted with the polymyxin B_1 scaffold: (Red) the hydrophobic *N*-terminal fatty acyl chain; (Purple) five Dab residues, positively charged at physiological pH; (Green) exo-cylic linear tripeptide sequence; (Blue) the hydrophobic motif at positions 6 and 7 and (Orange) the heptapeptide cyclic ring

N-terminal substituent [2]. The optimal fatty-acyl chain length for aliphatic groups appears to be C_7 to C_9 (as per the native peptides), as longer or shorter chain *N*-terminal analogs display reduced antimicrobial activity. This is consistent with the observation that LPS binding affinity appears to correlate with the length of the *N*-terminal fatty-acyl chain [7, 11]. While planar aromatic groups such as biphenyl are well tolerated, most sterically bulky or extensively branched *N*-terminal substituents are not as reflected by the poor antimicrobial activity of these compounds [2]. Likewise, *N*-terminal substituents with significant hydrophilic character also lead to decreased activity [2]. Overall, the available SAR data indicate that a hydrophobic substituent at the *N*-terminus of the polymyxin molecule is indispensable for antimicrobial activity. Intriguingly, there has been a recent report describing *des*-fatty-acyl polymyxin analogs, which display selective antimicrobial activity against *P. aeruginosa* [12].

20.1.2 Five Positively Charged Dab Residues

The critical involvement of the positively charged Dab residues (at physiological pH) in conferring the antimicrobial activity of the polymyxins has been well documented [13]. The key features of the Dab residues that are important for lipid A binding and antimicrobial activity include: a) the cationic character of the side chain groups; b) the length of the Dab side chain; and c) the specific order of the Dab residues within the primary sequence which confers the proper spatial distribution of the positive charges for electrostatic interactions with the phosphates of lipid A. To date, attempts to substitute or modify the Dab residues or reduce the number of positively charged positions have met with variable success [14]. In general, apart from Dab^1 and Dab^3, the remaining Dab residues (Dab^5, Dab^8, Dab^9) within the cyclic heptapeptide ring are indispensable for the antimicrobial activity of the polymyxins.

20.1.3 The Exo-Cyclic Linear Tripeptide Sequence

The heptapeptide cyclic ring of the polymyxin molecule is bridged to the fatty-acyl chain by an exo-cyclic linear tripeptide segment (Fig. 20.3). The first two amino acids in this sequence are highly conserved across the naturally occurring polymyxins with an L-Dab residue being found at position 1 and an L-Thr residue at position 2. Position 3 can see structural variation with L-Dab, D-Dab or D-Ser being found at this position [2]. Functionally, this segment in most cases contributes two positive charges towards the binding interaction with LPS. Moreover, the molecular model of the polymyxin-LPS complex indicates hydrogen bonds between: a) the amide nitrogen of Dab^3 and the hydroxyl side chain of Thr^2, and b) the main chain carbonyl of Dab^4 and the amide nitrogen of Thr^2, which bends the tripeptide towards the heptapeptide core (Fig. 20.2). A number of studies have explored the SAR of the linear tripeptide segment by examining the effects of amino acid deletions and substitutions [7, 14, 15]. The available SAR data relating to the tripeptide segment demonstrate that it represents an integral feature of the polymyxin structure. Two main SAR principles can be drawn from the data in the literature. Firstly, the tripeptide segment can only be truncated by one amino acid position (i.e. deletion of the Dab at position 1) from the *N*-terminus with a negligible loss of antimicrobial activity. Secondly, only conservative amino acid substitutions (substitution with an amino acid residue with similar functionality and size) appear to be tolerated without losing antibacterial activity.

20.1.4 The Hydrophobic Motif at Positions 6 and 7

The amino acid residues at positions 6 and 7 in the polymyxin heptapeptide ring (Fig. 20.3) form a hydrophobic motif that is generally conserved across the naturally occurring polymyxins and appears to be important for antibacterial activity and plasma protein binding [2]. The position 6 amino acid in particular is highly conserved across all polymyxins and is always either a hydrophobic phenylalanine or leucine residue. Furthermore, the amino acid residue at position 6 is always the D-stereoisomer. This is critical as it acts as a β-turn forming element, allowing the heptapeptide cyclic ring to adopt the necessary confirmation for interacting with the lipid A (Fig. 20.2) [2]. The residue displayed at position 7 can vary in structure with leucine, isoleucine, valine, nor valine and threonine being found at this position in the naturally occurring polymyxins [2]. While the introduction of less hydrophobic groups such as alanine at position 7 is tolerated without significant loss of antibacterial activity [5, 16], gross structural modification at positions 6 and 7, such as replacement of the native amino acid residues with β-turn mimetics appears to impact negatively on the antimicrobial activity [5, 16].

20.1.5 The Heptapeptide Cyclic Ring

The amino group of the side chain of the Dab residue at position 4 is acylated by the *C*-terminal Thr residue to form a 23-membered cyclic ring (Fig. 20.3). The molecular model of the polymyxin B-LPS complex (Fig. 20.2) shows how the precise 23-atom size of the heptapeptide ring acts as a scaffold for electrostatic and hydrophobic LPS contact points. The available SAR data demonstrates that the 23-atom size of the native polymyxin ring provides the most ideal structural configuration for potent antimicrobial activity, and that deletions or expansion of the ring size impact negatively on antimicrobial activity [2, 17]. As already discussed above for the Dab residues and the hydrophobic motif at positions 6 and 7, the side chain functionality of the amino acid residues in the heptapeptide cyclic ring are highly conserved across the naturally occurring polymyxins and generally intolerant to significant modification. The threonine residue at position 10 is also highly conserved in the native polymyxins and appears to make hydrophilic contacts with the sugar molecules of lipid A. However, in contrast to the other residues in the heptapeptide cyclic ring it appears to be more tolerant to structural modification [5, 16].

Our better understanding of polymyxin SAR is now utilized to design and develop new polymyxin lipopeptides with improved efficacy and toxicity profiles including the targeting of polymyxin-resistant Gram-negative pathogens. However, this is no trivial task. As highlighted above the whole molecular scaffold of the polymyxin molecule contributes to its antibacterial activity and is generally not amenable to significant structural change. This leaves a narrow window for exploring structural modification of the polymyxins in order to improve their pharmacological properties. In the following section we discuss the recent developments in the field of polymyxin drug discovery [18, 19] and provide a perspective on each of these in terms of the SAR knowledge base discussed above.

20.2 Preclinical and Clinical Development of Novel Polymyxins-Like Antibiotics

20.2.1 Monash University Lipopeptides

The increasing use of polymyxin B and colistin as a 'last-line' therapy for infections caused by multidrug-resistant Gram-negative pathogens has seen the emergence of resistance to the polymyxins in the clinical setting [1, 2]. This is very problematic as it implies that no antibiotics are available for the treatment of life-threatening infections caused by these Gram-negative 'superbugs'. Our novel lipopeptide discovery program at Monash University (Melbourne, Australia) is the first to use the aforementioned polymyxin SAR based mechanistic model (Fig. 20.2) to design novel polymyxin-like lipopeptides that specifically target polymyxin resistant Gram-negative bacteria [20]. The most common mechanism of polymyxin resistance is through covalent modification of one or both of the lipid A phosphates of LPS with a positively charged sugar (4-amino-4-deoxy-L-arabinose) or phosphoethanolamine group, which removes the negative charge of the phosphate groups [1, 21–23] and inserts a positive charge at these sites. According to our polymyxin SAR based mechanistic model (Fig. 20.2) these modifications to the LPS would disrupt the electrostatic interactions between the phosphate groups and the positively charged amino groups of the Dab residues in the polymyxin molecule. This would significantly weaken polymyxin-LPS binding. Therefore, we hypothesized that incorporating residues with side chains of increased hydrophobicity at positions 6 or 7 would help overcome the disrupted polymyxin-LPS electrostatic interactions by enhancing the polymyxin-LPS hydrophobic interactions. This lead to the design and synthesis of the polymyxin B analogue FADDI-002 (Fig. 20.4), which contains the non-natural amino acid L-octylglycine at position 7. Modeling of the FADDI-002-LPS interaction showed that compounds with these modifications were able to form a stabilized complex [20], which forms the basis of the ability of polymyxins to insert into the Gram-negative outer membrane [24].

Our molecular design strategy was validated when lipopeptide FADDI-002 showed significantly increased antimicrobial activity against polymyxin-resistant Gram-negative clinical isolates of *P. aeruginosa, A. baumannii* and *K. pneumoniae* (MICs of 2–16 μg/mL, vs colistin with MICs >128 μg/mL) [20]. In light of this promising activity we expanded our on SAR-based design strategy and synthesized a series of lipopeptides which incorporated various non-natural lipidic groups at positions 6 or 7 and the *N*-terminus (e.g. FADDI-003, FADDI-016, FADDI-017, FADDI-019, FADDI-020, Fig. 20.4) [20]. These lipopeptides showed very promising

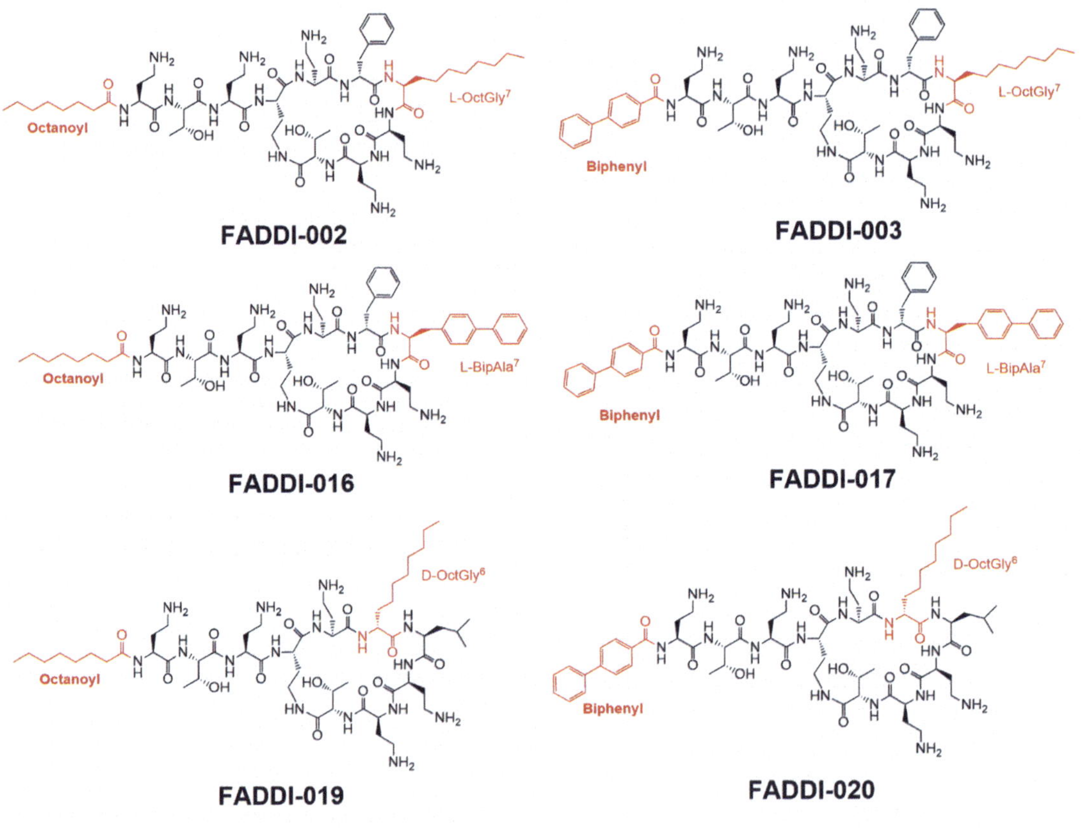

Fig. 20.4 Chemical structures of the novel polymyxin analogues by Monash University. The modifications that have been made to the polymyxin scaffold are highlighted in red

activity against polymyxin-resistant strains while also maintaining their activity against polymyxin-susceptible strains. Notably, against polymyxin-resistant clinical isolates of *P. aeruginosa, A. baumannii* and *K. pneumoniae*, these lipopeptides had MICs of 2–8 μg/mL, whereas polymyxin B or colistin was not inhibitory even at 128 μg/mL. In most cases the increase in antibacterial activity against polymyxin-resistant isolates is greater when the modification is at position 6. The structure of the side chain at positions 6 and 7 does have a small effect on activity with straight chain aliphatic groups giving the best result. Interestingly, for these position 6 and 7 modified peptides, replacement of the flexible aliphatic *N*-terminal octanoyl group with a rigid, aromatic biphenyl group did not have a negative effect on the antibacterial activity [20]. However, decreasing the length of the *N*-terminal fatty-acyl group did lead to decreased antibacterial activity with these peptides. The stereochemistry of the residue at positions 6 and 7 was important for antibacterial activity. The D-stereoisomer gave better activity than L-stereoisomer at position 6, while at position 7 the L-stereoisomer gave better activity than the D-stereoisomer. These observations are consistent with our current understanding of the position 6 and 7 SAR for the polymyxins [2]. Surprisingly, several of these novel polymyxin lipopeptides also displayed antibacterial activity against the problematic Gram-positive vancomycin-resistant *E. faecium* and methicillin- or vancomycin-resistant *S. aureus* (MICs of 4–8 μg/mL, vs polymyxin B or colistin with MICs of >32 μg/mL) [20]. This was unexpected as Gram-positive bacteria are usually intrinsically resistant to the native polymyxins [25]. Scanning and transmission electron microscopy images revealed that treatment with 4 μg/mL of lipopeptide led to the formation of blebs and protrusions (evidence of cell lysis) on the bacterial cell envelope of a polymyxin-resistant clinical *P. aeruginosa* isolate (colistin MIC >128 μg/mL; FADDI-003 MIC 4 μg/mL) [20]. Notably, a similar blebbing effect was observed with polymyxin-susceptible Gram-negative bacterial cells treated with polymyxin B and colistin [26], which would suggest a similar mechanism of action. Fluorescent dansyl-polymyxin displacement assays [27] revealed significantly higher binding affinities to isolated LPS (up to 27-fold) for FADDI-002 and FADDI-003 compared to polymyxin B and colistin [20].

A proof-of-concept study using a neutropenic mouse lung infection model demonstrated ($p < 0.045$) better *in vivo* efficacy of lipopeptide FADDI-002 against a polymyxin-resistant clinical isolate of *P. aeruginosa* compared with colistin [20]. After a single-dose treatment (40 mg/kg S.C), the bacterial burden in the lungs from the mice treated with FADDI-002 was 4.75 ± 0.80 log CFU/lung, which was significantly lower than 6.71 ± 0.46 log CFU/lung for the mice treated with colistin and 7.39 ± 0.17 log CFU/lung for the control group. In rats, lipopeptides FADDI-002 and FADDI-003 had substantially lower total clearances (0.66–1.30 mL/min/kg) and volumes of distribution (195–313 mL/kg), and longer half-lives (166–204 min), compared to colistin (5.2 mL/min/kg, 496 mL/kg and 74.6 min, respectively) [28]. Similar to colistin urinary recoveries of our lipopeptides were negligible (<1%) [28]. The results of preliminary animal studies suggest that our lipopeptides have at least similar tolerability to polymyxin B and colistin in rodents. There was no detectable hemolysis of human red blood cells after exposure to the examined lipopeptides, polymyxin B and colistin at concentrations up to 32 μg/mL.

Nephrotoxicity is the major dose-limiting factor for polymyxin B and colistin therapy [29]. The kidneys of mice subcutaneously treated with lipopeptides FADDI-003 or FADDI-019 (accumulated dose 105 mg/kg) were subjected to histopathological examination and compared to the kidneys of mice treated with an identical concentration of polymyxin B or a saline control [20]. Micro- and macro-morphological examination of kidney sections from the lipopeptide FADDI-003 treated mice revealed no significant lesions in the cortex, medulla and papilla regions. The kidneys of the lipopeptide FADDI-003 treated mice essentially resembled the kidneys of mice treated with the saline control and no histological grade was given. Micro-examination of the kidneys of mice treated with FADDI-019 showed mild tubular dilation and degeneration, and no tubular casts were identified. No macromorphological changes were evident, and the micromorphological changes observed in the kidneys was too mild to be graded. In comparison, the kidneys from the polymyxin B treated mice displayed damaged tubules, with marked tubular dilation and degeneration. It should be noted here that, the lower nephrotoxicity of the lipopeptide may be due to their high plasma protein binding (>90%), which would in turn reduce the exposure of the kidneys [20].

Overall, the results from this work support the use of our SAR-based mechanistic model to aid the design of novel polymyxins. It also lays a strong foundation for the further development of novel polymyxin lipopeptides that target polymyxin-resistant Gram-negative 'superbugs'.

20.2.2 Northern Antibiotics/Spero Therapeutics

Work originating from Northern Antibiotics (Helsinki, Finland) has focused on developing polymyxin analogs with reduced nephrotoxicity. Their design strategy involved generating analogues of polymyxin B with only three positive charges (compared to the five carried by polymyxin B and colistin) through modification of the exo-cyclic linear tripeptide sequence (Fig. 20.3) [14, 30–37]. The idea being that reducing the number of positive charges in the polymyxin scaffold would reduce its nephrotoxicity. This design strategy is based on the low toxicity observed for colistin methanesulfonate, the clinically used pro-drug of colistin [14]. In colistin methanesulfonate the amino groups of the Dab residues have been derviatised with negatively charged methanesulfonate groups, which blocks the amino groups and prevents them from being positively charged at physiological pH. However, this modification of the Dab residues renders the polymyxin molecule totally inactive. Therefore, by removing only some of the positive charge from strategic positions in the polymyxin scaffold you may be able to generate compounds with the right balance between antibacterial activity and nephrotoxicity. The most promising lead compound reported was NAB739, which shared an identical cyclic heptapeptide ring to that of polymyxin B, and a modified linear segment where Dab^1 has been removed and Dab^3 has been replaced with D-Ser (Fig. 20.5) [14]. These modifications afford a polymyxin analogue that carries only three positive charges at physiological pH. The *in vitro* antibacterial activity of NAB739 was evaluated against a large panel of clinically relevant Gram-negative isolates [14, 32, 34]. NAB739 displayed good activity against *E. coli* (66 strains tested in total) with MIC_{90} values (1–2 μg/mL) comparable to that of polymyxin B [14, 32]. Against *K. pneumoniae* (50 strains tested in total), the MIC_{90} of NAB739 was 2 μg/mL, versus polymyxin B with an MIC_{90} of 1 μg/mL [32]. Notably, the MICs of NAB739 against carbapenemase-producing (including KPC-, OXA-48-, VIM- and IMP-producing strains) *E. coli* and *K. pneumoniae* ranged from 1 to 4 μg/mL, whereas those of polymyxin B ranged from 1 to 2 μg/mL [34]. NAB739 was less active against *A. baumannii* (49 strains tested in total) with an MIC_{90} of 8 μg/mL, compared to that of polymyxin B with an MIC_{90} of 2 μg/mL [32]. Similarly, poor activity was observed against *P. aeruginosa* (49 strains tested in total), with the MIC_{90} of NAB739 being 16 μg/mL, whereas that of polymyxin B was 2 μg/mL [32]. Notwithstanding, its poor direct activity against *A. baumannii*, sub-inhibitory concentrations of NAB739 were shown to sensitize the *A. baumannii* strains to rifampicin, clarithromycin, and vancomycin by facilitating their entry into the bacterial cell [14]. NAB739 was not active against polymyxin-resistant strains of *E. coli* and *K. pneumoniae*, *Staphylococcus aureus* and *Candida albicans* [14, 32, 34]. NAB739 showed *in vivo* efficacy in an *E. coli* mouse peritoneal infection model, producing a 4.0 log_{10} reduction in bacterial load compared to the saline control within 6 h, when administered two times in 2-h interval at 1 mg/kg [37]. Based on *in vitro* studies, the toxicity of NAB739 appears to be lower than polymyxin B and colistin [31, 36, 37]. The binding affinity of NAB739 for rat kidney brush border membranes was approximately sevenfold lower than polymyxin B [14]. Compared to polymyxin B, NAB739 was eightfold less toxic in non-polarized porcine renal proximal tubular LLC-PK1 cells [37]. It should be noted that these cells express a functional megalin receptor, which is believed to be involved in the uptake of polymyxins [38]. In human renal proximal tubular HK-2 cells, NAB739 was 26-fold less toxic than polymyxin B and 7.5-fold less toxic than colistin sulfate [36]. Generally, the pharmacokinetics of NAB739 in rats was similar to colistin sulfate, however, some differences were notable, particularly with respect to kidney clearance rates and urinary recovery [30]. Following a single intravenous bolus of 1.0 mg/kg, the serum half-life of NAB739 in rats averaged 69.0 min (colistin 75 min), with a corresponding total body clearance and volume of distribution of 2.63 mL/min/kg (colistin 5.22 mL/min/kg) and 222 mL/kg (colistin 496 mL/kg), respectively [30]. Approximately, 19% of the dose was eliminated within 24 h via the urine unchanged, compared to the urinary recovery of colistin sulfate of just 0.2% [30]. The high urinary recovery of NAB739 may mean it has therapeutic potential in the treatment of urinary tract infections. To this end, Vaara et al. showed in a mouse pyelonephritis model that NAB739 was able to reduce the bacterial load of *E. coli.* in the kidneys, urine and bladder of at a significantly lower dose (tenfold lower) than polymyxin B [39]. Toxicokinetic studies in cynomolgus monkeys showed that NAB739 dosed at 24 mg/kg/d for 7-days was better tolerated than polymyxin B at the same dose based on analysis of biomarkers for kidney damage such as blood urea nitrogen and creatine [40]. As previously observed in rodents, the urinary recovery for NAB739 after intravenous infusion was significantly higher than polymyxin B in the cynomolgus monkeys [40].

Apart from NAB739, two additional Northern Antibiotics compounds are noteworthy, NAB7061 and NAB741 (Fig. 20.5), which do not possess potent direct antibacterial activity, however, they retained the ability to permeabilize the Gram-negative outer membrane [14, 33]. Similar to the potential application of NAB739 as a sensitizing agent against *A. baumannii,* Northern Antibiotics purports that NAB741 and NAB7061 may be useful for combination therapy to facilitate the access of hydrophobic antibiotics and the

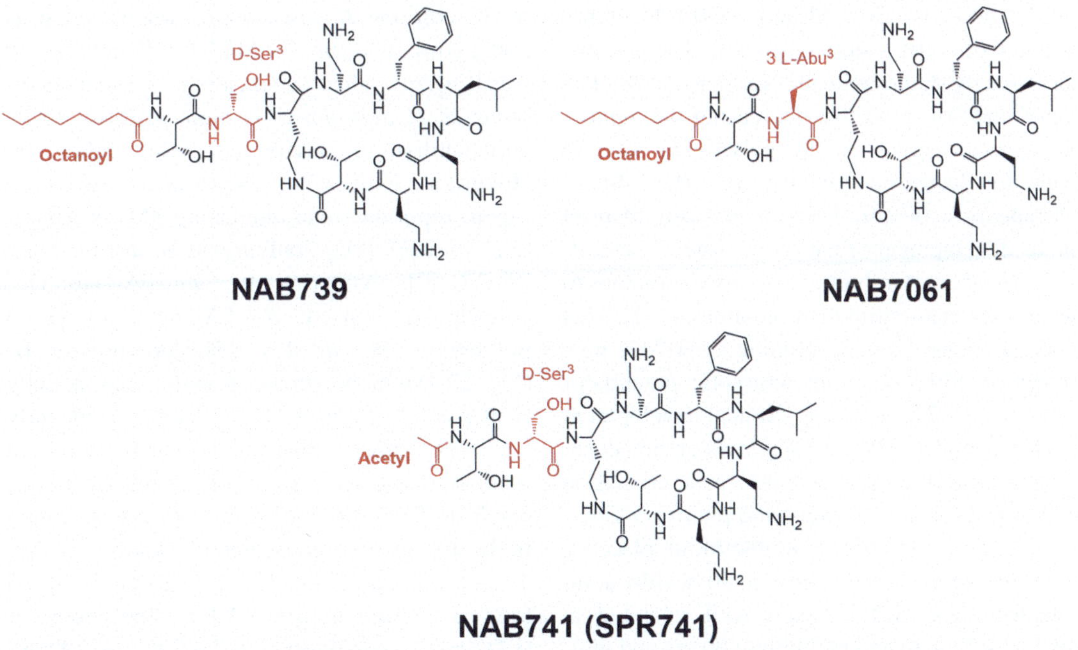

Fig. 20.5 Chemical structures of the novel polymyxin analogues by Northern Antibiotics/Spero Therapeutics. The modifications that have been made to the polymyxin scaffold are highlighted in red

large hydrophilic antibiotics such as vancomycin, which normally cannot permeate through the Gram-negative cell wall and gain access to target site inside the bacterial cell. To this end, they reported data showing that at concentrations of 4 μg/mL NAB7061 was shown to effectively decrease the MICs of rifampicin and clarithromycin against *E. coli*, *A. baumannii* and a polymyxin-resistant *K. pneumoniae* strain [14, 33–35]. Moreover, in an *E. coli* peritoneal mouse infection model, the combination of NAB7061 (5 mg/kg body weight, twice, at an interval of 2 h) and erythromycin (10 mg/kg) was more effective at reducing the bacterial load than either antibiotic alone [37]. Similar to NAB739, the *in vitro* toxicity of these two permeabilizer compounds appears to be lower compared to polymyxin B. NAB7061 displayed a fivefold lower affinity for isolated rat kidney brush border membranes compared to polymyxin B [14]. The cytotoxicity of NAB741 was shown to be 13-fold lower compared to polymyxin B [14]. In terms of pharmacokinetics, NAB7061 displayed a half-life 66 min, whereas NAB741 had a half-life of 33 min (after a single intravenous dose of 1 mg/kg) [30, 33]. The renal clearance of NAB7061 and NAB741 is ~30-fold and ~400-fold higher than that of colistin sulfate [30, 33]. The preliminary toxicity studies with the Northern Antibiotics compounds suggests that decreasing the number of positive charges on the polymyxin scaffold leads to decreased toxicity. In 2015, Spero Therapeutics (Boston, USA) a company focused on the development of antibiotic drugs, licensed-in the Northern Antibiotics polymyxin analogs to develop them as antibiotic potentiators [41]. This work is focused on developing NAB741, now known as SPR741 as an antibiotic potentiator [42, 43]. SPR741 is now in clinical development and has completed Phase I single ascending dose-escalation and multiple ascending dose-escalation studies to evaluate its safety and pharmacokinetics [44]. The randomized, double-blind, placebo-controlled phase I study enrolled 96 healthy adult volunteers and SPR741 was well tolerated at single doses up to and including 800 mg and multiple daily doses up to and including 600 mg every 8 h for 14 consecutive days [44]. A Phase 1b trial involving 27 healthy volunteers has also been conducted investigating the pharmacokinetic compatibility and tolerability of SPR741 when co-administered with β-lactam antibiotics [45]. No change in the PK or tolerability of SPR741 was observed when administered as a single dose of 400 mg in combination with either piperacillin/tazobactam, ceftazidime, or aztreonam. A Phase II clinical trial investigating its efficacy as a potentiator in combination with another antibiotic is now being planned.

20.2.3 Hokuriku University Polymyxin B Nonapeptide Derivatives

Polymyxin B nonapeptide (PMBN) which lacks the *N*-terminal fatty acid tail (*des*-fatty-acyl) and the Dab[1] residue (Fig. 20.6), is significantly less active compared to polymyxin B. However, it has significantly less acute toxicity and nephrotoxicity than polymyxin B [10, 15, 46–49]. Despite its apparent lack of antibacterial activity, PMBN retains an outer membrane permeabilizing activity [10, 46–48]. Interestingly, the MIC of PMBN for *E. coli* and *K. pneumoniae* was reported as 500 μg/mL whereas its MIC for *P. aeruginosa* was 8 μg/mL, clearly indicating the outer membrane of *P. aeruginosa* is more sensitive to its permeabilizing activity [9]. Researchers at Hokuriku University (Kanazawa, Ishikawa, Japan) reported some interesting PMBN derivatives (*des*-FA [Dap[1]]polymyxin B, *des*-FA-Dab[1] [Ser[2]-Dap[3]]polymyxin B, *des*-FA-Dab[1]-Thr[2] [Dap[3]]polymyxin B, *des*-FA-Dab[1]-Thr[2] [Ser[3]] polymyxin B, *des*-FA [Trp[1]]polymyxin B) (Fig. 20.4) with potent anti-pseudomonas activity (MICs of 0.5–1 μg/mL) [12, 50] and significantly less acute toxicity than polymyxin B. In rodent models, the acute toxicity of polymyxin B can result in death through respiratory arrest, potentially due to neuromuscular blockade [51, 52]. These compounds displayed up to an eightfold lower acute toxicity [*des*-FA [Dap[3]]polymyxin B (LD_{50} = 23.5 μmol/kg), *des*-FA-Dab[1] [Ser[2]-Dap[3]] polymyxin B (LD_{50} = 40.9 μmol/kg), *des*-FA-Dab[1]-Thr[2] [Dap[3]]polymyxin B (LD_{50} = >50 μmol/

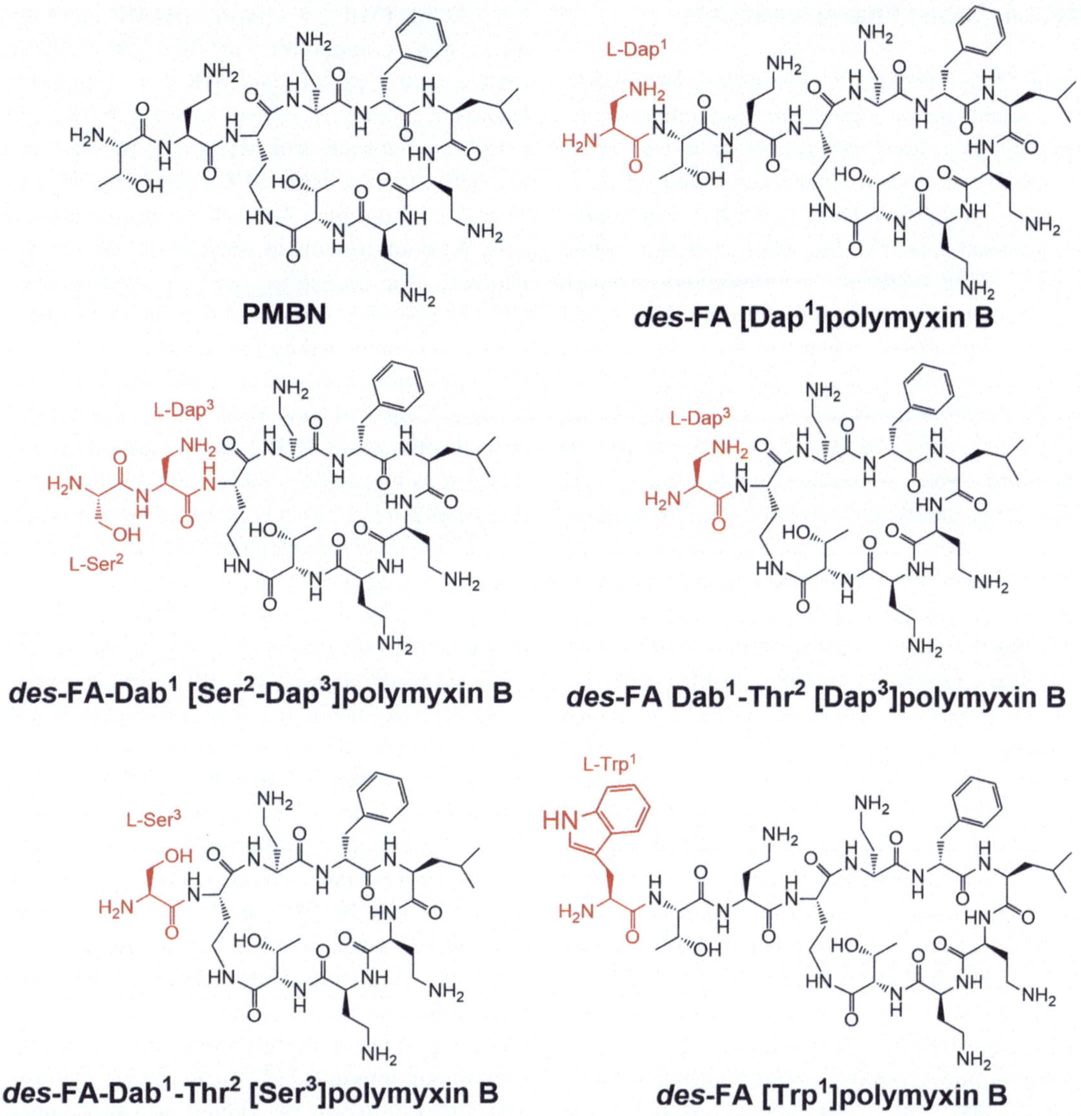

Fig. 20.6 Chemical structures of the novel polymyxin analogues by Hokuriku University. The modifications that have been made to the polymyxin scaffold are highlighted in red

kg), *des*-FA-Dab^1-Thr^2 [Ser^3]polymyxin B (LD_{50} = >50 μmol/kg), *des*-FA [Trp^1]polymyxin B (LD_{50} = 19.0 μmol/kg)] compared to polymyxin B (LD_{50} = 4.8 μmol/kg). Compared to PMBN (LD_{50} = 31.5 μmol/kg), some of these compounds displayed less acute toxicity, which highlights the positive impact of the modifications made to the residues presented at positions 2 and 3 in the exo-cyclic linear tripeptide sequence of PMBN [12, 50]. However, to date no information has been provided on the potential of these compounds for nephrotoxicity. Another notable aspect of PMBN, is that it is 25-fold less active at activating histamine release from rat mast cells compared to polymyxin B [53, 54]. Therefore, it follows that the development of aerosolized formulation of the aforementioned novel PMBN analogs may hold promise for inhalation therapy of *P. aeruginosa* lung infections in cystic fibrosis patients.

20.2.4 Cubist Pharmaceuticals

Cubist Pharmaceuticals (Lexington, MA, USA) had established a significant research program investigating novel *N*-terminal modified polymyxin B and colistin analogues based on intellectual property developed by BioSource Pharmaceuticals [55]. This work centered around novel semi-synthetic methodology which involved enzymatically removing the *N*-terminal fatty-acyl groups of polymyxin B or colistin mixtures to provide a single 'polymyxin core' of which the *N*-terminus was derviatised with novel aryl-urea groups. The strategy behind these modifications was to reduce nephrotoxicity by decreasing the hydrophobicity of the *N*-terminal fatty-acyl group, i.e. have enough hydrophobicity at the *N*-terminus to maintain antibacterial activity but not enough to cause nephrotoxicity. Over 200 novel analogues were prepared and tested. The lead compound to come out of this program was the polymyxin analogue CB-182,804, which contained an *N*-terminal 2-chlorophenylurea group (Fig. 20.7) [56]. Cubist screened CB-182,804 versus colistin against 455 Gram-negative strains selected from various surveillance programs which also included strains with acquired resistance to colistin, carbapenems and/or broad-spectrum cephalosporins [57]. Overall, CB-182,804 had a comparable *in vitro* MIC profile to colistin. Against *P. aeruginosa* (n = 100), including MDR strains resistant to carbapenems and/or aminoglycosides and/or fluoroquinolones, CB-182,804 was slightly more potent (MIC_{50} = 0.5 μg/ml and MIC_{90} = 2 μg/ml) than colistin (MIC_{50} = 1 μg/ml and MIC_{90} = 2 μg/mL). Likewise, against *Acinetobacter* spp. (n = 81), CB-182,804 (MIC_{50} = 1 μg/mL and MIC_{90} = 4 μg/mL) was comparable to that of colistin (MIC_{50} = 0.5 μg/mL and MIC_{90} = 4 μg/mL). However, against *E. coli* (n = 80), CB-182,804 (MIC_{50} = 1 μg/mL and MIC_{90} = 2 μg/mL) was less active than colistin (MIC_{50} = 0.25 μg/mL and MIC_{90} = 0.5 μg/mL). Against organisms intrinsically resistant to colistin, such as indole-positive *Proteae, Pr. mirabilis* and *S. marcescens*, CB-182,804 was also not active. In an independent study conducted by Quale and co-workers at the Department of Medicine at SUNY Downstate Medical Center in New York, the *in vitro* antimicrobial activity of CB-182,804 versus polymyxin B was screened against 5000 Gram-negative clinical isolates (*E. coli* (n = 3049), *K. pneumoniae* (n = 1155), *Enterobacter* spp. (n = 199), *A. baumannii* (n = 407), *P. aeruginosa* (n = 679)) from New York City, a region with a high prevalence of multi-resistant strains [58]. The results of this study showed that the MICs of CB-182,804 were generally twofold higher than polymyxin B and cross-resistance with polymyxin B was observed. It was also observed that the combination of CB-182,804 and rifampin had a synergistic effect, improving antimicrobial activity against polymyxin-resistant strains (*Enterobacter* spp. (n = 199); CB-182,804 MIC_{90}, = > 8 μg/mL vs CB-182,804 + rifampin MIC_{90} = 0.5 μg/mL).

In vivo studies in neutropenic mice lung and thigh infection model showed that CB-182,804 had comparable or slightly improved *in vivo* efficacy to polymyxin B [55]. In an *in vitro* cytotoxicity assay utilizing rat kidney proximal tubule cells, CB-182,804 displayed significantly reduced cytotoxicity (EC_{50} = >1000 μg/mL) compared to polymyxin B (EC_{50} = 318 μg/mL).

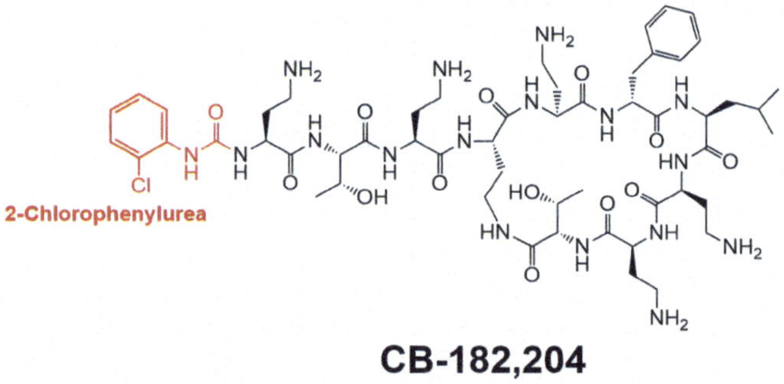

Fig. 20.7 Chemical structure of the novel polymyxin analogue by Cubist Pharmaceuticals. The modifications that have been made to the polymyxin scaffold are highlighted in red

Interestingly, the cytotoxicity observed in this assay, appeared to be significantly influenced by small variations in the chemical structure of the *N*-terminal aryl-urea group. The 3-chlorophenylurea *N*-terminal analogue (a shift in the position of the chloro-group by one carbon from the *ortho*- to the *meta*- position of the phenyl ring), was significantly more cytotoxic (EC_{50} = 619 μg/mL) than CB-182,804 (EC_{50} = >1000 μg/mL). While no data has been presented on its *in vivo* nephrotoxicity in rodent models, the *in vivo* nephrotoxicity of CB-182,804 was evaluated in female cynomolgus monkeys using clinically relevant doses. The comparative 7-day repeat dose safety study revealed that CB-182,804 was less nephrotoxic than polymyxin B with administration of CB-182,804 at 9.9 mg/kg/day (TID) showing similar renal tubular histological changes (increased renal tubular degeneration) to polymyxin B when dosed at 6.6 mg/kg/day (BID). At 6.6 mg/kg/day (BID or TID), CB-182,804 had limited to mild renal tubular histological changes comparable to the background changes observed in the vehicle control. This *in vivo* study also revealed that CB-182,804 had a different pharmacokinetic profile to polymyxin B, with CB-182,804 having decreased serum protein binding (30% vs 56% for polymyxin B), a two to threefold increase in plasma clearance, a twofold increase in the volume of distribution, less systemic exposure with a 2.5 fold decrease in AUC and a twofold lower C_{max} than polymyxin B. These pharmacokinetic differences to polymyxin B were viewed as being potentially exploitable at a therapeutic level, with CB-182,804 potentially having decreased toxicity and enhanced efficacy through greater tissue distribution. On the back of this nephrotoxicity and pharmacokinetic data in monkeys, CB-182,804 was taken into a phase I clinical trial in 2009, but did not progress any further and Cubist has since discontinued this program. No information has been made public as to the outcomes of the phase-I clinical trial. However, considering that Cubist's primary focus was on the development of anti-infectives and has successfully progressed other antibiotic candidates through clinical trials, one can only conclude that the phase-I clinical trial did not produce the desired results. Cubist reported no further work with these compounds and in 2014 the company was acquired by Merck Pharmaceuticals.

20.2.5 Pfizer Polymyxin Analogues

Pfizer (New York City, USA) had also instigated a discovery research program trying to alleviate polymyxin nephrotoxicity through modifications of the Dab residues and the *N*-terminus of polymyxin B. This work was first reported in 2012 in a patent application [59], followed by a peer-reviewed journal publication on their program in 2013 [60]. Initial work focused on trying to decrease nephrotoxicity by modulating the basicity of polymyxin core through the elimination of cationic charge or lowering the pKa of the dab residues. Through this work it was discovered that substitution of the Dab^3 with a diaminopropionic acid (Dap) residue to give lipopeptide 5a (Fig. 20.8), resulted in a twofold improvement in MIC values compared to polymyxin B against *P. aeruginosa* and *A. baumannii* strains, which also included polymyxin-resistant strains. Screening of lipopeptide 5a for *in vitro* nephrotoxicity utilizing human renal proximal epithelial cells showed a twofold decrease in cytotoxicity relative to polymyxin B. Further modifications to the *N*-terminal fatty-acyl group of lipopeptide 5a with novel biphenyl groups lead to the discovery of the lead compound in the program 5x (Fig. 20.8), which contains the *N*-terminal hetero-aromatic group, *N*-phenyl pyridone [60]. Similar to the Cubist lead polymyxin compound CB-182,804, the design strategy here was to decrease the hydrophobicity of the *N*-terminal fatty-acyl group to ameliorate nephrotoxicity without losing too much potency. These compounds were prepared via a total synthesis approach but could also be obtained utilizing a semi-synthetic approach [59].

The *in vitro* antimicrobial profile of lipopeptide 5x against susceptible Gram-negative strains was essentially the same as polymyxin B [*P. aeruginosa* (n = 96), MIC_{90} = 2 μg/mL; *A. baumannii* (n = 96), MIC_{90} = 2 μg/mL; *E. coli*

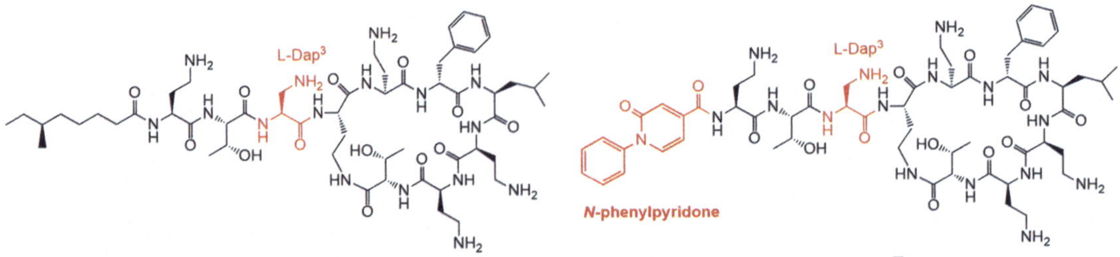

Fig. 20.8 Chemical structures of the novel polymyxin analogues by Pfizer. The modifications that have been made to the polymyxin scaffold are highlighted in red

(n = 101), MIC_{90} = 2 μg/mL; *K. pneumoniae* (n = 101), MIC_{90} = 1 μg/mL] [60]. Lipopeptide 5x also had a two to fourfold improved potency *in vitro* against polymyxin resistant subpopulations of *P. aeruginosa* and *A. baumannii* [60]. Most importantly, screening lipopeptide 5x for *in vitro* nephrotoxicity utilizing human renal proximal epithelial cells saw >5-fold decrease in cytotoxicity relative to polymyxin B [60]. A 7-day exploratory toxicity study in rats utilizing lipopeptide 5x demonstrated a lower incidence of necrotic kidney lesions relative to polymyxin B [60]. Dosing of lipopeptide 5x at 8 mg/kg/day (BID) for 7 days in rats was well tolerated and no significant histological kidney damage was observed whereas polymyxin B at the same dose was not tolerated, hence it's *in vivo* nephrotoxicity could not be assessed. However, polymyxin B dosed at 4 mg/kg/day (BID) for 7 days was tolerated, and resulted in minimal histological changes to the kidneys in all of the rats tested. To further evaluate the therapeutic potential of lipopeptide 5x, a 7-day exploratory toxicity study in dogs of lipopeptide 5x versus polymyxin B was carried out [60]. Unfortunately, the promising results observed with lipopeptide 5x in the rat study did not translate to dogs, with minimal kidney lesions being observed at the lowest dose of 5x, 5 mg/kg/day (BID). Higher doses of 5x at 11 and 20 mg/kg/day (BID) were tolerated but resulted in more significant kidney lesions in every animal. The highest dose of polymyxin B that was examined in dogs was 6 mg/kg/day (BID), which resulted in moderate to significant kidney lesions in every animal. The PK/PD profile of lipopeptide 5x was also examined in a neutropenic mouse thigh infection model against two *P. aeruginosa* strains in a direct comparison with polymyxin B. However, when matched for *f*AUC/MIC values required for similar efficacy targets, lipopeptide 5x (*f*AUC/MIC; EI_{80} = 157.55 EI_{50} = 87.92, Stasis = 85.26, 1 $\log_{10}$ decrease = 109.63) did not perform as well as polymyxin B (*f*AUC/MIC; EI_{80} = 59.00 EI_{50} = 37.38, Stasis = 37.07, 1 $\log_{10}$ decrease = 44.95). The variation observed with the animal nephrotoxicity data, and the inferior PK/PD profile of lipopeptide 5x relative to polymyxin B, were considered significant barriers to further exploration of its therapeutic potential [60]. To date no further work has been published on lipopeptide 5x and Pfizer has since ended its polymyxin discovery program.

20.2.6 Cantab Anti-Infectives/Spero Therapeutics

UK based biotech company Cantab Anti-Infectives (Hertfordshire, UK) has also been trying to develop novel polymyxin compounds to address the nephrotoxicity issues of the polymyxins [61–65]. This work has focused on replacing the *N*-terminal fatty-acyl group and Dab[1] of polymyxin B with a range of structurally diverse hydroxy or amino functionalized acyclic/cyclic acyl groups to afford compounds such as CA-2, CA-6, CA-14 and CA-824 (Fig. 20.9). These compounds can be derived semi-synthetically from polymyxin B, through enzymatic cleavage of polymyxin B at Dab[1] or Dab[3] [63]. In the initial *in vitro* MIC screening experiments versus polymyxin B and colistin, against *E. coli* (n = 4), *P.*

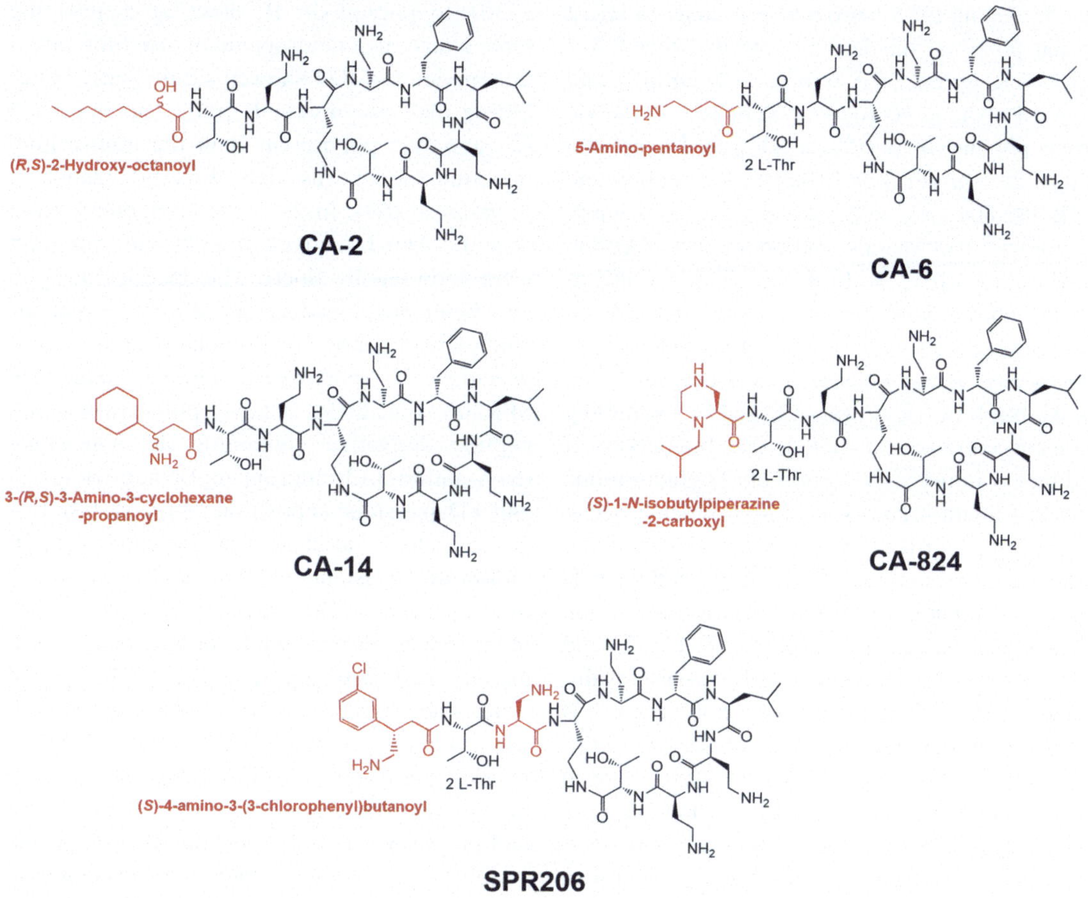

Fig. 20.9 Chemical structures of the novel polymyxin analogues by Cantab Anti-infectives/Spero Therapeutics. The modifications that have been made to the polymyxin scaffold are highlighted in red

aeruginosa (n = 4), *K. pneumoniae* (n = 4) and *A. baumannii* (n = 4), these compounds had MICs that were generally in the same range as polymyxin B (0.25–0.5 μg/mL) with CA-14 showing the best spectrum of activity. In some cases the antimicrobial activity of CA-14 was slightly better than polymyxin B and colistin [61]. Interestingly in these experiments, Northern Antibiotics' NAB739 and Cubist's CB-182,804 discussed in the previous sections above, where also used as positive controls and showed comparable antimicrobial activity (MICs) to CA-2, CA-6 and CA-14. The *in vitro* antibacterial activity of CA-2 and CA-6 was further evaluated against a larger panel of Gram-negative isolates [*E. coli* (n = 100), *P. aeruginosa* (n = 100), *K. pneumoniae* (n = 100) and *A. baumannii* (n = 100)]. Here the MIC_{90} values for CA-2 and CA-6 were 2–16 fold higher than the MIC_{90} values obtained for polymyxin B [61]. Assessment of *in vivo* efficacy in a neutropenic mouse thigh infection model of *E. coli*, showed that treatment with CA-2 (−4.48 log_{10}CFU) and CA-14 (−4.05 log_{10}CFU) gave a comparable reduction in the bacterial load to polymyxin B (−4.2 log_{10}CFU) when dosed at 10 mg/kg, with CA-6 (−3.38 log_{10}CFU) being less efficacious. However, at the lower dose of 3 mg/kg, these lipopeptides were not as efficacious as polymyxin B [61]. In a neutropenic mouse thigh infection model of *K. pneumoniae* these compounds gave a comparable reduction in the bacterial load (CA-2 = −2.22 log_{10}CFU, CA-6 = −1.92 log_{10}CFU and CA-14 = −2.30 log_{10}CFU) to colistin (−2.60 log_{10}CFU) when dosed at 10 mg/kg [61].

Screening for *in vitro* nephrotoxicity in HK-2 renal proximal tubule cells revealed that CA-2 (IC_{50} = 82 μg/mL), CA-6 (IC_{50} = 154 μg/mL), and CA-14 (IC_{50} = 60 μg/mL) were less cytotoxic than polymyxin B (IC_{50} = 11 μg/mL), colistin (IC_{50} 28 = μg/mL) and Cubists lead compound CB-182,804 (IC_{50} = 22 μg/mL), but were more cytotoxic than Vaara's lead compound NAB739 (IC_{50} = 176) [61]. To further evaluate the potential nephrotoxicity of CA-2, CA-6 and CA-14, the lipopeptides were screened for *in vivo* nephrotoxicity versus colistin in a 7-day rat study [61]. Nephrotoxicity was assessed by examining the concentrations of the key renal biomarkers of kidney injury; *N*-acetyl-beta-D-glucosamine (NAG), albumin and cystatin [61]. When dosed at 8 mg/kg/day BID CA-2, CA-6 and CA-14 all showed a two to threefold reduction in the levels of NAG, albumin and cystatin relative to colistin. The pharmacokinetic profile of CA-2 and CA-6 in rats versus polymyxin B was also evaluated [61]. Compared to polymyxin B ($t_{1/2}$ = 1.94 h), CA-2 had a half-life ($t_{1/2}$ = 1.34 h) that was slightly less, and a 1.5-fold increase in C_{max} and AUC. CA-6 had a half-life life ($t_{1/2}$ = 0.56 h), which was ~4 times less than polymyxin B, while its C_{max} was 2.5-fold greater than polymyxin B. Both CA-2 and CA-6 had smaller volumes of distribution (488 and 289 mL/kg) than polymyxin B (1120 mL/kg). CA-2 and CA-6 also had lower clearance (251 and 386 mL/h/kg) than polymyxin B (429 mL/h/kg).

More recently, Cantab presented *in vitro* and *in vivo* efficacy data for their novel polymyxin analog CA-824, in which the *N*-terminal fatty-acyl group and Dab[1] of polymyxin B has been substituted with a (*S*)-1-*N*-isobutylpiperazine-2-carboxyl group (Fig. 20.9) [63–65]. Against clinical isolates of *E. coli* (n = 30), *P. aeruginosa* (n = 30), *K. pneumoniae* (n = 36) and *A. baumannii* (n = 30), CA-824 had comparable MIC_{50} and MIC_{90} values to polymyxin B and less *in vitro* toxicity (IC_{50} = 148 μg/mL) against HK-2 proximal tubular cells when compared to polymyxin B (IC_{50} = 15 μg/mL) [65]. In a neutropenic mouse thigh infection model CA-824 showed comparable killing of a carbapenem resistant reference isolate *A. baumannii* NTNC 13301 to polymyxin B, however against the same isolate in a neutropenic mouse lung infection model, CA-824 showed significantly better killing than polymyxin B [64]. The improved efficacy over polymyxin B in the mouse lung infection model was also observed against *P. aeruginosa* [64]. In 2017 the compounds from Cantab Anti-Infectives polymyxin program were acquired by Spero Therapeutics and are now being developed as part of Spero's potentiator platform [66]. To this end, Spero is progressing the polymyxin clinical candidate SPR206 (Fig. 20.9), a novel polymyxin nonapeptide derivative containing an *N*-terminal (S)-4-amino-3-(3-chlorophenyl)butanoyl group and a Dap residue at position 3 [45, 67, 68]. It is designed to be used as a single agent to treat multidrug resistant (MDR) and extensively drug-resistant (XDR) bacterial strains, including carbapenem-resistant *P. aeruginosa*, *A. baumannii*, and Enterobacteriaceae [45]. Against Enterobacteriaceae species (541 clinical isolates, including carbapenem-resistant *K. pneumoniae* and *E. coli*), SPR206 displayed *in vitro* activity that was 2 to 4-fold greater than colistin and polymyxin B [68]. SPR206 also displayed potent *in vitro* activity compared to polymyxin B and colistin against the non-fermentative Gram-negative bacilli *P. aeruginosa* [($MIC_{50/90}$, 0.25/0.5 μg/mL), 2-fold lower than colistin ($MIC_{50/90}$, 0.5/1 μg/mL) and polymyxin B ($MIC_{50/90}$, 0.5/1 μg/mL)] and *A. baumannii* [($MIC_{50/90}$,0.12/0.25 μg/mL), 2 to 8-fold more potent than polymyxin B ($MIC_{50/90}$, 0.25/1–2 μg/mL) and 4- to 32-fold more potent than colistin ($MIC_{50/90}$, 0.5/4–8 μg/mL)] [67]. In 2018, Spero Therapeutics announced that SPR206 had successfully completed IND enabling studies and planned to take it into Phase I clinical trials in 2019 [45].

20.3 Conclusions

In the wake of our increasing understanding of polymyxin SAR, recent medicinal chemistry efforts have yielded some interesting novel polymyxin lipopeptides with promising activity and toxicity profiles compared to polymyxin B and colistin. The novel position 6 and 7 modified polymyxin lipopeptides from Monash University are unique with respect to their design, which specifically targets polymyxin resistance. This is important, as polymyxin resistance may become a greater issue in the future with the increasing clinical use of the polymyxins. The Monash compounds also highlight the value in using an SAR-based mechanistic model of polymyxin antibacterial activity to help aid the design of superior polymyxin lipopeptides. While the novel polymyxin compounds developed by Northern Antibiotics, and Hokuriku University lack the desired spectrum of antibacterial activity against clinically important Gram-negative pathogens, they also appear to lack the nephrotoxic side effects of the clinically used polymyxins. Hence, their clinical value may lie as antibiotic potentiators to be used in combination therapy with other antibiotics that have trouble penetrating the Gram-negative outer membrane. To this end, Spero Therapeutics has taken one of Northern Antibiotics polymyxin analogs into early stage clinical development as an antibiotic potentiator; however, it still remains to be seen if the antibacterial efficacy using the potentiator approach can be achieved in humans.

Cubist Pharmaceuticals and Pfizer both made significant attempts to develop less nephrotoxic analogues of polymyxin B and colistin. Both preclinical programs collected significant amounts of *in vitro* and *in vivo* nephrotoxicity and efficacy data on their lead lipopeptides against problematic Gram-negative strains, with Cubist taking their lead candidate into Phase I clinical trials. However, the fact that neither of their lead lipopeptides is being pursued any further and their polymyxin programs abandoned, highlights the immense difficulty in finding the right balance between efficacy, toxicity and PK/PD properties when it comes to developing new polymyxin antibiotics. In light of these setbacks, it will be interesting to see if the clinical candidate SPR206 from Spero Therapeutics, can be successfully translated into the clinic.

This collective body of pre-clinical and clinical work highlights how structurally intertwined the activity and toxicity of the polymyxins are and how difficult it is to try and structurally separate them through chemical modification of the polymyxin scaffold. Moving forward, the aforementioned pre-clinical and clinical drug development programs have provided valuable insights into not only polymyxin SAR but also polymyxin structure-toxicity relationships (STR). They have highlighted that the *N*-terminal fatty-acyl chain and the positively charged Dab residues represent nephrotoxicity 'hot-spots' around which medicinal chemistry efforts should be focused in order to reduce toxicity. In this respect, there is an urgent need to further develop our understanding of the molecular mechanisms and targets underlying the renal uptake, disposition and toxicity of the polymyxins. This would allow for the development STR-based mechanistic models of polymyxin nephrotoxicity to help aid the design of superior polymyxin antibiotics.

References

1. Velkov T, Roberts KD, Nation RL, Thompson PE, Li J (2013) Pharmacology of polymyxins: new insights into an 'old' class of antibiotics. Future Microbiol 8(6):711–724
2. Velkov T, Thompson PE, Nation RL, Li J (2010) Structure-activity relationships of polymyxin antibiotics. J Med Chem 53(5):1898–1916
3. Mares J, Kumaran S, Gobbo M, Zerbe O (2009) Interactions of lipopolysaccharide and polymyxin studied by NMR spectroscopy. J Biol Chem 284(17):11498–11506
4. Pristovsek P, Kidric J (1999) Solution structure of polymyxins B and E and effect of binding to lipopolysaccharide: an NMR and molecular modeling study. J Med Chem 42(22):4604–4613
5. de Visser PC, Kriek NM, van Hooft PA, Van Schepdael A, Filippov DV, van der Marel GA et al (2003) Solid-phase synthesis of polymyxin B1 and analogues via a safety-catch approach. J Pept Res 61(6):298–306
6. Okimura K, Ohki K, Sato Y, Ohnishi K, Sakura N (2007) Semi-synthesis of polymyxin B (2-10) and colistin (2-10) analogs employing the trichloroethoxycarbonyl (Troc) group for side chain protection of alpha,gamma-diaminobutyric acid residues. Chem Pharm Bull (Tokyo) 55(12):1724–1730
7. Sakura N, Itoh T, Uchida Y, Ohki K, Okimura K, Chiba K et al (2004) The contribution of the N-terminal structure of polymyxin B peptides to antimicrobial and lipopolysaccharide binding activity. Bull Chem Soc Jpn 77(10):1915–1924

8. Chihara S, Ito A, Yahata M, Tobita T, Koyama Y (1974) Chemical synthesis, isolation and characterization of α-N-fattyacyl colistin nonapeptide with special reference to the correlation between antimicrobial activity and carbon number of fattyacyl moiety. Agric Biol Chem 38(3):521–529
9. Tsubery H, Ofek I, Cohen S, Fridkin M (2001) N-terminal modifications of polymyxin B nonapeptide and their effect on antibacterial activity. Peptides 22(10):1675–1681
10. Vaara M (1991) The outer membrane permeability-increasing action of linear analogues of polymyxin B nonapeptide. Drugs Exp Clin Res 17(9):437–443
11. Okimura K, Ohki K, Sato Y, Ohnishi K, Uchida Y, Sakura N (2007) Chemical conversion of natural polymyxin B and colistin to their N-terminal derivatives. Bull Chem Soc Jpn 80(3):543–552
12. Katsuma N, Sato Y, Ohki K, Okimura K, Ohnishi K, Sakura N (2009) Development of des-fatty acyl-polymyxin B decapeptide analogs with pseudomonas aeruginosa-specific antimicrobial activity. Chem Pharm Bull (Tokyo) 57(4):332–336
13. Barnett M, Bushby SR, Wilkinson S (1964) Sodium sulphomethyl derivatives of polymyxins. Br J Pharmacol Chemother 23:552–574
14. Vaara M, Fox J, Loidl G, Siikanen O, Apajalahti J, Hansen F et al (2008) Novel polymyxin derivatives carrying only three positive charges are effective antibacterial agents. Antimicrob Agents Chemother 52(9):3229–3236
15. Kimura Y, Matsunaga H, Vaara M (1992) Polymyxin B octapeptide and polymyxin B heptapeptide are potent outer membrane permeability-increasing agents. J Antibiot (Tokyo) 45(5):742–749
16. Kanazawa K, Sato Y, Ohki K, Okimura K, Uchida Y, Shindo M et al (2009) Contribution of each amino acid residue in polymyxin B(3) to antimicrobial and lipopolysaccharide binding activity. Chem Pharm Bull (Tokyo) 57(3):240–244
17. Tsubery H, Ofek I, Cohen S, Fridkin M (2000) Structure activity relationship study of polymyxin B nonapeptide. Adv Exp Med Biol 479:219–222
18. Rabanal F, Cajal Y (2017) Recent advances and perspectives in the design and development of polymyxins. Nat Prod Rep 34(7):886–908
19. Brown P, Dawson MJ (2017) Development of new polymyxin derivatives for multi-drug resistant Gram-negative infections. J Antibiot (Tokyo) 70(4):386–394
20. Velkov T, Roberts KD, Nation RL, Wang J, Thompson PE, Li J (2014) Teaching 'old' polymyxins new tricks: new-generation lipopeptides targeting Gram-negative 'superbugs'. ACS Chem Biol 9(5):1172–1177
21. Clements A, Tull D, Jenney AW, Farn JL, Kim SH, Bishop RE et al (2007) Secondary acylation of klebsiella pneumoniae lipopolysaccharide contributes to sensitivity to antibacterial peptides. J Biol Chem 282(21):15569–15577
22. Moskowitz SM, Ernst RK (2010) The role of pseudomonas lipopolysaccharide in cystic fibrosis airway infection. Subcell Biochem 53:241–253
23. Moskowitz SM, Ernst RK, Miller SI (2004) PmrAB, a two-component regulatory system of Pseudomonas aeruginosa that modulates resistance to cationic antimicrobial peptides and addition of aminoarabinose to lipid A. J Bacteriol 186(2):575–579
24. Hancock RE (1997) Peptide antibiotics. Lancet 349(9049):418–422
25. Li J, Nation RL, Milne RW, Turnidge JD, Coulthard K (2005) Evaluation of colistin as an agent against multi-resistant Gram-negative bacteria. Int J Antimicrob Agents 25(1):11–25
26. Koike M, Iida K, Matsuo T (1969) Electron microscopic studies on mode of action of polymyxin. J Bacteriol 97(1):448–452
27. Soon RL, Velkov T, Chiu F, Thompson PE, Kancharla R, Roberts K et al (2011) Design, synthesis, and evaluation of a new fluorescent probe for measuring polymyxin-lipopolysaccharide binding interactions. Anal Biochem 409(2):273–283
28. Li J, Milne RW, Nation RL, Turnidge JD, Smeaton TC, Coulthard K (2003) Use of high-performance liquid chromatography to study the pharmacokinetics of colistin sulfate in rats following intravenous administration. Antimicrob Agents Chemother 47(5):1766–1770
29. Yousef JM, Chen G, Hill PA, Nation RL, Li J (2012) Ascorbic acid protects against the nephrotoxicity and apoptosis caused by colistin and affects its pharmacokinetics. J Antimicrob Chemother 67(2):452–459
30. Ali FE, Cao G, Poudyal A, Vaara T, Nation RL, Vaara M et al (2009) Pharmacokinetics of novel antimicrobial cationic peptides NAB 7061 and NAB 739 in rats following intravenous administration. J Antimicrob Chemother 64(5):1067–1070
31. Vaara M (2013) Novel derivatives of polymyxins. J Antimicrob Chemother 68(6):1213–1219
32. Vaara M, Sader HS, Rhomberg PR, Jones RN, Vaara T (2013) Antimicrobial activity of the novel polymyxin derivative NAB739 tested against Gram-negative pathogens. J Antimicrob Chemother 68(3):636–639
33. Vaara M, Siikanen O, Apajalahti J, Fox J, Frimodt-Moller N, He H et al (2010) A novel polymyxin derivative that lacks the fatty acid tail and carries only three positive charges has strong synergism with agents excluded by the intact outer membrane. Antimicrob Agents Chemother 54(8):3341–3346
34. Vaara M, Siikanen O, Apajalahti J, Frimodt-Moller N, Vaara T (2010) Susceptibility of carbapenemase-producing strains of Klebsiella pneumoniae and Escherichia coli to the direct antibacterial activity of NAB739 and to the synergistic activity of NAB7061 with rifampicin and clarithromycin. J Antimicrob Chemother 65(5):942–945
35. Vaara M, Vaara T (2010) Structure-activity studies on novel polymyxin derivatives that carry only three positive charges. Peptides 31(12):2318–2321
36. Vaara M, Vaara T (2013) The novel polymyxin derivative NAB739 is remarkably less cytotoxic than polymyxin B and colistin to human kidney proximal tubular cells. Int J Antimicrob Agents 41(3):292–293

37. Vingsbo Lundberg C, Vaara T, Frimodt-Moller N, Vaara M (2010) Novel polymyxin derivatives are effective in treating experimental Escherichia coli peritoneal infection in mice. J Antimicrob Chemother 65(5):981–985
38. Nielsen R, Birn H, Moestrup SK, Nielsen M, Verroust P, Christensen EI (1998) Characterization of a kidney proximal tubule cell line, LLC-PK1, expressing endocytotic active megalin. J Am Soc Nephrol 9(10):1767–1776
39. Vaara M, Vaara T, Vingsbo LC (2017) Polymyxin derivatives NAB739 and NAB815 are more effective than polymyxin B in murine Escherichia coli pyelonephritis. J Antimicrob Chemother 73(2):452–455
40. Vaara M, Vaara T, Tyrrell JM (2017) Structure-activity studies on polymyxin derivatives carrying three positive charges only reveal a new class of compounds with strong antibacterial activity. Peptides 91:8–12
41. Spero completes $30 million financing [press release]. June 8, 2015
42. Corbett D, Wise A, Langley T, Skinner K, Trimby E, Birchall S et al (2017) Potentiation of antibiotic activity by a novel cationic peptide: potency and spectrum of activity of SPR741. Antimicrob Agents Chemother 61(8):e00200-17
43. Zabawa TP, Pucci MJ, Parr TR Jr, Lister T (2016) Treatment of Gram-negative bacterial infections by potentiation of antibiotics. Curr Opin Microbiol 33:7–12
44. Spero therapeutics announces positive phase 1 clinical data from potentiator platform [press release]. October 3rd, 2017
45. Spero therapeutics announces positive top-line data for two product candidates from its potentiator platform [press release]. May 23rd, 2018
46. Vaara M, Vaara T (1983) Polycations sensitize enteric bacteria to antibiotics. Antimicrob Agents Chemother 24(1):107–113
47. Vaara M, Vaara T (1983) Polycations as outer membrane-disorganizing agents. Antimicrob Agents Chemother 24(1):114–122
48. Vaara M, Vaara T (1983) Sensitization of Gram-negative bacteria to antibiotics and complement by a nontoxic oligopeptide. Nature 303(5917):526–528
49. Vaara M (1992) Agents that increase the permeability of the outer membrane. Microbiol Rev 56(3):395–411
50. Sato Y, Shindo M, Sakura N, Uchida Y, Kato I (2011) Novel des-fatty acyl-polymyxin B derivatives with Pseudomonas aeruginosa-specific antimicrobial activity. Chem Pharm Bull (Tokyo) 59(5):597–602
51. Viswanath DV, Jenkins HJ (1978) Neuromuscular block of the polymyxin group of antibiotics. J Pharm Sci 67(9):1275–1280
52. Singh YN, Marshall IG, Harvey AL (1982) Pre- and postjunctional blocking effects of aminoglycoside, polymyxin, tetracycline and lincosamide antibiotics. Br J Anaesth 54(12):1295–1306
53. Bairamashvili DI, Voitenko VG, Gushchin IS, Zinchenko AA, Miroshnikov AI (1989) The histamine-liberating action of polymyxin B and its analogs. Biull Eksp Biol Med 107(4):447–449
54. Voitenko VG, Bayramashvili DI, Zebrev AI, Zinchenko AA (1990) Relationship between structure and histamine releasing action of polymyxin B and its analogues. Agents Actions 30(1–2):153–156
55. Leese RA, inventor; Biosource Pharm, Inc., assignee (2010) Antibiotic composition for the treatment of Gram-negative infections. U.S patent US20100160215
56. Keith DD, Borders D, Curran W, Leese R, Mahamoon A, Jarolmen H et al eds (2010) Poster F1-1619 synthesis and activity of CB-182,804, a novel polymyxin analog active against clinically relevant Gram-negative bacteria. In: 50th Interscience conference on antimicrobial agents and chemotherapy, 2010 September 12–15, Boston, MA, USA
57. Sader HS, Rhomberg PR, Jones RN eds (2010) Poster F1-1626 Antimicrobial activity of a novel polymyxin analog (CB-182,804) tested against clinical strains of Gram-negative bacilli, including colistin resistant organisms. In: 50th interscience conference on antimicrobial agents and chemotherapy, Boston, USA
58. Quale J, Shah N, Kelly P, Babu E, Backer M, Rosas-Garcia G et al (2012) Activity of polymyxin B and the novel polymyxin analogue CB-182,804 against contemporary Gram-negative pathogens in New York City. Microb Drug Resist 18(2):132–136
59. Chen JM, Li ZB, Magee TV, Martinez CA, inventors; Pfizer Inc, assignee (2012) Polymyxin derivatives useful as antibacterial agents. patent WO2012168820
60. Magee TV, Brown MF, Starr JS, Ackley DC, Abramite JA, Aubrecht J et al (2013) Discovery of Dap-3 polymyxin analogues for the treatment of multidrug-resistant Gram-negative nosocomial infections. J Med Chem 56:5079–5093
61. Saadi M, Duperchy E, Brown P, Dawson MJ, Wadman SN, inventors; Cantab Anti-Infectives, assignee (2013) Polymyxin derivatives patent WO2013072695
62. Brown P, Dawson M, Simonovic M, Boakes S, Duperchy E, inventors; Cantab Anti-Infectives, assignee (2014) Polymyxin derivatives and their use in combination therapy together with different antibiotics. patent WO2014188178
63. Brown P, Dawson M, Simonovic M, Boakes S, Duperchy E, Stanway S, et al., inventors; Cantab Anti-Infectives, assignee (2015) Polymyxin derivatives and their use in combintion theapy together with different antibiotics. patent WO2015135976
64. Boakes S, Duperchy E, Brown P, Teague J, Payne LJ, Dawson MJ eds (2015) Novel polymyxin derivative CA824: efficacy in neutropenic mouse thigh and lung infection models. In: 55th interscience conference on antimicrobial agents and chemotherapy and 28th international congress of chemotherapy meeting, San Diego USA
65. Wiederhold NP, Jorgenson JH, Boakes S, Collins M, McElmeel, Cushion MT et al (2015) Anibacterial activity of novel polymyxin B derivatives against

Gram-negative bacteria and intial toxicity assessment. In: 55th Interscience conference on antimicrobial agents and chemotherapy and 28th international congress of chemotherapy meeting, San Diego, USA
66. Spero therapeutics acquires next generation antibacterial candidates from Pro Bono Bio for treatment of multidrug-resistant, Gram-negative infections [press release]. January 31st, 2017
67. Arends S, Rhomberg P, Lister T, Cotroneo N, Rubio A, Flamm R et al (2018) In vitro activity evaluation of a next-generation polymyxin, SPR206, against non-fermentative gram-negative bacilli responsible for human infections. In: ESCMID/ASM conference on drug development to meet the challenge of antimicrobial resistance, Lisbon, Portugal
68. Arends S, Rhomberg P, Lister T, Cotroneo N, Rubio A, Flamm R et al (2018) Activity of an investigational polymyxin-B-like compound (SPR206) against a set of enterobacteriaceae organisms responsible for human infections. In: ESCMID/ASM conference on drug development to meet the challenge of antimicrobial resistance, Lisbon, Portugal

Conclusion

21

Roger L. Nation

Abstract

This chapter briefly reviews the progress that has been made in understanding the key chemical, microbiological and pharmacological properties of the polymyxins since they were resurrected for clinical use around the start of the current century. Discussed are some of the key outcomes of the first three international conferences on the poymyxins, including publication of a consensus statement on a framework for redevelopment of these last-line antibiotics and international consensus guidelines for their optimal use in patients.

Keywords

Polymyxins · Colistin · Polymyxin B · Progress in understanding how to use in patients · Consensus statements and guidelines

The mismatch between rising rates of multi-drug resistance among important Gram-negative pathogens and the dry drug development pipeline for new antibacterials was the driving force for the resurrection of the 'old' polymyxin antibiotics for treatment of life-threatening infections. Since their discovery in the late 1940s and introduction into the clinic about a decade later, colistin and polymyxin B have been on a remarkable journey. As reviewed previously [1–3], there is no doubt about the excellence of the science that was conducted back in the middle of the twentieth century leading to the discovery of the polymyxins and to the uncovering of some of their key chemical, microbiological, pharmacological and clinical properties. A difficulty imposed upon the researchers of the time was the relatively crude nature of the experimental methods that were available to explore the significant complexities of the chemistry and biological behavior of these antibiotics. With the increased clinical use of the polymyxins since around the beginning of the current century and with the benefit of modern techniques, there has been a very substantial increase in the scientific and clinical knowledge around the polymyxins. This is reflected in the number of papers on the polymyxins published annually; between the years 2000 and 2018 there was greater than a tenfold increase in papers per year and the publication rate continues to increase (see Chap. 1, Fig. 1.7).

Because colistin and polymyxin B are long off patent and multiple generic brands exist, the vast majority of the preclinical and clinical research undertaken over the last 10–15 years was investigator initiated and funded by government or other

R. L. Nation (✉)
Drug Delivery, Disposition and Dynamics,
Monash Institute of Pharmaceutical Sciences,
Monash University, Parkville Campus,
Melbourne, VIC, Australia
e-mail: roger.nation@monash.edu

J. Li et al. (eds.), *Polymyxin Antibiotics: From Laboratory Bench to Bedside*, Advances in Experimental Medicine and Biology 1145, https://doi.org/10.1007/978-3-030-16373-0_21

public grant bodies. In essence, the research conducted across the first several years of the twenty-first century has been to generate preclinical and clinical scientific information that is required by regulatory agencies today for approval of new drugs. Thus, the polymyxins have been undergoing 'redevelopment' to generate the information needed to enable clinicians to use them optimally; that is, to maximize bacterial killing, minimize emergence of resistance and decrease the potential for toxicity in patients. As the world continues to confront the 'Bad Bugs, No Drugs' scenario, the polymyxins have been at the vanguard in relation to the need to resurrect 'old' antibiotics. Undoubtedly, many of the issues confronted in the 'redevelopment' of the polymyxins will provide valuable lessons for the other 'old' antibiotics that follow.

At the dawn of the 'new' era for the polymyxins as they became increasingly needed for the treatment of infections caused by multidrug-resistant bacteria, it was clear that important information required to use them safely and effectively was not available. Many of these gaps and hindrances were summarized within the 'Prato Polymyxin Consensus' [4], which was an important outcome of the *1st International Conference on Polymyxins* held in Prato, Italy in 2013. In essence, the 'Prato Polymyxin Consensus' provided a roadmap of high-priority issues to be addressed in the ongoing efforts to optimize the clinical use of the polymyxins. The *2nd International Conference on Polymyxins* was held in San Diego, USA in 2015; a key outcome of that conference was the identification of the need for 'international consensus guidelines for the optimal use of the polymyxins'. Pleasingly, such guidelines have been prepared by an international group of experts; the guidelines have been published [5] and will be subject to wide dissemination. The *3rd International Conference on Polymyxins* was held in Madrid, Spain in 2018. That conference provided the forum for presentation of research that filled ongoing gaps in knowledge, and identified key areas where yet more work is required.

The chapters in this book provide a summary of the history of the polymyxins across the last six to seven decades, with a primary focus on presenting the progress that has been made over the last several years towards understanding how to effectively and safely use this important class of antibiotics. As is evident from the chapters, very substantial progress has been made in understanding the complexities of the polymyxins and how their clinical use may be optimized. However, it is also clear from the material presented in the chapters that there is still some work to do. Undoubtedly, the gain in knowledge over the next few years will lead to further improvements in the ability of clinicians to use the polymyxins safely and effectively.

References

1. Evans ME, Feola DJ, Rapp RP (1999) Polymyxin B sulfate and colistin: old antibiotics for emerging multiresistant gram-negative bacteria. Ann Pharmacother 33(9):960–967
2. Li J, Nation RL, Milne RW, Turnidge JD, Coulthard K (2005) Evaluation of colistin as an agent against multi-resistant Gram-negative bacteria. Int J Antimicrob Agents 25(1):11–25. https://doi.org/10.1016/j.ijantimicag.2004.10.001
3. Falagas ME, Kasiakou SK (2005) Colistin: the revival of polymyxins for the management of multidrug-resistant gram-negative bacterial infections. Clin Infect Dis 40(9):1333–1341. https://doi.org/10.1086/429323
4. Nation RL, Li J, Cars O, Couet W, Dudley MN, Kaye KS, Mouton JW, Paterson DL, Tam VH, Theuretzbacher U, Tsuji BT, Turnidge JD (2015) Framework for optimisation of the clinical use of colistin and polymyxin B: the Prato polymyxin consensus. Lancet Infect Dis 15(2):225–234. https://doi.org/10.1016/s1473-3099(14)70850-3
5. Tsuji BT, Pogue JM, Zavascki AP, Paul M, Daikos GL, Forrest A, Giacobbe DR, Viscoli C, Giamarellou H, Karaiskos I, Kaye D, Mouton JW, Tam VH, Thamlikitkul V, Wunderink RG, Li J, Nation RL, Kaye KS (2019) International consensus guidelines for the optimal use of the polymyxins: endorsed by the American College of Clinical Pharmacy (ACCP), European Society of Clinical Microbiology and Infectious Diseases (ESCMID), Infectious Diseases Society of America (IDSA), International Society for Anti-infective Pharmacology (ISAP), Society of Critical Care Medicine (SCCM), and Society of Infectious Diseases Pharmacists (SIDP). Pharmacotherapy 39(1):10–39

Index

J. Li et al. (eds.), *Polymyxin Antibiotics: From Laboratory Bench to Bedside*, Advances in Experimental Medicine and Biology 1145, https://doi.org/10.1007/978-3-030-16373-0

The manufacturer's authorised representative in the EU is Springer Nature Customer Service Centre GmbH, Europaplatz 3, 69115 Heidelberg, Germany. If you have any concerns regarding our products, please contact ProductSafety@springernature.com

Printed and bound by CPI Group (UK) Ltd, Croydon, CR0 4YY
30/09/2026
02225440-0020